The Bureau of Reclamation: From Developing to Managing Water, 1945-2000
Volume 2

Andrew H. Gahan and William D. Rowley

Bureau of Reclamation
U.S. Department of the Interior
Denver, Colorado
2012

TABLE OF CONTENTS: VOLUME 1

VOLUME 2

LIST OF ILLUSTRATIONS: VOLUME 1

CHAPTER 3

CHAPTER 4

CHAPTER 5

CHAPTER 6

CHAPTER 7

LIST OF ILLUSTRATIONS: VOLUME 2

CHAPTER 8

CHAPTER 9

CHAPTER 10

CHAPTER 11

CHAPTER 12

CHAPTER 13

CHAPTER 14

PHOTOGRAPHY OF WALTER J. LUBKEN

PHOTOGRAPHY OF ANSEL ADAMS

PHOTOGRAPHY OF BEN GLAHA

COMMISSIONER'S INTRODUCTION

On June 17th this year the Bureau of Reclamation's water and hydropower development in the American West turned 110 years of age. Because of that long tradition of Congressionally mandated development, Reclamation is the largest single electricity supplier in the West and the largest wholesale water supplier in the country.

The story of Reclamation is deeply entwined in the history of development of the American West in the twentieth and twenty-first centuries. One third of the West's population and about 10,000,000 acres of the West's irrigated land, about one-third of the irrigated land, use water from Reclamation projects. And, Reclamation-generated hydropower played an important role in electrification of western rural areas and development of industries, especially during and after World War II.

Throughout its history, Reclamation has been an innovator in the engineering and science of dam and canal design and construction, hydraulic modeling, hydroelectricity production and delivery, water delivery, conservation, and multipurpose uses of water. Reclamation's masonry dams represent a distinguished lineage from Pathfinder and East Park through Theodore Roosevelt, Arrowrock, Owyhee, Hoover, Grand Coulee, Friant, and Shasta, to Morrow Point. Reclamation's embankment dams share an equally distinguished lineage and include Belle Fourche, Anderson Ranch, and San Luis.

Reclamation's rich history is filled with colorful personalities and the unique character of the West. It is a history marked with engineering and construction innovation and wonder that have resulted in water and hydroelectric development, resource management, and resource preservation. This volume traces Reclamation's story from the end of World War II to the beginning of the twenty-first century. I hope you find this study as useful and informative as I do.

While Reclamation's mission always focuses on its two primary responsibilities to deliver water and hydroelectricity to the American public in the West, there are many subsidiary benefits of Reclamation projects which do not come directly from those responsibilities. Water bodies in the West naturally attracted recreationists from the earliest days of projects, and today extensive and varied recreation activities occur on projects. Operation of

Reclamation impoundments provides flood control and drought relief benefits. The U.S. State Department regularly uses Reclamation's technical expertise in international activities and in training foreign engineers and technicians. In addition, agencies from around the U.S. regularly find Reclamation's experience useful in developing water conservation, supplemental supply, and water augmentation programs.

More traditional roles continue for Reclamation. For instance, more than 180 Reclamation projects deliver agricultural water that produces a significant percentage of the value of all crops in the United States, including about 60 percent of vegetables and 25 percent of the fruit and nut crops. Yet Reclamation is an evolving institution, and it is important to understand Reclamation's past in order to permit intelligent management decisions in the present for the future. Reclamation has been moving away from new construction activities and into water management on its existing facilities. The Congress and Executive Branches are also developing new initiatives assigned to Reclamation. For instance, Reclamation now has partnerships on several rural water projects designed to deliver culinary water to rural areas that do not have good drinking water. Reclamation also provides staff and expertise to the Secretary of the Interior's negotiating teams working with Native Americans to quantify and deliver settlement water to tribes. The Department of the Interior's proposals for Reclamation's budget in Fiscal Year 2013 designated some 5 percent for the WaterSMART program so that Reclamation can work with states, tribes, local governments, and non-governmental organizations to develop sustainable water supplies by improving water conservation and fostering appropriate decisions about water use. Over 10 percent of the Fiscal Year 2013 proposed budget is designated for various environmental and river restoration initiatives. Understanding the evolution of Reclamation's programs, the environmental movement, and the various administrations' policy positions explains why shifts in emphasis like these occur in Reclamation programs.

Initiatives begun in the late 1980s and early 1990s continue to cause Reclamation's staffing level to trend downward, and the staffing mix has changed in recent years. In 2010, for instance, Reclamation staff was about 29 percent smaller than in 1993 and includes a much higher percentage of computer and non-engineering specialists than previously.

Water users, under contract with Reclamation, operate and maintain many projects. As Reclamation enters into additional partnerships with benefi-

ciaries of project water and electricity and shifts increasingly away from construction development projects toward water management activities, Reclamation staffing levels are expected to shrink further in the twenty-first century. These sorts of changes are manifestations of the natural organic evolution of Reclamation as it changes to meet shifting public perceptions and needs in the West.

Michael L. Connor
Commissioner
Bureau of Reclamation

AUTHORS' PREFACE AND ACKNOWLEDGEMENTS

During the second half of the twentieth century the Bureau of Reclamation underwent transitions reflected in this volume: *From Developing to Managing Water, 1945-2000*. Change over time marked these years in Bureau of Reclamation history. Beginning with the turbulent postwar years, Reclamation encountered a new urbanized and industrialized West, with multiple-purpose and competing water needs. Overseas, Reclamation engineers played prominent roles in spreading America's technical expertise to a war-torn world. Projects abroad allowed Reclamation to expand its own horizons at home and view water projects from multiple perspectives. Reclamation engineers and planners focused on entire river basins to achieve greater efficiency in water resource management, to ensure ample supplies of water for agricultural and urban needs, and to maximize hydropower production from Reclamation dams.

By the end of World War II, the Bureau of Reclamation was the world's foremost dam builder, a major producer of hydroelectricity, and water supplier to irrigation projects and urban centers. Almost fifty years of experience in dam building and hydroelectric development in the United States made the Bureau of Reclamation a world-renowned developer of water resources with monumental undertakings such as Hoover, Grand Coulee and Shasta dams to its credit. When the United States assumed a major role in "world rehabilitation" following the devastation of the world war, the nation called upon the Bureau of Reclamation for assistance. Overseas tasks involved the Bureau of Reclamation in "the revolution of rising expectations" amongst emerging nations formerly under colonial rule. In addition, the Bureau of Reclamation became a key player in American Cold War efforts to defeat the appeal of international communism as a path to economic development. To say the least, international activities form an important chapter in the history of the Bureau of Reclamation in the latter half of the twentieth century.

From 1945 to 1968, the construction record of the Bureau of Reclamation was impressive. Reclamation dams and powerplants helped continue the rapid pace of growth in the American West's urban centers, which began during World War II and still continues. Water stored in Reclamation reservoirs assisted in expanding agricultural production in the West and meeting the demands of urban customers. Reclamation powerplants played prominent roles in supplying electricity that fueled the phenomenal growth of western industries and metropolises. What one observer has termed the "go-go years," this period witnessed the Bureau of Reclamation involved in construction activities throughout all seventeen western states building structures both large and small and affecting nearly every community from the largest urban center to the smallest hamlet.* This era sealed Reclamation's reputation as one of the greatest construction organizations in the world, as it successfully met construction challenges and perfected engineering techniques and practices that were emulated throughout the world.

By 1968 signs appeared on the horizon that foreshadowed the end of this booming construction period. Since the end of World War II, the United States experienced a period of unprecedented economic growth that went hand-in-hand towards achieving funding for Reclamation's construction activities. By the mid 1960s the social policies of President Lyndon B. Johnson's "Great Society," the growing quagmire in Vietnam, and inflation placed a tremendous burden on the nation's economy, resulting in greater competition for scarce federal funds. In addition a new environmental ethos emerged in American society that raised questions about harmful effects of human activities on the natural world. One area of concern took direct aim at dam building and Bureau of Reclamation water projects. As a result of this societal shift, Congress passed a number of environmental laws that required Reclamation to include environmental considerations when planning and constructing water projects. Though Congress continued to approve Reclamation projects during this time, environmental regulations and decreasing budget appropriations slowed down the construction progress the Bureau of Reclamation had enjoyed since the end of World War II.

Reclamation's work during the "era of big dam building" did not occur without rumblings from a burgeoning environmental movement that demanded and received modification of proposed dams on the upper Colorado River. By the 1970s an emboldened environmental movement forcefully criticized Bureau of Reclamation planning and projects after the passage of the National Environmental Policy Act of 1969 and establishment of the Environmental Protection Agency the following year. EPA oversight and a growing public suspicion of the benefits of dams in preference for wilderness surrounding wild and scenic rivers curtailed the drive to build more dams. Reclamation's engineering reputation received a severe setback when Teton Dam in southeastern Idaho collapsed in June 1976, resulting in hundreds of millions of dol-

* Marc Reisner, *Cadillac Desert: The American West and Its Disappearing Water*, Revised and Updated (New York: Penguin Books, 1993), 145.

lars in property damage and the loss of eleven lives. On the heels of the Teton failure, President James (Jimmy) Carter's so-called "hit list" in 1977 took direct aim at the nation's water resources development agencies with the goal to rein in what he considered wasteful federal spending. These developments further eroded public support for dam building, and Reclamation once again found itself on the verge of another period of transition.

From 1980 to 2000, the Bureau of Reclamation began the process of transitioning itself from a construction organization into a water management agency. This transition was full of fits and starts, and Reclamation encountered new issues in its efforts to serve western water users. The West continued its phenomenal growth, and the region's limited water supply was hotly contested between urban and agricultural water users. In addition, Native Americans, long neglected in the traditional uses of western water, demanded greater control over their water resources. Environmental regulations required the readjustment of water diversions for fish and wildlife enhancement and water quality controls, placing further strains on the West's limited water supply. Because of its dams and conveyance systems, Reclamation found itself involved with water distribution issues often refereeing disputes among various water users.

In some instances, this meant a revision of Reclamation's commitment to the nineteenth-century ideal of the small farm (the much argued 160-acreage limitation rule) in favor of larger units of agricultural production characteristic of economies of scale in the late twentieth century. In other instances, the Bureau of Reclamation assumed the role of water arbiter and facilitator amongst competing interests for the scarce water resources of an arid environment. All meant a transition away from its origins and functions at the beginning of the twentieth century as a government service organization under the name of the U.S. Reclamation Service to a new organization with new purposes as the Bureau of Reclamation attempted to recreate itself in the closing years of the last century.

The Bureau of Reclamation played a prominent role in the phenomenal growth of the American West during the last half of the twentieth century. In terms of water resources development, this expansion was never easy because of competition for the West's limited water supply. From 1945 to 1968, Reclamation's success in securing funding for water projects came from a close partnership among the Bureau, western water users, and their elected representatives. Scholars term this relationship an "iron triangle." According to political scientist Daniel McCool, "an iron triangle is an informal political alliance that forms to influence a specific public policy to its advantage." McCool maintains that iron triangles influence "the allocation of government goods and services" whereby elected representatives receive credit for meeting constituents' needs, government agencies achieve expanded budgets and influence, and "interest groups get what they want from government." Reclamation's growth and achievements during the "dam-building era" were in no small part the result of an effective and powerful triangular alliance.[†] After 1968 as federal budgets tightened and environmental concerns gained in importance in American society, the reclamation "iron triangle" lost much of its effectiveness. Dam building slowed from the lack of prime dam sites in western America and the general public's lack of enthusiasm for water development projects.

These developments coincided with an ideological shift in American culture. From its inception in the Progressive Era, utilitarian conservation values drove Bureau of Reclamation activities. The utilitarian doctrine expressed by Theodore Roosevelt and Gifford Pinchot argued for proper scientific management of natural resources for the greatest good, for the greatest number, over the longest time. In terms of water resources development, this meant utilizing the West's limited water supply to its fullest extent; even to the point of begrudging "wasted water" flowing to the sea. In the West, water insured progress and growth, and Reclamation dams, canals, and powerplants assured water usage to its fullest potential for the benefit of society. By the 1980s, a more urban and environmentally conscious western population challenged the utilitarian conservation ethic espoused by Reclamation and most western water users, forcing a diversification of water use to include greater recognition of the effects of dams on fish and wildlife and the natural environment. The Bureau of Reclamation refocused its programs and personnel to respond to these changing values. Reclamation did not abandon its traditional constituency of western irrigators, but instead developed procedures and policies to meet increasingly diverse demands for the West's limited water supply. It is a challenge accepted by the women and men of the Bureau of Reclamation today.

The inclusion of the final chapter, "Selling Reclamation," attempts to analyze the modes of representation employed by the Bureau of Reclamation in its various campaigns to explain its mission and accomplishments to the

† Daniel McCool, *Command of the Waters: Iron Triangles, Federal Water Development, and Indian Water* (Tucson: University of Arizona Press), 5.

American public. The images Reclamation produced reveal the many changes in Reclamation's mission over the past one hundred years. More importantly, the photographs, works of art, and films provide visual evidence of the transformation of the American West from a nineteenth-century arid wasteland to a region of great urban centers and desert that does, in places, indeed bloom.

The second volume of the history of the Bureau of Reclamation offers a discussion and examination of the eventful years in the latter part of the twentieth century. Like many projects, this volume is a joint effort, and we would like to take this opportunity to express our gratitude to those who helped in its production. A special "thank you" goes out to the many librarians and archivists who gave invaluable assistance, especially those at the National Archives and Records Administration in Lakewood, Colorado, and College Park, Maryland; the Library of Congress in Washington, D.C.; the American Heritage Center at the University of Wyoming in Laramie, and the Mathewson Knowledge Center at the University of Nevada, Reno. We would like to acknowledge the assistance of Richard Ives, head of the Bureau of Reclamation Office of Foreign Affairs, who allowed us access to office files that helped to tell this important, but rarely examined, aspect in Reclamation history. The authors would also like to thank those who graciously reviewed and offered beneficial comments on portions of the manuscript, in particular Professors C. Elizabeth Raymond of the University of Nevada, Reno; Donald J. Pisani, emeritus, University of Oklahoma; and Donald C. Jackson of Lafayette College. In addition, we are grateful to Dr. Don Fowler, professor emeritus at the University of Nevada, Reno, who took time out of his schedule to discuss his experiences as part of the Glen Canyon Salvage Program. The authors hope that the following pages add some clarity and a degree of insight to the often labyrinth-like road map followed by the Bureau of Reclamation in the last dynamic fifty years of the twentieth century.

Finally we are grateful to Dr. Brit A. Storey, lead historian of the Bureau of Reclamation, for his unwavering support of this project. His editorial comments, critiques, and overall direction of the project were indispensible and greatly appreciated.

Andrew H. Gahan and William D. Rowley

SENIOR HISTORIAN'S PREFACE AND ACKNOWLEDGEMENTS

As we publish Volume 2 of the history of Reclamation, we are in the 110th year of Reclamation's historic work. Volume 2 covers from the end of World War II until 2000 and is the last volume in this project.

Reclamation's construction program remained very active into the early 1990s, but construction has slowed as political, environmental, and budgetary challenges to Reclamation's programs, as well as continued evolution of Reclamation's programs and the West, have occurred. Reclamation is now moving into the role of manager of the water, hydroelectric, and recreation resources it has developed as it works to perform its primary missions of water and hydropower deliveries while complying with the multifaceted and evolving, sometimes conflicting, legal and political direction that all large Federal bureaus receive.

During an active construction period after World War II, Reclamation saw some one hundred new projects placed in construction. Reclamation built these new projects while parallel work programs completed the large Depression Era projects which largely languished during World War II due to lack of essential manpower, budget, and construction matériel. Among the new postwar projects were Pick-Sloan Missouri Basin Program projects like the Armel Unit in Colorado, the Kirwan Unit in Kansas, the Canyon Ferry Unit in Montana, the Ainsworth and Farwell units in Nebraska, Jamestown Dam and Reservoir in North Dakota, the Angostura Unit in South Dakota, and the Owl Creek Unit in Wyoming. There were numerous other projects around the American West also. Beginning in the 1960s, Congress began to authorize some water projects, like the Norman Project in Oklahoma and the Canadian River Project in Texas, primarily for municipal water supply. Other projects, like the Cachuma Project in California, Congress authorized for both irrigation and municipal and industrial water supply. Authorizations during the post-World War II period tended to be smaller projects. However, a few large, spectacular construction projects like the Central Arizona Project and the Third Powerhouse at Grand Coulee Dam, while out of the ordinary in this period, joined the inventory of Reclamation's other major projects.

Most of Reclamation's large irrigation projects came into being before World War II although Reclamation built them out after the war. Thus, the major Depression Era projects, the Central Valley Project in California; Colorado-Big Thompson Project in Colorado; Boulder Canyon Project in Arizona, California, and Nevada; and the Columbia Basin Project in Washington; combined with the older Minidoka Project in Idaho, all had major construction and additions after World War II, and they represent about 50 percent of Reclamation's irrigated acreage—about 5,000,000 acres in an average water year. The other 50 percent of Reclamation irrigated acreage is included in the 180-plus other Reclamation projects, the majority of which Reclamation built after World War II. Irrigated agriculture uses 80 to 90 percent of Reclamation-developed water while some 10 to 20 percent of Reclamation-developed water supplies municipal and industrial uses for about one-third of the population of the American West.

A personal note on my career at Reclamation from 1988 to 2013. Reclamation gave me great opportunities to develop its history program and, as my first supervisor Jim Maxon put it, "go out there and show historians that Reclamation now is doing history." I am grateful that Reclamation chose me, i.e., gave me the opportunity, to develop the history program at a bureau of such importance to development of the twentieth century American West. I regret only that I was unable to sell development of a technological history of Reclamation to the executive leadership of Reclamation. Reclamation is generally known for its dam construction and hydroelectric developments, but the bureau had to gather the knowledge of the engineering and other communities and develop an intricate, innovative, science-based constellation of new engineering and technology to serve as the foundation upon which its dam design and construction developed and depended. Without these innovations and inventions the work would not have been possible. The list is extensive and includes: dealing with the tremendous pressures of high head hydroelectric systems, gates, and valves; spillways; adapting concrete to specific natural conditions and needs and testing it; laboratory modeling of hydraulic problems; electricity transmission; underwater paints; construction techniques; tunneling; lining canals while water flowed through them; water quality; urban and rural water conservation; desalinization; new construction techniques; effective fish ladders for both strong and weak swimmers; evolution of pipe design and pipe laying for water carriage over long distances; maximum probable flood forecasting; etcetera, etcetera, etcetera. These foundational chapters of Reclamation's history still hide from us in the shadows of Arrowrock, Belle Fourche, Buffalo Bill, Davis, East Park, Flaming Gorge, Folsom, Friant, Glen Canyon, Grand Coulee, Hoover, Hungry Horse, Owyhee, Parker, Pathfinder,

San Luis, Shasta, Stony Gorge, Theodore Roosevelt, and all the other spectacular dams that populate Reclamation's historic past, present, and future.

My thanks, in particular to all those who have in any way assisted and supported Reclamation's history program during the years that I have been here. I am somewhat reluctant to create a list since it is inevitable that someone of importance will be left out, but there are a few people who particularly provided policy and program support that cannot be ignored.

Reclamation's history program has enjoyed the support of every commissioner since I began to forward a program of historical research and publication in the early 1990s. Every living commissioner, except one, generously took time from busy schedules to do oral history interviews with me. The publication (six volumes), oral history (over 200 interviewees and over 900 hours of tape), and project history activities developed by the history program could not have occurred without the support, particularly, of commissioners Daniel P. Beard, Eluid L. Martinez, John W. Keys III, Bob Johnson, and Michael Connor and of various other executive staff, especially Deputy Commissioner Joe D. Hall. I was very gratified that both commissioners Keys and Johnson expressed their desire to have this volume published during their term in office—though I was unable to meet their hopes.

Since about 1991 I've been most fortunate in having supportive supervisors who have provided me the opportunity to hire part-time students, particularly in the summer, to do various projects. Our students have developed over 180 brief histories of projects which are now available on Reclamation's history website, they have provided editorial assistance in the layout of oral histories and project histories, and they have gladly undertaken special short term research projects. Some thirty people were included in this activity, and many of them are now out in the world as public historians and academics. Particularly, Robert Autobee, Adam Eastman, Lara Godbille, Stephen Bogener, Andy Gahan, Leah Glaser, Toni Rae Linenberger, Christopher McCune, Zachary Redmond, Jedediah Rogers, Wm. Joe Simonds, Eric Stene, Garrit Vogesser, and Roy Wingate came to Reclamation for some part of their careers.

Dr. Andrew Gahan worked for Reclamation while providing Professor William D. Rowley research and editorial assistance for both volumes of this history of Reclamation. Dr. Gahan also took over research and writing responsibilities when Professor Rowley's suddenly increased teaching schedule, necessitated by the recent economic crises at virtually all American public universities, precluded him from completing the work. Professor Rowley has, however, actively remained in the publication and editorial process to the last in spite of my pushy prodding to see outlines, chapters, editing and revision, and proofing.

Others who particularly assisted with publication of Volume 2 include:

- Dianne Powell and Cynthia Fields Cunningham in the Denver office library were of particular assistance.
- The Senate History Office's on-line Biographical Directory of the Congress of the United States;
- Carter Grant and John Lonnquest in the History Office of the U.S. Army Corps of Engineers;
- Margaret Schoneman in Reclamation's Ephrata office;
- Winetta Owens in the Mid-Pacific Region;
- Danica Rice and Emme Woodward in Reclamation's Lower Colorado Region;
- Richard Ives, the head of Reclamation's Native American affairs and international affairs offices, and Mary Mascarenhas and Leanna Principe of his staff;
- Professor Emeritus Don Fowler at the University of Nevada-Reno;
- Jim Maxon, John Lambert, Ronald (Rusty) Schuster, and Richard Rizzi my supervisors at various points in all this;
- Kathryn Ehler, Jaclyn Zechman, and Barry Waryanka in Reclamation's printing office.
- Rita Sudman and Curtis Leipold at the Water Education Foundation in Sacramento.
- Clark Bishop in the Power Resources Office who provided the data on Reclamation's generating plants.
- The Pacific Northwest, Mid-Pacific, Great Plains, and Upper Colorado regions invested the time needed to provide comments that improved this volume.
- Professors Donald J. Pisani and Donald C. Jackson who have provided peer review, planning, and editorial assistance to Reclamation and the authors at various stages of development of this two volume history of Reclamation.
- Patricia Cox in the Phoenix Area Office.

In addition, Charles Brown deserves recognition for his contributions to the success of the history program because he did graphics work and laid out all the history program's publications in preparation for publication, except this last volume which is the work of Network Typesetting Inc. of Highland Park, New Jersey.

In good bureaucratic fashion, each person on Reclamation's executive team was asked to provide comments on the manuscript as we began serious editing. Comments from those reviews improved the manuscript and caught occasional errors.

Last, but by no means least, Reclamation has let me allow our authors great freedom in their research and writing. Commissioner John Keys III, though no one had ever told me this, commented during his opening remarks at Reclamation's centennial history symposium that Reclamation's executive staff made this decision after some hard discussions. I have watched other bureaus' history initiatives founder because some manager or executive felt "I don't want someone to write or say something that reflects negatively on the (fill in the bureau/department)." That attitude is the kiss of death for good history. The truth always is that there are the good and the bad in the past of any large bureau, just as there are good and bad managers and good and bad decisions in any bureau. If we edit the past to include only the good, why bother? For if that is the approach we are not providing staff the background information to help them make good decisions of integrity in the present. AND, knowledgeable readers, particularly including historians, will out-of-hand reject the work as simply a public relations effort dressed up in the duds of "history." The other side of this coin is that, while Reclamation has supported and published this work, because of the nature of the intellectual process and historical method, the selection of facts and their interpretation are the authors' and do not necessarily represent the official views and policies of the Bureau of Reclamation—and they may not be cited as such.

Brit Allan Storey, Ph.D.
Senior Historian
Bureau of Reclamation

CHAPTER 8:

RECLAMATION ADJUSTS TO POSTWAR AMERICA

Introduction

By September 1945 the Bureau of Reclamation looked to new challenges after a decade and a half dominated by the Great Depression and World War II. With the conclusion of these tumultuous years, Reclamation congratulated itself on its new and growing importance to the nation. In the previous twenty years, the Bureau successfully mobilized engineering and organizational skills to supervise the building of some of the world's largest concrete dams. Monumental dam construction not only provided new water supplies and hydroelectric energy, but it also served the political and ideological goals of the New Deal. The construction of vast water and hydroelectric systems was part of a public works program designed by the administration of Franklin D. Roosevelt to free the nation from the grip of the Great Depression. When World War II brought an end to the Depression, Reclamation's great dams energized war industries that helped lead to urbanization and industrialization of the Far West.

During these turbulent decades, however, the Bureau of Reclamation struggled to come to grips with two competing ideologies within its ranks. Commissioner Elwood Mead (1924-1936) kept alive the idealism of the 1902 Reclamation Act of providing opportunity to small farmers even in the midst of building the big dams of the 1930s. However, his successor John Page (1936-1943), with a strong engineering background, expressed less concern for the social-agricultural mission of Reclamation, turning greater attention to the larger benefits-costs received from the sale of hydroelectric power. One source described the difference between the two commissioners as Page representing "a reassertion of the engineering dominance within Reclamation that dated back to the time of Newell and A. P. Davis." Engineers, according to this view, seemed more impressed with the revenues delivered by turbines and dynamos than with defending and championing "the social aspects of Reclamation policies" embodied in the 160 acre rule to ensure that Reclamation water served only the interests of the small farmer.[1]

---

1 Donald C. Swain, "The Bureau of Reclamation and the New Deal, 1933-1940," *Pacific Northwest Quarterly* 60 (January 1969): 146.

8.1. Hoover Dam, then known as Boulder Dam, in 1936. Photographer: Ben Glaha.

8.2. Grand Coulee Dam on June 14, 1948, discharged a flood of 590,000 cfs. Photographer: F. B. Pomeroy.

Nevertheless, the Bureau of Reclamation could feel justifiably proud on emerging from depression and war as one of the leading engineering and construction organizations in the world. Mead and Page brought Reclamation through the Depression and into the early years of the war, garnering professional and public acclaim for Reclamation's accomplishments. When Harry Bashore became commissioner in 1943, he showed greater dedication to the social and land reform ideals of federal reclamation. He had the complete support of Secretary of the Interior Harold Ickes, one of the most aggressive proponents of the New Deal, who encouraged this renewed idealism in Reclamation by urging strident enforcement of the 160 acre rule for farms receiving federal water. Bureau of Reclamation officials followed up in testimony before Congress emphasizing their commitment to longstanding policies to foster the creation of small farms on Reclamation projects. Reclamation rhetoric especially focused on this important social commitment in arguing for appropriations that competed with the Army Corps of Engineers and even proposals for the creation of independent river basin development.[2]

8.3. John C. Page served as Commissioner of the Bureau of Reclamation from 1936 to 1943.

[2] Alexander J. Field, "The Most Technologically Progressive Decade of the Century," *American Economic Review* 93 (September 2003): 1399-1413; Robert D. Leighninger Jr., *Long-Range Public Investment: The Forgotten Legacy of the New Deal* (Columbia, South Carolina: University of South Carolina Press, 2006); *Sacramento Bee*, February 10, 1944, quotes Commissioner Bashore's testimony before Congress arguing that Reclamation should develop the Kern River as part of the Central Valley Project and not simply as a flood control project under the Corps that would allow reservoir water to be exempt from the 160 acre limitation and therefore encourage further land monopolization in the Central Valley; Kathleen B. Freeland, "Examining the Politics of Reclamation: The 1944 Acreage Limitation Debate in Congress," *The Historian* 67 (Summer 2005): 217-33.

8.4. Elwood Mead, Commissioner of the Bureau of Reclamation from April 1924 to his death in January 1936.

Moreover, the Bureau of Reclamation was eager to get back to work on projects left unfinished and neglected during the war. Reclamation officials looked forward to continuing construction on the Central Valley Project in California, beginning the irrigation phase of the Columbia Basin Project in Washington, and implementing a long list of proposed projects in the upper and lower basins of the Colorado River and on the Missouri River. At the same time, Reclamation faced several obstacles during this period of postwar readjustment. Californians tenaciously fought off Bureau of Reclamation efforts to enforce the 160 acre rule in the Central Valley, vilifying Reclamation Commissioner Michael Straus in the process. Attacks on Straus reflected the larger struggle in American society between those wishing to return to the reform practices of the New Deal and those determined to put an end to the so-called "socialist tendencies" of New Deal policies. This debate not only led to controversy in California's Central Valley, but also renewed the bitter divide over public versus private power. In addition, anti-New Deal forces locked horns with the Truman administration's attempts to establish Tennessee Valley Authority (TVA)-like programs for the Columbia and Missouri river basins. And finally by the mid-1950s, critics questioned the entire Reclamation program, arguing that its activities were nothing more than pork barrel spending that subsidized Western growth and brought no appreciable benefits to the rest of the nation.

8.5. Harry W. Bashore, Commissioner of the Bureau of Reclamation from August 1943 to December 1945.

Postwar Transitions

As late as 1945, dedicated New Deal visionaries held out great hopes for what full development of the Columbia River basin promised to the people of the Pacific Northwest. With its immense bounty of natural resources, the area offered power, water, and land—all the basics for river basin development in cooperative enterprises that promised to foster a sense of regional cultural identity and prosperous community life. A leading advocate of this regional idealism was Leland Olds, chairman of the Federal Power Commission. Like many others, Olds believed the Columbia River basin uniquely suited to federally-sponsored regional growth. This meant full utilization of the region's natural resources, best represented by hydroelectric development of the Columbia River, on a foundation of family farms. In the summer of 1945 Olds shared his vision of the Pacific Northwest's future while addressing a meeting of the Columbia River Development League in Wenatchee, Washington. In his talk entitled "Building a Regional Culture," he saw the challenge not only in terms of utilizing the material resources of the region, certainly hydroelectricity as the basis, but also in terms of tapping "the spiritual resources of the people." For Olds, the great dams along the Columbia River and its tributaries represented the material structures of culture just as the great cathedrals of religion repre-

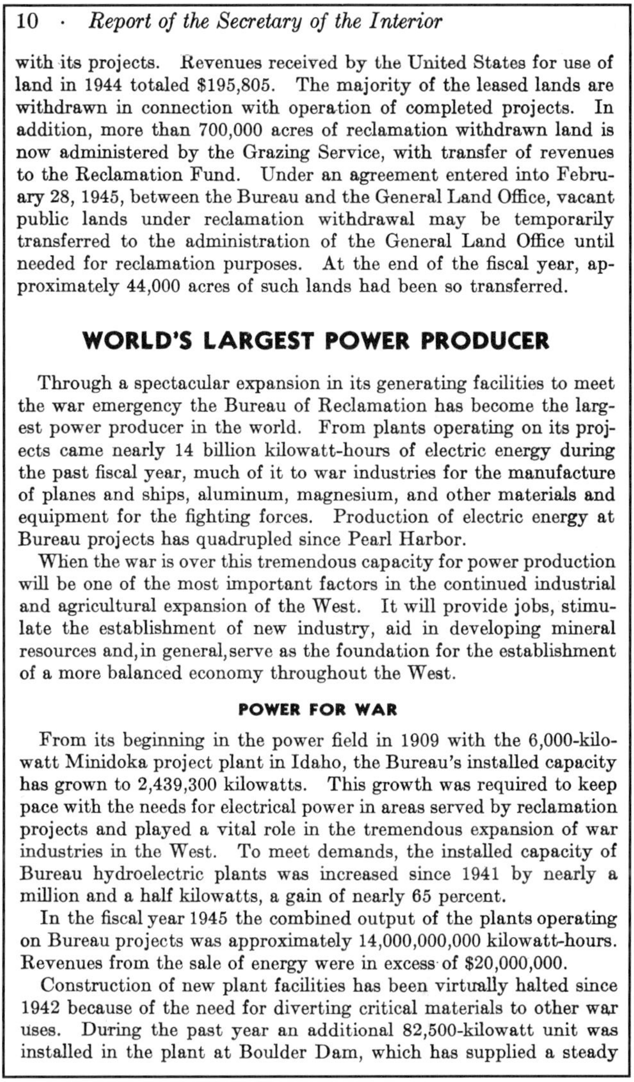
10 · *Report of the Secretary of the Interior*

with its projects. Revenues received by the United States for use of land in 1944 totaled $195,805. The majority of the leased lands are withdrawn in connection with operation of completed projects. In addition, more than 700,000 acres of reclamation withdrawn land is now administered by the Grazing Service, with transfer of revenues to the Reclamation Fund. Under an agreement entered into February 28, 1945, between the Bureau and the General Land Office, vacant public lands under reclamation withdrawal may be temporarily transferred to the administration of the General Land Office until needed for reclamation purposes. At the end of the fiscal year, approximately 44,000 acres of such lands had been so transferred.

WORLD'S LARGEST POWER PRODUCER

Through a spectacular expansion in its generating facilities to meet the war emergency the Bureau of Reclamation has become the largest power producer in the world. From plants operating on its projects came nearly 14 billion kilowatt-hours of electric energy during the past fiscal year, much of it to war industries for the manufacture of planes and ships, aluminum, magnesium, and other materials and equipment for the fighting forces. Production of electric energy at Bureau projects has quadrupled since Pearl Harbor.

When the war is over this tremendous capacity for power production will be one of the most important factors in the continued industrial and agricultural expansion of the West. It will provide jobs, stimulate the establishment of new industry, aid in developing mineral resources and, in general, serve as the foundation for the establishment of a more balanced economy throughout the West.

POWER FOR WAR

From its beginning in the power field in 1909 with the 6,000-kilowatt Minidoka project plant in Idaho, the Bureau's installed capacity has grown to 2,439,300 kilowatts. This growth was required to keep pace with the needs for electrical power in areas served by reclamation projects and played a vital role in the tremendous expansion of war industries in the West. To meet demands, the installed capacity of Bureau hydroelectric plants was increased since 1941 by nearly a million and a half kilowatts, a gain of nearly 65 percent.

In the fiscal year 1945 the combined output of the plants operating on Bureau projects was approximately 14,000,000,000 kilowatt-hours. Revenues from the sale of energy were in excess of $20,000,000.

Construction of new plant facilities has been virtually halted since 1942 because of the need for diverting critical materials to other war uses. During the past year an additional 82,500-kilowatt unit was installed in the plant at Boulder Dam, which has supplied a steady

8.6. The Secretary of the Interior's 1945 annual report bragged on Reclamation as the largest single power producer in the world.

sented the energy of religious spiritual beliefs.[3]

8.7. Harold L. Ickes served as Secretary of the Interior from March 4, 1933, to February 15, 1946.

Olds explained that nearly fifteen years earlier he had visited the state of Washington upon hearing that the people of the Pacific Northwest dreamt of building a regional culture "as a living purpose in their hearts." His traveling companion was the Irish cooperative leader and poet, George Russell whose pen name was simply AE, and author of *The National Being: Thoughts on Irish Polity* (1916). Similar to others of the time seeking answers to the economic and social confusion caused by the Great Depression, Olds and his Irish friend believed that answers to the problems of the time lay in developing strong regional traditions and cultures that would outweigh the perils of an overly centralized industrial economy. As Olds put it,

> We were traveling through a land to warn people of the fate of great over-centralized city civilizations, which divorced men from the soil. We were urging the building of a balanced rural-industrial social order, infused with the spirit of cooperation, as the noblest of undertakings.

On this trip, he found in the cities of the Pacific Northwest—Missoula, Moscow, Pullman, Seattle, Portland, and Eugene among others—a desire to identify with a strong regional culture among educators, state officials, businessmen, and labor leaders. All sought, he said, "Something more than a mere regional reflection of the cultural trends in the Eastern metropolis. They

---

[3] Leland Olds Address to the Columbia River Development League, Wenatchee, Washington, August 4, 1945, RG 48, Records of the Department of the Interior, Entry 779, Box 15, National Archives and Records Administration, College Park, Maryland; hereafter cited as RG 48; the Columbia River Development League was a group of project boosters who favored construction of Grand Coulee Dam and utilizing pumping as the means to bring water from the reservoir to the irrigated fields, see Paul C. Pitzer, *Grand Coulee Dam: Harnessing a Dream* (Pullman, Washington: Washington State University Press, 1994), 47-8.

8.8. Hoover Dam's Nevada Powerplant. January of 1996. Photographer: Andrew Pernick.

8.9. U.S. Army Corps of Engineers construction at Folsom Dam required careful preparation of the dam's foundation. October of 1952. Photographer: L. R. Murphy. Courtesy of the U.S. Army Corps of Engineers.

8.10. Relocation of a county road over Spring Creek during construction of Shasta Dam in June of 1949. Photographer: W. H. Colby.

wanted an indigenous culture which would make its unique contribution to the cultural mosaic of the nation."

His words resounded with a belief that Americans must proudly embrace the rich and distinctive forms of art, music, literature, and handicrafts found in their own various geographic and cultural regions of the United States. In the South voices of the Nashville School had already made themselves heard as they recognized the paralyzing themes of the Old South and sought to move away from the faceless and anti-communitarian impulses of the business oriented New South. Some from this region called themselves "the traditionalists" and organized many of their ideas around the landmark work by W. J. Cash entitled, *The Mind of the South.* All was an indication that some saw hope in the emergence of a promising new regionalism in the United States after the collapse of the national economy in the Great Depression. For Olds, the Pacific Northwest was no exception. It too held the promise of developing further into a distinctive regional district driven by the newfound potential for power and river development anchored around the Columbia River system.[4]

[4] Olds voiced a civic minded regional religion that verged on anti-modernism despite his praise of the technology of dams, irrigation, and the wide distribution of electrical power; see W. J. Cash, *The Mind of the South* (New York: A. A. Knopf, 1941). The historical context of this

Olds expressed to his audience his belief in strong regional cultures, whose development promised "the salvation of our democratic civilization." This movement, he added, "will contribute greatly to that material and spiritual strength of democracy through which alone we can hope to solve the pressing problems of the postwar world." Olds warned that only if the people of the region used their resources cooperatively would they avoid the wreck of civilization that selfish exploitation causes. The challenge was to hold to a vision and persevere "in spite of all the obstacles placed in your way by those who worship Mammon." He believed there was not a region in the United States or in the world that possessed more potential for achievement than the Pacific Northwest with its equable climate, abundance of water, fertile soil, and minerals. Olds declared that the people of the Columbia River basin had the opportunity to create "a veritable Garden of Eden, provided greed does not enter the Valley."

8.11. Michael W. Straus, Commissioner of the Bureau of Reclamation from December 1945 to February 1953.

Evil, of course, lurked in the forces of avarice that sought to control the natural resources of the Pacific Northwest. For New Dealers like Olds, the archetypical enemies of the good society—private power, land monopolies, and corporations—must be guarded against to secure for the region a safe and prosperous future. The Columbia Basin Project was foremost in his hopes to achieve that goal because, "The waters of the Columbia River and its tributaries, properly controlled and used, will provide the key to the development." The harmonious development of irrigation, navigation, and power was essential for that future. Of the three, "the greatest contribution which the Columbia River will make to the development of your regional civilization will unquestionably come from its enormous potentialities in the way of hydroelectric power." Without that power, Olds warned, the Pacific Northwest would remain a colonial economy "tributary to the industrial East."

regionalist revival is explored in Robert L. Dorman, *Revolt of the Provinces: The Regionalist Movement in America, 1920-1945* (Chapel Hill: University of North Carolina Press, 1993); for the opposite notion see William Leach, *Country of Exiles: The Destruction of Place in American Life* (New York: Pantheon Books, 1999).

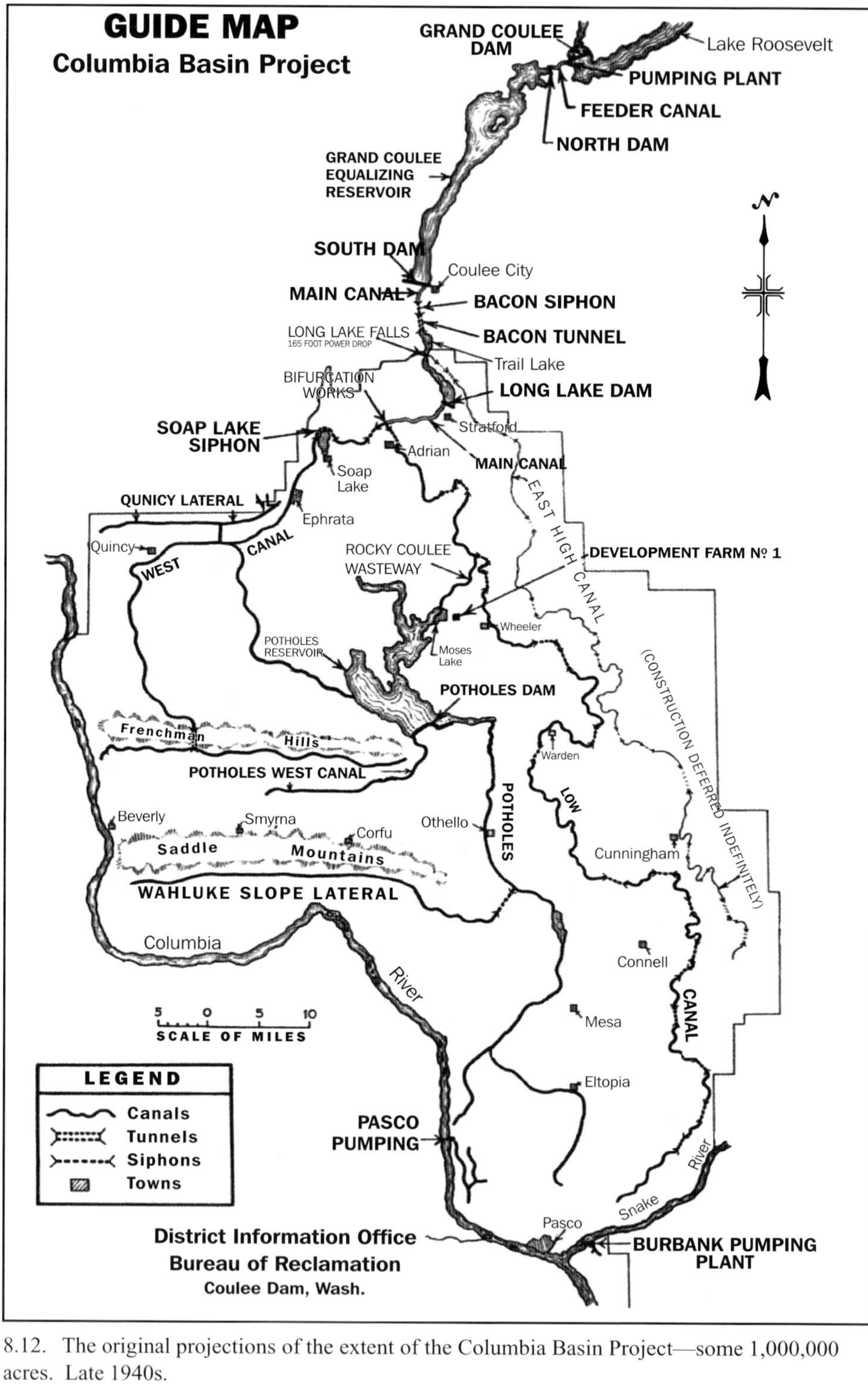

8.12. The original projections of the extent of the Columbia Basin Project—some 1,000,000 acres. Late 1940s.

Noting that war industries doubled the use of power in the Pacific Northwest, Olds projected that building new dams meant not only the additional benefits from navigation, irrigation, and flood control, but also increasing the industrial diversification of the region. Olds foresaw a string of dams on the Columbia River and its tributaries, bringing important additions to fuel economic growth. These dams would work in conjunction with the Army Corps of Engineer's Bonneville Dam and Reclamation's Grand Coulee Dam, building a vast network of electricity. Olds's vision reflected the basic public power argument on why government needed to take the initiative if private power interests refused. The point was that the availability of low-cost power from these facilities in the end created market-stimulating expansion of agriculture, industry, and trade.[5]

Olds's statements rested on an ideological foundation that stressed the role of the federal government to assist in regional growth. In one sense, it was an effort to reinvigorate reform-minded New Dealers pushed to the sidelines by the wartime emergency. On the other hand, visions of the future, such as those expressed by Olds, held out great hope for the Bureau of Reclamation. It assured the people of the Pacific Northwest that government, i.e., Reclamation, had not forgotten its obligation to the region. Taken on a wider scale, Olds's enthusiastic outlook foretold a great period of transition for the American West in which water resource development projects had a tremendous part to play.

Perhaps not with the same ideological vigor expressed by Leland Olds, the Bureau of Reclamation also looked forward to the postwar period. Yet, the challenges of new administrative directions and their successful implementation paled in comparison to the issues the Bureau of Reclamation faced in the shifting sands of agency functions and jurisdictions in the postwar years. The Army Corps of Engineers continued its competition with the Bureau of Reclamation on water development projects, as did the Department of Agriculture in its desire to play a larger role in project planning. Lurking behind the ambitions of these agencies was another "threat" to what the Bureau of Reclamation regarded as its domain. Some believed the extensive Columbia River basin lent itself to a TVA-like river authority, which meant a single government entity coordinating all aspects of river basin development in the Pacific Northwest. All of these questions came into play for the Bureau of Reclamation in the critical transitional years following World War II.

---

5 Leland Olds Address, in RG 48, Entry 779, Box 15.

At the end of World War II, the Bureau of Reclamation faced difficult external challenges and internal administrative changes. During the war, the War Materials Board had drastically curtailed construction on projects redirecting resources to essential war needs. The war denied Reclamation resources for the initiation of new projects and delayed the full development of projects already underway, most notably Parker Dam on the Colorado River along the California-Arizona border and the expansion of irrigated agriculture on the Columbia Basin Project.[6] However, the war helped to transform the public image of the Bureau of Reclamation. Reclamation's major dams, Hoover, Coulee, and Shasta, sent critical hydroelectric power supplies to aircraft factories, shipyards, and munitions producers, creating a formidable wartime economy in the West. These monumental dams and other projects reinforced the Bureau of Reclamation's image as the principal water developer in the West not only for farms, but increasingly for hydroelectricity to serve a new urban West. The completion of Hoover Dam in the midst of the Great Depression in 1935, Grand Coulee Dam in 1941 and progress on California's Central Valley Project that included the building of Shasta Dam (completed in 1944) meant that the Bureau of Reclamation had increased power supplies in eleven western states 84 percent by the end of 1944. These gains made the Bureau of Reclamation one of the largest single producers of electricity in the world.

With the end of the war, the Bureau of Reclamation began to complete unfinished projects. Construction started on the $4,688,000 Kortes Dam and Powerplant in Wyoming, while the Bureau also awarded contracts for Granby Dam in Colorado and the Ram Horn and Prospect Mountain tunnels on the Colorado-Big Thompson Project.[7] In 1946 Congress authorized funds to build Davis Dam on the California-Nevada border 67 miles below Hoover Dam—supporting further development of Colorado River resources. Construction

---

6 United States Department of the Interior, Bureau of Reclamation, *The Colorado River: "A Natural Menace Becomes a Natural Resource": A Comprehensive Report on the Development of the Colorado River Basin for Irrigation, Power Production, and Other Beneficial Uses in Arizona, California, Colorado, Nevada, New Mexico, Utah, and Wyoming* (Washington, D.C.: U.S. Government Printing Office, March 1946) notes that construction of the Parker Dam was temporarily halted by an order of the War Production Board, 164; work on the Friant-Kern and Madera Canals on the Central Valley Project was also put on hold by the War Production Board, see also Mary Montgomery and Marion Clawson, *History of the Legislation and Policy Formation of the Central Valley Project* (Berkeley, California: United States Department of Agriculture, Bureau of Agricultural Economics, March 1946), 161.

7 "Columbia Basin Program Speeded," *Reclamation Era*, 32 (July 1946): 148; Sidney D. Lawson, "The Meaning of Power Utilization," *Reclamation Era*, 32 (December 1946): 266-7; "Water Planning Pays Off," *Reclamation Era*, 32 (July 1946): 152-3.

8.13. Kortes Dam on the Kortes Unit of the Pick-Sloan Missouri Basin Program.

of the dam brought a new community to the lower Colorado as Reclamation and the Utah Construction Company set about building living quarters for the dam's construction workers. Where once only a few prospectors and ranchers lived, a new community of 2,200 people took up residence. In addition to producing more hydroelectric power, the dam, undertaken in accordance with the U.S.-Mexican treaty of 1944, created Lake Mohave and eased the delivery of water to Mexico. With completion of Davis Dam in 1953, it joined Parker Dam and Imperial Dam in regulating the flow of the lower Colorado River below Hoover. With the addition of the Davis Powerplant to Parker Dam, the Bureau of Reclamation boosted power production along this section of the river to 3,500,000 kilowatts. According to the *Reclamation Era*, the Colorado River and its tributaries produced twenty billion kilowatt hours annually.[8]

Amidst the flurry of activity at war's end came the resignation of Secretary of the Interior Harold Ickes and President Truman's appointment of Julius Krug to fill the post in early 1946. Ickes's departure cleared the way

[8] John A. Leveritt, "Camp Life at Davis Dam," *Reclamation Era*, 32 (September 1946): 296-7; Oscar Buttehdahl, "Corralling the Colorado," *Reclamation Era*, 32 (October 1946): 218-20; Bureau of Reclamation, *The Colorado River*, caption below a picture of a home, "Opportunities will be provided for many new farm homes for veterans and others," 102.

8.14. Davis Dam on the Colorado River. Parker-Davis Project.

for the president to consider changing the name of "Boulder" Dam. Truman largely was responding to a joint resolution from the newly elected Republican-dominated Congress in the fall of 1946 that declared Boulder Dam should "hereafter be known and referred to as Hoover Dam." In a spirit of cooperating with Congress, President Truman wrote to Secretary Krug, "I am of the opinion that if the present trip of Mr. Hoover turns out successfully [the president had appointed former President Herbert Hoover to head the Famine Emergency Commission to Europe] you should rescind the action of Harold Ickes in regard to the Boulder Dam." Upon completion of the dam, Ickes had no right, according to Truman, to "arbitrarily" overrule a previous congressional resolution that named the dam Hoover. Truman noted former President Hoover's service to the country, his role in planning for the use of Colorado River water, and his work in bringing food relief to war-torn Europe warranted recognition. And while Truman said he could not agree politically with Hoover, he deemed Ickes's efforts to overrule a resolution of Congress as "petty and should not be countenanced by this Administration." The president's openness to the idea of changing the name of Boulder Dam back to Hoover Dam was the result of his desire to cooperate with congressional Republicans, but it may have been prompted as much out of his dislike for Roosevelt's secretary of the interior.

While the president faced opposition from members of his own party on the official name change, the resolution was approved and Truman signed the legislation on April 30, 1947.[9]

8.15. Secretary of the Interior Julius Krug served from March 18, 1946, to December 1, 1949.

Another issue under consideration by Congress in the postwar readjustment was the question of veterans and what privileges to extend to them in terms of taking up lands on Reclamation projects. After World War I, Congress had also considered granting exclusive privileges to reward returning veterans on the Reclamation projects. Congress's generosity, however, was not overwhelming. Despite proposals that veterans be granted free water and land, Congress agreed to only "preference rights" for veterans when they applied for farms on the projects. At the end of World War II the same issue reappeared in Congress. Some, especially Arizona Congressman John R. Murdock, chairman of the House Irrigation and Reclamation Committee, wanted homesteads on Reclamation projects for servicemen as part of the Servicemen's Readjustment Act of 1944, more widely known as the GI Bill of Rights. Congressman Murdock projected more than four thousand new "homestead farms" available for veterans by 1951 and wanted Reclamation to launch a campaign entitled, "Veterans—Here's Your Farm." Nevertheless, Congress failed to include a veterans' benefit for farms on Reclamation projects. In early 1945 the Bureau of Reclamation and western congressmen tried again and sought soldiers' benefits in waivers of construction costs and programs of technical assistance. In the end Congress approved the traditional offering of preferential treatment of veterans when farms on projects became available. While Congress and surveys conducted by the Department of the Interior and the Bureau of Reclamation overplayed

---

9 Personal and Confidential Memo from Harry S. Truman to Julius A. Krug, February 18, 1947, Julius A. Krug Papers, Box 69, Manuscript Division, Library of Congress, Washington, D.C.; hereafter cited Krug Papers; Senate Joint Resolution, 80th Congress, January 29 (legislative day, January 27), referred to the Committee on Public Lands, Library of Congress; letter to editor from Harold Ickes complaining of an editorial favoring name change to "Hoover Dam," *New York Times*, April 7, 1947; "...Democratic Opposition...," *New York Times*, April 23, 1947; "It's Hoover Dam," *Reclamation Era*, (June 1947): 143; Eugene Lyons, *Herbert Hoover: A Biography* (Garden City, New York: Doubleday & Company, Inc., 1964): 385; "Hoover Dam Bill Signed by Truman," *New York Times*, April 30, 1947.

soldier aspirations for farm ownership in the face of demographic trends to leave the farms for urban life, the number of returning servicemen expressing interest in Reclamation farms exceeded farm availability on the various projects. These included the newly-opened Columbia Basin Project where lotteries occurred to distribute farms. The Columbia Basin Project especially excited many in Congress and in the Department of the Interior as a project offering great opportunities to returning veterans and a partial fulfillment of the small farm ideology that originally inspired western reclamation at the beginning of the century.[10]

8.16. On February 8, 1956, officials, at Antelope Union High School, drew 40 names to go on the priority list of veterans preference applicants for farm units for sale in the Wellton-Mohawk Division under Gila Project Public Announcement #2.

The end of the war also renewed the stiff competition between the Bureau of Reclamation and the Army Corps of Engineers, in many cases caused by their similar yet often conflicting missions. This rivalry became a focus of the Hoover Commission, created by Congress in 1947 to study the reorganization of the executive branch. In its report, the commission criticized the detrimental competition and duplication of effort when both Reclamation and the Corps sought funding to build water development projects in the same river basin. It noted that after enactment of the Flood Control Act of 1936 "administrative confusion" occurred. The Act gave to the Corps the primary responsibility for flood protection on the main streams and the development and improvement of the upper watershed to the Department of

---

10 Fred W. Johnson, Commissioner, Harold L. Ickes, Secretary of the Interior, "Governmental Aids to Land Acquisition by War Veterans, 1796-1994," RG 48, Entry 768, Box 12, "Soldiers Preference for Land;" see also John R. Murdock, "Veterans—Here's Your Farm," *Reclamation Era*, 32 (May 1946): 95-6; Brian Q. Cannon, "Creating a 'New Frontier' Opportunity: World War II Veterans and the Campaign for Western Homesteads," unpublished manuscript read at the Pacific Coast Branch of the American Historical Association, Honolulu, Hawaii, August 2003.

8.17. Cover of "The Commission on Organization of the Executive Branch of the Government: Department of the Interior. A Report to the Congress."—i.e., the First Hoover Commission. March 1949. Courtesy of the National Archives and Records Administration.

the Interior and the Bureau of Reclamation. According to the Commission, this distinction was not at all clear-cut and muddled relations between the Corps and Reclamation. The Hoover Commission observed that the Flood Control Act expanded the Corps' original responsibility not only to improve navigation and flood control but also to include the construction of hydroelectric facilities. This, the commission maintained, conflicted with the Bureau of Reclamation's longstanding commitment to irrigation development and its by-products (hydroelectric power production, farm improvements, and the prevention of land monopolies), often causing fierce competition between bureaus. The commission noted, "Now we are witnessing the spectacle of both agencies contending for the authorization, construction, and operation of projects in the same river basins, for example, in the Central Valley, the Columbia, and Missouri Basins."[11]

In terms of the economics of Reclamation projects, the Hoover Commission brought more bad news to the Bureau of Reclamation. The commission reported on the inability of projects to repay their capital costs. Even with the subsidy of no-interest loans on construction costs, the commission noted that "these projects, on the average, do not pay off." Their original costs, with

---

11 *The Hoover Commission Report: On Organization of the Executive Branch of the Government* (New York: McGraw-Hill Book Company, Inc., 1949), 282; see also David Arlin Kathka, "The Bureau of Reclamation in the Truman Administration," PhD diss., University of Missouri, Columbia, 1976, 129-31.

a few exceptions, were too high for agriculture to bear, and the Hoover Commission concluded, "It is simply accepted that the national advantage of more farm homes and more national productivity are advantages which will offset Government losses." At this point, the commission echoed previous investigations into the economics of federal reclamation. It asserted that drastic changes must occur to make Reclamation projects economically viable and to control the overall costs to the Treasury. Most significantly, however, the commission addressed duplication of efforts by the Corps and the Bureau of Reclamation. It recommended merging the two agencies into a Department of Water Development under the Department of the Interior.

In 1947, however, the recommendation fell on deaf ears. The War Department was in no mood to give up its cherished Army Corps of Engineers. Also, the Pentagon was grudgingly responding to other suggestions from the Hoover Commission to reorganize all military responsibilities into a single

8.18. Covers of the 2 volumes of "Commission on Organization of the Executive Branch of the Government: Water Resources and Power. A Report to the Congress."—i.e., the Second Hoover Commission 2. June 1955. Courtesy of the National Archives and Records Administration.

Department of Defense.[12] Furthermore, western Republican critics in Congress, who were not necessarily against Reclamation projects per se, remained wary of any aspects of social planning these efforts included. They based their criticism on ideology, seeking to squash any "socialist" tendencies in Reclamation left over from the New Deal.

The transformation of the Bureau of Reclamation into a major public power producing agency not only offered new opportunities, but also led to the resurgence of old adversaries. Long-standing foes of public power continued to heap criticism upon New Deal public works projects, especially the Tennessee Valley Authority, which for these critics became a symbol of government overreaching. Now that the Bureau of Reclamation appeared to be much more than a government effort to promote small farm irrigation in the arid West, its public power operations, once the wartime emergency passed, became a target for those who guarded the interests of private power and decried the growth of the federal government during the New Deal. They saw Reclamation's production of power as one more step toward big government—a euphemism for "creeping socialism." Also Reclamation faced opposition and criticism even in its more traditional role of developing small farms on its projects. Critics began an overt attack on the 160 acre limitation rule, never evenly or rigorously enforced, which under Reclamation law banned water to acreage exceeding 160 acres per farm owner. Many saw the limitation as a restraint on ambition and a kind of leveling socialism on projects, an especially strong view among farmers in the Central Valley of California.[13]

Regionalization

During the war, a significant change to Bureau of Reclamation policies and procedures came with the decentralization of the Bureau as mandated by the 1939 Reclamation Project Act. One month after Commissioner John Page retired in August 1943, newly appointed Commissioner Harry W. Bashore announced the regionalization plan. Regionalization created seven regional directors who were to report directly to the commissioner, which in effect reduced the influence of the Chief Engineer's Office in Denver. The reorganization of Reclamation went forward amidst growing debate on the future of comprehensive multipurpose river basin development, as discussions raged about the creation of new TVA-like valley authorities. While removing

---

12 *The Hoover Commission Report*, 278, 197-8.

13 Lawrence B. Lee, "California Water Politics: Opposition to the CVP, 1944-1980," *Agricultural History* 54:3 (July 1980): 407.

8.19. Sinclair O. Harper was Reclamation's Chief Engineer, 1940-1944.

some centralized authority from the Chief Engineer, Sinclair O. Harper, decentralization offered the Bureau of Reclamation greater flexibility to assume a larger role in river basin development. Moreover, reorganization provided Reclamation a surer foothold in its competition with the Corps of Engineers.[14]

In the fluidity of the postwar years, uncertainty loomed over the manner in which water resources development in the seventeen western states occurred. At the same time, discussion on new independent river basin authorities threatened to take over water development projects from both the Bureau of Reclamation and the Corps of Engineers. The course of future events was uncertain, but Reclamation was clearly positioning itself to assume any new leadership mantle in river basin developments by implementing regionalization. With more administrative powers in the hands of regional directors, Reclamation prepared to claim it was on the ground and ready to go should river basin development projects capture further interest in Congress. After its wartime hiatus, articles in *Reclamation Era* confirmed the view that reorganization occurred in response to prospects for multipurpose river basin developments. One 1946 article asserted that when Reclamation undertook development of western rivers "on a basin-wide scale" regionalization became necessary. Departure from individual projects to multi-river basin tasks required that the seven regional offices have more individual authority to deal with localized problems. Later, in 1947, as battles in Congress over funding for Reclamation

[14] Lee, "California Water Politics," 408-9; see also "Reclamation Project Act of 1939," Public Law 206, August 7, 1939, in United States Department of the Interior, Bureau of Reclamation, *Federal Reclamation and Related Laws Annotated*, Volume I, Richard K. Pelz, editor (Washington, D.C.: United States Government Printing Office, 1972), 634-64 (this series of publications currently, in late 2012, includes 5 volumes, 2 supplements, and one volume on the Reclamation Reform Act, 1982-1988); Kathka, "The Bureau of Reclamation in the Truman Administration," 22.

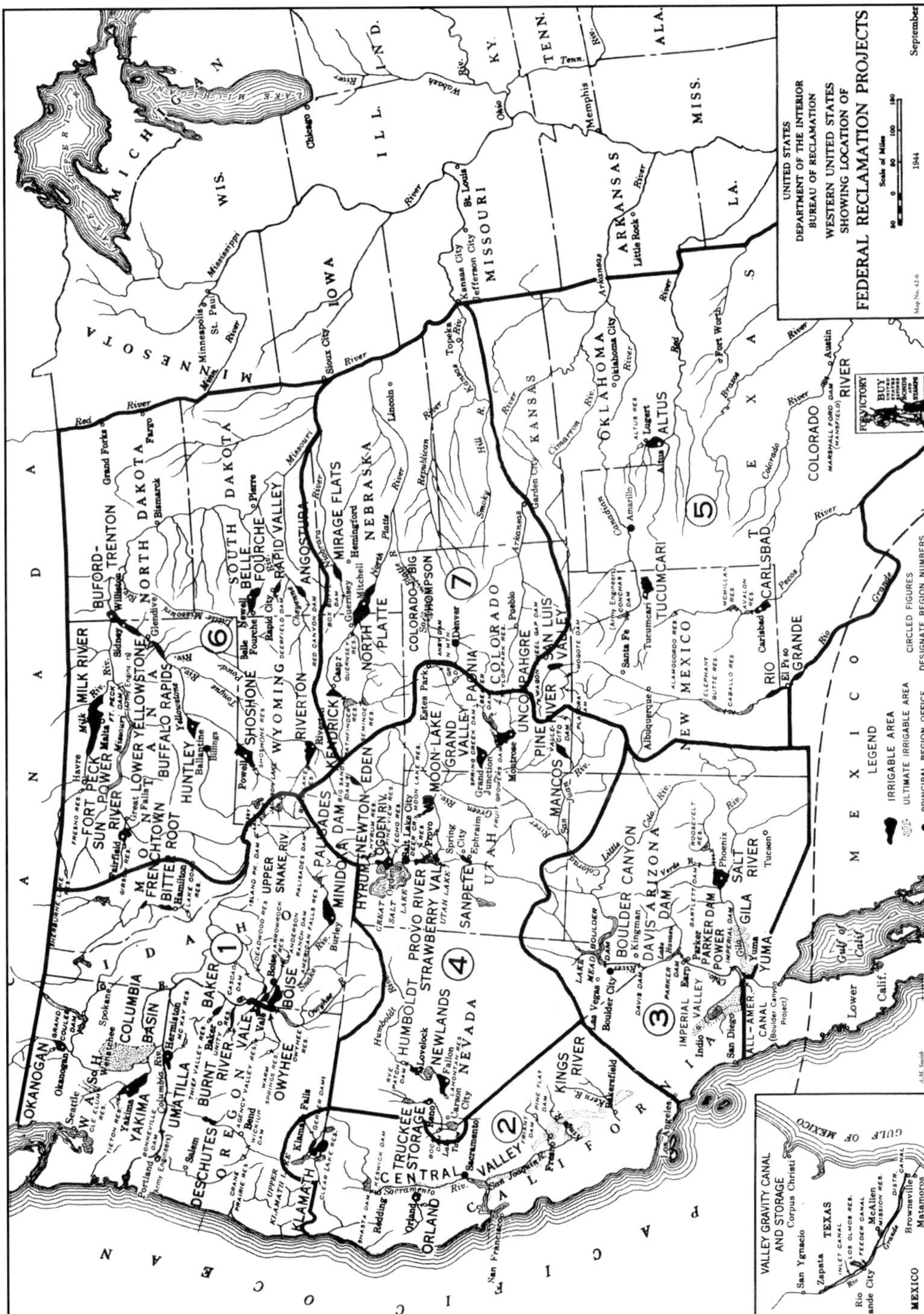

8.20. This map shows the regions as created in the period 1944 to 1946. Region 7 was created out of parts of regions 5 and 6, at the request of Colorado politicians, to meet the heavy planning demands of the Pick-Sloan Missouri Basin Program.

took place, *Reclamation Era* claimed that budget constraints forced decentralization of the Bureau's engineering offices. The Denver Office would continue to be the focal point of dams and major structures, but according to this official voice of Reclamation, "The field offices will be responsible for detail designs on major structures," that included "works appurtenant to dams, camps, roads, design of transmission lines and irrigation distribution systems."[15]

Still, questions remained over how regionalization would impact the Bureau of Reclamation's ability to fulfill its mission. In 1944 Chief Engineer S. O. Harper spoke to the Western Association of State Engineers in Denver on the reorganization of regional offices. He began by asserting, "I want now to puncture any fallacious idea that I am opposed to the regionalization of the Bureau." Harper insisted that he viewed it as a step forward and that in fact he had been instrumental in establishing that set up for the Central Valley Project in California. But he was critical of the diminished influence of the Chief Engineer's Office by splitting up of a single-headed organization that he believed operated with efficiency "and substituting for it a 5- or 6-headed group with no directing head." Harper maintained that the Bureau was a construction agency, and construction projects, he said, did not function well unless there was authoritative direction from the top and not direction by a committee or organization heads.[16]

Harper noted that he had seen this all before in his thirty-eight year career with the Bureau of Reclamation. In the name of efficiency, other attempts at creating a "commission" to oversee large construction had failed in the past. As an example, Harper recalled that the commission created to manage the Panama Canal construction utterly failed, and only after the job bogged down did President Theodore Roosevelt appoint General George W. Goethals "as a one-man czar; who brought order out of chaos and built the canal." According to the chief engineer, the Bureau of Reclamation was not immune to past efforts at decentralization, in 1914 and again in 1924, and only through the exertions of Arthur P. Davis and Dr. Elwood Mead, respectively,

[15] "The Changing Years," *Reclamation Era*, 32 (May 1946): 113; "Design Work Decentralized," *Reclamation Era*, 33 (October 1947): 227.

[16] S. O. Harper, "Farewell Remarks from the Chief Engineer," in *Denver 1944 Symposium of Conference of Regional and Branch Directors, Bureau of Reclamation, Denver, Colorado, November 18-21, 1944, with Selected Addresses Presented Before National Reclamation Association and Association of Western Engineers* (United States Department of the Interior, Bureau of Reclamation), 90, RG 48, Entry 779, Box 16; for background information on S. O. Harper see Lewis Nordyke, "River Doctor," *The Saturday Evening Post*, 218 (September 8, 1945): 19, 92-4.

did Reclamation return to a top-down hierarchal organization. Nonetheless, Harper recognized Reclamation's reorganization as a sign of the times brought about by the war and the bureaucratic centralization in Washington, D.C. He lamented, however, that the result of this state of affairs wasted time and energy of engineers who had better things to do than focus on personnel matters or other service functions, draining resources from Reclamation's construction objectives. Harper declared that the sole mission of Reclamation engineers was, "The building of lasting and enduring engineering structures and their operation for the benefit of the people in the West and the nation."

In what ultimately became a farewell speech, the retiring chief engineer harbored grave concerns about regionalization, and its impact on the ability of Reclamation to construct irrigation projects. His words caused a flurry of comments in subsequent months inside the Bureau and the Department of the Interior. Michael Straus, assistant to the secretary of the interior, noted in a memorandum to the secretary that Harper's engineering qualifications were unassailable, but warned of possible morale problems if Reclamation hired Harper as an outside consultant. Straus stated, "I have reason for believing that Mr. Harper was not in sympathy with the regionalization of the Bureau of Reclamation or with many of the newer policies of that Bureau, nor did he aid their implementation and adoption." Although Straus acknowledged that there was much to be said for the Bureau of Reclamation to "take care of its own," there was danger in employing former high-ranking individuals who still held tremendous influence.[17]

Nevertheless, regionalization allowed the Bureau of Reclamation the flexibility to focus on some of its ideological commitments and better explain its point of view. By granting more independence to regional directors, they paid more attention to problems that fell outside the realm of construction. Although constructing engineering works remained the central undertaking, there was also the need to work closely with state and local constituents on policy issues. This not only meant working to coordinate water development plans but also living up to the letter and intent of Reclamation law. Regionalization gave regional directors the freedom to concentrate on their localities and develop comprehensive plans, build personal relationships with water users, and work more closely with decision makers in the nation's capitol. Out of the reorganization came the Colorado River Storage Project from the Salt Lake City Office (Region 4), Reclamation's contribution to the Pick-Sloan

[17] Michael W. Straus to the Secretary, May 30, 1945, RG 48, Entry 779, Box 15.

Plan from Region 6, later the Upper Missouri Region, in Billings, Montana, and the Columbia Basin Project from the Boise Office (Region 1) with the goal of developing the agricultural potential of the project. All of these proposals sought to fulfill the Bureau of Reclamation's postwar ambitions in river basin development and revealed close cooperative efforts that met the needs of local water users.[18]

Valley Authorities and the Public v. Private Power Debate

Upon conclusion of World War II, the idea of instituting valley authorities as a means to achieve the fullest potential for water development in the West gained momentum within some circles in the Truman administration. Veteran New Dealers perceived valley authorities as the best way to spur regional growth. Leland Olds in particular characterized TVA as promoting healthy regional growth while balancing the forces of centralization concentrated in Washington, D.C. He denounced those who called river valley authorities and the TVA experiment "a trend towards socialism," the favorite charge of Republican congressmen against the TVA. River authorities, according to Olds, represented a trend toward decentralized management that was more familiar with the problems of a region. Also Olds believed that TVA, as well as any other river basin programs Congress might authorize, promoted the beneficial utilization of natural resources. The TVA experience encouraged cooperation of federal, state, and local agencies; farm organizations; agricultural colleges; experiment stations; chambers of commerce; private enterprise; labor organizations; and cooperatives. Furthermore, Olds claimed that river basin projects advanced the growth of private enterprise: "Figures are eloquent as to the extent to which the TVA has brought about an expansion of private enterprise in the Tennessee Valley. It provided the underpinning of prosperous development in a significant region of the South that encouraged the private enterprise in the life of the region."[19]

In the American West two regions appeared ideal for valley authority development—the Missouri River basin and the Columbia River basin. Both had the tacit support of President Truman. Valley authorities extended the progressive ideals of the New Deal in the production of public power to help

---

[18] Bureau of Reclamation, *The Colorado River*; Pitzer, *Grand Coulee*, 268; Harold L. Sylten, "Pioneering in the Missouri Basin," *Reclamation Era*, 32 (May 1946): 98-9; Oscar Buttehdahl, "Corralling the Colorado: Part II—Empire Builder," *Reclamation Era*, 32 (October 1946): 218-20, 229.

[19] Leland Olds, Address, in RG 48, Entry 779, Box 15.

build-up regional economies and complete the electrification of rural areas. The Rural Electrification Administration (REA) played a large role in this mission.[20] These idealistic goals, however, faced daunting resistance. Discussions about river authorities in the Pacific Northwest and the Missouri River basin renewed the debate over public versus private power development. More importantly, the heated conversations raised the larger question of how best to accomplish regional development. Should the people of the region be the catalyst behind their own economic and social growth, or should they rely upon a large federal entity to dictate how growth would occur? Unlike Olds, many TVA opponents saw valley authorities as an invasion by a powerful federal entity that acted against the will and desire of the people.[21] During this debate, the Bureau of Reclamation found itself in the difficult position of planning for extensive work on both the Columbia and Missouri river systems, but faced with the possibility that proposed valley authorities might repudiate its efforts.

As part of planning for the future during the war, Bureau of Reclamation leaders began investigating ways to fit Reclamation's expertise into multiple-use projects. With some items already earmarked for continued development—the Central Valley Project, the Columbia Basin Project, and the unfinished business on the lower Colorado River—Bureau planners sought larger challenges in river basin development. Bureau of Reclamation officials asserted that the organization was well positioned to render services and lend its experience to an entire region through generation of hydroelectric power, opening new irrigation opportunities, providing water for municipalities, and creating new recreational areas under one large comprehensive plan.

At a 1944 conference in Denver, Assistant Secretary of the Interior William E. Warne outlined to Reclamation regional officials his hopes for the Bureau's future in river basin development. He said that the West and the future of that region owed much gratitude to Reclamation visionaries such as F. H. Newell, A. P. Davis, F. E. Weymouth, Elwood Mead, R. F. Walter, J. L. Savage, S. O. Harper, F. A. Banks, John C. Page, the present Commissioner Harry W. Bashore, and many others. Warne noted that completion of Boulder Dam, which some said was impossible at the unheard of height of 726 feet,

---

20 P. R. DeLuna, "Bureaucratic Opposition as a Factor in Truman's Failure to Achieve a Columbia Valley Authority," *Historical Papers/Communications historiques* 10:1 (1975): 233; for more information on the REA see Leah S. Glaser, *Electrifying the American West: Stories of Power, People, and Place* (Lincoln: University of Nebraska Press, 2009).

21 William R. Arthur, "MVA—Its Background and Issues," *Congressional Record*, 29:1 (January 1050): 13; Mark W. T. Harvey, "North Dakota, the Northern Plains, and the Missouri Valley Authority," *North Dakota History* 59:3 (1992): 34.

8.21. The Friant-Kern Canal.

helped to win a worldwide war by producing electricity to feed West Coast war industries. He praised the "visionary undertaking" of Grand Coulee Dam which was once denounced as a "gigantic white elephant, but now its kilowatt and water storage capacity shamed detractors and the faint of heart who now welcome its power to the war muscle of the United States." Warne also noted the near completion of Shasta Dam with its multi-purpose functions of hydropower output, irrigation, and flood control in California. All three developments, according to Warne, begged the question, "Who will say today that a river is too broad to dam or a task to difficult to be completed by our engineers?" His words reflected a renewed vigor and energy emanating from Reclamation at the end of World War II for the continuation of its mission to develop the water resources of the West.[22]

It was not enough merely to celebrate the great edifices constructed by engineers. There were other, perhaps, greater benefits to recognize. Dams provided irrigation water, opportunities to generate hydroelectric power, and complementary uses involving flood control, municipal water supplies, abate-

22 William E. Warne, "Operation and Maintenance, Land use, and Settlement," in *Denver Symposium*, 1-5, RG 48, Entry 779, Box 16; "Water Planning Pays Off," *Reclamation Era*, 32 (July 1946): 152-3; for more information on the "white elephant" reference, see Pitzer, *Grand Coulee*, 247.

8.22. Work on the Central Valley Project after World War II included lining the Friant-Kern Canal in February 1947. Photographer: A. Ross.

ment of pollution, and finally recreational opportunities to fish, swim, camp, and picnic. Warne saw the end of World War II as the beginning of a third phase in water development in the West. He noted that the first occurred in the late nineteenth century—a time of pioneers who made simple stream diversions and built highline ditches to irrigate western valleys. The Reclamation Act of 1902 began the second period that brought the federal government into development of water resources in the West. During this period another 10,000,000 acres were added to the irrigated lands of the West, about half from the efforts of the Bureau of Reclamation. The greater resources of the federal government enabled the Bureau of Reclamation to build large storage dams that could regulate, conserve, and control the water supply. This second phase in the view of many represented the "zenith" of Reclamation's accomplishments, especially with the launching of the Central Valley and the Columbia Basin projects. In 1944 Assistant Commissioner Warne saw a third era: basin-wide developments based on multiple-purpose projects. The Roosevelt administration, while receiving much partisan criticism for the Tennessee Valley Authority, was well-satisfied with the development and touted it as a hallmark of progressive achievement in water management and power development that aided economic growth in the Tennessee River Valley. For many supporters, it offered a model for future water development that the West might well emulate.

Warne revealed that the Bureau of Reclamation had some fifteen additional basin reports that it would eventually present to Congress. Clearly he believed that basin development was the wave of the future and was the most economical way of approaching the full utilization of water resources in the West in the postwar period. He saw in all of this the foundation of a great postwar program, "one of transcendent importance to the West, and to the whole Nation." Ultimately the developments Warne envisioned offered opportunities for returning servicemen and demobilized war workers for settlement on family farms. His words still echoed the ideals of the original Reclamation mission to make farm homes available for families, an ideal now more relevant than ever in this new post-world-war future.[23]

Though Warne praised the accomplishments of the TVA, the establishment of valley authorities for western river basins was not exactly what Reclamation officials had in mind. Nevertheless the ideas expounded by Leland Olds on regional planning still had considerable support within the Roosevelt and later the Truman administrations. For the Bureau of Reclamation, these ongoing discussions on creating new TVA-like agencies with all encompassing powers to control water resource management in the river basins of the American West was a looming threat to its own postwar plans. Such agencies would surely limit opportunities for the Bureau, and Reclamation was not alone in harboring these fears. As it nurtured its own ambitions for the river basins of the West, the Army Corps of Engineers came to oppose independent river basin authorities. Usually at loggerheads, Reclamation and the Corps joined forces to oppose all efforts to create river authorities. P. R. DeLuna writes, "Often in conflict with each other, these two agencies were united in their opposition to the establishment of an organization similar to the T.V.A. in the Columbia Valley." Though DeLuna's statement only reflects activities for the Columbia River, a much more earnest effort by Reclamation and the Corps occurred to stop attempts to form a river authority for the Missouri River.[24]

In 1944 Congress passed the Flood Control Act that authorized the Pick-Sloan Plan. Named after General Lewis A. Pick of the Army Corps of Engineers and W. G. Sloan, assistant director of the Bureau of Reclamation's Region VI in Billings, Montana, the plan presented a comprehensive project

---

23 Warne, "Operation and Maintenance, Land use, and Settlement," 1-5, RG 48, Entry 779, Box 16.

24 DeLuna, "Bureaucratic Opposition as a Factor in Truman's Failure to Achieve a Columbia Valley Authority," 239.

for development of the Missouri River basin. In essence, the Pick-Sloan Plan divided the Missouri River basin between Reclamation and the Corps. The Army Corps of Engineers's primary responsibility was flood control and to improve navigation along the main stem of the river through a series of dams and levees. The Bureau of Reclamation took up irrigation and hydroelectric power development on the main tributaries of the upper Missouri River. Sometimes touted as a model of interagency cooperation, the Pick-Sloan Plan addressed flood control, enhanced agricultural production of the northern plains, and increased hydroelectricity production to spur the economic diversification on the upper Missouri River basin.[25]

8.23. Assistant Commissioner William E. Warne, 1946.

Initially Assistant Secretary of the Interior William Warne downplayed the Corps' plan in favor of Reclamation's report. Later, however, he acknowledged that negotiations reconciled the Bureau's plan with the Corps' studies. Warne called this an excellent example of cooperation between the Bureau of Reclamation and the Corps of Engineers. More importantly in the process of merging the two plans, both organizations agreed that there would be a ratio of benefits to costs in the range of 2.45 to 1. "Think of it," Warne wrote, "the benefits were virtually two and one-half times the costs, when a plan for the comprehensive use of all the water and related resources was considered." Cost benefits aside, another possible motivation for the Bureau and the Corps to come to terms on the Missouri River basin plan was President Roosevelt's growing impatience with the inability of these two rivals to reach an agreement on river basin development. According to Peter Carrels in *Uphill Against Water*, "To solve the impasse, President Roosevelt advised

[25] John Ferrell, "Developing the Missouri: South Dakota and the Pick-Sloan Plan," *South Dakota History* 19:3 (1989): 308; Mark W. T. Harvey, "North Dakota, the Northern Plains, and the Missouri Valley Authority," 35; Kathka, "The Bureau of Reclamation in the Truman Administration," 113; for more information on the Pick-Sloan Plan, see "Flood Control Act of 1944" in USDOI, BR, *Federal Reclamation and Related Laws Annotated*, Volume 1, 796-812.

formation of a new agency—one to be called the Missouri Valley Authority (MVA)—to rule the Missouri River."[26]

Although publication of the Missouri River Plan, or the Pick-Sloan Plan, for the multipurpose development of Missouri River basin resources seemed to bode well for a working relationship between the Bureau of Reclamation and the Corps of Engineers, the harmony existed mostly on paper. One observer wondered if two entirely separate agencies, one interested in irrigation and power and the other concerned with flood control and navigation could really cooperate to accomplish multiple-purpose development of a river basin. There also remained questions on how to divide responsibilities in reference to structure designs to avoid duplication of effort and policy goals that worked at cross-purposes. Corps methods ordained that flood control reservoirs be empty, while the opposite was true for the Bureau of Reclamation which wanted full reservoirs for irrigation and power. Concerns also arose over whether either organization had any interest in tackling other areas of resource management that valley authorities embraced such as soil erosion, reforestation, recreation, while balancing economic growth between industry and agriculture. And finally with the growth of the War Department in both size and prestige during the war, Reclamation officials wondered whether any real cooperation with the War Department could exist. In a 1944 letter to William Warne, one Reclamation official wrote, "I am all the more convinced that you can't 'cooperate' with the War Department. I think we should boldly move out ahead of them and assume leadership, if public interest in irrigation, power, etc., is to be served."[27]

Despite its comprehensiveness, the Pick-Sloan Plan still had its detractors within the Missouri River basin. Some critics were in fact avid proponents of a Missouri Valley Authority and included various MVA committees from St. Louis, Kansas City, Omaha, and Denver, along with representatives of organized labor and the Farmers' Union. Upstream opponents of the Corps' mainstream plan objected to its focus on flood control that only benefitted larger downstream urban communities. They complained that the Corps ignored the needs of northern plains farmers by not showing any concerns for

---

[26] Warne, "Operation and Maintenance, Land use, and Settlement," 3, RG 48, Entry 779, Box 16; see also Peter Carrels, *Uphill Against Water: The Great Dakota Water War* (Lincoln: University of Nebraska Press, 1999), 16.

[27] Phillip Dickinson to William E. Warne, February 20, 1944, William E. Warne Papers, Box 2 Correspondence, American Heritage Collection, University of Wyoming, Laramie, Wyoming, hereafter cited Warne Papers.

soil erosion that plagued the region. One observer, commenting on the devastating flooding on the Missouri River that occurred during the early 1940s, noted, "The richest land in the United States has been washed under the Eads Bridge in St. Louis at a rate of twenty acres a minute." In addition, Missouri Valley Authority proponents believed that federal control of the river that the Pick-Sloan Plan represented seriously challenged the water rights of the basin states. Similar to the vision of regional control extolled by Olds, MVA supporters argued that a valley authority was a much better method to serve all the needs of basin states through, in the words of historian Mark Harvey, "centralized allocation of Missouri River water."[28]

Other detractors saw valley authorities as threats to free enterprise and local government. The National Reclamation Association supported Commissioner Bashore and chief of the Army Corps of Engineers Major General Eugene Reybold's success at reconciling differences in the development of waters in the Missouri River basin under the Pick-Sloan Plan. The association believed that Pick-Sloan better served irrigation interests and other beneficial consumptive uses of water in the Missouri River basin. Association members were adamant about preventing any extension of TVA-like authorities to the West. Its resolutions declared all "such authorities are unnecessary, unwise, and undesirable" and pledged "to defeat all such measures." The National Reclamation Association expressed not only antipathy to river basin authorities, but also any New Deal-like centralized efforts for economic development. It asserted that river valley authorities, among other unwise measures, encroached upon states' rights, hindering their ability to enter into water compacts. Moreover valley authorities, the association argued, represented unwise centralization that resulted in too much public ownership of resources, creating unwholesome government monopolies that removed property from local tax rolls. These arguments echoed Republican Party objections to what they considered the New Deal's "socialist tendencies" as well as resurrecting private power's arguments against public power.[29]

---

[28] Wesley Price, "What You Can Believe About MVA," *Saturday Evening Post*, 218 (January 19, 1946): 124; Ernest Kirschten, "MVA: Stalled But Not Stopped," *Nation*, 163:7 (August 17, 1946): 184; see also Harvey, "North Dakota, the Northern Plains, and the Missouri Valley Authority," 34.

[29] Resolutions Adopted by the National Reclamation Association Thirteenth Annual Convention, Denver, Colorado, November 15, 16, 17, 1944, RG 48, Entry 779, Box 15; see also Kathka, "The Bureau of Reclamation in the Truman Administration," 44.

Most opponents of river valley authorities saw the Pick-Sloan Plan as the best course of action to defeat the proposed Missouri Valley Authority. For years, competition between the Bureau of Reclamation and the Corps of Engineers stalled Missouri River development. Now when talk of a valley authority emerged, these two rivals joined forces to produce a comprehensive plan for the Missouri River basin. Some perceived this as more than just mere coincidence and called it a "marriage of convenience" or a "shotgun wedding," but all understood that the primary purpose was to ensure the presence of Reclamation and the Corps in the basin. Still some in Congress and the Truman administration continued to nurture the vision of one Missouri River Basin Authority similar to the TVA.[30]

In 1947 Assistant Secretary of the Interior C. Girard Davidson sent a report to Secretary Julius A. Krug that portrayed governmental participation in natural resource development as holding the key to economic growth. Davidson traced the conception of this activity back to the conservation priorities established by the Theodore Roosevelt administration. Since that time, he claimed, twenty "different federal resource agencies" operated independently providing "no centralized responsibility for resource conservation and development." Davidson praised the TVA as not only the salvation of an impoverished region, but also as a model for natural resource development. He targeted the Pacific Northwest and the Columbia River basin as the future site of a successful river basin authority. The plan envisioned by the assistant secretary proposed to transfer flood control, navigation, irrigation, and power development performed by the Corps of Engineers, the Bureau of Reclamation, and the Bonneville Power Administration to a regional authority. Sounding very similar to Leland Olds, Davidson claimed that while the TVA sought to rejuvenate "an exhausted people" and "an exhausted land," the Pacific Northwest offered the opportunity to build a "new economy." For Davidson the conclusion was: "Just as the big and tragic problems of the Tennessee Valley could be solved only through the sort of teamwork and integrated effort provided by the TVA, so can the challenging and provocative problems of the Pacific Northwest be solved."

---

30 William R. Arthur, "MVA—Its Background and Issues, *Congressional Record*, 29:1 (January 1950): 13; Kirschten, "MVA: Stalled But Not Stopped," 184; Price, "What You Can Believe About MVA," 124; DeLuna, "Bureaucratic Opposition as a Factor in Truman's Failure to Achieve a Columbia Valley Authority," 239; Harvey, "North Dakota, the Northern Plains, and the Missouri Valley Authority," 33; Kathka, "The Bureau of Reclamation in the Truman Administration," 113.

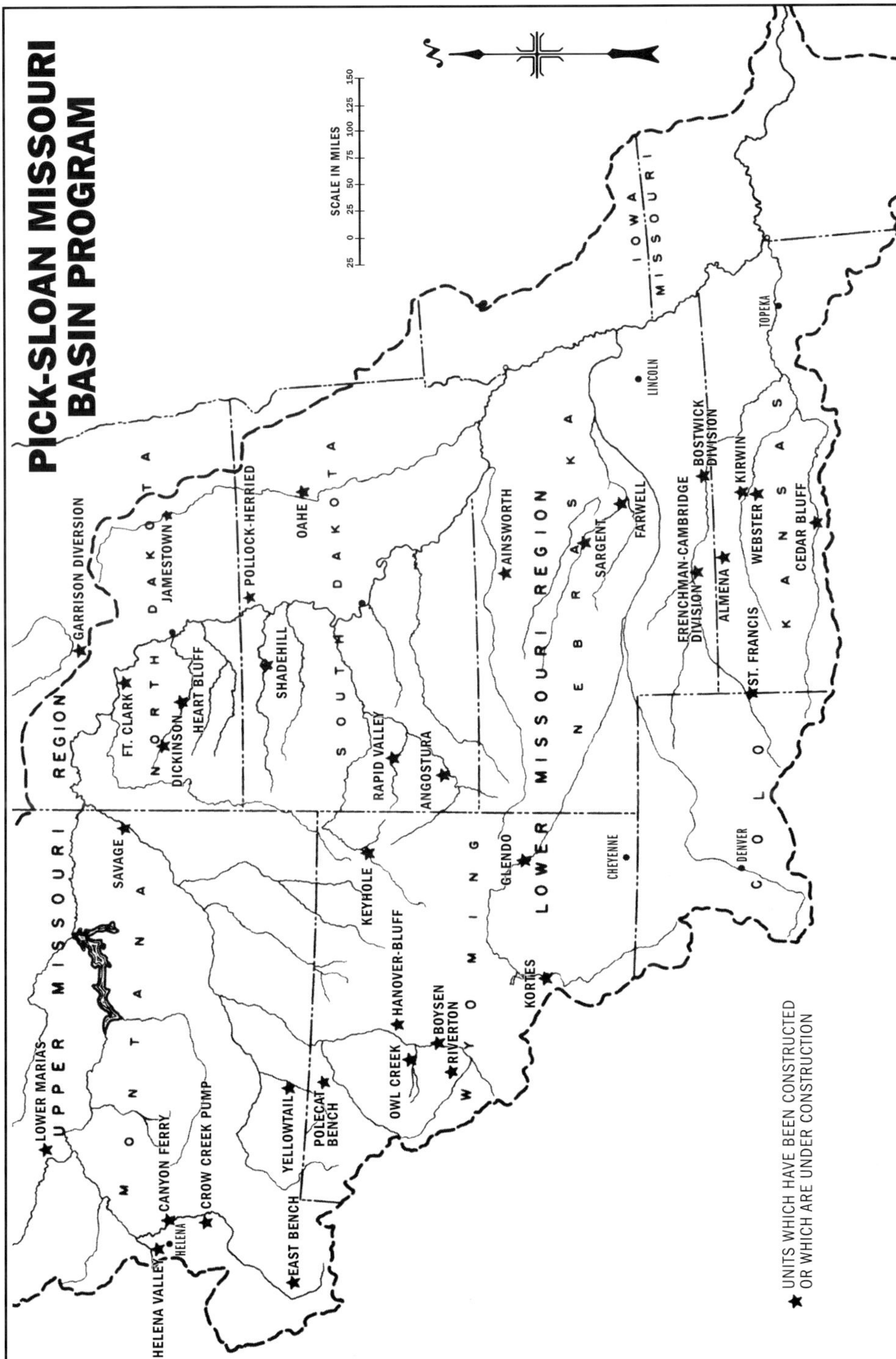

8.24. 1977 map of the Missouri River Basin showing Pick-Sloan Missouri River Basin Program units.

While coordinated development of all of the natural resources of a region—land, forests, fish, water, parks, minerals, heat, and energy—struck a responsive chord in many, others saw the erosion of their power and influence. The Bureau of Reclamation was no exception. Admittedly, the imposition of a river basin or regional authority, Davidson noted, contained the seeds of "possible conflict between the regional agency and the strongest of the existing Federal resources agencies." States too feared the loss of power. Practical considerations or concerns for bureaucratic prerogatives in designated spheres of power trumped much of the idealism expressed in Davidson's report, which spoke to the promises of new regional authorities. All of which is to say that established agencies including state governments came to fear new "TVAs" in the West.[31] Support for valley authorities within the Department of the Interior was not universal. Regional TVA-like agencies threatened the power and influence of the Department of the Interior, and some department officials sought to view the issue pragmatically. One memo cautioned that "it would be premature to commit ourselves to regional authorities." It advised that only time would tell whether regional agencies were the wave of the future and suggested that in the meantime the Department refrain from supporting autonomous river basin authorities. Indeed, some observers perceived a complete transformation of the executive branch stemming from the creation of valley authorities. One *Saturday Evening Post* article noted in 1946,

> If authorities take over planning and management of our natural resources, the executive branch of the United States Government is in for a major overhaul. Great chunks would be torn from the Department of Agriculture. The entire Department of the Interior might be razed … and a historic mass of law and judicial interpretation would become obsolete.[32]

Observations, such as this one, provided ample reasons for wariness by government bureaucracies to the idea of establishing valley authorities in the American West.

---

31 C. Girard Davidson, Assistant Secretary, Memorandum to Secretary Krug, "Special on Regional Resource Development," 1947, Krug Papers, Box 4; for information on C. Girard Davidson see Phyllis Komarek De Luna, *Public versus Private Power during the Truman Administration: A Study in Fair Deal Liberalism* (New York: Peter Lang, 1997), 31.

32 Evelyn N. Cooper to Assistant Secretary C. Gerard Davidson, March 18, 1947, Krug Papers, Box 45; Price, "What You Can Believe About MVA," 24; William L. Lang, "Failed Federalism: The Columbia Valley Authority and Regionalism," in *The Great Northwest: The Search for Regional Identity*, William G. Robbins, editor (Corvallis: Oregon State University Press, 2001), 70-4.

Curtailment of Columbia Valley Authority discussion after the war testified to the strength of the political opposition to new river valley authorities and flagging interest in New Deal era programs among members of Congress, farmers, and regional business people in eastern Washington. They feared the social rhetoric of New Dealers like Leland Olds and Secretary of the Interior Ickes. Opposition forces capitalized on these concerns. They cast river authorities as the manifestation of a growing "federal octopus," poised to strangle state powers as well as private enterprise. Of course, already established agencies and service bureaucracies did not take kindly to the creation of new authorities or new agencies that would infringe upon their realms of activity. Indeed, some political scientists drew lessons from the massive governmental undertaking involved in the war effort to suggest that the creation of new authorities and leadership czars to accomplish tasks did not work as well as mobilizing existing governmental structures. Amidst the opposition and dwindling power of Depression-era politics, the dreams of the New Dealers for a new social experiment in a cooperative community on the Columbia Basin Project dimmed. What remained was the initiative and independence of the Bureau of Reclamation to renew its commitment to construct dams, hydroelectric facilities, and water delivery systems for both urban and agricultural purposes in an era of expansion in the decades following World War II.[33]

8.25. Secretary of the Interior Oscar Chapman, December 1, 1949, to January 20, 1953.

Shifts in the political tides manifested themselves in a series of events, and the Pacific Northwest provided the backdrop for changes in water development policies in the 1950s. Gone were the days of the New Deal when public power advocates could rely on a general consensus that public power was a

---

[33] Luther Gulick, *Administrative Reflections from World War II* (Birmingham: University of Alabama Press, 1948), 102; Brian Q. Cannon, *Remaking the American Dream: New Deal Rural Resettlement in the Mountain West* (Albuquerque: University of New Mexico, 1996); Paul K. Conkin, *Tomorrow a New World: The New Deal Community Program* (Ithaca, New York: Cornell University Press, 1959).

national benefit that brought affordable electricity to all Americans. In 1950 Secretary of the Interior Oscar Chapman warned an audience in the public power state of Nebraska that attacks upon public power were gaining momentum: "they will be both direct and indirect and under cover," but he said, "as long as the people and their government remain alert, and fully informed, these attacks will fail." Indeed, power industry publications intensified their attacks referring to the "pinko power policies" of the Department of the Interior and the Bureau of Reclamation. These assaults undermined support for public power even among the faithful of the Pacific Northwest. The withering attacks on the proposed Hells Canyon High Dam project on the Snake River, a major tributary of the Columbia River, in the 1950s ultimately demonstrated how the new political climate emerging after the war defeated a prize project of the Bureau of Reclamation.[34]

For the Bureau of Reclamation, opportunities for the expansion of hydroelectric power and irrigation beckoned in the Pacific Northwest's flowing rivers and desert landscapes. In 1946 the Bureau's long range planning agenda included building a high dam in the region's last remaining major prime damsite—Hells Canyon on the Snake River. Several obstacles loomed to foil Reclamation's plans. First was the Corps of Engineers, as elsewhere, an ever-present rival. Second were private power interests, principally in the form of Idaho Power Company, which stood ready to attack any plans for a new major public power producing dam in its domain. On the other side, President Truman's administration saw Hells Canyon as part of the larger Columbia River Valley Authority (CVA) to oversee all dam building, the production of electricity and its distribution in the region. Following the president's lead in June 1949, Secretary Julius Krug testified before Congress that one unified agency was the best means to develop the resources of the region. He noted that the administrative centers of various resource agencies—the Bureau of Reclamation, the Bureau of Land Management, the U.S. Forest Service, and the Soil Conservation Service—were remote from the Pacific Northwest and their policies sometimes worked at cross purposes. Under a Columbia River Valley Authority, Krug suggested, one administrative authority would effectively administer the entire region. Others argued that a CVA would mean

---

34 Oscar L. Chapman, "From Nebraska to the Sea," *Reclamation Era*, 36 (March 1950): 45-7; Robert E. Firth, *Public Power in Nebraska: A Report* (Lincoln: University of Nebraska Press, 1962); George A. Dondero, U.S. Representative from Michigan and Chairman of the Public Works Committee, "Wanted: A New Federal Policy," *Public Utilities Fortnightly*, 42 (September 9, 1948): 333-42.

the complete federalization of the region and the removal of local voices from critical policy decisions.[35]

Truman's CVA vision sought to lay the foundations for the Pacific Northwest's future in the spirit of the New Deal's undertaking in the Tennessee River valley. But, as already noted, the postwar political climate proved a minefield for the extension of New Deal river basin developments and planning enterprises that in any way resembled the scope and scale of the Tennessee Valley Authority. Longtime guardians of the public purse argued that public dam building—whether it was a project of a government agency, i.e., the Bureau of Reclamation, or a larger river basin authority—was too costly, unnecessary, and subject to charges of inefficiency and monopolization of resource development to the exclusion of private enterprise. In response to such attacks, Assistant Secretary of the Interior Warne sought to alleviate fears by portraying the high dam Hells Canyon undertaking as "self-liquidating" based on potential power sales in a region that faced imminent shortages. By "self-liquidating," he meant that the dam's electrical revenues would pay for its construction and even the expansion of irrigation in the deserts of Idaho. His urgency was all the more heightened because his statement came in the aftermath of the disastrous flood that destroyed the community of Vanport, Oregon, near Portland in late May 1948. Warne argued large dams on the upper Snake River could have prevented the flood. Yet, neither the political boost the president received from his unexpected victory in the presidential election of 1948 nor the devastating Vanport Flood the previous spring provided the momentum necessary to achieve a Columbia River Valley Authority.[36]

The Truman administration's inability to achieve valley authority legislation on the Columbia or Missouri rivers was due to a combination of factors. The Bureau of Reclamation and the Corps of Engineers opposed it. For many, the president's half-hearted commitment to the idea undermined some strong supporters for a valley authority in both the Missouri and Columbia river basins. In addition, the heated debate over public versus private power and accusations in the early McCarthy Era, that the administration was soft on communism along with its schemes that smacked of social planning, planted seeds of suspicion about the proposal. More importantly, international events such as the Berlin Airlift and the advent of the Korean War consumed much

---

35 Statement of Secretary of the Interior, Julius Krug, before the Senate Public Works Committee, June 1, 1949, Krug Papers, Container 45; see also Pitzer, *Grand Coulee*, 244.

36 William E. Warne to Julius A. Krug, June 9, 1948, Krug Papers, Box 69; see also Lang, "Failed Federalism," 66-79.

of Truman's attention during his second term. While the Truman administration continued to publically support valley authorities, the president made little effort to see it through, and opposition forces in the region and the election of Republican Dwight Eisenhower to the presidency in 1952 finally put an end to plans and legislation for a regional authority centered on water resource development.[37]

Reclamation's Leadership Issue and the 160 Acre Rule

Amid the lively debates over valley authorities in the postwar period, the leadership of the Bureau of Reclamation and the Department of the Interior underwent enormous adjustments after the long tenure of Secretary of the Interior Harold Ickes ended in 1946. Likewise the appointment in 1945 of a new commissioner of the Bureau of Reclamation, Michael Straus, to succeed Harry Bashore opened a new era. Reclamation's leadership set about completing many projects left unfinished during the war, in addition to confronting and defending its role as the chief water developer of the West. Yet, the appointment of a man perceived to be a non-engineer to the leadership position in the organization faced severe criticism. Other than David W. Davis (1923-1924), commissioners of the Bureau of Reclamation usually had been engineers who had come through the Reclamation ranks. On the other hand, what better choice to lead the charge into this new era than a man with a background in public relations, a newspaper man, and a publicist. Formerly an assistant secretary of the interior, Straus seemed to be assuming a lesser administrative role by moving into the commissioner's office. In reality, however, he entered an office that commanded over sixty percent of the Department of the Interior's expendable budget.

Not unexpectedly, Straus emerged as a controversial figure in his role as commissioner from 1945 to 1953. As a New Dealer and disciple of cantankerous Secretary of the Interior Ickes, he raised the ire of Republican members of Congress and some Democrats. With his career origins in the Chicago newspaper business along with other major dailies and the International News Service in Washington, D.C., Straus was familiar with the ways of Washington at the beginning of the New Deal. With this background, Secretary Ickes invited him in 1933 to manage public relations of the Department of the

37 William L. Lang and Robert C. Carriker, editors, *Great River of the West: Essays on the Columbia River* (Seattle: University of Washington Press, 1999); Lang, "Failed Federalism."

Interior, becoming Ickes's point man with Congress. In 1943 Ickes appointed Straus First Assistant Secretary of the Department of Interior before his "lateral" move to Commissioner of the Bureau of Reclamation in 1945.[38]

The personality and leadership qualities of Commissioner Straus drew attention in the press especially after congressional committees openly criticized his efforts at publicizing Reclamation. To some skeptics, Straus was an outsider who had no intimate connection with the Bureau of Reclamation and no real understanding of its mission. Nevertheless, jockeying for advantage to defend and expand the role of Reclamation in these years of opportunity was made to order for Straus's unique talents. But what passed for talents to some, others saw as flaws. Even within the Department of the Interior, Assistant Secretary William Warne tried to calm the waters as indicated in a 1948 note to Secretary Krug, "I have begun an active campaign on my own to tone down Reclamation's single track and aggressive reactions to many things. I believe that a calmer approach and a little more flexibility would help Reclamation's official relations." One of the "many things" was Commissioner Straus's clear statement about Reclamation's intentions of enforcing the 160 acre limitation rule on the Central Valley Project (CVP) in California. Warne's candid communication was one indication that deep within the Department of the Interior there was growing inclination to back away from the forced sale of "excess lands" required under strict adherence to the 160 acre rule in the original 1902 Reclamation Act.[39]

Reclamation's attempts to enforce the 160 acre rule and Straus's stringent advocacy of it were cause for concern by some who looked forward to the continuation of water resource development in the West. In 1944 the National Reclamation Association's meeting in Denver revealed the Association's own ideas about future water developments in the postwar West. Its deliberations foreshadowed some of the disputes between it and the Bureau of Reclamation over the next decades. Prominently featured in the list of resolutions adopted by the Denver conference was the removal of the excess lands provision in national Reclamation law. One resolution asked that the excess lands provision not pertain to Reclamation projects utilizing partial water supplies whether from surface or underground sources. But failing a general revision of Reclamation law on this issue, the meeting called for immediate enactment of the "Elliott Amendment" to exempt the Central Valley of California from

---

38 Kathka, "The Bureau of Reclamation in the Truman Administration," 43.

39 William E. Warne to Julius A. Krug , January 12, 1948, Krug Papers, Box 69; William E. Warne, *The Bureau of Reclamation* (New York: Praeger Publishers, 1973), 85.

the excess land provisions. The Elliott Amendment eventually caused the failure of the Rivers and Harbors appropriation bill at the end of 1944 because the Senate refused to accept it.[40]

California Congressman Alfred J. Elliott's amendment to the 1944 Rivers and Harbors Bill was the first earnest effort by Californians to exempt the Central Valley Project from the 160 acre limitation rule. The issue had been simmering ever since the federal government took over the CVP in 1937. Though the Bureau of Reclamation never forcefully enforced the land limitation rule on other Reclamation projects, leadership within the Bureau and the Department of the Interior appeared to make a stand in California. And since the 160 acre rule became part of Reclamation law in 1902, Congress had repeatedly affirmed its support. The rule promised equitable distribution of water while limiting the harmful effects of land speculation and monopoly. It centered on the very idea of the family farm. Historian Donald Worster maintains, "The acreage limit was clearly a family, not an individual, standard, and it applied in all cases, whether the land to be watered had been in private ownership for a hundred years or whether it was newly segregated out of the public domain was immaterial."[41]

However, the Bureau of Reclamation's lackadaisical enforcement of the rule on other Reclamation projects offered CVP farmers hope that acreage limitation would not apply to them. After all, Congress removed the restriction on the Colorado-Big Thompson Project in Colorado and two Reclamation projects in Nevada. In addition, the rule was not vigorously enforced in the Imperial Valley as part of the Boulder Canyon Project Act. All in all, there was a growing and understandable sense that Reclamation had little intention of strictly adhering to the 160 acre limitation rule. By 1943 the Central Valley appeared to be an exception. An exception that Secretary Ickes and newly designated Reclamation Commissioner Harry Bashore hoped would reinvigorate Reclamation's idealism concerning the virtues of the family farm.[42]

---

[40] Resolutions Adopted by the National Reclamation Association Thirteenth Annual Convention in Denver, Colorado, November 15, 16, 17, 1944, RG 48, Entry 779, Box 15.

[41] Donald Worster, *Rivers of Empire: Water, Aridity, and the Growth of the American West* (New York: Oxford University Press, 1985), 161; for a more in depth discussion of the 160 acre rule see Robert de Roos, *The Thirsty Land: The Story of the Central Valley Project* (New York: Greenwood Press, Publishers, 1968; originally published by Stanford University Press, 1948), 73-90.

[42] Lee, "California Water Politics," 407; Montgomery and Clawson, *History of the Legislation and Policy Formation of the Central Valley Project*, 144, 162-3; Freeland, "Examining the Politics of Reclamation," 222.

Land ownership in California reflected the novelty of the Central Valley Project compared with other Reclamation projects. By all accounts, the Bureau of Reclamation was providing an irrigation infrastructure to a valley with little or no public lands. Indeed, a Department of Agriculture report observed, "Land ownership in the Central Valley is heavily concentrated. Studies by the Bureau of Agricultural Economics indicate that nearly half the land in Madera, Tulare, and Kern Counties is owned by three percent of all landowners." For New Dealers Ickes, Bashore, and subsequently Michael Straus, this "concentration" of land begged for enforcement of the 160 acre rule. Observance of the limitation law meant increasing the number of family farms in the Central Valley, distributing benefits to a greater extent, and achieving, in this showcase project, a major Reclamation ideal. On the other hand, the valley's large farm interests saw this move as a governmental attempt to take private property to serve an outmoded ideal.[43]

When Congressman Elliott introduced his amendment to the 1944 Rivers and Harbors Bill, big farmers in California, along with the National Reclamation Association, saw it as a measure to protect private property rights in the Central Valley. While the land limitation provision meant that an individual farmer could receive water for only 160 acres, or 320 for a married couple, a farm with more than 160 acres could still stay intact and even receive water for excess lands as long as those parcels were sold off in ten years. The controversy over the land limitation rule turned into an ideological debate, and for the most part, Elliott's colleagues in the House agreed with the Central Valley farmers. But this was not so in the United States Senate. A myriad of authors have noted the sentiment in the Senate on the issue that supported the ideal of the small farmer reflected in the 1902 Reclamation Act and the recognition in the Senate that a 160 acre farm in California produced considerably more wealth.[44]

While the controversy over the 160 acre rule simmered, the National Reclamation Association ardently defended the principle that hydropower revenues from Reclamation dams be committed to retire the costs of irrigation works. It rejected any idea of subsidizing urban power users to the detriment

---

[43] It was generally recognized that the majority of farms in the Central Valley easily fell into the limits set by the 160 acre rule, but it was the large landholdings in the southern San Joaquin Valley that were the target of Reclamation's strict enforcement of the limitation law; see Montgomery and Clawson, *History of the Legislation and Policy Formation of the Central Valley Project*, 146; Worster, *Rivers of Empire*, 243-7.

[44] Kathka, "The Bureau of Reclamation in the Truman Administration," 50-1; Montgomery and Clawson, *History of the Legislation and Policy Formation of the Central Valley Project*, 166.

of rural water users on irrigation farms. In the postwar years, the Bureau of Reclamation pursued the strategy of supplying cities and industry with power, using the revenues to subsidize the high cost of irrigation development. The cost offset, or subsidy, helped underwrite many water projects in the postwar period. This practice drew caustic attacks from Reclamation critics who saw it as a drain on the national treasury. They argued that despite assertions that power revenues paid for irrigation works the fact was that power revenues went into irrigation projects that could not possibly pay for themselves. Both the 160 acre rule and power subsidies for irrigation projects sparked continuing controversies. For critics of the Bureau of Reclamation, Commissioner Straus became a lightning rod for everything that was wrong with the federal reclamation program.[45]

In the spring of 1952 *The Saturday Evening Post* ran an article highly critical of Commissioner Straus under the title, "Our Most Arrogant Bureaucrat." The title set the tone of criticism throughout, asking, "What is an old-time Chicago newspaperman and an ex-New Deal publicist doing in a job which is largely an engineering operation?" The answer was not flattering. The article asserted that, "Straus runs the Bureau of Reclamation for its maximum publicity and propaganda effect, applying to dam building and power production some of the same drum-beating showmanship that P. T. Barnum once gave to a gaudier but somewhat less colossal enterprise." The article went on to say that many in Congress and notably the private-power interests regard Straus as "Washington's most arrogant bureaucrat." Nevertheless, the article also noted that to his coworkers at the Bureau—apart from some professional engineers—he appeared, "forthright, conscientious, honest administrator who has put push and drive into the Bureau and made it bigger and better known than ever before." A third "I like Mike" group warmed to his "bluff charm" but remained, "uneasy about his casual attitude toward the taxpayers' dollar and his free-wheeling bulldozing in general." In these first paragraphs of the story, it is not difficult to see where the writer is going as he constructs a story of a blustery personality governing and promoting a western water empire in seventeen states and acting "like a crusading satrap of the Truman Administration."[46]

---

[45] National Reclamation Association Resolutions, RG 48, Entry 779, Box 15; Lawson, "The Meaning of Power Utilization," 266-7.

[46] Paul F. Healy, "Our Most Arrogant Bureaucrat," *Saturday Evening Post*, 224 (April 19, 1952): 46.

Yet as noted earlier, the Truman administration never successfully developed a comprehensive water-policy legislative agenda. No doubt this was a central reason Commissioner Straus assumed such a large presence on the public scene as he sought to preserve Reclamation's authority, influence, and functions vis-à-vis would-be competitors: the Corps of Engineers, river basin authorities, and the state of California that expressed interest in taking over the Central Valley Project if the Bureau failed to yield on the 160 acre limitation rule. In 1949, authorization for the Corps to construct the Pine Flat Dam on California's Kings River threatened the Bureau of Reclamation's major-player status in the Central Valley. It was a veiled attempt by California interests to circumvent the 160 acre rule. In this case, however, Congress thwarted the effort when it stipulated in the legislation that the water and power distribution fall under Reclamation law.[47]

When he succeeded Harry Bashore as commissioner, Straus and other New Dealers struggled to move the Bureau of Reclamation in the direction of a "new school of thought." Straus argued that the engineering leadership of the Bureau, especially after the death of Commissioner Mead, paid too much attention to the engineering side of Reclamation, neglecting its commitment under Reclamation law to the social reform goals of supporting the welfare of and opportunities for small family farms. He argued that the Bureau must enforce policies against land monopolization and seriously pursue the 160 acre limitation provisions of Reclamation law.[48] With the support of Interior Secretary Ickes, Straus tried to turn the Bureau's ideology into action.

Straus saw California's Central Valley as a starting point for reform. Here, Reclamation leaders stepped to the forefront in defense of family farms and the fair and equitable distribution of water and public power. Straus's appointment of Richard Boke as director of Region II in Sacramento, which included the Central Valley Project, underscored the "New School" policies. Boke easily melded with Straus's plans. According to one source, Straus was "Impressed with the fact that Boke was a 'card carrying ecologist,'" which could mean in Straus's view that Boke's interest lay in the health of the entire community and not simply in the sturdy construction of dams. Like Straus, Boke was not an engineer. To some congressional Republicans, these key figures in the Bureau of Reclamation reflected not a "New School" but the old school of the New Deal Era that they hoped to root out after their victory in

---

[47] Lee, "California Water Politics," 413; de Roos, *The Thirsty Land*, 68-9.

[48] Charles Coate, "'The New School of Thought': Reclamation and the Fair Deal, 1945-1953," *Journal of the West* 22:2 (April 1983): 59.

8.26. Senator Patrick McCarran. Courtesy of the U.S. Senate History Program.

the 1946 congressional elections.[49] The Democratic victories in the fall election of 1948 that kept the Truman administration in power and regained Congress for the Democrats foiled Republican plans to purge the Bureau of Reclamation of its resurgent New Dealers.

Nevertheless, the critical tone of the 1952 *Saturday Evening Post* article foreshadowed a Republican sweep in the fall elections. The article portrayed Straus and his appointees as "publicity workers" for the Bureau with Straus always expecting, "each employee to double informally as a pitchman for his reclamation policies." Straus's strategy, however, appeared to gain allies. Not only did the publicity seem to win over much of the western constituency of the Bureau, always eager for more water projects, but Straus impressed congressional committees with his forthright and folksy testimony. All of which resulted in increased budgets for Reclamation projects. One agency, however, stood in the way of Straus's success with Congress. The *Saturday Evening Post* applauded the Bureau of the Budget's efforts to inform Congress that not all of Straus's proposals were economically feasible. Even President Truman sometimes had to veto bills for water projects because they threatened to disrupt budget targets. Suffice it to say that Straus's administrative style and policies made him a prime target should the reins of power change in Washington.[50]

Also noteworthy, Straus drew the attention and wrath of a growing anti-communist crusade led by Senator Joseph McCarthy to expose communists and "fellow travelers" in government. The accusations served to undercut Straus's New Deal ideology that, according to these critics, smacked of socialism. They portrayed Reclamation's recent intentions to enforce the 160 acre rule as an attack on private property. Shortly after Truman's 1948 election, Commissioner Straus came under suspicion. This occurred despite the

49 Kathka, "The Bureau of Reclamation in the Truman Administration," 40.
50 Healy, "Most Arrogant Bureaucrat," 145-6.

popularity of the Bureau of Reclamation and in western states where Truman's campaign played up support for Reclamation projects and pointed derisively to the lack of support from the "do nothing" Republican Congress. To further stoke the fire, the House Un-American Activities Committee revealed that the Commissioner's wife, Nancy, belonged to organizations listed by the Attorney General as subversive. This included the League of Women Shoppers, the Washington League for Democratic Action, and the Southern Conference for Human Rights. While these organizations ceased to exist by 1949 and Straus's spouse had resigned from them before the Attorney General listed them, some senators, Patrick McCarran of Nevada among them, insisted upon an FBI investigation of Straus. The investigation was also to include the regional director of the Central Valley Project, Richard Boke. After Oscar Chapman took charge as secretary of the interior in 1949, succeeding Charles Krug, the new secretary, consistent with President Truman's policies, refused to release FBI reports on Straus to the Senate. This caused no end of trouble with Senator McCarran who served on the Senate Appropriations Committee.

8.27. Senator Sheridan Downey. Courtesy of the U.S. Senate History Program.

Infuriated Republicans in the House and Senate, including the Democrats McCarran and Senator Sheridan Downey of California, saw all of this as valid reasons to support the rider to the Department of the Interior's appropriation bill that disqualified both Straus and Boke from their positions because they were not engineers. The president reluctantly signed the bill to keep the Department of the Interior running and enable it to meet payroll for its employees. In 1949 when Democrats took control of Congress, Truman reinstated both Straus and Boke. Despite the controversy that seemed to surround Michael Straus, the Bureau of Reclamation grew considerably under his direction. Straus maintained an average expenditure rate of nearly two hundred million dollars annually while commissioner. During his tenure from 1945 to 1952, the Bureau of Reclamation produced an incredible construction record by completing thirty-six dams in the West. According to William Warne in *The Bureau of Reclamation*, Straus oversaw the Bureau staff's growth "to its highest peak and its construction work reached its greatest volume."[51]

The Transition: From Truman to Eisenhower

Overall, the Truman administration concluded its water policy in the West with a mixed record marked by the blustery and confrontational style of Straus. At the outset, the administration sought a policy of rational development of water resources and some satisfactory solution to the bureaucratic conflict in the development of these resources. For various political and bureaucratic reasons the Hoover Commission's recommendations for a single water development agency never came to pass nor did the proposed creation of valley authorities for the Columbia and Missouri river basins to address comprehensively the management of water resources. Furthermore, the ambitions of the Bureau of Reclamation under Straus sometimes outstripped the administration's, and its watchdog Bureau of the Budget's, plans to fund water development especially in California's Central Valley and the proposed Central Arizona Project.

For example, in February 1949 when Frank Pace, director of the Bureau of the Budget, recommended against the Central Arizona Project, and President Truman accepted the recommendation, Interior Secretary Krug and Commissioner Straus received the news with profound disappointment. Krug told Straus that the Department of Agriculture's studies, which played a role in the Bureau of the Budget's decision, exaggerated costs for bringing lands into production. He suggested that an effort be made to obtain support in the state for the project, "which is so absolutely essential to maintaining agriculture." He then referred to "the more or less abstract studies of statistics relating to benefits" that should not confuse the primary issue which is, "with only a little exaggeration, that the entire economy of central Arizona will crumble within the next few decades unless a supplemental source of water can be made available." Krug expressed to Straus his opinion that the "dire and urgent need of Arizona" could be met without threatening the

---

51 Warne, *The Bureau Reclamation*, 19, 77-9; see also, Kathka, "The Bureau of Reclamation in the Truman Administration," 27; "Reclamation Revelations—1948-1949," *Reclamation Era*, 12 (December 1948): 223-4, 239; Keynote Comments of Michael W. Straus Commissioner of Reclamation, Opening of the Annual Programming Conference of the Bureau of Reclamation at Boulder City, Nevada, August 1, 1949, Krug Collection, Box 69.

interests of California and other Colorado River basin states for the waters of the Colorado River.[52]

Another impediment to the Bureau of Reclamation's ambition came in a proposal that Reclamation build a steam plant fueled by natural gas to produce electricity for the Atomic Energy Commission's laboratories at Los Alamos, New Mexico. In this case, the Bureau of the Budget declared that it was, "not within the purview of the Bureau of Reclamation" to diversify the manner in which it produced power beyond hydroelectricity. The Bureau of the Budget took a particularly dim view of Secretary of the Interior Krug's effort to push the project. By the end of his presidency, Truman's own vacillation about and bureaucratic squabbling over water projects in the West continued. In addition, conservative opposition frustrated administration efforts to expand public power, and a stalemate prevailed on the enforcement of the 160 acre limitation on farms receiving federal water. In the end, the Bureau of Reclamation, with the approval of Commissioner Straus and the acquiescence of Secretary of the Interior Oscar Chapman, moved toward compromise on the issue when it was ruled that 320 acres were allowable for husband and wife. Also excess landowners could accept water for 320 acres of their land without being required to sell off excess acreage. If the sale requirement was enforced, the landowner was allowed ten years to dispose of the acreage over 320 acres. As one historian put it, "Although the Bureau's actions resulted in an increased technical compliance with the law, these provisions achieved far less change in landholding patterns than Reclamation's founders had envisioned," and concluded that while the letter of the law survived Truman's Fair Deal unchanged, its principle and practice did not fare so well.[53]

Dwight D. Eisenhower's election to the presidency in 1952 revived the power and influence of the economy-minded Republican Party in domestic affairs. The new Eisenhower administration placed great emphasis upon reducing the role of the federal government in the lives of everyday Americans. The Bureau of Reclamation faced adjustment to this new political landscape. A new Secretary of the Interior, Douglas McKay from the state of Oregon, and a new commissioner of Reclamation, Wilbur A. Dexheimer, brought less

---

52 Julius A. Krug to Michael Straus, February 10, 1949, Krug Papers, Box 69.

53 Julius Krug to President Truman, August 6, 1948, Krug Papers, Box 69; Coate, "'New School of Thought,'" 61-2; Clayton R. Koppes, "Oscar Chapman: A Liberal at the Interior Department, 1933–1953," unpublished dissertation, University of Kansas, 1974, 119; for more information on "technical compliance see Lee, "California Water Politics," 414; Kathka, "The Bureau of Reclamation in the Truman Administration," 65.

ambitious visions for western dam building to their offices, especially if the resulting hydroelectric production and distribution threatened private power interests. Dexheimer prudently chose a lower profile in his leadership style when it became clear that attacks upon big government might also include the Bureau of Reclamation and its longstanding programs to develop water resources and particularly hydroelectric power.

Even with all the changes and controversies that the Bureau of Reclamation faced during the immediate postwar years, there was still time to reflect on achievements. In 1952 among the growing disenchantments with New Deal ideology, Reclamation celebrated its golden jubilee. It was time to look back not only on the accomplishments of Hoover and Grand Coulee dams, but also the gains made in water resource management. Reclamation officials felt justifiably proud of the role Reclamation played in transforming the American West. For these individuals, the past fifty years revealed much more than just increases in crop production or electricity for an emerging urbanized West. They pointed to concrete evidence of Reclamation's ability to fulfill its original mission. One observer wrote, "Time and again, the placing of water on semiarid areas in the West has brought about the establishment of a new town, or resulted in a phenomenal economic boost to an erstwhile slumbering area."[54]

The celebration of the golden jubilee was also an opportunity to face some of the criticism that detractors had heaped upon the Bureau of Reclamation in recent years. In 1952, assistant commissioner of the Bureau of Reclamation, Kenneth Markwell, attempted to define the Bureau's ideological position. In a paper delivered to the American Society of Civil Engineers celebrating Reclamation's golden jubilee, Markwell touted the Bureau's achievements. He wrote, that Reclamation strove "to ensure a livelihood to as many families as possible and prevent concentrating ownership in the hands of a few persons." Markwell expressed the crux of the matter by simply stating Reclamation was merely enforcing the law as Congress had intended. He also attacked those who questioned the Bureau's motives,

> In encouraging and assisting, through the Bureau of Reclamation, independent individuals or family ownership and operation of family-sized farms, the various Congresses and Presidents of the United States have, since 1902, broadened

---

[54] United States Department of the Interior, Bureau of Reclamation, "Facts about Reclamation's Golden Jubilee 1902-1952," 1952, 5-7, RG 115, Records of the Bureau of Reclamation, Office of the Chief Engineer, ACC# 8NS-95-090, Box 12, National Archives and Records Administration, Denver, Colorado; hereafter cited RG 115.

the basic free enterprise foundation of our nation. The reason why corporation farms call this socialism I leave to your imagination.

Markwell's comment suggests that there were larger issues at stake as the Bureau of Reclamation sought higher, albeit more idealistic, goals. He turned more to the Progressive goals inherent in the 1902 Reclamation Act that called for a utilitarian approach to resource management. It sought not only to improve the conditions for people living in the arid West, but, Markwell argued, that by doing so land reclamation attained greater prosperity for the nation as a whole. Markwell concluded that "through the wise provisions of the Reclamation Act … irrigated lands have been developed to the extent that although they comprise less than six percent of arable land, they provide the livelihood for eleven percent of the country's population."[55]

As an arm of government public service, created in the context of the early twentieth-century Progressive Era, the Bureau of Reclamation constantly faced changing times and policies that challenged its original charter and goals. The new political setting in Washington after the 1952 election allowed for renewed attacks on Reclamation just at a time when it entertained expansionary designs on the upper Colorado River and the Central Arizona Project on the lower Colorado River. The announced fiscal conservatism of the Eisenhower administration did not shield the Bureau of Reclamation from charges of extravagance and being used by politicians as an example of an unnecessary federal service bureaucracy. Longtime New Deal critic Raymond Moley emerged with a long slate of criticisms that primarily centered on profligate federal spending and he placed the Bureau of Reclamation near the top of the offending list.

The Bureau of Reclamation's plans for development of the upper Colorado River presented an excellent target for critics biding their time while Reclamation basked in the light of the successful developments on the lower Colorado River—Hoover Dam, Parker Dam, and Davis Dam providing power and water to southern California, along with the fully operational All-American Canal to the Imperial Valley. By the mid-1950s the novelty and sheen of these accomplishments faded as the Eisenhower years revived ardent rhetoric about economy in government. The time appeared ripe for a rollback of government including Reclamation's water and power projects. Critics of

---

55 Kenneth Markwell, Assistant Commissioner, Bureau of Reclamation, Washington, D.C., "Fifty Years of Reclamation Progress," A Paper for the Summer Convention, ASCE, Denver, Colorado, June 1952, RG 115, ACC# 8NS-115-95-090, Box 12.

Moley's stripe rose to the occasion ignoring praiseworthy accomplishments in the realm of public works. Pushing aside the achievements of Reclamation's big dam engineering feats during the New Deal and the resulting enormous power output that helped place the nation quickly on a war footing when World War II engulfed the United States, no amount of past good works shielded the Bureau from these critiques.

Writing a series of pamphlets for the conservative American Enterprise Association, Moley laid bare what he saw as the economic failures of federal reclamation from its origins to the mid-twentieth century. In his 1955 publication *What Price Federal Reclamation?* Moley recounted the origins of federal reclamation in the 1902 Reclamation Act. He noted the enthusiasm of Reclamation advocates who argued that advances of interest free money for development of Reclamation projects would duplicate the successes of private efforts in irrigation of arid lands achieved by the Mormons in Utah and private irrigation developers in Colorado and southern California. Originally the payback period for the federal projects was ten years, but as Moley pointed out ten years was not enough and in some instances was extended to forty years by the 1920s. He believed the extensions merely increased exponentially the subsidy offered by the federal treasury to this uneconomical undertaking. Moley argued that federal reclamation was a failure by the 1920s: "The Bureau of Reclamation was living on borrowed time, and the farmers were living on borrowed money." In the nick of time, according to Moley, President Franklin D. Roosevelt saved the program with the infusion of new money to build dams for public works and "novel bookkeeping devices, and economic formulas" to justify the program to Congress.[56]

Finally, he noted, the Reclamation Project Act of 1939 gave authority to the Department of the Interior to develop repayment contracts under "variable repayment," that extended the repayment period and insured that the least efficient projects stayed in business. The 1939 Act, Moley argued, "practically" put the federal government "into permanent possession of water rights," which was being bitterly contested in the Central Valley of California. In addition, the Department of the Interior interpreted language in the Act to justify power revenues to subsidize irrigation projects, which Moley contended Congress never approved. Another innovation was justification of projects on the basis of benefit/cost ratios creatively devised even before the 1939 Act but extensively employed ever since. Moley regarded the methodology as an exercise in fictive

---

56 Raymond Moley, *What Price Federal Reclamation?* (Washington, D.C.: American Enterprise Association, Inc., 1955), ii.

economics that "opens the way to almost unbelievable abuses in fabricating a case for the feasibility of a project." He used the words, "fictitious, indirect and highly speculative" to describe how the benefits of projects were inflated while estimates of project costs fell well below the real construction costs. He ridiculed the "multiplier" benefit of projects, extolled in a never-ending stream of Bureau of Reclamation publications and public relations bulletins. Such claims could not stand the burden of proof, he claimed. From a historical perspective, this was particularly discomforting because Commissioner Mead had used such arguments in defense of Reclamation in some of its darkest days in the 1920s to defend and even save Reclamation from a looming threat in Congress to do away with the program.[57]

Moley's main points stressed the uneconomical aspects of federal reclamation arguing against all claims for community and social values and benefits of the program for rural America. The American Enterprise Association, the forerunner of the American Enterprise Institute for Public Policy Research that promoted free competitive enterprise and saw government expenditures as the bane of free market economic systems, sponsored his work. Moley focused on an irony at the core of the Reclamation Program. The investment in arid land irrigation programs represented a negative return on investment and if the same amount of investment could be made in "the relatively cheap process of irrigation in humid areas in order to stabilize the growing of crops" the return on investment would overwhelmingly fill the profit column. To continue the irony, Moley noted that the "Golden Jubilee" brochure of the Bureau in 1952 asserted that between 1943 and 1959 "water should be assured for 3,111,400 additional thirsty acres." This development, Moley declared, incurred additional subsidies in interest "not paid by those who are benefited" all to add to the nation's supply of food and fiber not needed. Taxpayers, he said, asked why are we subsidizing the production of yet more crops when we already are subsidizing the production of crops through the federal farm subsidy programs. For Moley and his cohorts at the American Enterprise Association the whole picture failed to make sense. Yet, Moley declared that all the claims of congressmen from the Reclamation states, their supporting organizations, and the Department of the Interior were "pouring out" what he regarded as bogus arguments for more irrigation

---

[57] Moley, *What Price Federal Reclamation?*, iii-iv. Moley's claim that the federal government retained water rights in the Central Valley stems from what became known as the 9e contract. According to David Kathka, "In the Central Valley, the 9e contract granted water rights to irrigation districts for a limited period of time. The water rights in perpetuity remained with the Bureau." See Kathka, "The Bureau of Reclamation in the Truman Administration," 75.

of arid lands. He was not arguing for the "termination of reclamation" but hoped that the process of economic and political education might bring about a rationalization of the Reclamation program. Moley appeared to be simply asking the Bureau of Reclamation to redirect its efforts from watering an arid West "to support food and fiber that we cannot use" to more economical environments.[58]

Beyond the broad criticism of Bureau of Reclamation undertakings, Moley turned his attention to the immediate Reclamation plans to proceed with development of the waters of the upper Colorado River. While the promises of the project were widely advertised by Reclamation and ardently supported by Utah's U.S. Senator Arthur V. Watkins, Moley eagerly sought to engage Senator Watkins over the virtues and especially the economics of the undertaking. He did this in a forum again provided by the American Enterprise Association in 1956 that presented a point-by-point debate between the Utah senator and Moley on the virtues and failings of the Colorado River Storage Project (CRSP) before Congress. Moley described the CRSP as so costly "as to be wholly indefensible, despite the trick bookkeeping incorporated in the bill and habitually practiced by the government in reporting such projects." While Moley castigated the Colorado River Storage Project for its economic shortcomings, he also lamented the fact that the major debate about the project centered on the preservation of wilderness rather than on its astronomical costs. Moley found it a "bitter reflection that so many commentators and others have interpreted this controversy only as a battle to preserve the sanctity of a beautiful national monument and that so few have comprehended the enormity of the engineering folly and financial wastes involved." Still the opposition, he noted with satisfaction, had stopped Echo Park Dam in a compromise deal in Congress. Moley believed that the true meaning of conservation was to "conserve the solvency of the nation," that such extravagant and unneeded projects threatened by placing undue burdens on the nation's taxpayers for the benefit of regional interests.

Moley's discussion of the economic failings of the Colorado River Storage Project were as lengthy and caustic as were his previous discussions about the entire failure of arid land reclamation in the previous year. He noted that both the president and the secretary of the interior recommended authorization and construction of two dams, Glen Canyon and Echo Park, along with eleven other "participating projects." He pointed out that the aim was to add

[58] Moley, *What Price Federal Reclamation?*, vi-vii.

hundreds of thousands of acres of productive land, while at the same time, Secretary of Agriculture Ezra Taft Benson sought ways to relieve the "burden of surpluses through retiring land from use" in the Soil Bank Program. Moley found that irony almost unbearable.[59]

8.28. Senator Arthur Watkins. Courtesy of the U.S. Senate History Program.

What especially upset Moley was the argument that power production from the large dams would pay not only for their construction but for all of the costs of Reclamation projects on lands that raised mostly low market price forage crops. Such erroneous projections, he believed, did not take into consideration supplies of coal resources in the region that could fuel coal powerplants producing electricity at lower costs than could be provided by hydropower. In any event, the point was that coal fired plants offered the possibility of producing low cost electricity to compete with the Bureau of Reclamation's hydropower dams resulting in overall lower returns for power produced. Cheaper electricity from coal plants was not the only threat to the price of hydroelectricity. Moley argued that the future seemed to hold the bright promise of "atomic energy plants" that offered even cheaper electricity. All of this made the pursuit of hydroelectric power under the guise of land reclamation something no more productive than a dog chasing its tail. But the pursuit was not harmless play. It was a costly undertaking for which every American taxpayer would pay dearly.

Utah's Senator Watkins had a decidedly different view. For him the Colorado River Storage Project was the fulfillment of promises made dating back to the 1922 Colorado River Compact. Delay in honoring it meant the continued enjoyment by California and Mexico of water rightly assigned to the upper basin states. According to Watkins, development of the lower Colorado

59 Raymond Moley and Arthur Watkins, *The Upper Colorado Reclamation Project* (Washington, D.C.: American Enterprise Association, Inc., 1956), 42-4.

8.29. Major components of the Colorado River Storage Project. Note that the Central Utah Project and San Juan-Chama Project exported water from the Colorado River Basin.

River basin was widely acknowledged to be an easier task than the challenges facing water development in the upper basin. The topography and the inaccessibility of major storage sites for the regulation of the river presented horrendous challenges in the upper basin region. A major reason the states of the upper basin signed the Colorado River Compact in 1922 lay in the pledges of the states in the lower Colorado River basin to support the eventual expansion of water projects in the upper basin. Senator Watkins asserted that the time for this development was "now."

Only through the Colorado River Storage Project would the people of the upper basin, where waters for the river originate, be able to assert their rightful claims to the waters of the Colorado River. In contrast to Moley, Senator Watkins believed that the benefits of the project justified the costs. In fact, the price tag was in his view almost completely reimbursable. The power generated over the years from the proposed big dams at Echo Park and Glen Canyon went a long ways toward this goal plus contributing to the repayment of irrigation costs. In addition, the flood control provided by the dams contributed untold benefits. Interestingly, the project did not involve the controversies over public versus private power that plagued other proposals in the postwar period. Private companies shied away, and probably wisely so, from such investments in remote hydroelectric damsites. Senator Watkins also cited the benefits that would accrue to the Navajo Tribe starting them "on the road to independence and self-sufficiency."[60]

Not only would local entities benefit from the project but every section of the nation profited. This followed an argument that Moley had earlier derided as a long time chimera of the Bureau of Reclamation to justify expenditures on questionable projects. Watkins recounted the argument in detail:

> Eighty-one percent of the construction cost of the Colorado River Storage Project will be spent in markets outside the Upper Colorado Basin, for labor and materials with which to build the Project. This means that practically every state in the union will benefit from expenditures resulting from such construction.

Also what was noted as "reimbursable costs" will be paid for by water users and power sales. And with more families on the land and power to the cities there will be increases in individual incomes and wealth in the region. All of

---

60 *Ibid.*, 42-4, 53-4, 2, 11.

8.30. The compromise location for a dam at Glen Canyon for the Colorado River Storage Project was considerably downstream of Dinosaur National Monument, but it was still in the Upper Basin of the Colorado River.

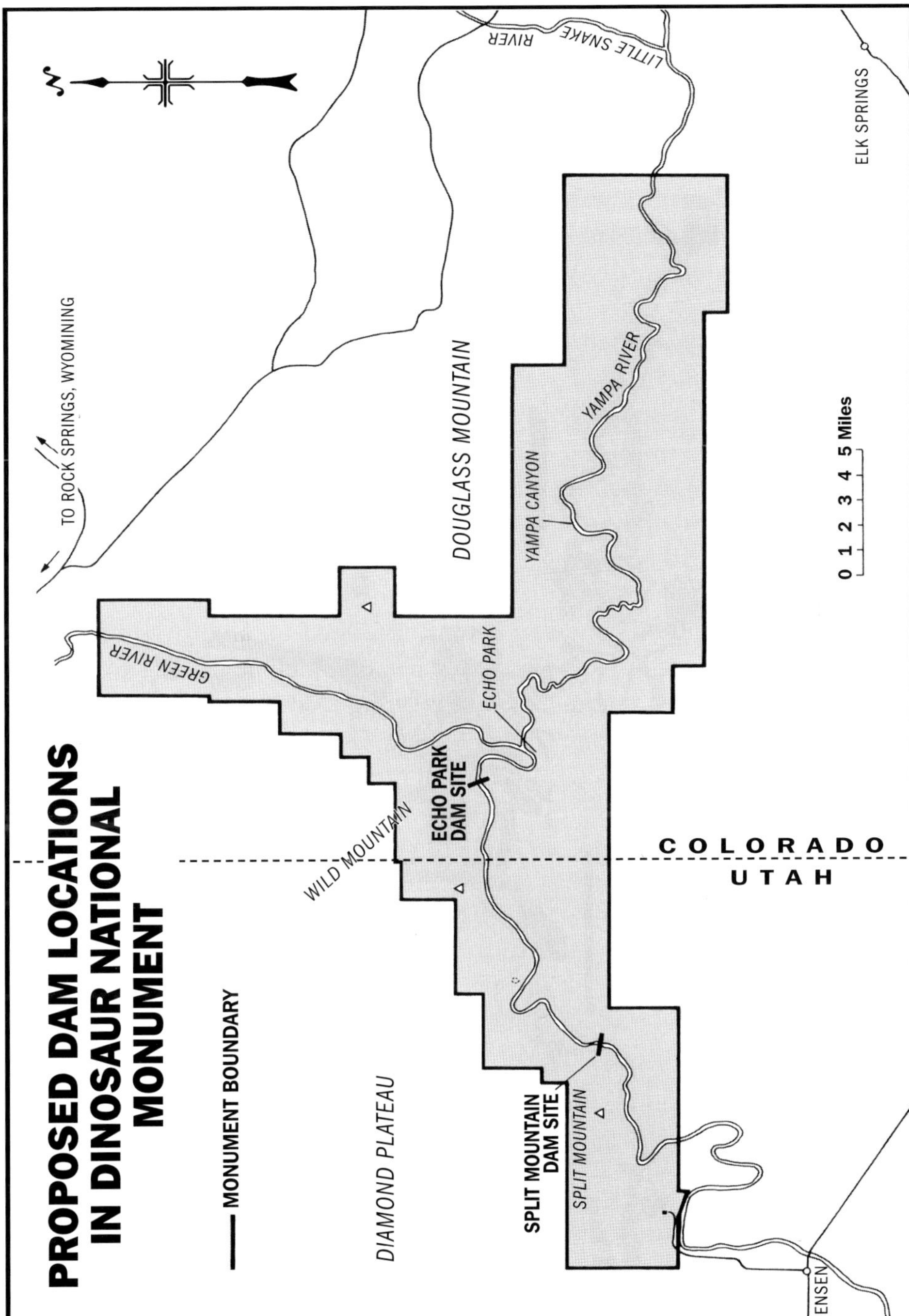

8.31. Echo Park, Split Mountain, and Glen Canyon were dams which figured prominently in the several-years-long debate over the Colorado River Basin Projects Act, which finally became law in 1968. After Mark W. T. Harvey's *A Symbol of Wilderness: Echo Park and the American Conservation Movement* (Albuquerque: University of New Mexico Press, 1994).

this redounded to the benefit of the U.S. Treasury in terms of the income taxes collected. New incomes and burgeoning population as a result of workers seeking jobs in war industries in the West laid a foundation for later manufacturing in the region—manufacturing that required power from dams. This alone justified the interest free loans extended to Reclamation projects that some have called a subsidy. "This is not a subsidy. It is a form of payment for indirect benefits received by the public from the results of reclamation," declared the senator.[61]

While Moley raised the specter of Reclamation compounding the agricultural surplus situation in the United States in the 1950s, Senator Watkins said the Colorado River Storage Project would have no effect on current agricultural resources. First of all, the crops raised in the high mountain valleys were not the same as the principal staple crops raised in the Midwest and South—corn, wheat, tobacco, and cotton. In these valleys specialty crops of alfalfa, forage, vegetables, and fruits prevail. Furthermore, the senator noted that by 1975 the United States would be a nation of over 200 to 250 million people. This meant a need for another 100 million acres of productive land in the United States of which he estimated 20 million could be obtained by draining and clearing land in the eastern humid and sub-humid regions of the United States. Only about 6 million acres of land remain irrigable in the arid West and the Colorado River Storage Project will only account for about 132,360 acres when they are brought into full production by 1980. For all of these reasons the Senator believed: "The project must be built beginning now. It cannot possibly add to the current surplus. It is even doubtful that it can meet the increased food demand by the time it is in full production."[62]

Conclusion

During the immediate postwar period (1945-1952), the Bureau of Reclamation faced many internal and external challenges. Within Reclamation, there was optimism that the construction activity that occurred during the Depression would once again restart. Indeed through much of World

---

61 *Ibid.*, 19-20, 23, 25; see also Harvey, *Symbol of Wilderness*, notes that the war stimulated the economic development of the states in the upper Colorado Basin: "These economic developments focused ever-increasing attention on water and power supplies, and elevated the value of the Colorado River," 35; Harvey also cites Gerald Nash, *The American West Transformed: The Impact of the Second World War* (Bloomington: Indiana University Press, 1985), 17.

62 Moley and Watkins, *The Upper Colorado River Project*, 30-3, 35.

War II, Bureau planners conducted studies and surveys throughout the West to ensure readiness once the wartime emergency ended. By war's end, they had developed comprehensive plans for the major river systems in the West, especially for the Colorado, Columbia, and Missouri river basins. The end of World War II saw a transformed West more urbanized and industrialized. Reclamation's great dams Boulder, Grand Coulee, and Shasta had won acclaim for their contributions to the war industries that emerged to service the war, and that experience emphasized the greater importance of hydroelectric power. A new West was emerging as a result of the war, and the Bureau of Reclamation was positioning itself to aid in that transition.[63]

Yet irrigation and development of agricultural lands were still central to the Bureau of Reclamation's vision. The proposed million-plus acres on the Columbia Basin Project alone spoke of the Bureau of Reclamation's commitment to advancing the agricultural potential of the American West. In California, Reclamation personnel looked forward to expanding the Central Valley Project by injecting new and stable water resources into that already established agricultural enterprise. Opportunities were also available in the upper Colorado River basin where the expansion of farming could occur in the states of Utah, Colorado, Wyoming, and New Mexico. The Bureau's portion of the Pick-Sloan Plan on the Missouri River basin promised new opportunities for farm families in that immense river basin. In short, developing vibrant irrigation communities still commanded an important place on Reclamation's agenda in the years immediately following World War II.

That said, the Bureau of Reclamation faced new and old criticisms. The most vocal was an emerging movement among conservative forces in the United States to curtail and even turn back what they perceived as the socialistic tendencies of FDR's New Deal. For these individuals, Reclamation's stance on development of public power and enforcement of the 160 acre limitation rule meant attacks on free enterprise and private property rights. Others viewed the Bureau of Reclamation as an inefficient bureaucracy and a drain on the national treasury whose accomplishments never measured up to its promises. Of course, the criticisms were not new and were largely overcome by the

---

63 For more information concerning the transformation of the American West after World War II see Nash, *The American West Transformed*; Richard W. Etulain, "Gerald D. Nash and the Twentieth-Century American West," in *The American West in 2000: Essays in Honor of Gerald D. Nash*, Richard W. Etulain and Ferenc Morton Szasz, editors (Albuquerque: University of New Mexico Press, 2003), 186-98.

prospect of hydroelectric energy from new dams and new water supplies for a thirsty West. What was new was the brief climate of virulent anti-communism or a "Red Scare" that threw a blanket of suspicion over long-time public servants and the first spark of an environmental movement that saw dams and reservoirs as destroyers of natural and scenic river systems.

In the immediate postwar period, however, the Bureau of Reclamation succeeded in developing power and enthusiastic constituents in both Congress and communities throughout the West. On the world stage, the emerging Cold War between the United States and the Soviet Union offered new challenges for the Bureau of Reclamation. The Truman administration sought Reclamation's technical expertise believing it uniquely suited to serve American foreign policy goals. This new mission launched the Bureau of Reclamation into various overseas tasks of dam building, hydropower development, and construction of improved irrigation works.

CHAPTER 9:

RECLAMATION AND THE POSTWAR WORLD, 1945-1969

Introduction

By the mid-twentieth century, the Bureau of Reclamation's Denver offices and laboratories received a steady stream of visitors from abroad. Mostly engineers and some administrators, they came to observe firsthand the Bureau of Reclamation's various engineering challenges in the American West and to transfer technology and know-how back to their numerous homelands. Water resource development promised to underwrite agricultural and industrial advances in countries throughout the world. What better place to study the pathways to successful water resource development than in the American West? Narratives of American history argued for the view that the nation's progress and its "mastery of the North American wilderness [i.e., the American West] ought to serve as a model of modernity for all humankind."[64] The parade of progress in the American West, due in no small part to the technical and engineering accomplishments of the Bureau of Reclamation, altered land and waterscapes, expanded irrigation acreage, provided urban water supplies, and most dramatically made possible the production of millions of kilowatts of hydroelectric power.

The world took notice of the progress from wilderness and desert to furrowed agricultural landscape and cityscapes with imposing skyscrapers. Americans took pride in national achievements that some called "high modernism" in the form of monumental dam structures and multiple purpose regional development projects. After the devastation of World War II, a repository of American technical, engineering, and administrative expertise existed with the potential to serve the cause of worldwide rehabilitation and modernization for underdeveloped nations, especially in the former colonial possessions of European powers. The American example provided a road map to a modern developed society under free institutions, and the Bureau of Reclamation was a chief construction agent along this well-marked highway to modernity. As the United States confronted the Soviet Union in the Cold War, it struggled to win friends and allies around the world by offering a pathway to development quite different from the pattern advanced by Soviet Com-

64 Michael Adas, *Dominance by Design: Technological Imperatives and America's Civilizing Mission* (Cambridge, Massachusetts: Harvard University Press, 2006), 73.

9.1. One of many foreign visitors to early Reclamation projects, L. Beata Neves, a Brazilian engineer, inspected construction on Theodore Roosevelt Dam in February of 1909. Photographer: Walter J. Lubken.

munism's style of state driven development. American policy makers eagerly pointed to the United States as a shining example of prosperous modernity. Certainly the "lessons of America's past demonstrated the route to genuine modernity," and the best and most humane way to move "stagnant" or traditional societies into the modern world. America's mission to this world developed into a commitment to block the spread of Communism during the Cold War, summoning Bureau of Reclamation planners into the Lower Mekong River Basin in Southeast Asia as a little noted sidelight of American military intervention in Vietnam in the 1960s.[65]

Reclamation in the World Setting

American engineering expertise, having achieved Hoover Dam, Grand Coulee Dam, Shasta Dam, and even river basin development with the Tennessee Valley Authority and a burgeoning Missouri River Basin Program, pre-

---

[65] James C. Scott, *Seeing Like a State: How Certain Schemes to Improve the Human Condition Have Failed* (New Haven, Connecticut: Yale University Press, 1998), 6; Michael E. Latham, *Modernization as Ideology: American Social Science and "Nation Building" in the Kennedy Era* (Chapel Hill: University of North Carolina Press, 2000), 4; Nguyen Thi Dieu, *The Mekong River and the Struggle for Indochina, Water, War, and Peace* (Westport, Connecticut: Praeger Publishers, 1999), 150-1.

sented sources of technical expertise ready to be tapped for service throughout the world. Asia, the Middle East, South America, and Europe now presented opportunities for the expansion and application of American enterprise, skill, talent, and organizational experience in the fields of water and hydroelectric development. Even earlier, American experts pioneered special intergovernmental agreements and private consultations in diverse places—Puerto Rico, Central America, Hawaii. Commissioner Elwood Mead's early twentieth century career, although outside of the Reclamation Service, took him to Australia where he engaged in irrigation developments and returned home prior to World War I with revamped ideas on the importance of governmental leadership in the promotion of irrigation projects. Consultations overseas began early in the century for John "Jack" L. Savage, legendary dam design engineer for the Bureau of Reclamation who was the principal designer of Hoover, Grand Coulee, and Shasta dams. Unlike Mead, whose early employment was with the Department of Agriculture and then the Commonwealth of Australia (1906-1915), Savage's career was almost entirely with Reclamation until his retirement in 1945 whereupon he entered into private consultations devoted to postwar reconstruction until his death in 1967.[66]

In the aftermath of Hoover Dam's much-celebrated construction in the mid-1930s, engineers from around the world visited the site. Their journeys often included requests to observe and study Reclamation's administrative structure, laboratories, and design operations at the office of chief engineer in Denver. In return, Americans traveled abroad. In 1937, Savage, chief designing engineer for Reclamation, addressed the Institute of Civil Engineers in London on the Boulder Canyon Project. Visits of foreign engineers to Denver brought information from around the world about potential dam, reservoir, and power sites, especially in China, where visions of a great dam on the Yangtze River took shape on the eve of World War II. In the immediate postwar period, Reclamation saw opportunities to put its expertise to work for a world in dire need of water and power development.

These included war-ravaged lands as well as former European colonies transitioning to nationhood. Not only did opportunities arise for grand dams, hydroelectricity, and water distribution projects, but also opportunities for achieving larger social and political goals associated with the "revolution of rising expectations"[67] in the postwar world. The development of

---

66 "John Lucian Savage Dies at 88; Designed World's Great Dams," *New York Times*, December 29, 1967.

67 Adlai Stevenson, "Will India Turn Communist," *Look*, 17 (July 14, 1953).

9.2. John L. "Jack" Savage, Reclamation's chief design engineer, worked at Reclamation from 1903 to 1945 except for a brief period when he worked as a consultant.

water resources offered the prospect of Americans playing a major role in the improvement of the material life of millions throughout the world and, most importantly, served American policymakers as they sought to stem the tide of international Communism in the beginning years of the Cold War. Reclamation's close identification with the New Deal's economic recovery programs during the Great Depression, particularly its contribution to the Tennessee Valley Authority project, marked it as an agent of social and economic development. Many postwar planners for economic development saw the American experience during the New Deal in the Tennessee Valley as a model for modernized liberal development. As one source notes:

> That the TVA appeared so easily and often in the context of so many basic discussions regarding modernization demonstrates the importance of the liberal development ideas it represented to the overall conception of how modernity could best be cultivated in a changing world.[68]

With its experience and expertise, the Bureau of Reclamation appeared as the proper vehicle to bring that development to a world beyond the borders of the United States. This is to say nothing of the original social goals of Reclamation whose centerpiece was promotion of irrigated farm communities with assured water supplies.

Americans found their own irrigation and water development efforts from the late nineteenth century onward a possible model for worldwide development to bring under cultivation marginal, mostly arid lands in the Middle East, Asian subcontinent, and even areas of Australia. Some projects sought to expand upon ancient irrigation systems as evidenced in India, Ceylon, and

[68] David Ekbladh, *The Great American Mission: Modernization and the Construction of an American World Order* (Princeton, New Jersey: Princeton University Press, 2010).

9.3. Hoover Dam, then known as Boulder Dam, in 1941.

Egypt. In their own development process in the nineteenth century, Americans saw the importance of drawing upon the wisdom and experience of faraway places and times to guide the development of their own water projects. In the 1870s, Congress asked the author of *Man and Nature* (1864), George Perkins Marsh, about his observations of irrigation in the Mediterranean. The late-nineteenth-century United States Geological Survey sent engineers abroad to gain knowledge of how other societies in varied environments constructed water storage and distribution systems. Likewise, inquisitive foreign visitors came to the U.S. to observe conditions and prospects for irrigation in the American West. John Wesley Powell's ill-fated and short-lived Irrigation Survey of the West from 1889 to 1891 noted the journey of Herbert W. Wilson to Egypt, Arabia, Italy, and France where he inspected irrigation projects, dams, storage facilities, and canals.

Wilson's investigations in India provided the material for his U.S.G.S. 1891 publication Water-Supply and Irrigation Paper, "Irrigation in India." He noted that American engineers might see the similarity between the climate and topography of the great northern plains of India and the arid American West, including especially the Central Valley of California. In the 1902 preface of the second edition, prompted by passage of the Reclamation Act, Wilson asserted the belief that American engineers would find much to learn from projects in India that combined the building of a dam, its reservoir, the diversion of water from a running stream and its storage in a reservoir located at a considerable distance from the canal head.[69]

Americans as well as their foreign contemporaries understood the advantages of sharing knowledge and experience related to the technical challenges of water development. Australia's future prime minister, Alfred Deakin, visited the American West in the early 1880s to study arid land irrigation after having read Powell's 1878 *Report on the Lands of the Arid Region of the United States*. In considering techniques employed elsewhere, Reclamation quickly concluded that models of direct diversions from flowing rivers and streams for irrigation used in India and Egypt were not appropriate in the American West. Rather it adopted plans for storage reservoirs to capture spring snow melt storing it to insure favorable flows of water through the long, dry summer months. Reclamation's achievements within its first ten years of work drew international visitors as announced in its annual report for 1911:

> There has been an almost continual series of investigations of the work and its results by men both from this country and abroad. Nearly every foreign country having large areas of arid lands has been represented by visitors who have studied the works on the ground, and particularly the methods and analyses of cost. Official and unofficial representatives from Great Britain and its colonial possessions … from various portions of the German Empire, from Austria, Russia, Spain, and other European countries, and from Mexico and South America. These men have been interested not only in irrigation but in the control and conservation of flood waters.[70]

---

69 Herbert M. Wilson, *Irrigation in India*, Water Supply and Irrigation Paper No. 87 (Washington, D.C.: United States Department of the Interior, United States Geological Survey, United States Government Printing Office, 1903), 7-10.

70 United States Department of the Interior, Reclamation Service, F. H. Newell, Director, *Eleventh Annual Report of the Reclamation Service, 1911-1912* (Washington, D. C.: United States Government Printing Office, 1913), 15.

As it built impressive high dams, i.e., (1904-1910) Shoshone Dam, (renamed Buffalo Bill in 1946),[71] Arrowrock Dam 1915, Owyhee Dam 1932, and finally in 1935 Hoover Dam, Reclamation gained an enviable international reputation. Its experiments and refinements of the trial-load method of analyzing projected dam designs and test modeling at universities (Colorado State College and University of California at Berkeley) developed a body of public knowledge. The accumulated dam-building information (design, engineering practice, materials, and management) appeared in Reclamation bulletins, pamphlets, and publications. Reclamation made knowledge open and accessible to the world engineering community. That information included laboratory reports, modeling experiments, and reports on design—all shared at professional meetings and in professional publications. For its large projects, especially Hoover Dam, Reclamation made available publications entitled *Technical Record of Design and Construction*. By making knowledge gained in building dams, spillways, penstocks, and even hydroelectric plants into "public knowledge," Reclamation promoted its national as well as international reputation.[72]

In 1920 the *Nineteenth Annual Report of the Reclamation Service* summarized the importance of Reclamation's work on a world scale:

> Irrigation development of hitherto largely unused lands is becoming more and more prominent in Australia, South Africa, Canada, Brazil, [the] Argentine, Russia, and other countries, and the works of the Reclamation Service have for many years attracted engineers and economists from all over the world. There can be no doubt that much of the stimulus for extended reclamation development of the arid regions of the world has been the direct result of first-hand study of the irrigation problems in the United States, and particularly that as exemplified by the work of the Federal Government.[73]

---

71 Robert E. Bonner, *William F. Cody's Wyoming Empire: The Buffalo Bill Nobody Knows* (Norman: University of Oklahoma Press, 2007), 204.

72 United States Department of the Interior, Bureau of Reclamation, *Thirty-First Annual Report of the Bureau of Reclamation, 1931-1932*, Elwood Mead, Commissioner (Washington, D.C.: United States Government Printing Office, 1932).

73 United States Department of the Interior, Reclamation Service, *Nineteenth Annual Report of the Reclamation Service, 1919-1920*, Arthur P. Davis, Director (Washington, D. C.: United States Government Printing Office, 1920), 43-4.

DEPARTMENT OF THE INTERIOR
JOHN BARTON PAYNE, SECRETARY
UNITED STATES RECLAMATION SERVICE
ARTHUR P. DAVIS, Director

NINETEENTH ANNUAL REPORT

OF THE

RECLAMATION SERVICE

1919-1920

WASHINGTON
GOVERNMENT PRINTING OFFICE
1920

9.4. The title page of Reclamation's *Nineteenth Annual Report.*

At the end of World War II, Reclamation possessed undisputed credentials to assume a leading role in the rehabilitation of nations. Emerging nations, eager to move beyond the limitations of colonial pasts, saw water power development as the key to the future, and the United States saw the Bureau of Reclamation with its technical know-how as an ambassador to the world. As the United States chose to support, and even celebrate decolonization, it also made an effort to supplant British influence in the Middle East and even in India.[74] Reclamation's close identification with the New Deal in the prewar years of the Great Depression gave it a broad mantle not only as a water development agency but as an organization familiar with the social and economic needs of rural communities. Equally, or more so, Reclamation commanded an unmatched reputation for hydroelectricity development and in the process the delivery of electricity served to modernize communities both rural and urban. Its engineers worked closely with the Tennessee Valley Authority (TVA) and saw the social and economic uplift that multiple purpose river development offered. Reclamation's close association with the public works efforts of the New Deal in the 1930s positioned it to assume a developmental role in the wider world should the opportunity arise to export the spirit and energy of America's public work programs to build democracy overseas.

---

[74] Emily S. Rosenberg, *Spreading the American Dream: American Economic and Cultural Expansion, 1890-1945* (New York: Hill and Wang, 1982), 194.

While the Bureau of Reclamation was well-positioned to respond to the needs of emerging nations in the postwar period, it remained an open question as to how much energy and political will could be mustered in postwar America to apply Reclamation's know-how to the problems of international development. At home in the American West the Columbia River, the Colorado River, the Missouri River basin, and California's Central Valley Project demanded attention. Beyond the unfinished work remaining in these projects, their very existence served as examples of comprehensive river basin developments and showcased the successes of American technical and organizational knowledge.

When the war ended in the late summer of 1945, the United States emerged from the conflict with its industry and cities intact. During the war American industry and agriculture amazed the world and the U.S. itself with an extraordinary ability to produce the weaponry for victory and at the same time sustain a high standard of living for the civilian population. Eager now for peacetime growth and determined not to revert to the depressed economic conditions of the 1930s, the United States was on the threshold of historic domestic economic growth. Also, its business community stood poised to extend outward to many regions of the world. In these theaters, the first challenges were to meet the needs of war-ravaged regions and secondly, to seize the opportunities presented in an emerging post-colonial world. In 1947 President Harry S. Truman asked Secretary of the Interior Julius A. Krug for a report on "National Resources and Foreign Aid." After reviewing the report, the president wrote to the many individuals, within the Department of the Interior, thanking them for contributions "[you] undertook on short notice … to [do] this work which is important to world-wide human rehabilitation and economic recovery." Most importantly, American interest in aid to foreign development became tied to the emerging Cold War with the Soviet Union and the growing fear of Communist influence throughout the world.[75]

The Possibilities of China

China presented a test case for American good intentions and ambitions. In the 1930s, Chinese engineers visiting damsites and studying at the Denver facilities of the Bureau of Reclamation raised the possibilities of large river basin development projects in China. Groups in the United States concerned with the welfare of China (China International Famine

---

75 "Jack Returns," *The Engineers Bulletin* (July 1937): 5, John L. Savage Collection, American Heritage Center, University of Wyoming, Laramie, Wyoming; hereafter cited as Savage Papers; President Harry Truman to William E. Warne, October 24, 1947, Warne Papers, Box 2.

Relief Commission and the Rockefeller Foundation) reinforced the suggestions and saw the New Deal's TVA accomplishment in the American South as a model for China's modernization. American dam builders and engineers within Reclamation expressed interest, if not enthusiasm. In 1937 the Japanese invasion of mainland China complicated matters. And while American involvement in World War II after the attack on Pearl Harbor in December of 1941 suggested an indefinite postponement of Chinese reclamation investigations, the Chinese government, in 1943, invited Reclamation's John Savage to explore design possibilities for what would eventually be known as the Three Gorges Dam on the fabled Yangtze River. Even in the midst of war, Nationalist Chinese government officials looked ahead to the postwar era and sought American know-how for development of water and power resources. Savage's investigations in China occurred within the sound and sight of clashing armies, revealing both his and Reclamation's commitment to participate in water development programs in the postwar rehabilitation efforts. For American dam builders the underdeveloped world offered new frontiers

9.5. Jack Savage in China looking at potential dam sites on the Yangtze River near the end of World War II. Left to right: Hong-bin Li; General Qi-wei Wu, Commander of Yangtze Defense Headquarters; Y. H. Huang; John "Jack" L. Savage. Boatman unidentified. 1944.

9.6. Left to right: Hong-bin Lee; Huaj-yun Hsu; Zhong-xi Chan; Col. W. A. Dexheimer (future commissioner of Reclamation); John "Jack" L. Savage; Bai-heng He; Fu-shi Sun; Li Zhou; Y. H. Huang during Savage's visit to China toward the end of World War II.

and challenges beyond the now much-dammed American West as well as the opportunity to harness water energy for a brighter postwar future.[76]

Ventures overseas presented problems, especially in wartime. The Department of State guarded its authority in the domain of foreign affairs, requiring all foreign investigations by Reclamation to occur under Department of State direction and approval. The Division of Cultural Relations established within the Department of State in 1938 took charge. In 1941 and early 1942 Reclamation's chief design engineer, John Savage, found himself on assignment in Australia under State Department auspices working on Australian water and reservoir issues. His itinerary also originally included India and possibly China for similar duties, but the Japanese attack on Pearl Harbor made the entire Pacific Ocean a war zone preventing his travel beyond Australia. By 1943 the British Indian government renewed the invitation for his visit and officials in the Chinese government likewise revived their suggestion that Savage continue on to China. China wanted Savage to consult on building a

[76] David Ekbladh, "'Mr. TVA': Grass-Roots Development, David Lilienthal, and the Rise and Fall of the Tennessee Valley Authority as a Symbol for U.S. Overseas Development, 1933-1973," *Diplomatic History* 26 (Summer 2002): 339.

series of dams in the Yangtze River system. Since completion of Hoover Dam and visits from Chinese engineers in the mid-1930s, the prospect of damming China's largest river system had captured the imagination of Savage and other engineers in Reclamation's Denver Office.

The Chinese invitation offered to assume the expenses of the trip and any preliminary design work. Savage's enthusiasm showed in a note to the State Department when he welcomed, "An opportunity to be of service to the valiant people of China," and saw it as, "a signal honor." He also noted that the trip provided "for the renewal of friendships with a dozen or more Chinese engineers who supplemented their technical education in the United States with practical experience in the laboratories and design sections of the Bureau of Reclamation." Savage's plans for travel in late 1943 and through 1944 did not end with India and China. He also had in hand inquiries from a Zionist organization in New York to do studies of irrigation and hydroelectric development in Palestine. He characterized this work as having "altruistic aspects of international concern" because of its prospects of providing a "home-land for large numbers of people of the Jewish race." If this additional travel were approved, he saw himself proceeding to India then to Chungking and finally to Palestine. He acknowledged that complications might arise with the Palestine work because a private American organization was paying for his travel and not foreign governments through the State Department. In the event of complications he volunteered, "to give gratis the necessary time for the field work in Palestine (estimated at about a month)." He hoped this would facilitate the international arrangements.[77]

No facilitation occurred. The Palestine visit raised the delicate question of a United States government official serving as technical consultant to a Zionist organization in Palestine with expenses paid by the organization. Within the State Department, the Division of Near Eastern Affairs quickly nixed Savage's acceptance of the invitation by a "Zionist organization" to act as a consulting engineer for irrigation and hydroelectric development. It was pointed out that these developments were "closely linked with the question of the economic absorptive capacity of that country." The question was highly controversial among the competing groups interested in the future of Palestine. For this reason it was considered "inadvisable" that Savage should visit. He should only go there at the request of the Palestine governmental authority.

---

[77] John L. Savage to Haldore Hanson, Department of State, September 3, 1943, RG 59, Records of the Department of State, Decimal File 1940-1944, Box 5882, National Archives and Records Administration, College Park, Maryland; hereafter cited RG 59.

The exchange on Palestine demonstrated the tight supervisory role that the State Department asserted over any venture by Reclamation, any other U.S. bureau, and U.S. officials overseas. Eventually Savage did work in Palestine, but as a private consultant. His service did not escape the attention of *Reclamation Era*. In the summer of 1946 the publication noted that proposals for two irrigation projects in Palestine had been "reviewed on the ground by John L. Savage, former chief designing engineer of the Bureau."[78]

Regardless of the outcome of Savage's plans to visit Palestine, by the end of 1943 he was on his way to the Punjab in India and then to the Yangtze in China. William Warne described Savage, "as excited as a kid," before departure on a trip to last over a year in the Far East with his preliminary investigation of the Three Gorges project in China commanding the bulk of his attention. He conducted studies on the Yangtze as the struggle between the Chinese Government of Generalissimo Chiang Kai-shek and Japanese armies raged near his encampments. At this point the American engineer envisioned five project sites on tributaries of the Yangtze rather than one large dam at Three Gorges. Upon arrival in China, he inspected the gorge above Ichang and became convinced that it was a feasible damsite. Savage telegraphed Commissioner Bashore from China that the National Resources Commission of China was seriously considering the Yangtze Gorge Project. He described it as a dam of Shasta height and a powerplant that could produce ten and one half million kilowatts. He said the Chinese desired a cooperative arrangement for assistance from the Bureau and TVA. Present in China also was Donald M. Nelson, chairman of the important War Production Board that directed the allocation of materials and energy for the war effort. He told Savage that he was "enthusiastic" and that President Roosevelt would be "extremely interested."[79]

He went on to explain the organizational makeup under which the construction might occur. Generally the method employed by Reclamation for major projects was to give the construction over to a private company. While this procedure involved a bidding process, he urged that contracts not be awarded on the criteria of lowest bid alone. Only companies with experi-

---

78 Internal Memo: "Possible Visits of Mr. John L. Savage, Engineer of the United States Bureau of Reclamation to India, China, and Palestine," September 9, 1943; Paul Alling to RC, September 14, 1943, RG 59, Decimal File 1940-1944, Box 5882; "Reclamation Abroad: New Promise in the Holy Land," *Reclamation Era*, 32 (July 1946): 151.

79 William E. Warne to Philip Dickinson, January 8, 1944, Warne Papers, Box 2; Savage to Bashore, n.d., RG 59, Decimal File 1940-1944, Box 5882.

9.8. Donald M. Nelson, Chair of the War Production Board during World War II. Courtesy of the Franklin Delano Roosevelt Library, National Archives and Records Administration.

ence and records of success in building large projects should be allowed to bid. Looking at the situation in China, Savage was not totally convinced that Reclamation's practice of hiring private contractors was workable. In the challenges facing China with the Three Gorges project he foresaw that the TVA approach with government-employed design and construction teams would be more practicable. It was to be understood that much of "this personnel happens to be largely U.S. government personnel," but he was not suggesting that the U.S. government employ people in China, but rather, that the government of China undertake the project and employ Americans where required. Of course, thousands of Chinese must be brought into the effort. Chinese officials replied that they could not give a commitment in terms of the number of workers, engineers, and scientists that China could commit to the project. Instead of providing definitive figures, they indicated that any agreement must include a "sufficient number" of Chinese personnel. The entire arrangement, of course, was subject to final approval of the Generalissimo, "the reason being self-explanatory," the Chinese communication emphasized.[80]

[80] John L. Savage to C. C. Chien, Vice Chairman, National Resources Commission, September 17, 1944; C. C. Chien to John L. Savage, September 26, 1944, RG 59, Decimal File 1940-1944, Box 5882.

During his May 25 to November 24, 1944, visit to China, Savage filed several reports. They addressed the feasibility of the proposed dams and power projects and the administrative and financial arrangements necessary for Reclamation's participation in planning and construction. At home Reclamation officials and most importantly the Department of State carefully reviewed his suggestions and outlines for action. The Chinese Minister of Economic Affairs, Won Wen-hao, expressed enthusiasm for "the Yangtze Gorge Project" to American Secretary of State Edward R. Stettinius, Jr. He saw it as helping Chinese industrialization, solving the problem of flood control and navigation of the Yangtze River, and improving the livelihood of the people. He thanked the secretary of state, "for sending Dr. Savage to our assistance" and noted, "As the detail design of the Yangtze Gorge Project has yet to be done, we are negotiating with your Bureau of Reclamation to do this work, which will be automatically under Dr. Savage's supervision." Finally the minister wrote that the American government would earn "the ever-lasting gratitude of the Chinese people" for helping build the project.[81]

As dramatic as the prospect appeared for Reclamation to participate in the China project, Acting Commissioner J. Kennard Cheadle cautioned that Reclamation must reserve the right to give its own domestic program precedence over any work for the Chinese Government. He saw delays occurring because of shortages of technical manpower. Still he offered the reassurance that Reclamation endeavored to obtain whatever manpower priorities for the Chinese work proved consistent with American policies, which meant that, if called upon, Reclamation would serve U.S. foreign policy objectives. The Department of State, however, saw another set of problems. It declared that if Reclamation employees were to operate overseas, they must do so under the Department of State's aegis or oversight and on its payroll.

That assertion caused some consternation in Reclamation circles and was an indication that the State Department intended to protect its administrative turf in the conduct of foreign relations even in the specialized fields of technical aid and construction projects. At the close of 1944 the Financial and Monetary Affairs Division (FMA) within the State Department made it clear that there should be no financial commitment on the part of the United States. Moreover, news that "Jack Savage of the Reclamation Bureau" was negotiating a contract for Reclamation "to design and construct" projects did not infer any financial commitment on the part of the United States. State's memo empha-

[81] Won Wen-hao, Minister of Economic Affairs in Chungking, to Edward R. Stettinius, December 15, 1944, RG 59, Decimal File 1940-1944, Box 5882.

sized that "projects concerning the economic development of China are to be handled for the State Department by FMA, in consultation with CA [Cultural Affairs] and other appropriate Divisions." The State Department also made it clear that Savage had been in China under a program of cultural cooperation that it administered and approved.[82]

With so many avenues opening for the extension of American aid to the immediate postwar world, the State Department hastened to recommend and achieve legislation in Congress giving it exclusive control over all foreign projects whether technical or cultural. As Michael W. Straus, assistant secretary in the Department of the Interior explained it to Secretary Ickes, State Department policy and congressional legislation prohibits any foreign activity unless the United States personnel involved therein is transferred to the State Department payroll. Ultimately it meant that the State Department must approve activities before it accepted personnel on to its payroll.[83]

To avoid this, Reclamation hoped for passage of a bill pending in Congress to allow it and other agencies to loan personnel and enter into contractual arrangements with foreign governments. In a letter to Savage, Straus said its passage depended considerably on the attitude of the Department of State toward the bill. During the spring of 1945 Straus also noted that "after long and distinguished service" Savage was taking retirement from Reclamation, but he assured Savage that "change in official relationship in no way alters our interest in your plans and desire for their success." Savage's retirement plans, of course, included employment as a consultant for Reclamation on the Three Gorges Project in China. Straus declared that the Department of the Interior led by Secretary Ickes and the Bureau of Reclamation supported the plans for the Three Gorges Project on the Yangtze as presented in Savage's preliminary report. In fact, he had given the report to Lauchlin Currie, assistant to the president and advisor to the White House on Chinese affairs, after a discussion over lunch on the project.

The other news in this communication with Savage was not so encouraging. The State Department in a letter signed by Under Secretary Joseph Grew turned down the proposed Reclamation-Chinese National Resources

[82] Memo from C. F. Remer, "The Yangtze Gorge Project," December 16, 1944; Joseph C. Grew, Acting Secretary of State, to Wong Wen-hao, Minister of Economic Affairs in Chungking, February 10, 1945, RG 59, Decimal File 1940-1944, Box 5882.

[83] Michael W. Straus to Secretary Harold Ickes, June 12, 1945, RG 48, Entry 779, Box 15.

Committee engineering contract for the Yangtze project. In the opinion of the State Department the Yangtze Gorge was "economically unfeasible and unwarranted for an indefinite time in the future, and therefore the preliminary engineering report contract should not be entered into at the present time." A disappointed Straus believed that some of the reasoning in the rejection letter was "strangely reminiscent of the many arguments opponents made to Boulder Dam, Grand Coulee Dam, Shasta Dam, etc., in this country." With opposition from the State Department and an existing law directing that all foreign work must be done under its auspices, the Gorges project seemed blocked, but the Department of the Interior did not give up. Indeed, Savage proposed several routes to follow including transferring personnel to the State Department from Reclamation and TVA as well as having the Natural Resources Commission of China contract with large private companies. Straus could not embrace any of these suggestions but concluded, "I want very much to see the bureau dominant in this engineering development."[84]

Opposition from the Department of State did not prevail against the determination of the Department of the Interior and Reclamation (and probably elements within the White House) to honor the agreement to develop plans for the Chinese Three Gorges Project. By November 1945 Secretary of the Interior Ickes announced a cooperative agreement between Reclamation and China on a "comprehensive basin-wide development program in the Yangtze River Valley." The agreement signed by the National Resources Commission (NRC) of China directed Reclamation to make final studies and prepare specifications for the Yangtze Gorge Project and five tributary projects. There was as yet no commitment by either the Americans or the Chinese to finance the project, but Reclamation assumed the design work in Denver under the supervision of Savage as a special consultant to Reclamation. In December 1945 a *New York Times* story contained full page illustrations of the proposed dam and enormous locks on the river to enable ships to navigate the Yangtze and bypass the "Ichang Dam." The story noted that the proposed dam dwarfed the Grand Coulee and offered power, flood control, and irrigation, while facilitating navigation. All in all, the development was "A Super TVA." A follow-up story appeared in the paper in May 1946 with the news that China had made a $250,000 payment to Reclamation to cover the initial costs for developing plans to harness the Yangtze River. Planning work began at the Engineering and Research Center in Denver with the assistance of fifty NRC engineers from China. Secretary Ickes, prior to his resignation in February 1946, esti-

[84] John L. Savage to Michael Straus, June 4, 1945; Straus to Commissioner of Reclamation Bashore, June 7, 1945, RG 48, Entry 779, Box 15.

mated that planning work would cost $500,000. With support for the project continuing to come from the Department of the Interior under the new Secretary Julius Krug, Savage left Denver on August 26, 1946, to consult in Australia, China, and India. He returned March 29, 1947, but the civil war in China and the collapse and retreat of the Chinese Nationalists, in the face of the Communist revolutionaries, to Taiwan or Formosa caused suspension of any assistance to the Chinese on the Gorges project during the summer of 1947. All but four of the Chinese engineers in Denver returned to China. In October of 1949 the People's Liberation Army declared victory in China, which ended any American role in China's postwar development.[85]

In the aftermath of a devastating world war, exclusion of the United States from mainland China did nothing to discourage American desires for "world rehabilitation" through economic and technical assistance. In fact, the "fall of China" underlined the urgency of such efforts. At the same time, Secretary of State General George C. Marshall and the Truman administration were formulating the Marshall Plan, a massive program to aid the recovery of war-torn Europe. Of course, threats of Communist expansion in Europe also played a major role in American desire to rebuild Europe after the war, but so did a realization that poverty throughout the world created opportunities for conflicts everywhere.

In 1946, the Truman Doctrine announced to the world the United States's commitment to stop the expansion of Soviet Communism by providing military and economic aid to Greece and Turkey. The event marked a dramatic move by the United States to "contain" Soviet expansion by not only military aid, but also American technical and economic aid to thwart the appeal of Communism throughout the world. A *New York Times* story in late 1946 saw dam building in particular as forestalling "the floods of war." TVA-like projects, according to the author, could raise living standards and promote peace in "backward lands." He referred to the "starving East," and to achieve a last-

---

[85] Department of the Interior, Bureau of Reclamation, Press Release, November 28, 1945; Itinerary for J. L. Savage, Consulting Engineer, August 22, 1946, as noted in George O. Pratt, "Foreign Activities of the Bureau of Reclamation, 1953," typescript, 44; Bureau of Reclamation, "Status Report: Yangtze Gorge and Tributary Project, China, prepared for National Resources Commission of China," August 15, 1947, in the files of the Office of International Affairs, Bureau of Reclamation, Washington, D.C.; *New York Times*, December 28, 1945 and May 17, 1946; T. H. Watkins, *Righteous Pilgrim: The Life and Times of Harold L. Ickes, 1874-1952* (New York: Henry Holt and Company, 1990), 832; "Chronology of Activities Involving the People's Republic of China," RG 115, ACC# NRG-115-05-088, "Case File on Yangtze Gorge and Three Gorges Dam, China, 1943-1996," Box 1.

ing peace, stability and well-being must be brought to Asia, Africa, and the Middle East as well as Europe. Otherwise, the only option was chaos through the loss of vast sections of the world to Communism. "My purpose here," said the author, "is to show that it is feasible for us to take the lead now in developing an international policy toward backward regions which will increase their effective use of their own resources—a policy which is constructive, practical and worthy of democracy." The order of the day, in the view of this journalist, was to build infrastructure in the developing world the way the New Deal built it inside the United States during the 1930s.

Since American technical skills afforded the United States the highest standard of living, these same skills were applicable to internationally sponsored projects "comparable to our own TVA" and offered soil conservation, irrigation, reforestation, power development, and flood control. He cited Reclamation's efforts in the Yangtze Valley with the work of Dr. Savage "one of the great dam builders of the age" as an example of plans that could be brought to fruition for the benefit of world peace, asserting, "Long after the present quarrel between Communist and Nationalist was a footnote in history books, long after the Great Wall of China had crumbled away, the Yangtze Dam would continue to spread its abundance over a smiling China." And he concluded, "It is possible that our grandchildren would be considerably prouder that we sent our engineers to China after World War II than our marines." The same efforts should be made in India and elsewhere especially with the shining example of the experience of the Bureau of Reclamation and its record of accomplishments in making arid lands productive in the American West. Reclamation itself announced in 1946 that it had completed at its Denver laboratories a model built on a scale of 1 to 80 of the proposed 482-foot high Bhakra Dam in India's Punjab Province. It had undertaken the effort for a private international engineering firm commissioned by the Indian government to prepare designs.[86]

A World in Need

The abrupt end of the China project closed off one field of endeavor for Reclamation's technical expertise. Still there was the rest of the developing world. A larger world-wide vision and ambition depended upon American resolve to extend aid overseas in advisory capacities or in directly building

---

86 Willard R. Espy, "Dams for the Floods of War," *New York Times*, October 27, 1946; Caption under photo entitled, "Model of a Giant," *Reclamation Era*, 32 (July 1946): 163.

projects to help meet the demands of what became a popularized term in the late 1950s: "the Revolution of Rising Expectations." Development of water resources, of course, was important for "human rehabilitation." It involved food and electrical energy production, protection from floods, improved transportation—all embodied in the various missions of Reclamation in the American West during the decades since its creation in 1902. In the conditions of the postwar world, Reclamation did not have to seek out these projects. Requests flowed into its office for aid and assistance. An early recognition of these demands showed in the addition of an item to the organization charts of Reclamation in June 1945. The new line stated that, "the responsibility for assisting representatives of foreign governments" should be placed in the office of the engineering assistant. In the first year following the war, Reclamation highlighted its international reputation in its publications noting that it had attracted wide-spread requests for technical assistance and training. One Department of the Interior official stated, "In its program for international cooperation, the Bureau of Reclamation is adding to the world-fame of its engineering accomplishment." He made the good points that (1) foreign undertakings would enrich opportunities for Reclamation to learn from others as it helped others; (2) operations in other countries open outlets for foreign trade; and (3) foreign activities paved the way for greater understanding of problems throughout the world. Altogether he sketched the role of the United States as a good neighbor in the world. Over a year later *Reclamation Era* announced the appointment of the author by the commissioner of Reclamation to assignment in Greece to work as an "irrigation advisor in the rehabilitation of irrigation works on 800,000 acres in northern Greece, as part of the American mission for aid to that country." The use of the word "rehabilitation" reflected at this point Reclamation's perception of its mission overseas rather than the mission and goal of "economic development" characteristic of programs as the Cold War became a world-wide struggle for the United States.[87]

The administration pondered a response to the question raised by the fast-pace of developing events: would the American government commit to overseas programs of development at a time when the mood of the American electorate indicated a desire for a retreat from world responsibilities? The first priority had been waging war from which they now expected a rapid demobilization of military forces, but now their government hinted that it continue with international obligations that might include commitments of foreign aid

87 William E. Corfitzen, "Ideas Go Traveling," *Reclamation Era*, 32 (May 1946): 102-3; "Corfitzen Goes to Greece," *Reclamation Era*, 33 (October 1947): inside back front cover.

in many forms including technical assistance. While Commissioner Bashore expressed Reclamation's readiness to make international commitments, he advised that, "The Bureau's own program for continued river-basin development in the western United States will continue to have first priority and the major attention of the Bureau's staff."[88]

Where National Interests Directed

American government under the Truman administration shortly made a series of decisions that positioned the United States to play a permanent role on the world stage both militarily and economically. The policies won hard-fought victories in a Congress eager to be done with saving the world. The Truman Doctrine for the defense of Greece and Turkey (1947), the Marshall Plan for the recovery of Europe (1948), and the approval of the North Atlantic Treaty Organization (NATO) along with participation of the United States in the formation of the United Nations at the end of the war all confirmed "the internationalization" of the United States—a very different nation than the one that concluded World War I with a refusal to participate in the League of Nations. Moreover, Congress approved legislation in 1948 to permit agencies of the government to operate overseas without being under the jurisdiction and administration of the Department of State. Previously Public Law 63 passed by the Seventy-Sixth Congress, only permitted federal agencies to loan personnel on a reimbursable basis to foreign countries through the Department of State. The passage of the U.S. Information and Educational Exchange Act of 1948 (P.L. 80-402) facilitated Reclamation's interaction with foreign governments, but did not relieve the State Department of its responsibilities and even oversight of the actions and programs of American governmental personnel overseas.[89]

Before the Korean War (1950-1953) revealed in no uncertain terms American willingness to meet military threats from Communist expansion, U.S. economic development programs made their appearance in many countries. How much of a commitment from Reclamation to international development largely depended upon its domestic agenda. From the beginning of this critical era, which pointed to a division of Reclamation's energies between

---

88 Press Release, November 28, 1945, Department of the Interior, Bureau of Reclamation, typescript in files of the Office of International Affairs, Bureau of Reclamation, Washington, D.C.; Pratt, "Foreign Activities," 4.

89 "Bureau Engineers Loaned to Foreign Government," *Reclamation Era*, 33 (March 1947): 70-1.

9.9. Harry W. Bashore served as Commissioner of the Bureau of Reclamation, August 1943-December 1945.

domestic and foreign endeavors, cautionary notes arose. Commissioner Bashore's position was well-known that domestic duties for the Bureau of Reclamation commanded priority. Beyond some plans for overseas projects that were at best tenuous, Reclamation could not ignore a stream of demands upon it from foreign engineers as the world settled in to recover from war and started to build a new future. Engineers and foreign officialdom frequented Reclamation damsites and studied the operations of the Denver Office under the chief engineer. In addition, Reclamation obliged requests from every part of the globe for copies of its technical bulletins accumulated from almost a half century of work. The Department of State's Economic Cooperation Administration (ECA) provided funds to Reclamation's foreign relation activities by paying the expenses of visiting engineers from around the world.[90]

With Denver as the center of Reclamation administration in the West, it was perhaps no accident that an Inter-American Conference on Conservation of Renewable Natural Resources took place in the Mile High City in Septem-

[90] Julius A. Krug, *Annual Report of the Secretary of the Interior: Fiscal Year Ended June 30, 1946* (Washington, D.C.: United States Government Printing Office, 1946), 120; Oscar L. Chapman, *Annual Report of the Secretary of the Interior: Fiscal Year Ending June 30, 1949* (Washington, D.C.: United States Government Printing Office, 1949), 12; "ECA Sponsors Greek Engineering Visits to Reclamation Area," *Reclamation Era*, 35 (June 1949): 137.

ber of 1948. The tenor of the proceedings reflected a postwar neo-Malthusian concern for the improvident consumption of resources by the ravages of the recent world-wide conflict and the impending demands upon resources presented by the growth of world population. A Reclamation official from the Central Valley Project in California noted that river basin projects in the United States were instructive for similar works in the Mediterranean countries, Central America, and the Middle East. In terms of river basin development and the utilization of resources the writer was not only concerned with water utilization for irrigation and power purposes, but also noted that land utilization questions must be addressed: small and large land ownership questions, agricultural practices, erosion, and grazing. Land and water issues were inseparable. Other U.S. delegates from resource agencies noted the "huge drain of the recent war" and the threat of another war on the horizon—all occurring amongst postwar shortages in many basic commodities.

Reclamation's William Warne, who was among those representing the Bureau's interests at the conference, presented a wide-ranging view of the expertise available from the almost half century history and experience of Reclamation dealing with the challenges of water development. The conference signaled an affirmation of American interest in world resource issues and assurances that the United States was not to turn inward upon conclusion of the worldwide war struggle.[91]

The course pursued by the Truman administration in the postwar years confirmed the commitment of the United States to a foreign policy engaged with the world. After President Truman won a spectacular victory in the 1948 election, he announced in his inaugural address in January 1949 the Point IV Program: (1) to work with the United Nations; (2) to help the recovery of the world economy; (3) to offer aid to nations struggling against aggression; and finally (4) a program of technical and scientific aid to help underdeveloped areas of the world. All of the points underscored American commitment to international involvement. The last point, Point IV, contained important tasks

---

91 "Are We Becoming a 'Have-Not' Nation?" *Current Events*, 45 (March 25-29, 1946): 201-2; William E. Warne, "Legal and Economic Problems Associated with Excessive Withdrawal from Ground-Water Sources," 297-304; Hugh G. Calkins, Bureau of Reclamation, Department of the Interior, Sacramento, California, "Human and Land Problems in Three Regions," 385-91; E. I. Kotok, Forest Service, U.S. Department of Agriculture, Washington, D.C., "The Ecological Approach to Conservation Programs," 472 in *Proceedings of the Inter-American Conference on Conservation of Renewable Natural Resources*, Denver, Colorado, September 7-20, 1948 (Department of State, Publication 3382, International Organization and Conference Series II, American Republics 4: Division of Publications, Office of Public Affairs).

9.10. President Harry S. Truman announced his Point Four initiative during his inaugural address on January 20, 1949. Courtesy of the Harry S. Truman Library, National Archives and Records Administration.

and opportunities for the Bureau of Reclamation. Its meaning and relevancy for Reclamation was unmistakable when the president asserted, "We must embark on a bold new program for making the benefits of our scientific advances and industrial progress available for the improvement and growth of underdeveloped areas." The president noted the limitation on American material resources but asserted, "Our imponderable resources in technical knowledge are constantly growing and are inexhaustible. I believe that we should make available to peace-loving peoples the benefits of our store of technical knowledge in order to help them realize their aspirations for a better life." Truman emphasized that this should be in cooperation with other nations and denounced, "The old imperialism—exploitation for foreign profit," and said that what he envisioned was, "a program of development based on concepts of democratic fair-dealing."[92]

92 United States President, *Inaugural Addresses of the United States from George Washington,*

The new Congress soon approved President Truman's Point IV Program to go into effect in 1950 underwritten by the Economic Cooperation Administration. Point IV extended American technical aid to the undeveloped world and allowed Commissioner of Reclamation Michael Straus to proclaim that Reclamation's reach stretched around the world. Much of American technical aid overseas from 1950 to 1953 operated under this program administered by the Technical Cooperation Administration within the Department of State. After 1953 the Eisenhower administration initiated subsequent broad foreign aid programs. In any event, Reclamation's overseas work still occurred under close oversight and even assignment by the Department of State. Straus's 1955 book, *Why Not Survive?*, made the case for worldwide resource development to meet what some were calling the "revolution of rising expectations." If these expectations were to be met, of course, increasing energy from water development and food production was essential—all longtime domestic concerns of the Bureau of Reclamation in the United States.[93]

Opening the Door to Ceylon (Sri Lanka)

An early example of postwar interest by the Bureau of Reclamation in cooperative international water development occurred in Ceylon—an island with an ancient tradition of irrigation works.[94] By 1946 the Chief of the Hydrology Division in Denver noted suggestions from the Department of State that Reclamation become involved in designing two projects in Ceylon (renamed Sri Lanka in 1972). One project, the Gal Oya Project, required a reservoir and dam, irrigation plans, flood control provisions, and development of hydroelectricity. Its watershed was in the northeastern portion of the island that received rainfall in the monsoons. The early fascination with a water project in Ceylon illustrates the eagerness with which American interests sought to move into former British Empire possessions, but, as was the practice, under the guidance of the Department of State. In June 1948 (the year Ceylon achieved independence) the United States's embassy in Colombo noted in communications to the Department of State that the new Prime Minister of the Ceylonese government urgently requested American loans and engineers to represent its interests in negotiations with the Morrison-Knudsen Company on

---

1789 to Harry S. Truman, 1949 (Washington, D.C.: United States Government Printing Office, 1952).

93 Michael W. Straus, *Why Not Survive?* (New York: Simon and Schuster, 1955). See footnote 67 for the source of the phrase "revolution of rising expectations."

94 R. L. Brohier, *Ancient Irrigation Works in Ceylon*, 3 parts (Colombo: The Ceylon Government Press, 1934, reprint 1949).

the Gal Oya Dam and irrigation scheme. The Ceylonese wanted Reclamation engineers to review designs for two dam projects and prepare supplemental plans and specifications with funds to be provided in advance by the Government of Ceylon. The communiqué emphasized that a civilian appointee was preferable, but an army engineer was acceptable. Furthermore, the message noted that an American failure to respond might result in a request to the British all of which would complicate future operations of an American company on the project.[95]

In September of 1946 the secretary of the interior (with the approval of the Department of State) asked the comptroller general to approve a contract between Reclamation and the government of Ceylon. The comptroller general, according to the Budget and Accounting Act of 1921, made determinations on the legality of expenditures by the various agencies of the executive branch of government. In the letter to the comptroller, the Department of the Interior asserted that "it would be desirable to arrange for compliance with that request, particularly, inasmuch as State has expressed the opinion it will tend to promote good foreign relations and accordingly serve the best interest of the United States to perform this work." The secretary of the interior not only justified the agreement on the grounds that it promoted good foreign relations, but that the experience gained by Reclamation's design engineers, as well as any economic and technological advances it achieved in connection with the proposed research, will be applied to domestic reclamation problems. In addition, the Ceylonese government stood ready to finance the enterprise, which prompted the letter to emphasize, "Thus American reclamation projects may benefit from the technical work proposed to be performed without cost to the United States or expenditure of appropriated funds." Interior further asserted that the agreement was justified under the Reclamation Act of 1902, and subsequent acts of Congress, authorizing use of Reclamation moneys to aid various entities of local government that promoted land reclamation.

Clearly the Department of the Interior had searched Reclamation law diligently to justify foreign operations, especially if foreign governments financed those endeavors. In reply the comptroller general rejected the arguments and stated flatly that no Reclamation laws justified aid by Reclamation even in planning and design consultations to foreign governments. The law of March 4, 1921, cited by Interior, that authorized Reclamation to receive

95 John R. Riter, Chief, Hydrology Division in Denver, Memorandum for Files, March 1946; Jester to Secretary of State, June 16, 1948, RG 115, ACC# NRG-115-00-265, General Correspondence 1933-1989, Box 4.

moneys from states, municipalities, corporations, associations, firms, districts, or individuals for investigation, "does not specifically include foreign governments." The ruling was passed on to the Division of the Budget and to the Solicitor General's Office. The comptroller general further argued that "such possible benefit is too remote and indirect to support a conclusion that the proposed activities for Ceylon would be authorized on the basis that they might result in some possible benefit to projects in the United States." Moreover, the office of the comptroller general claimed that the Bureau of Reclamation lacked any definite statutory provision or the expressed will of Congress to make agreements to work in Ceylon or any foreign country. The Department of the Interior eventually notified the State Department of the decision and requested that the State Department inform the Government of Ceylon through the British Embassy of the comptroller general's decision.[96]

Two years later the way cleared for Reclamation to assign an engineer to Ceylon. Congress passed the Information and Educational Exchange Act of 1948 (Smith-Mundt Act) authorizing activities in foreign countries and freeing Reclamation from many of the details of transferring personnel to the Department of State for assigned service overseas. While the law opened the way for Reclamation to assign personnel directly to foreign aid technical projects, it did require close ties with the State Department in all activities, if not virtual oversight by State. Ultimately the Department of State bore responsibility for Americans overseas, especially Americans on assignment by the U.S. government. Overall the law indicated that Congress and the administration sought to expand American activities overseas to aid impoverished nations and, of course, to combat the appeal of Communism to the needy throughout the world. The Marshall Plan announced and approved by Congress a year before in 1947, of course, was the most outstanding example of Congress's willingness to shoulder additional burdens of foreign aid in the postwar years.

From Denver to Ceylon (Sri Lanka)

Now Reclamation was in a position to loan construction engineer Paul von der Lippe out of the Denver Office to the government of Ceylon. The State Department continued to play a key role in directing Bureau participation in overseas projects and considered it extremely desirable to place a Bureau

---

96 Lindsay C. Warren, Comptroller General, to Secretary of the Interior, October 8, 1946; Warner W. Gardner, Acting Secretary of the Interior, to James F. Byrnes, Secretary of State, October 31, 1946, RG 115, ACC# NRG-115-00-265, General Correspondence 1933-1989, Box 4.

of Reclamation engineer in Colombo by July 15, 1948. As an engineer on foreign assignment von der Lippe proved to be particularly observant. Already by 1947, advanced parties of American engineers in Ceylon reported back to Denver on the successful expansion of a hydraulic laboratory for river modeling as their work progressed. Beginning in late 1950 most of the American work in Ceylon occurred under a Point IV agreement with the United States.[97] Von der Lippe, however, worked directly for the Ministry of Agriculture and Lands in Ceylon, and he and his Ceylonese colleagues worked from Bureau drawings produced in Denver to develop blueprints and specifications for ditches, grouting, and drainage galleries in the concrete dams under construction. Sometimes he asked for clarification on specifications that had been determined in Denver. Beyond advice on technical matters, his experience reflected the challenges of American construction projects overseas. The work of construction was not being done by Reclamation but by American private contractors, i.e., International Engineering Company headquartered in Denver and Morrison-Knudsen of Boise, Idaho.

The contractors, while completing dams and shoring up reservoirs, were not confident of how the Ceylonese planned to use, operate, and maintain facilities in the future. D. J. Bleifuss, engineer for International Engineering Company, voiced concern that there was no assurance that any developed operating scheme would be adhered to in the future. He feared there would be a tendency to use reservoir capacity for irrigation at the expense of flood control, especially after a series of dry years. From his observations of the site under consideration, he believed it was a mistake to base design on one particular scheme of operation. Rather, Bleifuss suggested, the design of Gal Oya Dam should be based only on consideration of dam safety, and that this assumed some flexibility in the implementation of various operating schemes. Finally he expressed an engineering dictum that, "No project can be made absolutely fool-proof, but a project should be as fool-proof as possible." A conference at the resident engineer's office in Inginiyagala, Ceylon, September 7, 1948, endorsed the plan for the greatest possible spillway capacity to handle floods and provide for reservoir safety, but the resident Ceylonese engineer, R. Kahawita, desired a more limited spillway. To resolve the impasse, all parties agreed to let Ceylon's Director of Irrigation resolve the disagreement. On other matters von der Lippe noted that advice had been received from Dr. Savage in Denver regarding the downstream slope of the dam confirming that it should remain as designed.

---

97 Robert V. Burns to John L. Savage, February 10, 1947, Savage Papers, Box 1; "U.S., Ceylon Agree on Point Four Aid," *New York Times*, November 16, 1950.

Other notes from the conference indicated the need for facilitating production of electricity to supply government buildings and construction personnel with air conditioning.

Bureau of Reclamation Chief Engineer L. N. McClellan corresponded directly from Denver with von der Lippe about details of the Gal Oya Dam and powerplant. He said the undertaking was similar to the Green Mountain Dam and Powerplant in Colorado as he sought to apply Reclamation's experience in building western dams to the Ceylonese project. The estimated cost of $13,500,000 for the Ceylonese project was comparable to the Green Mountain undertaking as were the 426 construction drawings with a cost of approximately $500 per drawing for direct labor. The division of costs between powerplant and dam on the Gal Oya job would be 80 percent for the dam and 20 percent for the powerplant. Based upon Reclamation's experience at Green Mountain Dam, the chief engineer believed the International Engineering Company's estimate that drawings should be increased from 350 to 426 at an increased cost from $231,000 to $290,000 "is reasonable." Reclamation's experience building Green Mountain Dam proved useful half a world away.[98]

Encountering Ceylonese Culture

Ater a year on the job in Ceylon, in November 1949 von der Lippe wrote a general report on his experiences for the American Embassy in Colombo. Written primarily for State Department officials, it was forwarded on to the secretary of state and to the Bureau of Reclamation. Von der Lippe's observations encompassed more than his job description as an American Technical Adviser to the Ministry on Water and Lands. He noted that Ceylon achieved its independence February 4, 1948, and since then had launched ambitious projects, but without proper planning. He had repeatedly recommended to the Ministry to establish a project planning board for water resources for all of Ceylon because it was moving actively into too many projects too fast. The State Department heartily agreed, but it also saw opportunity for American enterprise to open operations in new nations emerging from the disintegration of the British Empire after World War II.

The American engineer familiarized himself with Ceylonese customs and traditions during construction of the Gal Oya Dam and Powerplant.

---

98 Paul von der Lippe to Chief Engineer, Bureau of Reclamation, October 10, 1949 and September 17, 1949; L. N. McClellan, Chief Engineer, to Paul von der Lippe, October 5, 1949, RG 115, ACC# NRG-115-00-265, General Correspondence 1933-1989, Box 4.

He did not live on site but made field trips from Colombo at least once a month to the project site. His role also required him to act as a liaison between the American contractor (Morrison-Knudsen and American International) and the Ministry of Irrigation and Water, his immediate employer. Monthly reports from von der Lippe to the Ministry contained suggestions for improvements and changes he thought necessary. His year of experience gave perspective on the working environment that he willingly shared with State Department officials in the Embassy, presumably at their request. He pointed out that some British officials still remained in key technical posts. Both the Director and Deputy Director of the Irrigation Department were British having been in Ceylon since before the war and seemed to be out of touch with modern engineering methods. On top of this handicap, officials were overworked and in need of additional "real" engineers. One of the few qualified engineers, "a Ceylonese gentleman," had recently departed for Denver to participate in the design planning for the Gal Oya and Walawe multiple-purpose projects. Von der Lippe's reference to "multiple-purpose" indicates his understanding of the various roles that water development would play for Ceylon in dam building, reservoir storage, hydroelectric power, flood control, and irrigation.

When "the Ceylonese gentleman" returned from his work in Denver, he took charge of the Hydraulic Laboratory for Irrigation Development. His participation in the Denver design team's efforts and his previous training as an engineer made him a key figure and decision maker in the project. According to von der Lippe, the hydraulic laboratory in Colombo was fairly well equipped for soils and concrete testing. Earlier another engineer from this laboratory also gained experience with Reclamation engineers in Denver. Still the laboratory needed more qualified people if it were to serve in designing hydraulic structures and other river studies. While the laboratory was ably managed, it suffered from a lack of funds. Von der Lippe believed the Ceylonese government did not spend enough money on laboratory matters directly related to adequate planning that should give more attention to issues related to flood control. He noted the torrential rainfall during certain seasons brought destructive floods and cited Easter Day 1949 in Colombo when ten inches of rain fell in three hours.

Moves by the Ceylonese government to eliminate European engineers from key civil service positions since independence led to a large turnover and unfilled vacancies primarily because the government was reluctant to pay for engineering talent. Still von der Lippe believed that Ceylon would continue to need "European (white) engineers for many years to come." At the same time he complimented Ceylonese engineers for their ability "in certain phases of engineering and they do learn fast," he wrote. Yet he criticized them for being too theoretical, not practical, and lacking "Method" in their approach. The men on the Gal Oya project, he indicated, however, have done very well and will form an important nucleus of talent if the government encourages them to stay on. Von der Lippe's report also contained remarks on the relationship of the American companies with the cultural environment. For instance, when the Morrison-Knudsen Company hired a Canadian doctor to take charge of the local hospital, Ceylonese officials protested that native doctors should have been hired. Other items of controversy occurred over insurance questions, job site security, and customs duties on imports.

In struggling to meet various construction requirements, von der Lippe expressed appreciation for the Bureau of Reclamation's commissioner's office and the chief engineer's office in Denver for providing technical information. He made special mention of the large demand for the *Reclamation Manual* that caused him to place a large order for additional copies. Praise for the manual could not be overstated in terms of its helpfulness on the new projects launched in Ceylon. In addition, the Irrigation and Land Ministry was interested in the *Manuals on Organization* from the Bureau. And well they should be, according to von der Lippe, because their administrative procedures were "quite antiquated." Too much detail was performed by high level officials that could just as well be taken care of by clerks. The Ministry and its workings were badly in need of an administrative analysis or an efficiency expert to save it time and money.

Von der Lippe alluded to a new project under consideration. The Walawa Banga was a proposed multiple purpose project located in the southern part of the island. Already the drawings were in preparation by International Engineering Company in Denver. But he expressed reservations because it appeared too costly for the amount of irrigable land that could be brought into production. Ceylon's government developed a Six Year Plan to bring 12,000 acres of paddy land a year into production. Looking ahead, these same planners hoped to increase that acreage by 25,000 acres annually from 1954 to 1959. Von der Lippe thought this was too optimistic and argued that planners should focus on developing hydroelectricity on the island. He maintained that, at best, the Ceylonese could hope to gain 12,000 acres overall in paddy lands a year.

Von der Lippe described the agricultural methods employed in the "irrigation districts," as "antiquated, the same as have been used for a thousand years." In this respect he had at hand a work that he referenced: R. L. Brohier, *Ancient Irrigation Works in Ceylon*, 3 volumes, published in 1934 and 1935, that gave him a historical cultural perspective on the island's irrigation practices. He noted that Buddhist teachings tell the people to get along in the world with as little as possible and that most farmers held only one or two acres of paddy land. But he observed that farmers work hard in the fields with their mamootees and take great pride in raising their own rice. He emphasized the hard work because much had been said about the laziness of the Ceylonese and countered that it needed to be remembered that Ceylon has a humid tropical climate surrounded by the Indian Ocean and the Bay of Bengal. All in all, von der Lippe declared, it is "not an invigorating climate." Also, he observed, the diet of the island is deficient in minerals with the staple food of rice supplemented with dried fish and sometimes a little meat for the native curries. In addition, von der Lippe noted that Ceylon's colonial past under the Portuguese, the Dutch, and finally the British created perceptions of Ceylonese backwardness. None, of course, he argued, encouraged or permitted Ceylonese to obtain top positions. Ambition was largely unrewarded. He believed time was needed for the nation to realize what independence meant and predicted "many and costly blunders." In the final analysis, change, he concluded, was occurring too rapidly to avoid mistakes, but overall he defended the culture of the island and its friendly response to progress.

His remarks generally supported policies to provide technical assistance to Ceylon as quickly as possible. Both the Irrigation Department and the Gal Oya Development Board stood in need of immediate assistance and aid to achieve at least a portion of their ambitious goals. He felt the new commonwealth was under able leadership and deserved support and was optimistic about the soundness of the Ceylonese economy as a dollar earner. Prices for coconut and tea made these exports profitable, and the recent devaluation of the pound sterling boosted the prospects of rubber plantations on the island. Finally, he concluded, Ceylon was a good bet for future development. He based his conclusion on his faith that Ceylonese administration and planning of projects would mature with application of western technology through American know-how in the form of aid to government agencies and the energy and can-do attitude of American companies.

Von der Lippe's report reveals an American with a growing sensitivity to the long irrigation heritage and irrigation experience of the Ceylonese on their island. He exhibited a growing respect for the talents of native engineers who were expanding their education by acquiring the latest in techniques and knowledge from Bureau of Reclamation training sessions in Denver. Most importantly this American observer and in-country worker from the Bureau of Reclamation showed an awareness of the repression imposed by colonialism upon Ceylon. He realized that the new post-colonial period promised a better future for the country and, at the same time, acknowledged that the pace of change would probably come too fast and mistakes would be made, but they would be the errors of self government and not foreign rulers. His attitude reflected the anti-colonialism that informed American foreign policy in the post-World War II period—most pointedly even toward America's closest ally during the war, Great Britain. To brace up independence and insure the success of economic development projects related to water, von der Lippe believed, the technical knowledge and administrative procedures learned from the Bureau of Reclamation would play a key role. In addition, the presence of Reclamation assistance facilitated the activity and success of American companies under contract to the new government for construction projects. At this point von der Lippe's analysis did not reveal a Cold War mentality of competition with the Soviet Union. His interest seemed primarily in getting a job done efficiently and recognizing that it was important to have an understanding of local history and culture for the success even of technical projects. His remarks represent an internationalizing spirit at work amongst Americans overseas. As such, Bureau engineers were new agents of modernization displacing a previous colonial era.[99]

Reclamation in the Midst of New Foreign Policy Formulations

Intensification of the Cold War upon the outbreak of a hot war in Korea in the summer of 1950 heightened American concern about stemming the tide of Communist advances throughout the world. In late 1951, Congress passed the Mutual Security Act that created an independent Mutual Security Administration. It supervised military assistance to other countries and economic programs to enhance the security of friendly nations. The economic assistance portion of the act, while vigorously debated in Congress, survived

---

[99] Paul von der Lippe, "Impressions from First Year Assignment to Government of Ceylon," to the American Embassy in Colombo, Ceylon, November 23, 1949; Charles P. O'Donnell, First Secretary of Embassy, to the Secretary of State, "General Report by Mr. Paul von der Lippe on his Assignment in Ceylon," November 23, 1949, RG 115, ACC# NRG-115-00-265, General Correspondence 1933-1989, Box 4.

to provide funding for the expansion of American technical assistance overseas and further opened to Reclamation the opportunity to assist projects in foreign countries. By March 1951 Reclamation Commissioner Michael Straus created the Office of Foreign Activities. At the golden anniversary celebration of the creation of Reclamation in 1952, Chief Engineer Leslie N. McClellan declared, "The Foreign Activities Program of the Bureau has developed in the last two years from an unorganized extracurricular activity to a full-fledged Bureau-wide operation with a budget last year of nearly two million dollars." In the same year the Bureau held an international water conference along with the first International Reclamation Conference in Yakima, Washington with representatives from twenty nations attending.[100]

While its foreign activities appeared to feed the ambitions of the Bureau of Reclamation's expansion in the postwar period, it should be noted that Reclamation did not aggressively pursue foreign involvement. More correctly, the executive branch, with full knowledge of Reclamation's accomplishments in the American West over the past half century, encouraged Reclamation to direct resources to the world scene. The Bureau's growing domestic agenda, however, caused ambivalence within Reclamation about overseas commitments. World events crowded in upon an organization originally dedicated to internal improvements. In many ways this is the story of the United States in the world in the twentieth century—world events and America's own robust internal development made the nation an international player in the larger world community. Visitors and trainees continued to visit Denver, including Japanese engineers from occupied Japan interested in techniques of concrete engineering and hydroelectric development. Visits were still very much under the authorization and jurisdiction of the Department of State. It was not until 1961 and passage of Public Law 87-195, the Foreign Assistance Act, that the Bureau of Reclamation received the authority to enter directly into agreements with the U.S. Agency for International Development (USAID) to provide services to foreign governments. Reclamation, however, always asserted that it was not in the business

[100] Michael J. Hogan, *The Marshall Plan: America, Britain, and the Reconstruction of Western Europe, 1947-1952* (New York: Cambridge University Press, 1987); Doris M. Condit, *The Test of War, 1950-1953*, in Alfred Goldberg, editor, *History of the Office of the Secretary of Defense*, Vol. II (Washington. D.C.: Historical Office, Office of the Secretary of the Defense, 1988); J. B. Bingham, *Shirt-sleeve Diplomacy: Point 4 in Action* (New York: J. Day Company, 1954); *Reclamation's Golden Jubilee 1902-1952 Conference: A Summary of the Proceedings of Reclamation's 1952 Annual Meeting, Cody, Wyoming, July 28-August 1, 1952*, 26-7; Donald J. Pisani, "The Bureau of Reclamation and the West, 1945-2000," in *The American West in 2000: Essays in Honor of Gerald D. Nash*, Richard W. Etulain and Ferenc Morton Szasz, editors (Albuquerque: University of New Mexico Press, 2003), 56.

of training foreign consultants, its technical aid should only be called upon as a necessary instrument of American foreign policy, and that its work was of a planning nature that facilitated the participation of American companies in the construction phases of projects. Costs for training and technical assistance became a sensitive issue. To counter any criticism, Reclamation noted that reimbursement must come from the countries of the foreign visitors, the United Nations, or American foreign aid programs and not from money designated for the tasks of its now expanding domestic program.[101]

After World War II, Reclamation joined several international organizations: International Committee on Large Dams (ICOLD); International Commission on Irrigation and Drainage (ICID); International Conference on High Tension Electric systems (CIGRE) and others. Of course, membership in international organizations marked no new departure from past practices. From the early years of its creation the Bureau sent personnel to international meetings and believed the original Reclamation Act authorized its participation in international organizations.[102] International involvement resulted in a flow of information not simply from Reclamation to foreign countries but, as the world prospered after World War II, a flow back of information to Reclamation from the world in a genuine exchange of information gathered in international conferences and the publication of foreign technical bulletins in the pattern that Reclamation adopted in its early and continuing publication program. As a matter of administrative clarification, fields of foreign activities by Reclamation also include the territorial holdings of the United States. Although the trust territories of Puerto Rico, Palau, Guam, Saipan, U.S. Virgin Islands were under the administration of the Secretary of the Interior, work in them was not authorized by the legislation creating Reclamation. Activities in these places fell within the bureau's foreign activities program.

The Wider World Draws on the Talents of Reclamation

The global events of 1948 through 1950 (the Berlin Airlift in 1948, the Soviet Union's testing of an atomic bomb in 1949, the Korean War in

---

[101] Leslie N. McClellan, "Foreign Activities—Commitments, Opportunities, and Problems," *Reclamation's Golden Jubilee 1902-1952*, 26-7.

[102] Sam Guy, Chief of Foreign Activities Division, Statement to the Subcommittee on Public Lands of the House Committee on Interior and Insular Affairs, October 8, 1985, typescript copy in files of the Foreign Affairs Office of the Bureau of Reclamation, Department of the Interior, Washington, D.C.

June 1950) brought home to Americans the seriousness of the worldwide struggle against Communist expansion. The sudden entrance of the Communist Chinese forces into the struggle on the Korean Peninsula in late 1950 frustrated an almost complete victory by the United Nations's (predominantly American) forces. Although the United States entered the war under a United Nations mandate to stop North Korean aggression, it clearly shouldered the major burden not only in Korea but throughout the world against what was increasingly termed "the Communist Menace," embodied in the Soviet Union, Red China, and client states. While the nuclear option always loomed in the background, conventional military measures in localized conflicts often resulted in negotiated standoffs. Consequently, a major thrust of foreign policies from both sides in the Cold War was peaceful competition in terms of economic development to win the hearts and minds of millions throughout the world. Witness the aggressive efforts and eventual success of the Soviet Union to win the approval of Egypt to build the Aswan Dam on the Nile River by the late 1950s. This occurred after the U.S. and Britain withdrew offers because the new Egyptian government under Gamal Abdel Nasser disdained Israel, took a stance of neutrality in the Cold War, and, of course, nationalized the Suez Canal incurring an ill-considered military response by Britain, France, and Israel. Soviet engineers eagerly undertook the task. The Aswan Dam, although it created multiple environmental problems unrelated to the Cold War, became a monument to a lost opportunity for the West to use its technical expertise in the struggle against the spread of Communist influence.[103]

The leadership of Reclamation in this postwar period was under the direction of Commissioner Michael Straus (1945 to 1953). Secretary Ickes brought him into the Interior Department in 1933 and he became first assistant secretary of the Department of Interior in 1943. Unlike his predecessor, Harry Bashore, his background was not in civil engineering, but in chemical engineering and journalism. His strength was in public relations and overall policy formulation. His outlook served Reclamation well in the transition years from the crises of World War II and the New Deal depression years. His commissionership faced an expanding economy and world involvement in contrast to the struggling economic times of the Great Depression; its inward focus upon the domestic scene and foreign policies constrained by the isolationist bloc within Congress. Keenly aware of the challenges and opportunities presented on the foreign front, Straus believed the resources of Reclamation up to the

---

[103] Charles D. Smith, *Islam and the Search for Social Order in Modern Egypt: A Biography of Muhammad Husayn Haykal* (Albany: State University of New York Press, 1983); Charles D. Smith, *Palestine and the Arab-Israeli Conflict* (Boston: Bedford/St. Martin's, 2001).

challenges. If the United States needed the Bureau to play a role in its foreign policy strategies and to win hearts and minds in the underdeveloped world, Reclamation stood ready for the task.

While the Truman administration pursued overall strategies to aid underdeveloped nations to utilize their own natural resources for economic growth, Straus embraced aspects of the irrigation development that sought the same goals under the president's Point IV program. Straus also welcomed Reclamation's relationship with the Department of State through its Economic Cooperation Administration. He saw a "Reclamation Street" reaching 26,000 miles or around the entire world that he called the "arid belt." All was a part of the new international burdens that the United States must shoulder. Straus showed his journalistic flare in statements to the press and his support of administration policies: "But as America sends military aid to the defenders of liberty over land, sea, and air, so it is sending technical aid to help the teeming peoples of the under-developed areas win the war against want." Want and material deprivation loomed as the chief allies of the Communist menace, Straus warned, and as such should be attacked with the technical know-how of the United States. Straus emphasized that Reclamation's experience stood ready to serve the arid regions of the world: "Vast reaches of these under-developed areas lie in the arid belt stretching around the globe—where deserts await only sweet water to grow food and clothing."[104]

By late 1951 the Bureau of Reclamation was active in over ten countries around the globe: Chili, Ecuador, Formosa, Greece, India, Liberia, Malaya, Saudi Arabia, Thailand, Australia, and New Zealand. Reclamation portrayed its assistance to the Snowy Mountains Authority in Australia as helping to build a project similar to the TVA. Not only did Reclamation provide technical assistance on the ground in Australia, but Australian engineers received training at its Denver facilities. The general field of water resources drawing upon the expertise of Reclamation engineers included irrigation, drainage, hydroelectric power surveys, and problems of dam design and construction. Most of the activities occurred through the Department of State and its Mutual Security Administration in cooperation and funding from the Point IV Program of technical assistance to underdeveloped countries. Some countries objected to the term "underdeveloped" when receiving the aid but nevertheless welcomed the material assistance. Some, however, hired or borrowed Bureau of Reclamation engineers at their own expense. New Zealand hired

[104] Michael W. Straus, "Reclamation Street—26,000 Miles Long," *Reclamation Era*, 37 (May 1951): 95-8.

A. W. Simonds, an authority on grouting dam foundations. "Foundation grouting" is a highly technical phase of dam construction in which liquid cement mortar is injected into natural fissures in the foundation rock to prevent seepage after the structure was in service. He went to New Zealand in 1950 to work with engineers on the foundations of the Maraetai Dam, one of a series of dams being developed to provide hydroelectric power.[105]

By the 1950s Reclamation concluded that foreign activities must be a part of its agenda despite the pressing demands of its own domestic programs. Reclamation's well-established international reputation placed it at the beck and call of various administrations' policies to join in programs to develop water resources around the world. It was part of a strategy to compete for the loyalties of people by raising their standard of living and thwarting the Communist revolutions and any alignments with the Communist bloc in the developing Cold War. Of course, it was in Reclamation's interest as a government service bureau to make its know-how available to American foreign policymakers in the executive branch. The cooperation opened opportunities for enhanced budgets and generally a larger sphere of activity and importance. Also to have ignored the requests from foreign governments to send forth engineers, Reclamation risked losing valuable personnel to foreign governments on consulting jobs and to American firms doing international business. The situation created competition between private enterprise and government for experienced engineers at a time when their numbers fell short of the demand. The resulting hiring competition caused private companies to complain, but Reclamation usually pointed out that its technical assistance to foreign governments often resulted in equipment purchases in the United States and further opportunities for private companies to undertake construction projects after Reclamation's completion of the planning phase of a project. Clearly the seriousness of Cold War competition dictated that American foreign policy not rely entirely upon private enterprise to provide comprehensive programs of technical assistance. The activities of the Department of State and the ambitious Point IV Program reinforced this conviction.

The Point IV Program did not preclude agreements between foreign governments and the Bureau of Reclamation. In October 1951 Reclamation Commissioner Straus announced an agreement with the government of India in which the Bureau of Reclamation was to make the facilities of its engineering center available to assist the Indian government with water development

---

105 "Foreign Activities," *Reclamation Era*, 37 (December 1951): 291; "A. W. Simonds Aids New Zealand Development," *Reclamation Era*, 36 (April 1950): 84.

problems. Unlike Point IV such agreements did not involve costs for the United States or loans from USAID. The Indian government paid the Bureau of Reclamation for its services. Direct involvement of Reclamation with foreign governments oftentimes was an offshoot of the continuing program of hosting foreign visitors to the large dams—their hydroelectric facilities and the accompanying irrigation projects—in the United States and training visiting engineers at the Denver engineering center.

In the immediate postwar years, articles in *Reclamation Era* reveal a steady stream of foreign engineers training at the Denver office and Reclamation experts on loan to foreign governments. Engineers and farm experts from twelve nations were on hand to witness the drawing for Columbia Basin Project farm plots in Othello, Washington, when Reclamation turned on water to the first of 66,000 acres of what was announced as a one million-acre project made possible by Grand Coulee Dam. Other veterans of training in Denver often reported the benefits of their training as they confronted problems in their parts of the world. Arturo Carvajal, who had studied with Reclamation from 1945 to 1946, noted that cavitation at dams in Chile was especially severe because of the high velocity of escaping water from the country's high dams and deep reservoirs. In 1949 he reported that his work on cavitation problems in turbines at dams in both Chile and Argentina had been largely corrected by application of technology used in the construction of powerplants at Reclamation dams.[106]

American reclamation experts, i.e., Elwood Mead, had a long history of relations with Australia. The postwar period saw the Bureau of Reclamation offering technical assistance to the ambitious Snowy Mountains Hydro-Electric Scheme. From 1951 to 1961 Reclamation provided advisory services on the construction of a trans-mountain water diversion plan based largely on its experience planning and constructing the Colorado-Big Thompson Project. Completion of the Snowy Mountain Scheme boosted national pride in Australia—similar to the pride manifest in the United States upon completion of Hoover Dam some two decades earlier in the midst of the Great Depression. Snowy Mountain can be seen as among the world's most prominent power and irrigation projects. Over ten years a full-time Reclamation team provided

---

[106] "Gains for India Cited," *New York Times*, October 3, 1951; "Aides of 12 Nations See Coulee Project," *New York Times*, May 31, 1952; "India Studies Reclamation," *Reclamation Era*, 32 (July 1946): 143; "Bureau Engineers Loaned to Foreign Governments," *Reclamation Era*, 33 (March 1947): 70-1; Arturo Carvajal, "Cavitation Phenomena in Chile and in Argentina—Some Solutions," June 20, 1949, RG 115, ACC# 00-265. Box 4.

on site engineering advisory services and tests in the laboratories in Denver helped analyze structures and building plans. Also in the Denver Office over 110 engineers received training in engineering specialties related to the Australian "Scheme." No other institution in the world could have provided the services made available for this project. Triumphal celebration over its completion by 1961 added to the Snowy Mountain mystique in Australian history, but critics charged that the scheme dislocated people and exploited environments, and, of course, American technical aid and training was complicit in the undertaking. The latter criticism also became characteristic of environmental, social, and economic critiques of dam building in the United States well before the end of the twentieth century.[107]

The Commitments Continue

Increasingly, the parameters of the Cold War began to shape American response to its role in the world. This occurred not only in Europe with the Truman Doctrine to support the government of Greece against armed Communist subversion, the Marshall Plan, and ultimately NATO, but also throughout the developing world in efforts to promote technical aid programs and raise standards of living to ward off the appeal of Communist revolutionaries, i.e., Point IV. A major Democratic Party campaign statement for Truman's reelection in 1948, *The Vital Center* by historian Arthur S. Schlesinger Jr., first published in the election year and later republished in 1970, asserted that, "Dams were the American alternative to Communist land reform." While Communists violently demanded "crude redistribution" of land, American engineers offered "wonderlands of vegetation and power" in TVA-like developments that could "outbid all the social ruthlessness of the Communists for the support of the peoples of Asia."[108] By 1950 such idealism tapped the skills and talents of the rank and file within the Bureau of Reclamation. Not only was Reclamation deemed a successful domestic institution whose employment numbers had expanded to almost 20,000, but now the talents of those numbers were to be exported to a troubled world. Within the orga-

---

[107] Gilbert G. Stamm, Assistant Commissioner, Bureau of Reclamation, "Bureau of Reclamation International Technical Assistance in Development of Arid Lands," International Conference on Arid Lands in a Changing World, sponsored by the Committee on Arid Lands, American Association for the Advancement of Science, Tucson, Arizona, June 3-13, 1969, 4, RG 115, ACC# 8NS-115-95-089, Box 1; Jan Kociumbas, "Myths, Men and History: The Making of the Snowy Mystique," *Journal of the Royal Australian Historical Society* 88:1 (2002): 1-19.

[108] As noted and quoted in Nick Cullather, "Damming Afghanistan: Modernization in a Buffer State," *Journal of American History* 89:2 (September 2002): 524; see also Cullather, *The Hungry World*.

9.11. After U.S. Supreme Court Chief Justice Fred M. Vinson swore in Paul G. Hoffman as Administrator for Economic Cooperation on April 9, 1948, President Truman congratulated him on his new job which included the Marshall Plan. Courtesy of the Harry S. Truman Library, National Archives and Records Administration.

nization remained the nagging question of whether these new international demands threatened its performance on the domestic front.

During the 1950s, the world became increasingly defined as a playing field on which the free world contended with the forces of international Communism. The transition to the new presidency of Dwight D. Eisenhower and a negotiated armistice brought the Korean War to an end in 1953. The war's stalemated conclusion underlined the realization that the struggle against worldwide Communism required more than military confrontations. Although Congress merged the Technical Cooperation Administration (TCA), the administrative apparatus of Point IV into the Mutual Security Administration in 1952, the goals of Point IV, within the International Development Act of May 1950 remained intact. Still congressional appropriations were modest in contrast to funds allocated for military aid. The foreign aid budget in 1952 totaled \$6.5 billion with only \$155.6 million designated for the Point IV program.[109]

[109] Latham, *Modernization as Ideology*, 54; Dennis Merrill, editor, *Documentary History of the Truman Presidency*, "The Point Four Program: Reaching Out to Help the Less Developed

With the success of the Marshall Plan in Europe, American policy-makers gained confidence that economic development and technical assistance offered promising returns elsewhere in the world, especially to counter Soviet efforts to take advantage of the political instability connected with wars of liberation against the remnants of European colonialism. While the Eisenhower administration stressed aggressive actions against threats to stability (i.e., the overthrow of a government in Guatemala and the toppling of Prime Minister Mohammed Mosedec in Iran when his government moved to nationalize oil production), the Bureau of Reclamation continued programs of technical aid. In the Truman administration, Secretary of the Interior Oscar Chapman enthusiastically supported the cooperation of Reclamation with foreign aid programs. Long-time Reclamation official William E. Warne, formerly assistant commissioner of the Bureau of Reclamation and then assistant secretary of the interior in 1947, departed for Iran in late 1951 as an economic specialist. His presence in Iran signaled the growing interest of the United States in the Middle East and opened doors for opportunities for technical assistance in water development.

Afghanistan

In 1946 the Afghan government with its own funds revived work started by the Germans and Japanese in the 1930s on the Helmand Valley Project. Eager for postwar contracts, the American firm of Morrison-Knudsen entered into an agreement to plan, design, and construct the project with the Afghan government. Under Afghan financing the project included hydroelectric power, flood control, and irrigation for an estimated 500,000 acres. The costliness of the project by 1950 required help from the Export-Import Bank that extended loans of $39.5 million. American governmental assistance came in 1953 to counter the presence of Soviet technical and military advisors. American technical advice addressed project maintenance, irrigation farming practices, drainage, and other problems of the Helmand Valley and Arghandab Valley Authority (HAVA). With this assistance came teams of American personnel, some hired away from Reclamation, to the Helmand Valley Authority. Their presence continued until 1974. By the mid-1950s Morrison-Knudsen faced insurmountable difficulties with the project in terms of land distribution, drainage or waterlogged land issues, and saline lands. In 1955 President Eisenhower's secretary of state, John Foster Dulles, regarded the company as "one of the chief influences which maintain Afghan connections with the

Countries," Vol. 27 (Bethesda, Maryland: University Publications of America, 1999), xi.

9.12. In 1954 Reclamation published the "Development Plan for the Litani River Basin in the Republic of Lebanon" while working on behalf of the United States Foreign Operations Administration.

West." In 1960, however, Morrison-Knudsen withdrew from its endeavors in Afghanistan. A 1975 memo in Reclamation's Office of Foreign Activities notes that the assistance was designed to train Afghans to operate the project efficiently, but many of the efforts were largely "frustrated by Afghan resistance to land reapportionment, leveling, and assignment of trained Afghans to areas outside of HAVA." Instabilities in the Afghan government made for additional problems.[110]

After Floyd E. Dominy became commissioner in May of 1959, the Bureau of Reclamation "got into foreign affairs pretty deep," according to Dominy's remarks in his 1996 oral history. Problems in the Helmand Valley Project occasioned this deeper involvement. The State Department approached Reclamation, according to Dominy, to send a team to study and make recommendations to correct problems in this faltering project that had been variously supported by the Afghan government, the Export-Import Bank, and the Point IV program. Dominy's response was to demand a more independent role for Reclamation rather than its work overseas being done under the auspices of the State Department or Point IV-type of program. To meet the situation in Afghanistan, he determined to travel there with aides and then decide upon what efforts should be undertaken. The State Department or other agencies were to provide the money, but Reclamation under Dominy's leadership decided upon the course of action and what people should be sent to work in the field. L. W. Damours, chief of the Division of Foreign Activities in the Bureau of Reclamation visited Afghanistan in June and July 1959 and conducted discussions with the American Ambassador Henry A. Byroade and HVHA officials.

Damours's trip was a prelude to Dominy's visit to the Helmand Valley and Kabul. Significantly, P. R. Nalder, the project manager of the Columbia Basin Project, accompanied Dominy in December 1959. Decades later, reflecting on Reclamation's involvement in Afghanistan, Dominy made the point that the Department of State permitted the HVHA to be built by Morrison-Knudsen "on a cost-plus contract, with no assistance whatever from the Bureau of Reclamation." He berated the State Department for not bringing in "the one agency in the Federal government that knows how to do these things. You didn't consult with us. Now, it's a mess; and you don't know what to do about it." He said the only reason the State Department knocked on the door of Reclamation was because the Afghan government requested the aid of Reclamation believing that perhaps it ought to be consulted. Unbelievably, in Dominy's opinion, it was a foreign government that pushed the State Department to request the aid of Reclamation.[111]

---

110 Nick Cullather, "Damming Afghanistan," 522, 527; Project Review: Helmand Valley Project-Afghanistan (June 23, 1975), International Affairs Office files, Bureau of Reclamation, Department of the Interior, Washington, D.C.

111 Floyd E. Dominy, *Oral History Interviews*, Transcript of tape-recorded Bureau of Reclamation Oral History Interviews conducted by Brit Allan Storey, senior historian, Bureau

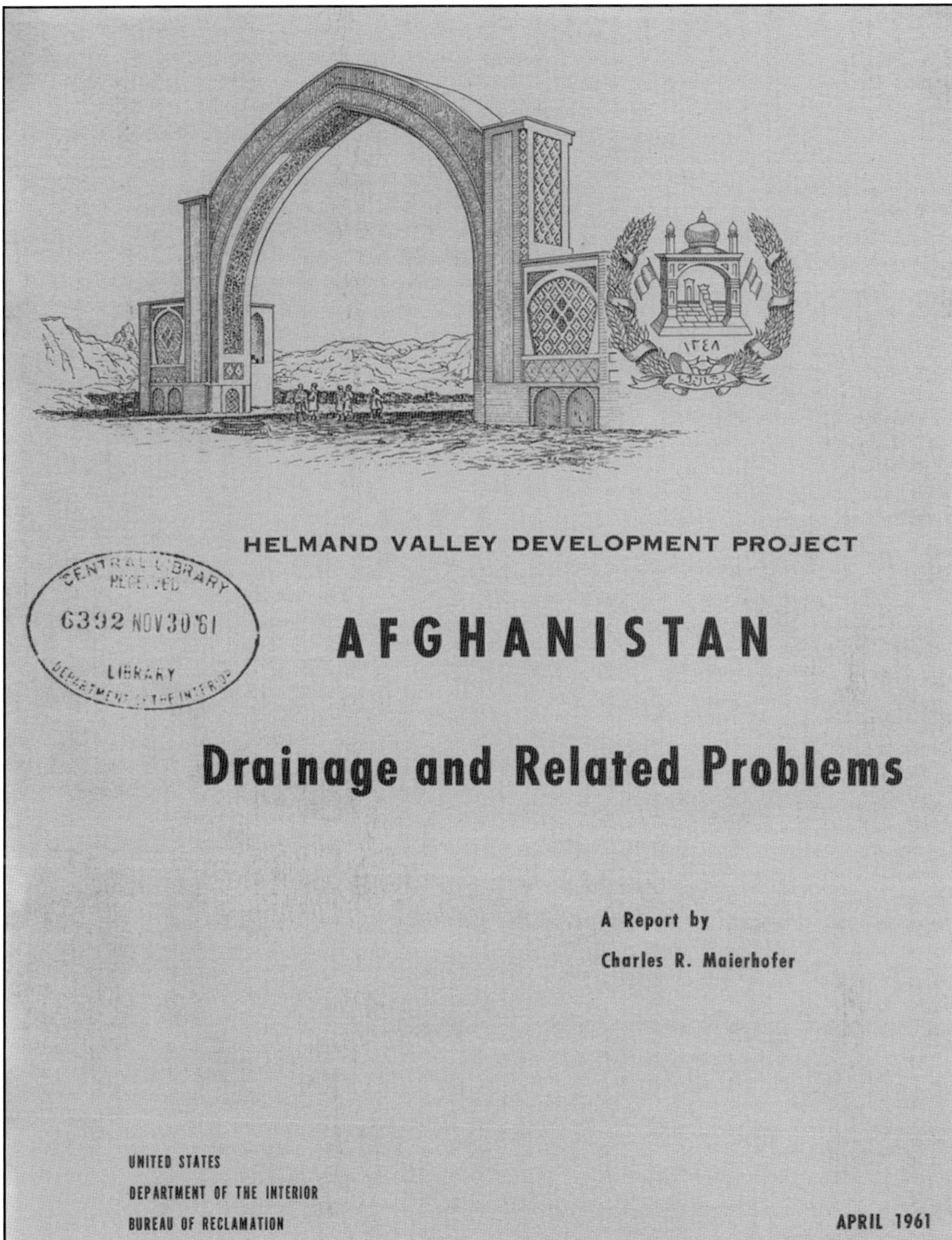
HELMAND VALLEY DEVELOPMENT PROJECT

AFGHANISTAN

Drainage and Related Problems

A Report by
Charles R. Maierhofer

UNITED STATES
DEPARTMENT OF THE INTERIOR
BUREAU OF RECLAMATION

APRIL 1961

9.13. Reclamation's April 1961 report on issues in the Helmand Valley of Afghanistan.

9.14. Reclamation technical staff working on drainage with Afghanis in the Helmand Valley.

The following year (1960) the Bureau of Reclamation signed, "A Participating Agency Service Agreement" (PASA) with the International Cooperation Agency (ICA). Four general objectives of the agreement emerged:

1) To assist and advise the HVA in the execution of the Authority's responsibility for policy development, planning, construction management, operation and maintenance of project works in the Helmand Valley for water storage, control, and distribution; power generation; irrigation and drainage.
2) To train Afghan personnel so that HVA can in the shortest possible space of time assume full responsibility for the operation and maintenance of the facilities of the Helmand Valley Development program.
3) To assist the HVA in the procurement and management of supplies and equipment, as well as the training of personnel concerned therewith.

---

of Reclamation, April 6, 1994 and April 8, 1996, at Bellevue Farm in Boyce, Virginia, 68-9; Cullather, "Damming Afghanistan," 524; "United States Agency for International Development in Afghanistan: Project History, June 1959-December 31, 1961," RG 115, ACC# 8NS -115-95-089, "International Affairs Report, 1914-1987," Box 1.

9.15. In December 1963 Commissioner Floyd E. Dominy posed for this portrait with Hoover Dam in the background.

4) To assist and advise in the establishment of a program for the settlement of people in new farm areas, including the provision of assistance to new settlers in the construction of farm laterals, surface and subsurface drainage.

The report from Reclamation on the first year of involvement in Afghanistan under the agreement emphasized operations and maintenance and the centralization and coordination of these efforts. It was important that operations begin with an inventory of maintenance needs based upon inspections. Upwards of twenty Reclamation technicians undertook the direction of these activities cleaning out of laterals, repairing erosion with rip-rap, and conducting intensive work on drainage problems. A crew of ditch riders received special training in the system of checks, turnouts, and structures for measuring and controlling water deliveries. The upbeat tone of the report lauded the progress underway.

Reclamation's project review of the Helmand Valley Project in March 1961 noted that Reclamation inherited a difficult situation in the making for over a decade. Troubles followed inadequate soil studies, poor selection of project lands, disputes between Morrison-Knudsen and the Afghans, and, especially, insecurities over land tenure. With no guarantee to ownerships rights and persisting cultural antagonism to a settled existence, nomads resisted settlement projects that placed them on the land and then later demanded their removal in order that the land be leveled for better application of irrigation water. Moreover, one report concluded, "The diligence, labor and knowledge required by a successful irrigation farmer is entirely foreign to the nomads."

Still these reports disclaimed a story in the *New York Times* earlier in 1960 that the Helmand Valley Project was a "Comedy of Errors." According to Reclamation's report, the article, although correct on many fronts, failed to appreciate the complexity of the project, and the limited resources of funds and skilled manpower available for the project. The report concluded that the road ahead would not be easy, painless, or inexpensive. Patience and money would be required over a period of years to maintain and rehabilitate the project, as well as train farmers, technical and administrative personnel, and Afghan government officials. Reclamation reported the project was basically well engineered and built and run by HVHA engineers and administrators who were conscientious and capable people. Still there was a need to develop policies and legislation adapted to the surrounding culture for the project to succeed.[112]

Ten years later an "End of Tour Report" by a Reclamation official in Afghanistan expressed both optimism and caution about future prospects in the country. He was convinced that the Helmand-Arghandab Valleys had the potential to be one of the most productive areas in the world. Water supplies were adequate, soils good, and the climate ideal. Still there were many problems, but they could be overcome with "strong direction on the part of the United States, adequate financing, and a determination on the part of the Afghans." All of these factors, however, were not operating in the current situation and without their implementation the report expressed, "strong fears that the present plan will not be successful." The writer concluded, "I think we should be prepared to go all the way and see it finished properly or we should terminate." A year later in 1971, just three years before Recla-

112 Helmand Valley Project Review Afghanistan, March 1961, Office of Foreign Affairs files, Bureau of Reclamation, Washington, D.C; Cullather, "Damming Afghanistan," 534; "Comedy of Errors," *New York Times*, March 13, 1960.

9.16. Reclamation published a 1964 report on the Blue Nile in Ethiopia for AID, the Agency for International Development.

mation withdrew its efforts in Afghanistan, another report on the Helmand Project lamented the persistence of severe problems that centered on social and cultural issues as one of the root causes for difficulties with the project. The report expressed amazement that such a project could possibly have been planned without detailed data on the nature of the involved population. Extension agents among the population simply had not been able to gain villagers' supportive responses to the implementation of new irrigation schemes nor did they understand the ideas and traditions of "the socio-cultural system of the country."[113]

9.17. Premier Süleyman Demirel of Turkey, with Vice President Lyndon B. Johnson in August 1962 in Ankara. Demirel trained with Reclamation in 1949 when he was project manager of Turkey's Electric Power Resources Administration.

The Bureau of Reclamation launched its Afghanistan work upon the request of the Department of State and with the approval of Commissioner Dominy. The timing coincided with the new decade of the 1960s and the abundant energies the John F. Kennedy presidency applied in the "Development Decade" that confronted new horizons and volatile events on the foreign scene. As one source noted, the increased competition "altered the significance, but not the fortunes, of the Helmand project in the 1960s." Afghanistan became the "arena for a tournament of modernization," and the topic of a novel by James A. Michener, one of the late twentieth century's most prolific authors. In his *Caravans* (1963), he plots the struggle of the modern versus the traditional and the religious in the face of efforts to build dams, roads, communications, and provide the gifts of electrical power to the countryside.[114]

---

113 A. L. Mitchell, Project Manager, Bureau of Reclamation, Laskar Gah, to Commissioner, April 30, 1970, International Affairs Office files; R. Scott, "Memorandum" to Albert R. Baron, Assistant Director, HAVR, April 3, 1971, International Affairs Office files, Bureau of Reclamation, Department of the Interior, Afghanistan File, Washington, D.C.

114 Cullather, "Damming Afghanistan," 531; James A. Michener, *Caravans* (New York: Random House, 1963); James A. Michener, "Afghanistan: Domain of the Fierce and the Free," *Reader's Digest*, 67 (November 1955): 161-72.

Southeast Asia

The 1960s saw the entanglement of the United States in Southeast Asia. With increasing military involvement of the U.S. in Vietnam, the Mekong River Valley became a field of increased activity for Reclamation. The Mekong River, one of the longest rivers in the world, arises in the Tibetan Plateau. It flows southeast out of China through Burma, Thailand, Laos, Cambodia, and Vietnam forming the Mekong Delta along its southern routes before emptying into the South China Sea at the southern tip of Vietnam. Since the 1940s American dam builders took an interest in the Mekong River. The concept of river basin development captured the imagination of policymakers as well as American construction companies, who looked to foreign aid funds to finance the design of river management undertakings based upon American models. One study suggested what it called, "The Mekong Blueprint," as one such example, defining river basin development there as a multi-purpose cascade of hydroelectricity, flood control, irrigation, improved navigation, and even tourism.[115]

For many alarmist voices in the United States the 1954 French defeat at Dien Bien Phu in Indochina made the entire lower Mekong Basin vulnerable to a Communist takeover. Thailand, however, appeared resistant to communist influence, partly because it had escaped colonial domination. In addition, the introduction of "advanced" technology, i.e., a major dam across the Mekong, promised modernization for the country and linkage to the modern West that would frustrate unpredictable experiments with communist modes of development. The lower Mekong River offered a splendid theater for the application of strategies based upon theories of how to move societies in underdeveloped nations toward modernization based upon the delivery of advanced technologies.

In the Cold War struggle, the U.S. enlisted American academics who believed they possessed road maps on how to move countries toward modernization. Central to the modernization paradigm was the application of technical aid to foster, in the words of prominent advisor to the Kennedy administration and Massachusetts Institute of Technology Professor Walter Rostow, "economic take-off." With technical and economic aid, underdeveloped countries might, for example, emulate the economic growth of the United

[115] Liesbeth Sluiter, *The Mekong Currency: Lives and Times of a River* (Utrecht, The Netherlands: International Books, 1993), iv.

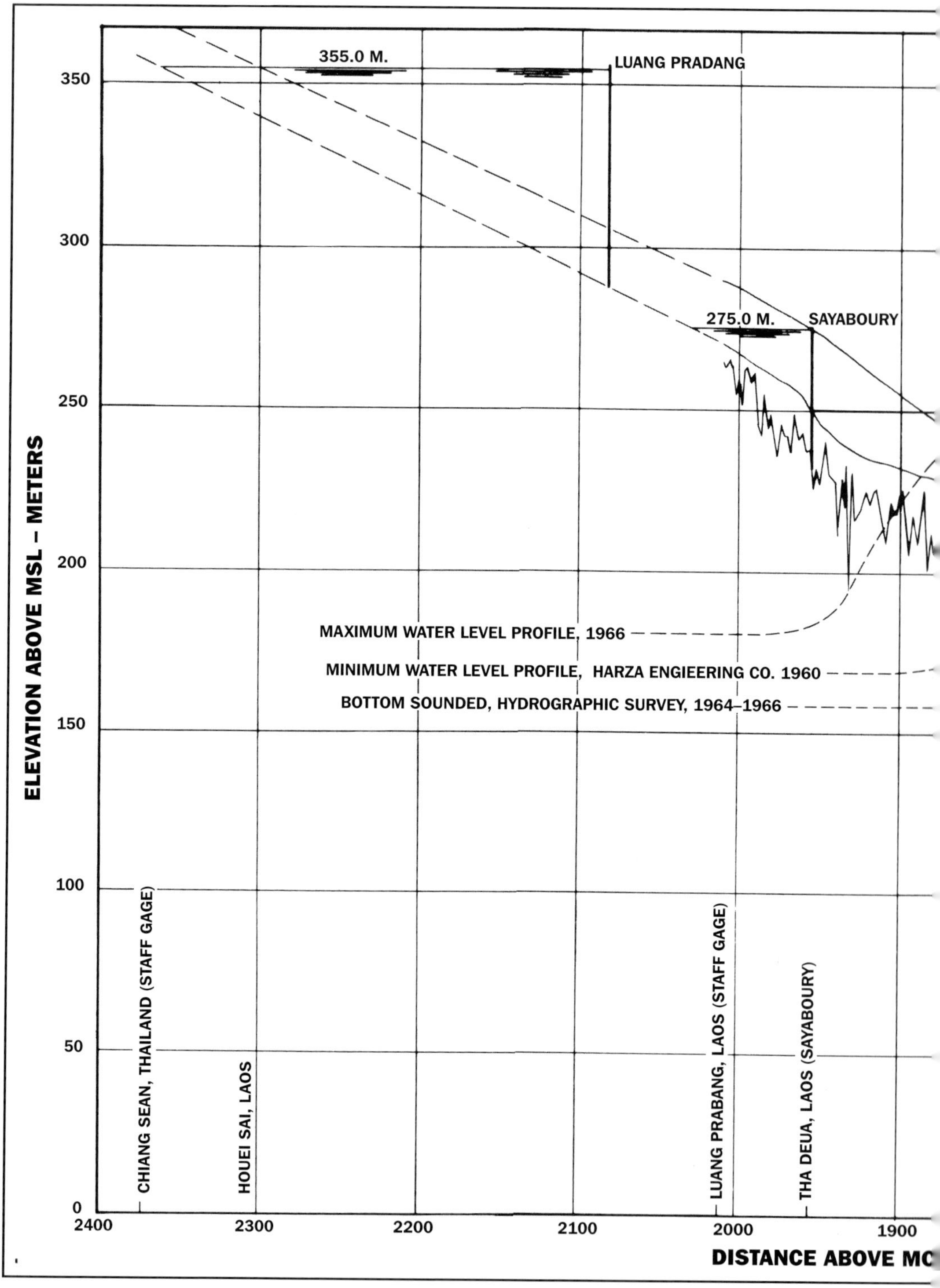

9.18. The "Pa-Mong Cascade" on the Mekong River was visualized in this Reclamation profile dated Octobe 4, 1968. Note that only two dams were under active study: Pa Mong and Sambor.

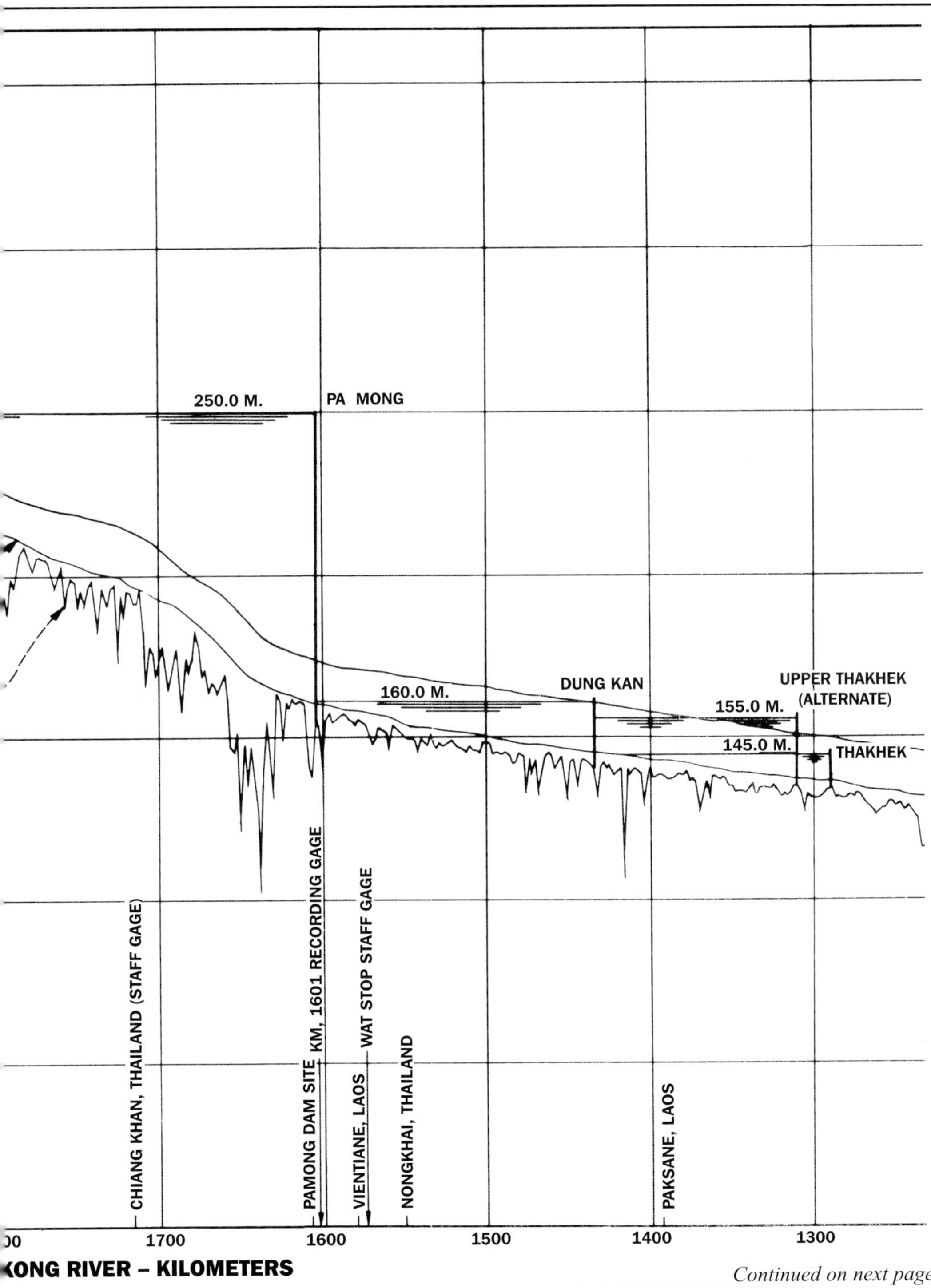

Continued on next page

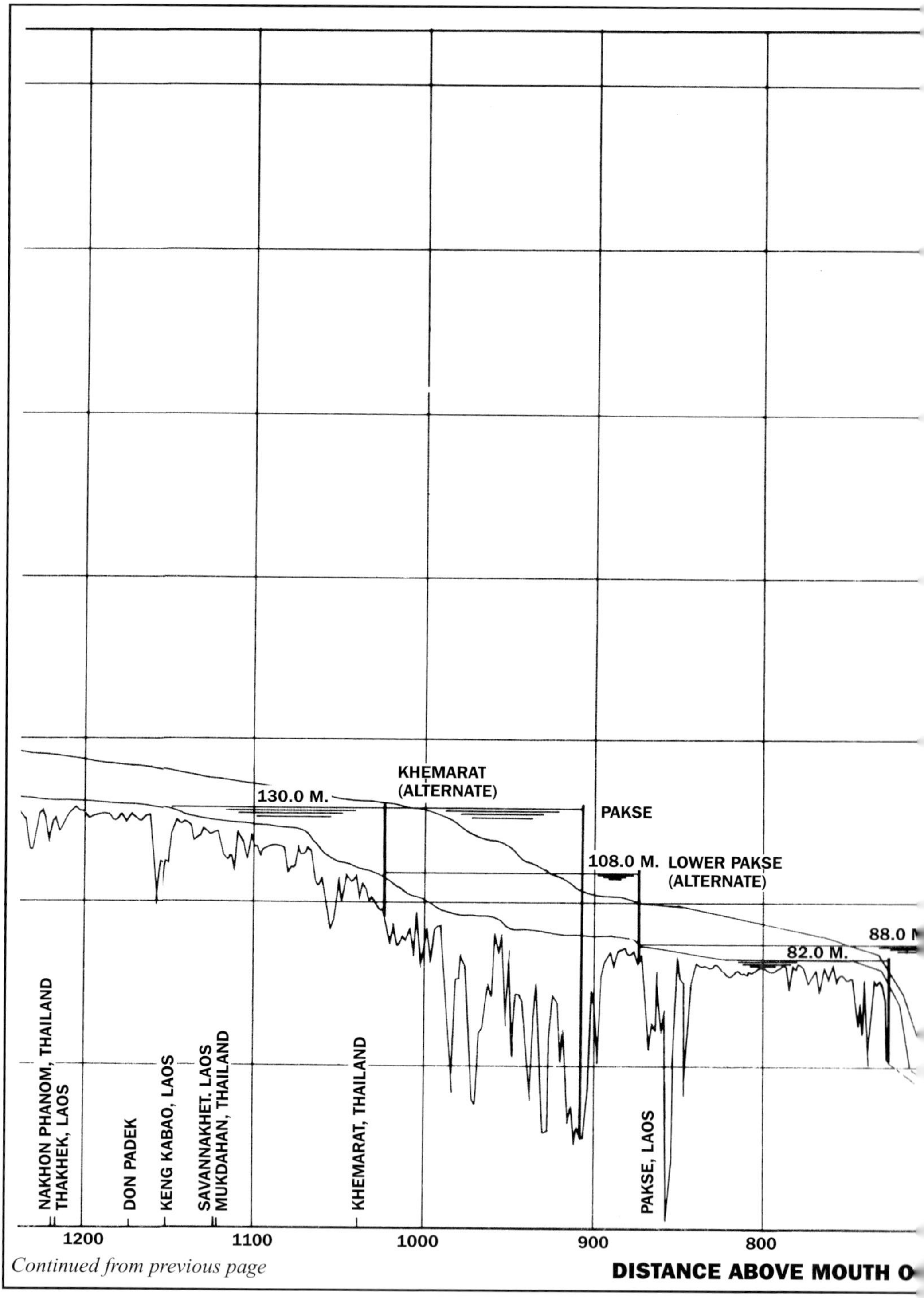

Continued from previous page

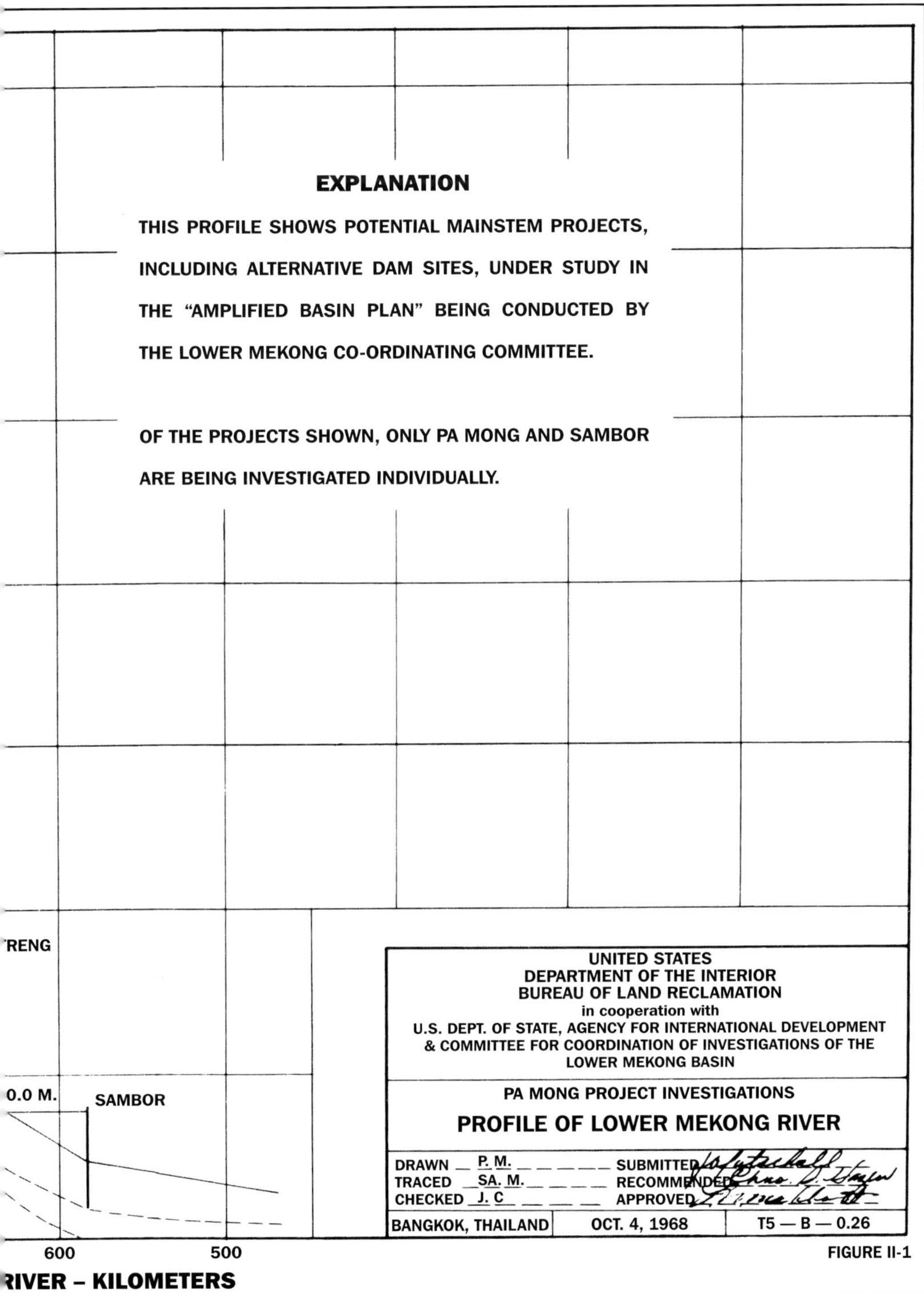

EXPLANATION

THIS PROFILE SHOWS POTENTIAL MAINSTEM PROJECTS, INCLUDING ALTERNATIVE DAM SITES, UNDER STUDY IN THE "AMPLIFIED BASIN PLAN" BEING CONDUCTED BY THE LOWER MEKONG CO-ORDINATING COMMITTEE.

OF THE PROJECTS SHOWN, ONLY PA MONG AND SAMBOR ARE BEING INVESTIGATED INDIVIDUALLY.

UNITED STATES
DEPARTMENT OF THE INTERIOR
BUREAU OF LAND RECLAMATION
in cooperation with
U.S. DEPT. OF STATE, AGENCY FOR INTERNATIONAL DEVELOPMENT
& COMMITTEE FOR COORDINATION OF INVESTIGATIONS OF THE
LOWER MEKONG BASIN

PA MONG PROJECT INVESTIGATIONS

PROFILE OF LOWER MEKONG RIVER

DRAWN P. M. — SUBMITTED [signature]
TRACED SA. M. — RECOMMENDED [signature]
CHECKED J. C — APPROVED [signature]

BANGKOK, THAILAND	OCT. 4, 1968	T5 — B — 0.26

FIGURE II-1

Outstanding Allies in Our Midst

A few years ago an extensive tour of Northwest power and irrigation projects was given to four Afghan dignitaries included in group standing in front of the airplane. The men from Afghanistan and the positions they held at that time are: His Excellency Hashim Maiwandwal, counselor of embassy and chargé d'affaires in Washington, D.C.; The Honorable Abdussattar Shalizi, honorary consul at San Francisco; His Excellency Abdullah Khan, general president of the Helmand River Valley Authority in his country; and Dr. A. Kayeum, vice president of the Helmand authority. The three American guides are Fred J. Huber, then vice president of International Engineering Co., Don Stoops, then a division chief in the U.S. Foreign Operations Administration; and Gordon K. Ebersole, then head of the Bureau's foreign training branch.

L. Rochanasiri "Rocky" Warindr, right, Administrative Officer of the Royal Irrigation Department of Thailand, stops while inspecting a dam to chat with Mr. Ebersole, author of the foregoing article.

A 6-month, in-service training program will be completed in early 1966 for Mrs. Parween Azize. Learning secretarial work at the Columbia Basin Project office in Ephrata, Wash., Mrs. Azize was bettering her qualifications as an employee in the Ministry of Finance in Kabul, capital city of Afghanistan.

9.19. A February 1966 Reclamation Era article on Reclamation's work in the Helmand Valley of Afghanistan.

9.20. In 1954 Reclamation designed the upper right abutment of the Klang Gates Dam, a water supply dam, for this location near Kuala Lumpur, Malaysia.

States and pass through stages of economic growth similar to the American path toward industrialization. The Rostow model projected a process for Americanizing much of the world. As such, "the rhetoric of modernization" could be seen as an updated form of imperialism and quite at odds with the spirit of a post-colonial world.[116]

As early as 1951 American advisors began appearing in Thailand for the Chao Phya Project and later the Yanhee Project. The World Bank invested $66 million in 1957 for construction of Bhumiphol Dam. By 1963 the 505 foot-high concrete dam stood completed. Its hydroelectric plant helped expand the domestic economy of Thailand and helped control water supplies that contributed to the doubling of rice yields in the central Thai plain in the 1960s. These successes opened the door to planning for larger scale projects in Thailand. The Lower Mekong Coordinating Committee envisioned the Pa Mong Project and, in 1963, invited the Bureau of Reclamation to conduct reconnaissance investigations. Soon a Reclamation team of forty was on

[116] Sluiter, *The Mekong Currency*, iv; Latham, *Modernization as Ideology*, 58-9, 66, 68.

9.21. With a Malay police officer standing guard, in November 1951 W. H. Irwin, the assistant chief geologist of the Bureau of Reclamation (left), and A. L. McClure, the senior engineer of the Water Works, Public Works Department, Federation of Malaya, examined a small gravel bar on the Klang River about 200 yards downstream from the Klang Gates damsite.

the ground assisted by 170 Thai and Laotian engineers and technicians who received "on-the-job" training in studies of land and water issues. The plan's broad outline called for a large storage reservoir in the Mekong River at Pa Mong with lateral basins on the adjoining Nam Lik River in Laos and on the Nam Mong in Thailand, a large hydroelectric powerplant, and in time large areas of irrigated land in northeast Thailand and on the Vientiane Plain in Laos. The multipurpose project envisioned development in irrigation, power, navigation improvement, and flood control. Power and irrigation, however, predominated. There would also be other benefits in domestic, industrial, municipal water supplies, enhancement of fisheries, recreation and in preventing salinity in the Mekong Delta. A preliminary report declared, "About 5,000,000 acres of land are being studied to determine their suitability for sustained crop production under irrigation."[117]

The proposed Pa Mong Project stands as an example of the optimism surrounding the deployment of American technical expertise (i.e., Reclamation

[117] Gilbert Stamm, "Bureau of Reclamation International Technical Assistance in Development of Arid Lands," International Conference on Arid Lands in a Changing World, Tucson, Arizona, June 3-13, 1969, 5.

engineers) to develop the resources of a pre-modern society. The expected outcomes offered economic development, democracy, and Westernization—all fortifying the society against the chimera of Communism. As American military involvement escalated in Vietnam, Commissioner Dominy took greater interest in the project. In 1965 President Lyndon B. Johnson gave an important speech related to the Mekong at Johns Hopkins University. The president coupled an announcement of an aggressive bombing campaign against North Vietnam with his decision to ask Congress for $1 billion to construct "the Mekong Cascade" that would "dwarf even our own TVA." He hoped this commitment to peaceful economic growth would be the ultimate "footprint" of the United States in the region rather than the scars of war. Following this announcement, the president dispatched longtime associate from the New Deal Era and former director of the TVA and chairman of the Atomic Energy Commission, David Lilienthal, to visit the region and tout the benefits of multipurpose, basin-wide river development. Author of *TVA: Democracy on the March* (1944) Lilienthal saw the accomplishments of the TVA as an example to export overseas to a world undergoing decolonization and crying out for programs of democratic development and modernization.[118]

9.22. The Klang Gates Dam in Malaysia on March 22, 1976.

[118] *Public Papers of the Presidents of the United States: Lyndon B. Johnson*, Volume 1 (Washington, D.C.: U.S. Government Printing Office, 1966), 394-99 as quoted in David Biggs, "Recla-

SI METRIC

TECHNICAL FEASIBILITY INVESTIGATION
INTERIM REPORT

ENLARGEMENT OF
KLANG GATES DAM
AND RESERVOIR

KUALA LUMPUR FLOOD
MITIGATION PROJECT
MALAYSIA

Prepared by
Resident Team
UNITED STATES
DEPARTMENT OF THE INTERIOR
Bureau of Reclamation
Kuala Lumpur, Malaysia
October 1976

9.23. To protect downstream Kuala Lumpur, Malaysia, Reclamation studied adding flood control to the Klang Gates Dam, October 1976.

In 1966, Commissioner Dominy delivered to the Lower Mekong Basin Coordinating Committee a report in French entitled: "*L'Etude Exploratoire Le*

---

mation Nations: The Bureau of Reclamation's Role in Water Management and Nation Building in the Mekong Valley, 1945-1975," *Comparative Technology Transfer and Society* 4 (December 2006): 234-5; David Biggs, *Quagmire: Nation Building and Nature in the Mekong Delta* (Seattle: University of Washington Press, 2010), 170-5, note that the author slipped into using Bureau of Land Management when referring to the Bureau of Reclamation; U.S. House of Representatives, Subcommittee on National Security Policy and Scientific Developments, *The Mekong Project: Opportunities and Problems of Regionalism* by Franklin P. Huddle (Washington, D.C.: United States Government Printing Office, 1972): 2; Ekbladh, " 'Mr. TVA': Grassroots Development," 336-44.

Projet De PaMong Laos-Thilande" or "Exploratory Study of Phase One of the Pa Mong Project." Its purpose was to determine the practicability both economically and technically of the Pa Mong Project. Such studies were not without precedent since the Department of the Interior had already undertaken and published in 1956 a previous report on the lower Mekong River Basin entitled: *Le Bassin du Bas-Mekong: Rapport de Reconnaissance prepare pour l'Administration de Cooperation Internationale*. The earlier study preceded by one year the formation of the Mekong Committee in 1957 to assess the possibilities of developing the water resources of the lower reaches of the river. In that year following organization of the Mekong Committee, an important delegation of visitors arrived including Lieutenant General Raymond A. Wheeler, former chief of engineers in the U.S. Army Corps of Engineers, Kanwar Sain, chairman of India's Central Water and Power Commission, John W. McCammon, former general manager of Quebec Hydro, G. Duval, director, SOGREAH, France, and Yutaka Kubota, president of Nippon Koei, Japan. Their various points of visitation included the Pa Mong site fifteen miles upriver from Vientiane, Laos. Bureau of Reclamation engineers guided the delegation to each site. Since 1952 Reclamation engineers had looked at Pa Mong as a possible "replacement" project for Three Gorges that was now off limits because of the victory of the Communists in China.[119]

9.24. President Lyndon B. Johnson spoke about the Mekong River during an address at Johns Hopkins University on April 7, 1965. Photographer: Frank Wolfe. Courtesy of the Lyndon B. Johnson Library, National Archives and Records Administration.

[119] Biggs, "Reclamation Nations," 232-3.

9.25. General Raymond Wheeler, U.S. Army Corps of Engineers. Courtesy of the History Office, U.S. Army Corps of Engineers.

The visit of Wheeler's international delegation represented a transition in American "modernization strategies" from containment in 1949 to what one historian describes as the "global imaginary of integration." The integration of the former colonial world into a world capitalist system served both capitalism and the political-military ends of containment. An America "looking outward" sought collaboration with allies to march in concert toward the goals of bringing modern power sources and advancements in agriculture within the framework of a free market capitalist economy. The world did not always share the enthusiasm for free trade that, of course, allowed American businesses access to former colonial countries now assuming nationhood. In some regions, i.e., the lower Mekong system, without the application of considerable military intervention the strategies could not go forward. Ultimately the defeat of American efforts in Vietnam by 1975 halted ambitious projects for the lower Mekong.[120]

But few envisioned an eventual retreat of the United States from the Mekong in the mid-1960s when the Bureau of Reclamation made its intensive studies of the Pa Mong Project. *Reclamation Era* featured a speech by President Johnson to the first International Conference on Water for Peace in Washington, D. C., in May of 1967. The president called "Water the Key to Sustaining Growth." He told representatives from 91 nations that water was necessary to sustain an ever-growing population; that it was the key and there was need for: "Water to drink; water to grow the food we must eat; water to sustain industrial growth." He reaffirmed his support of international coopera-

120 As noted in Biggs, "Reclamation Nations," 234; Christina Kline, *Cold War Orientalism: Asia in the Middlebrow Imagination, 1945-1961* (Berkeley: University of California Press, 2003), 25-7.

tion in water developments like those underway in the Mekong and the Indus river basins. Calling water use programs "the enduring servant of peace," he announced that he had directed the Secretary of State to establish a Water for Peace Office.[121]

Increasingly the work of Reclamation engineers in the lower Mekong began to take on greater political significance according to social scientists commissioned to examine the impact of dams, hydroelectric production, and new irrigation regimes upon local populations and societies. Beginning with a 1962 report by geographer Gilbert White and funded by the Ford Foundation, social scientists of varying views brought to the forefront questions about social and economic welfare that should be considered in any future funding of Pa Mong Dam and other dam projects in the "Mekong Cascade." Generally studies conducted by the Rand Corporation criticized Reclamation for a lack of sophistication in benefit-cost analysis. Even later studies in the 1960s began a critique not only of the social aspects of the dams, Bhumipol and Nam Ngum (Laos), but also the environmental problems stemming from these dams. In 1969 Reclamation sponsored a study on the social feasibility of the Pa Mong irrigation undertaking that indicated a wide gap between technical ability to build the project and acceptance by the local population. That study also noted that social aspects of river basin development projects usually have "not been done as part of the preconstruction feasibility study." A deteriorating security situation after 1968 caused Reclamation to reduce its visibility in preliminary and secondary investigations of the Pa Mong project and then in 1972 to withdraw entirely from Mekong Committee activities.[122]

Still, publications in the late 1960s waxed on enthusiastically about development of the lower Mekong. A *National Geographic* report in 1968 declared that even in the midst of the Vietnam War, "Men of many nations worked on the Giant Mekong Project, a plan that promises to revolutionize life for millions of Southeast Asians by harnessing the river for power and irrigation." The benefits envisioned by the Pa Mong Dam, the article continued,

---

121 "Water—The Key to Sustaining Growth," *Reclamation Era*, 53 (August 1967): 57-60.

122 University of Michigan Team for Pa Mong Research, *Pa Mong Resettlement Research Project: Final Report*, (Ann Arbor: University of Michigan Center for South and Southeast Asian Studies, 1982), 29, iii as noted in Biggs, "Reclamation Nations," 238; Kurt Finsterbusch, Jasper Ingersoll, and Lynn Llewellyn, editors, *Methods for Social Analysis in Developing Countries* (Boulder, Colorado: Westview Press, 1990); U.S. Department of Interior, Bureau of Reclamation, and U.S. Agency for International Development, U.S. Department of State, "The Social Feasibility of Pa Mong Irrigation: Requirement and Reality," by Jasper Ingersoll, typescript mimeograph, 1969.

9.26. President Lyndon B. Johnson addressing the International Symposium on Water for Peace at the Sheraton Park Hotel, Washington, D.C. Secretary of the Interior Stewart Udall is seated behind the president on the right of the picture. May 23, 1967. Photographer: Frank Wolfe. Courtesy of the Lyndon B. Johnson Library, National Archives and Records Administration.

"would dwarf the achievements of the Tennessee Valley Authority." After travelling to the site with Lyle W. Mabbott from Dubois, Wyoming, engineer on the Pa Mong Project and a long-time engineer with the Bureau of Reclamation with work experience on dams in Arizona and studies of the Blue Nile, the *National Geographic* writer asserted, "No area in the world ever planned such a regional economic development on so massive a scale." While the tone of the article generally lauded the undertaking, there was an element of caution about changing the water regime of the lower Mekong. It noted that some experts worried about the impact on the environment. The questions posed included: "What would happen when areas now flooded annually would be dry all year, thanks to flood control? Unless soon irrigated and fertilized, they might turn into dust or into soil hard as rock. It had happened in Pakistan." Would standing reservoir water bring more disease as had occurred in Egypt? Would fish die? Would unwanted plants multiply and clog reservoirs, turbines, and canals?[123]

[123] Peter White, "The Mekong River: Terror and Hope," *National Geographic*, 134 (December 1968): 743, 749, 771-2.

9.27. Eugene R. Black, president of the World Bank, with President John F. Kennedy. Courtesy of the World Bank.

Yet Bureau workers in cooperating with the Lower Mekong Committee saw great opportunity. In 1968 Eugene R. Black, former President of the World Bank, delivered a report to President Johnson on the possibilities of development in Southeast Asia that concentrated on the Lower Mekong River Basin. In a forward to the published study, President Black expressed his belief and hope that peaceful economic development and cooperation of nations would be the future of the region. Black for his part saw opportunity for development of the Mekong Basin and the eventual transformation of the Mekong Committee into a "Lower Mekong Basin Authority." Black noted that a detailed feasibility study, "completed by the largest team of U.S. Bureau of Reclamation engineers ever assembled overseas" would soon appear. He was convinced that this first dam across the main artery of the river would someday be built bringing cooperation to the troubled nations of the region.[124]

Optimistic technical prospects for dam building could not erase a growing American frustration with the Vietnam War or prevent eventual American military withdrawal from the region. Without American military presence, the Lower Mekong projects stalled. The general withdrawal by the United States in the region brought to an end major involvement by the Bureau of Reclamation in foreign theaters in the postwar period. While Reclamation appeared to be a partner in American foreign policy and stepped up to meet the demands placed upon it, the organization faced setbacks when the American anti-communist military effort in Vietnam failed. Increasingly budgets were strained at home by military expenditures in Southeast Asia to carry

[124] Eugene R. Black, *Alternative in Southeast Asia* (New York: Frederick A. Praeger, Publishers, 1969): ix, 134, 138-9.

9.28. While visiting Reclamation Mr. Chon Myong Kim (left) of Korea and Professor Nakasi (right) of Japan posed in front of a bulletin board posting photographs of international trainees and visitors.

on the Vietnam War. After the war, the burden of debt still persisted, and it crimped congressional appropriations for new water projects at home, which had been so generously funded during the 1950s and early 1960s.[125]

[125] Pisani, "The Bureau of Reclamation and the West, 1945-2000."

Reclamation Staff traveled by a variety of transport on international projects

9.29. Nicaraguan Mission, 1951.

9.30. Nicaraguan Mission, 1951.

9.31. Ethiopia, Blue Nile Project, 1960.

9.32. Nicaraguan Mission, 1951.

9.33. Nicaraguan Mission, 1951.

CHAPTER 10:

RECLAMATION IN AN ERA OF GUNS AND BUTTER: RIVERS, VALLEYS, AND CANYONS–1945 TO 1956

Introduction

By the early 1950s, the Truman administration found itself awash in a sea of criticism. Events of the early Cold War—the Soviet Union's atomic bomb, China's fall to Communism, and the outbreak of the Korean conflict in June 1950—unsettled the American public and undermined confidence in the administration in foreign and domestic affairs. In historical hindsight, the Marshall Plan, NATO, the Berlin Airlift, and the limited war strategy in Korea emerge as success stories, but at the time the picture was unclear to the American public. Frustrations dogging early Cold War foreign policy fueled criticism of the president's domestic Fair Deal that the opposition party identified as a warmed-over New Deal. Still the Republican opposition smarted from unexpected defeats, especially the return of Congress to the Democrats and Truman's come-from-behind victory in the 1948 presidential race. Even in the hands of an unlikely heir, Harry Truman, the ideals of the New Deal, including its activist view of government's role in providing economic security, seemed to win hands down against any threat to repeal basic programs.

While the foundations of the New Deal and government's regulatory power remained intact, perennial controversies about the Bureau of Reclamation continued to inflame politics—public versus private power, ownership of power transmission lines from public dams, and the Bureau's responsibility to enforce its 160 acre limitation rules. The press and a vocal wing in the Republican Party launched attacks against what it regarded as the socialistic economic planners left over from the New Deal. Also as Cold War fears intensified, the loyalty and patriotism of ex-New Dealers fell under suspicion in an atmosphere rife with accusations that communists held positions of power in the highest echelons of American government.

The volatile political climate of the early 1950s provided critics of the Bureau of Reclamation an especially inviting platform to target their discontent at Reclamation's leaders, its past accomplishments, and future goals. While political change was in the air, the election of Republican presidential candidate General Dwight D. Eisenhower in 1952 did not mark a complete reversal of the domestic reform achieved under the previous Democratic administrations. Despite Eisenhower's overwhelming victory, the new president was either unwilling or unable to dismantle the major underpinnings of the New Deal. Eisenhower's presidential campaign emphasized smaller government, encouragement of private enterprise, and local control rather than federal authority directed from centralized bureaucracies. On all of these scores the Bureau of Reclamation stood vulnerable to attack from those who demonized its bureaucratic power and huge (for the day) expenditures, along with its participation in the production of public power. At the same time, however, demands upon the federal budget increased as the United States fully committed itself to the Cold War through defense measures to bolster its positions against Communist expansionism throughout the world. Eisenhower's response to this dilemma was an attempt to rein in defense spending by instituting his "New Look" defense policy that relied on the United States' nuclear arsenal as America's first line of defense. Though the "New Look" did somewhat lessen spending on conventional forces, military expenditures never did noticeably drop, even after the 1953 truce in the Korean War. In addition, the American economy in the 1950s still supported increases in domestic spending, which included not only administration supported programs such as interstate highways but also water development projects.[126]

Economists labeled the joint effort to maintain both military spending and domestic prosperity "Guns and Butter."[127] For the Bureau of Reclamation, bolstering American economic strength on the domestic front meant full speed ahead for water resource development in the American West. A softening of anti-New Deal rhetoric during the Eisenhower years did not lessen the fervor of private power interests and anti-federal government forces to oppose Reclamation's proposal to build a gigantic dam in Hells Canyon on the Snake River. In addition, agribusiness interests in California and Washington tested the Bureau of Reclamation's longstanding commitment to the 160 acre limitation rule and the family farm ideal. By the end of the 1950s, another source of opposition to Reclamation plans came from concerns spearheaded by a coalition of small but vocal preservation groups.

---

[126] For information on Eisenhower's "New Look" see Derek Leebaert, *The Fifty-Year Wound: How America's Cold War Shapes Our World* (Boston: Back Bay Books, 2002), 148-9; Norman Friedman, *The Fifty-Year War: Conflict and Strategy in the Cold War* (Annapolis, Maryland: Naval Institute Press, 2000), 194-8; Gerard DeGroot, *The Bomb: A Life* (Cambridge, Massachusetts: Harvard University Press, 2005), 190, 201-3.

[127] For reference to the term "guns and butter" see Paul Anthony Samuelson, *Economics: An Introductory Analysis* (New York: McGraw-Hill, 1951), 20-2, 28-9.

10.1. During his first campaign for the presidency Dwight D. Eisenhower visited Hoover Dam on June 22, 1952.

The Excess Lands Issue in the Central Valley Project

In 1933 the Bureau of Reclamation took over the Central Valley Project (CVP), and from the beginning the 160 acre rule increasingly met bitter opposition from valley agricultural interests and their political representatives. A strict interpretation of provisions in the 1902 Reclamation Act made lands under a single ownership in excess of 160 acres ineligible for water supplies provided by dams, reservoirs, or ditches built by the Bureau of Reclamation. The 160 acre rule in the law stood as a constant reminder of the early goals of the federal reclamation policy to encourage establishment of small farmers on the land, and to permit government subsidized water to foster agricultural development, while at the same time preventing land monopoly. In the view of Secretary of the Interior Ickes the question involved "the age old battle over who is to cash in on the unearned increment in land values created by a public investment."[128] Nowhere did the Bureau's efforts to uphold the land

[128] Joseph L. Sax, "Selling Reclamation Water Rights: A Case Study in Federal Subsidy Policy," *Michigan Law Review* 64 (November 1965): 13; "Irrigation Warfare Renewed: Old Disputes Over U.S. Limitations on Size of Tracts Eligible for Water," *Business Week*, (November 17, 1945): 19-20.

limitation ideal come under heavier attack and scrutiny than in California's Central Valley. This is not to say that Bureau of Reclamation officials followed a policy of vigorous enforcement. Some, however, were more ardent about the issue than others. When the "Elliott Amendment," which sought to have 160 acre rule restrictions removed from the Central Valley Project, caught the attention of Congress in 1944, the Senate staunchly refused to abandon the rule. In addition, Secretary Ickes expressed a fierce commitment on the part of the Department of the Interior to the 160 acre limitation rule. The introduction of the Elliott Amendment along with Secretary Ickes's refusal to abandon the 160 acre limitation rule in the Central Valley Project established battle lines over this clause of Reclamation law during the postwar period.

Agriculture in California differed from other regions in the West, due in large degree, because of historical forces. Historians and other researchers note this peculiarity in California history as a result of the large land grants given to settlers during the Spanish/Mexican period. To work the landed estates efficiently required a large labor pool originally made of Native Americans. Beginning in the mid-nineteenth century, Anglo-American settlers not only overwhelmed the descendents of Spanish/Mexican residents, but also assumed the historical pattern of agricultural production in California. Over time corporations and business conglomerates, such as Miller and Lux and the Southern Pacific Railroad, acquired vast swaths of land throughout California. In addition California agriculture moved toward the production of cash crops—wheat, livestock, citrus fruit, etc.—abandoning subsistence farming in favor of "industrial farming." By the time the Bureau of Reclamation arrived in the Central Valley in 1937, the model of corporate agriculture was firmly entrenched, assisted by a pool of cheap labor. In his seminal study on farming practices in the Central Valley, *As You Sow*, Walter Goldschmidt noted that "4 per cent [*sic*] of all farms in California have 1,000 acres or more, and that these farms owned 66 per cent of the land under actual cultivation in the state." Little wonder that Reclamation attempts to enforce the 160 acre rule met fierce opposition.[129]

The secretary's determination to implement the 160 acre rule not only reflected his confrontational reform-minded spirit, but also his commitment to defend the utilitarian ideals in Reclamation law that sought to guarantee dispersal of government benefits as widely as possible. And Ickes was not alone, his interpretation of the land limitation rule carried considerable weight

---

129 Walter Goldschmidt, *As You Sow: Three Studies in the Social Consequences of Agribusiness*, forward by Gaylord Nelson (Montclair, New Jersey: Allenheld Osmun, 1978), 13.

in some spheres of California politics. At the height of the controversy from 1944 to 1946, the California Grange, the National Farmers Union, organized labor, the National Catholic Rural Life Conference, and some consumer groups rallied to support Reclamation and the Department of the Interior against the Elliott Amendment and subsequent attempts to repeal the 160 acre limitation measure.[130] Studies by New Deal social planners, especially from the Bureau of Agricultural Economics, supported their views by arguing that acreage limitation provisions in Reclamation law helped to serve the national interest by producing stable communities of small farmers and small businesses. In short, all of these efforts seemed to confirm the view that the Department of the Interior, the Bureau of Reclamation, and the persistent voices of New Dealers within the administration wanted to see only small farms developed in the Central Valley. They contended that these developments offered social benefits best illustrated in anthropologist Walter Goldschmidt's study. Goldschmidt compared what he called the "good society" of the Central Valley community of Dinuba, where small farms were prevalent, against the lack of community services and prevalence of large numbers of poor agricultural labors in Arvin, California, where large landholdings dominated. The study, done for the Bureau of Agricultural Economics at the University of California, Berkeley, first appeared in a 1946 report to the U.S. Senate Special Committee to Study Problems of American Small Business entitled, "Small Business and the Community." A year later a commercial press published Goldschmidt's work under the title *As You Sow* (1947).[131]

In California another early figure to raise a hue and cry against the Elliott Amendment and lifting the land limitation requirement was Paul S. Taylor, long-time agricultural economics professor at the University of California, Berkeley, and an advisor to Goldschmidt's Central Valley study. He wrote articles, letters to Congress, and appeared in public forums arguing for enforcement of the 160 acre rule to save the Central Valley Project from the domination of big growers and agribusiness. Taylor developed deep convictions on the subject during the 1930s because of his pioneering studies of Mexican farm workers and agricultural labor in California. His marriage to Dorothea Lang, noted photographer of Dust Bowl migrant workers' experi-

---

130 For information on groups supporting the 160 acre rule see Freeland, "Examining the Politics of Reclamation," 222; Kathka, "The Bureau of Reclamation in the Truman Administration." 60.

131 Senate Special Committee to Study the Survey Problems of American Small Business, *Small Business and the Community: Study in the Central Valley of California on Effects of Scale Operations*, 79th Cong., 2nd sess. (Washington, D.C.: United States Government Printing Office, 1946); see also Goldschmidt, *As You Sow*.

ences in California, reinforced his commitment to social justice causes, and made Taylor a prominent proponent of the idea of small farms in the valley. During the 1930s, both Taylor and Lang witnessed the use of local law enforcement, under the direction of the growers, to break up unions and intimidate workers. On one occasion Assistant Secretary of the Interior Michael Straus wrote to Taylor congratulating him on his remarks heard nationwide over the Town Hall of America radio program. Straus described Taylor's position on the issue as fitting squarely with that of himself, the Department of the Interior, and the Bureau of Reclamation. In the face of growing opposition, however, the problem was how to make the right political moves to protect and enforce the 160 acre rule in California. One failure in the process had already occurred. Unfortunately for the Bureau, the press, and even the Town Hall program, branded Reclamation's effort to enforce the 160 acre rule as "breaking up" of large holdings. These charges made the Bureau of Reclamation appear to be attacking established and legitimate business interests. It created a climate that invited amendments like the one Congressman Elliott championed in 1944, and most annoyingly to Straus, as riders to any legislation pertaining to the Department of the Interior.[132]

As already noted Reclamation was ambivalent about enforcing the law's requirements, and some Department of the Interior officials began to reconsider its usefulness. In the summer of 1944, Assistant Commissioner William Warne wrote, "This darn 160 acre law in the Central Valley project has caused us unending grief."[133] It quickly became apparent that the fight over the 160 acre limitation was emerging as a major issue that threatened Reclamation's Central Valley programs and highlighted the Bureau's competition with the Corps of Engineers. Warne noted that even the Department of Agriculture and the International Boundary Commission had joined the fray. The assistant commissioner suspected that those fighting against the land limitation clause might seek to scuttle the entire Reclamation program if Reclamation did not back down. Warne referred to them as "the venal interests that would sell the whole reclamation policy for a temporary advantage."[134] The Army Corps of Engineers labored under no such limitations in distributing water from reservoirs it built, which made the Corps an attractive alternative

---

132 Michael W. Straus to Paul S. Taylor, August 11, 1944, RG 48, Entry 779, Box 15.

133 Alfred R. Golzé, *Reclamation in the United States* (Caldwell, Idaho: Caxton Printers, 1961), 196-7; Swain, "The Bureau of Reclamation and the New Deal, 1933-1940," 145; Freeland, "Examining the Politics of Reclamation," 217-8; William E. Warne to Phil Dickinson, August 15, 1944, December 28, 1944, Warne Papers, Box 2, Correspondence.

134 William E. Warne to Phil Dickinson, March 23, 1944, Warne Papers, Box 2, Correspondence.

10.2. Millerton Lake behind Friant Dam with the Friant-Kern Canal heading south from high on the left abutment of the dam. April 12, 1967. Photographer: Wes W. Nell.

water provider for Central Valley agricultural interests. Moreover, it brought to the surface the underlying struggle between the Bureau and the Corps over which bureau would take a major role in building dams and developing California's water resources in the postwar years.

Despite the land limitation controversy, the Central Valley Project remained one of the major construction efforts the Bureau of Reclamation enthusiastically resumed at the end of World War II. Construction began in earnest to complete the Friant-Kern Canal to transport water from Millerton Lake on the San Joaquin River to the southern valley communities in Kern and Tulare counties. Reclamation was also working with the Bureau of Mines in establishing an electric steel plant in Redding, California, using the hydroelectricity produced at Shasta Dam. In 1947 the Bureau awarded a $5,888,000 construction contract for the 2½-mile Delta-Mendota Intake Canal which included the Tracy Pumping Plant. These activities reaffirmed Reclamation's commitment to the Central Valley Project wherein "every little foothill stream worth damming will have been dammed to put every drop of water to its fullest use," according to Martin H. Blote chief of Region II's Irrigation Operations Division. Moreover, the 160 acre rule debate had little effect on Reclamation's

10.3. Tracy Pumping Plant.

appropriations, allotting over $40 million for the Central Valley Project out of the $198 million 1948 budget construction program.[135]

As construction on the Central Valley Project progressed at a steady pace, the struggle among bureaucracies vying for congressional funding in conjunction with the land limitation controversy eventually spilled into California political campaigns. In 1945 California's liberal and movie-star celebrity congresswoman, Helen Gahagan Douglas, took up the cause of saving the CVP for the small farmer and requested tactical political advice from Secretary of the Interior Ickes. Ickes offered a number of arguments for Douglas to take to her constituents in opposition to the notorious Elliott Amendment. He noted that she should begin by citing her long-time support for "family-size farms" in the Central Valley of California. On another point, the draft letter spoke of the "people's dams" and the necessity not to give up the government's right to build transmission lines through the Central Valley to carry power produced by federal dams. As Ickes framed it, private power companies should not be put between the people and their dams. "Both the efforts to remove acreage restrictions on the federally irrigated land and to impose restrictions against federal transmission lines are efforts of monopolists—in one case, power monopolists; in the other, land monopolists," he wrote. The letter

[135] "Commissioner Authorized to Award Top Value Contract," *Reclamation Era*, 33 (August 1947): 187; "Reclamation's Construction Program," *Reclamation Era*, 33 (September 1947): 194; Martin H. Blote, Chief Irrigation Operations Division, Region II, Sacramento, "Water Runs Uphill," *Reclamation Era*, 33 (November 1947): 213.

emphasized that a struggle of vital importance was now underway in California and in Congress to protect the public's interest on these matters and, most importantly, the public must be kept informed to prevent monopolistic interests from winning the day.[136]

From 1945 to 1950, the 160 acre rule haunted California politics, and Helen Gahagan Douglas remained in the thick of it. When she ran for the Senate in 1950 Douglas attacked her opponent veteran Senator Sheridan Downy in the Democratic primary for his opposition to the small farm and his favoritism to the large agricultural interests in the Central Valley. Downey was a vocal opponent of the 160 acre rule who expounded on his distaste for acreage limitation in his passionate treatise *They Would Rule the Valley* denouncing it a threat to the American dream of land ownership which "would stymie initiative." Douglas, on the other hand, relied on New Deal ideology and the style and rhetoric of her campaign, according to Douglas's biographer, "captivated liberals everywhere." However, others claimed, "She rather overdid the 160 acre problem; she simply could not understand that workers streaming out of a factory in Los Angeles, for example, had no interest in water reclamation." But somehow she was convinced that workers should lend their support to keeping the Central Valley in the hands of small farmers. From the Republican side of the senatorial campaign, Richard M. Nixon used red-baiting tactics that said "socialistic planners and political demagogues" favored the 160 acre limitation and were responsible for delaying completion of the Central Valley Project by their insistence upon the 160 acre rule that would undermine legitimate property interests in the project.[137]

Change was in the air, however, as New Deal rhetoric began to lose much of its luster during the postwar years. Commissioner Harry W. Bashore's retirement in 1945, and the abrupt departure of Secretary of the Interior Ickes from President Truman's cabinet in 1946 signaled the beginning of a period of accommodation on the part of the new administration toward those seeking to subvert the 160 acre limitation. Before his resignation, however, Ickes sought to head off compromise. His appointment of Michael Straus as com-

---

136 Harold Ickes to Helen Gahagan Douglas, June 13, 1945, as copy of letter to be sent to Victor V. Bowker, RG 48, Entry 779, Box 15; Helen Gahagan Douglas to Harold Ickes, June 4, 1945, and Ickes's suggested reply, June 13, 1945, Helen Gahagan Douglas Papers, Box 17, Carl Albert Research Center, University of Oklahoma, Norman, Oklahoma.

137 Ingrid W. Scobie, *Center Stage: Helen Gahagen Douglas A Life* (New York: Oxford University Press, 1992), 205, 230, 234-5, 247-8; see also Sheridan Downey, United States Senator, *They Would Rule the Valley* (San Francisco: Sheridan Downey, 1947); Kathka, "The Bureau of Reclamation in the Truman Administration." 67.

missioner of the Bureau of Reclamation and the selection of Richard Boke to head Region II in California stand out as his last-ditch effort to leave behind personnel friendly to public power and the family farm. Both were "expansionists" who sought to protect Reclamation in interagency jurisdictional battles and advance its interests in obtaining federal funding from Congress. As he assumed his new position in 1945, Straus seemed to live up to Ickes's hopes of defending the 160 acre rule against all enemies, especially the reclamation associations in California.[138]

10.4. Richard Boke in 1950

With Ickes out of the way it did not take long for the postwar Republican-controlled Congress in 1946 to focus its attention and criticisms upon the Bureau of Reclamation, especially on public relations oriented Commissioner Michael Straus. The prospect of Reclamation enforcing the 160 acre restriction rule, or 320 acre rule for married couples, in the Central Valley fanned opposition among those who viewed Straus's position as a furtherance of New Deal programs. Hostility on this score soon appeared in congressional investigating committees and even House Resolutions to appoint a subcommittee to investigate the "publicity and propaganda of Federal officials." Both Straus and Boke became particular targets of a special House subcommittee to investigate the Bureau of Reclamation in the spring of 1948. The committee's report charged that Straus's purpose in appointing Boke was "to place the vast Central Valley Project in the hands of a propagandist for the Bureau's socialistic policies," namely the pursuance of the 160 acre limitation policies on the CVP.[139]

Labeling Straus and Boke as propagandists and socialists was a political device meant to lessen the commissioner's influence in the public realm and demean New Deal programs. For his detractors, Straus represented gov-

[138] Kathka, "The Bureau of Reclamation in the Truman Administration," 37-43.

[139] United States House of Representatives, House Subcommittee of Expenditures in Executive Departments, *Investigation of the Bureau of Reclamation*, H. Rpt. No. 2458, 80th Cong., 2nd sess., August 7, 1948 (Washington, D.C.: United States Government Printing Office, 1948), 10.

ernment efforts to place limits on the rights of property owners, whether they were large landowners in the Central Valley or private utilities. They viewed Reclamation's publicity programs as attempts to turn public opinion away from their perceptions of American ideals of free enterprise without government limitations. Straus, of course, viewed his mission differently. He saw Reclamation projects as vital components of the economic stability of the West. Speaking in 1947, before the National Reclamation Association, whose members were not friends of land limitations, Straus pleaded for their support in influencing Congress to consistently fund Reclamation construction projects in order to meet the needs of a growing and diversified American West. His promotional activities pursued two avenues: showcase Reclamation's accomplishments and build public support for Reclamation programs. No doubt, politics motivated Straus to defend and uphold the 160 acre rule, but his primary goal was to advance Reclamation. David Kathka writes, "A positive public image of reclamation," Straus knew, "reflected on the Truman Administration and the Democratic Party. His personal political party preference, however, did not dominate Bureau politics."[140]

Congressional investigations into Bureau of Reclamation promotional activities revealed the intensity of the debate over the 160 acre rule. One such effort that caught Congress's attention was a 1947 typescript publication entitled *They Subdued the Desert*. In it the Bureau of Reclamation presented a series of interviews with Reclamation project farmers throughout the seventeen western states who recalled their hardships and successes. The interviews contained in the book were collected by Barrow Lyons, Bureau of Reclamation chief information officer. In the introduction, Commissioner Straus expressed his belief that the interviews represented life experiences of a group of irrigation farmers "objectively recorded." On this score, Straus hoped their comments might "throw light upon phases of social and economic conflict, which will not be resolved for many years."

In general most of the interviewees favored Reclamation's commitment to the small farm, but they did not conceal criticism of the Bureau on other matters. Some testimonies expressed weariness with Reclamation restrictions on their desire to expand acreage and still receive government water. Others, like Robert Franklin Schmeise of Fresno, California, and presi-

---

140 Michael W. Straus, "Take Reclamation off the Merry-Go-Round," *Reclamation Era*, 33 (November 1947): 235-6, 246; see also Kathka, "The Bureau of Reclamation in the Truman Administration: Personnel, Politics, and Policy," 43; Pisani, "The Bureau of Reclamation and the West, 1945-2000," 53.

dent of the Associated Farmers of America, condemned Reclamation's enforcement of the 160 acre rule and its advocacy of public power. He declared that the Bureau was a "propaganda organization" possessing "dictatorial powers" that in effect restricted the pursuit of free enterprise. On the other hand, Joseph Claude Lewis, part owner of a 230 acre farm near Bakersfield, member of the State Democratic Central Committee, and defender of the 160 acre law and public power, applauded Reclamation efforts to stand by its principles and noted:

> Some conservatives in the Central Valley of California call Joseph Claude Lewis a Communist. That is a name frequently applied these days to anyone who advocates Government distribution of low-cost power and enforcement of the 160 acre limitation in the Reclamation laws, which would result eventually in dividing up some of the large California estates.[141]

These telling comments exemplified the complex social, political, and cultural conditions that Reclamation faced in the Central Valley of California during the postwar era.

Reclamation's *They Subdued the Desert* was an attempt to display the hardships and achievements of irrigation farming in the mid-twentieth century. It revealed the intense dichotomies within project communities wherein farmers sought, and to a limited degree achieved, the independence that the 1902 Reclamation Act promised, while at the same time, many interviewees felt constrained by Reclamation policy. Still some respondents appreciated the opportunity to remain in farming made possible through the work of the Bureau of Reclamation. Congressional critics, however, took a staunchly different view of the book and its message. Members of the U.S. House of Representatives investigating Reclamation's promotional activities interpreted statements from *They Subdued the Desert* to argue that the publication was "sheer propaganda." They railed that the book was designed to pit "class against class, liberal against conservative, and inject into the minds of readers ideologies sponsored by some of the planners with the Bureau." And all of this was done, asserted the committee's report, at government expense.[142]

---

141 U.S. Department of Interior, Bureau of Reclamation, *They Subdued the Desert: The Story of Irrigation as told to Barrow Lyons by the men who apply water, till the land and feed their flocks and herds*, Barrow Lyons, editor, typescript, August 1947, iii.

142 House Subcommittee on Publicity and Propaganda of the Committee on Expenditures in Executive Departments, *Investigation of the Bureau of Reclamation*, 80th Cong., 2nd sess., April

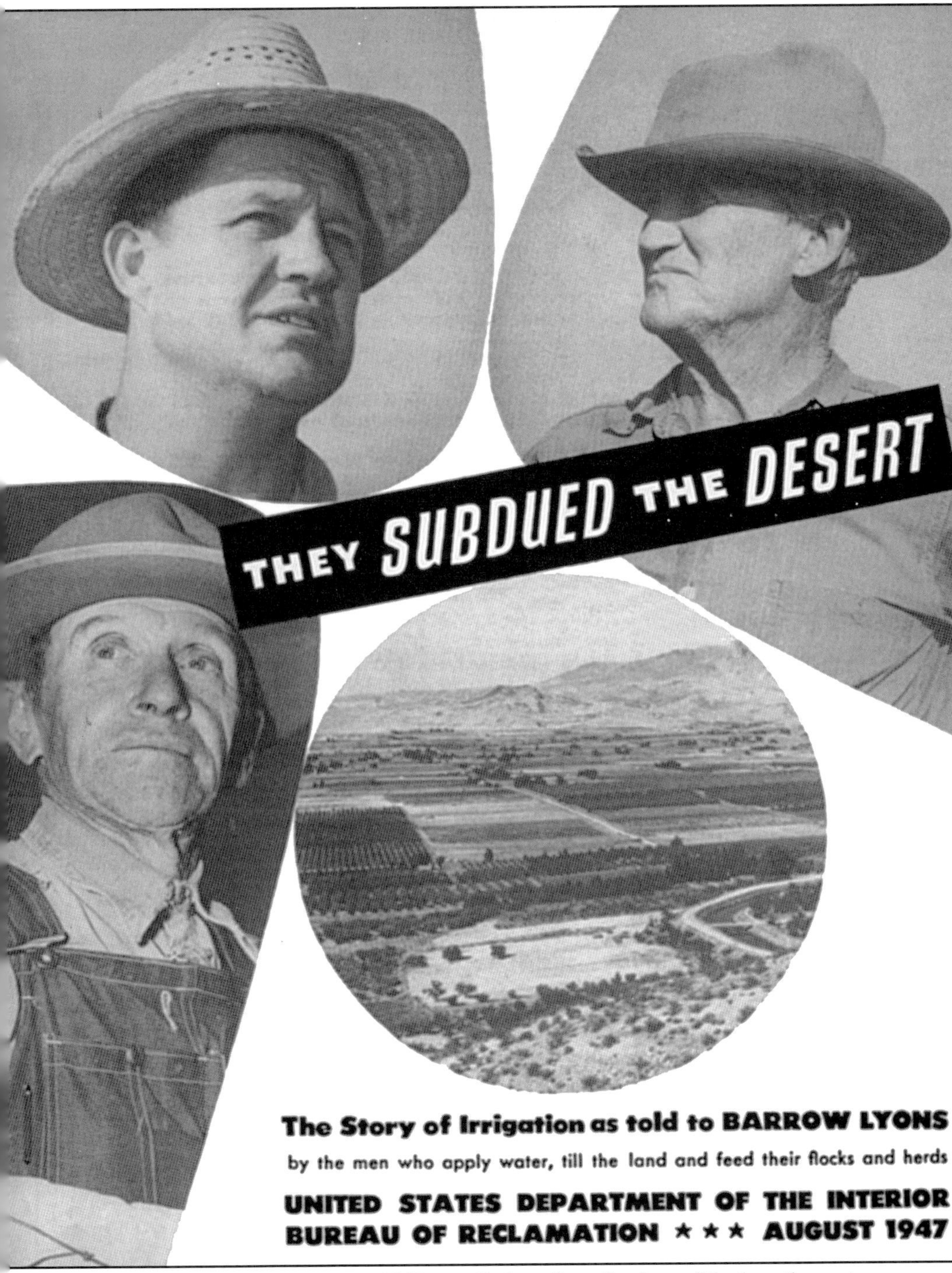

.0.5. The cover of Barrow Lyons' August 1947 compilation of recollections of Reclamation water users.

In addition, the presence of the Corps of Engineers in western water development created manifold complications for the Bureau of Reclamation. In the hopes of staving off the creation of a Missouri Valley Authority, both agencies successfully joined forces to produce a comprehensive plan for water resource development in the Missouri River basin. But this exemplar of inter-agency cooperation did little to diminish competition in the West's other river basins. In light of the debate over the 160 acre rule in California, the Corps offered Central Valley water users another option. First and foremost, water delivered from Corps dams was not subject to Reclamation law. Furthermore, Corps dams were primarily for flood control: a non-reimbursable benefit that lowered construction repayment costs. These two aspects were highly advantageous to water users and exceedingly popular. In his interview for *They Subdued the Desert*, Schmeise favored the Corps of Engineers "chiefly because it might provide irrigation water at a less charge to the users than the Bureau of Reclamation."[143]

These were no shallow threats to Bureau of Reclamation plans for the Central Valley. Reclamation officials understood that many in Congress, under the influence of water interest lobbyists, might look to the Corps of Engineers in future appropriations for water projects that also included water storage for irrigation. In part, Reclamation publicity efforts attempted to forestall such an eventuality. By emphasizing fundamental differences between Corps objectives and those of Reclamation, Bureau publicity activities sought to append Reclamation projects and programs to achievement of higher national aspirations: the family farm and the distribution of government benefits to the greatest number. Reclamation had allies who shared its utilitarian commitment and voiced concerns about the Corps' ambitions in the Central Valley. The 1948 book, by Robert de Roos, *The Thirsty Land: The Story of the Central Valley Project* asserted that the Bureau of Reclamation was "on the right track" in trying to implement the 160 acre limitation. He said, "I am opposed to the Army Corps of Engineers building irrigation and power dams and reservoirs." In this work, de Roos accused the Corps of being on the side of "the interests," and concluded that the Army should keep its hands off the CVP and "go away and shoot their guns." Others who believed that large-scale farming held the key to the future economic survival of agriculture congratulated the historical pattern and tendency of the Bureau of Reclamation to avoid or show flexibility in enforcing the

---

27-30, May 11-14, 17-20, 24-28, June 1-4, 7-9, 14, 15, 29, 1948, 12.

143 Lyons, *They Subdued the Desert*, 101.

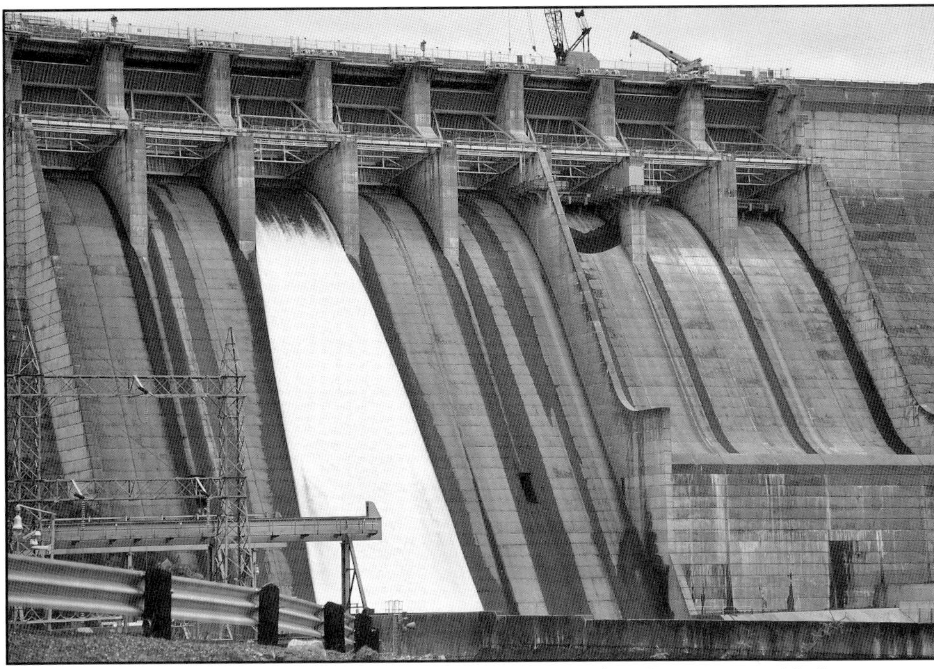

10.6. The U.S. Army Corps of Engineers built Folsom Dam, but Congress specified Reclamation would control water deliveries from the dam.

160 acre limitation and said that it was "a remarkable triumph in public administration."[144]

In 1945 there was some discussion about turning over Corps of Engineer projects in the Central Valley to the Bureau of Reclamation. In a May 1945 memorandum to Secretary of the Interior Harold Ickes, Assistant Secretary Michael Straus advised the secretary to send a formal request to the Bureau of the Budget asking "whether President Truman wanted to follow President Roosevelt's policy … to avoid duplication and conflict." During the heady days at the close of World War II, there appears to have been no effort to comply with Straus's desire to head-off conflict. Indeed the 1944 Flood Control Act gave the Corps a substantial presence in California authorizing the construction of flood control dams on the American River (Folsom Dam), the Kern River, and the Kings River. By the late 1940s the question of how to utilize the water behind these dams became embroiled in the land limitation debate. In August 1949 the dilemma between Reclamation and the Corps achieved some measure of resolution when President Truman announced the

[144] de Roos, *The Thirsty Land*, vii; see also Harry J. Hogan, *Acreage Limitation in the Federal Reclamation Program* (Arlington, Virginia: National Water Commission, 1972), 97.

Folsom Formula determining agency responsibilities. In short, the Folsom Formula stated that "multiple-purpose dams are the responsibility of the Bureau of Reclamation, and dams and other works exclusively for flood control are the responsibility of the Corps of Engineers."[145]

During the late 1940s, however, political pressures against Straus and the Bureau of Reclamation on the 160 acre rule continued to mount. It was becoming increasingly clear that the 160 acre limitation cause had lost much of its luster, and Commissioner Straus moved with the times in spite of his previous commitment. Years earlier in a December 1945 memorandum to Ickes, Straus privately acknowledged that "the 160 acres is an arbitrary figure which got into the law by an historical sequence and has become anachronistic." In 1947 Straus announced that Reclamation would accept what was called "technical compliance." For Straus "technical compliance" calmed the waters by giving the impression that "the family farm law was still intact," which permitted the Bureau of Reclamation to move forward in a variety of directions beyond the issues in the Central Valley. Technical compliance recognized the right of a spouse, and even children, to claim 160 acres each under one proprietor to expand and to comply with the Bureau of Reclamation's increasingly flexible interpretation of the law. In some cases, Reclamation officials stretched the concept of technical compliance to incredible lengths. For example, if land was held by a corporation, members of the corporate board could claim the 160 acres due water under federal reclamation. Straus tacitly recognized these interpretations of the law, including a dubious 1915 interpretation that projects that were paid off became exempt from the 160 acre rule. According to one source, these policies obfuscated "the spirit of the law."[146]

If Straus hoped that the announcement of technical compliance might calm the waters permitting the Bureau of Reclamation to move past the issues in the Central Valley, he was sorely mistaken. In 1948 two separate congressional subcommittees conducted investigations of the Bureau of Reclamation

---

[145] Memorandum for the Secretary, VIA the Undersecretary, May 17, 1945, in RG 48, Entry 779, Box 15; "What Is the Folsom Formula," *Reclamation Era*, 36 (January 1950): 15; "Flood Control Act of 1944," in United States Department of the Interior, Bureau of Reclamation, *Federal Reclamation and Related Laws Annotated*, Volume II of Three Volumes through 1958, Richard K. Pelz, editor (Washington, D.C.: United States Government Printing Office, 1972), 809-10.

[146] Michael W. Straus, Assistant Secretary, Memorandum for the Secretary, December 3, 1945, RG 48, Entry 779, Box 15; see also Lee, "California Water Politics," 404, 414; for information on "technical compliance" see Lee, "California Water Politics," 414; Kathka, "The Bureau of Reclamation in the Truman Administration: Personnel, Politics, and Policy," 65, 69-70.

primarily focusing on its publicity activities. The Subcommittee on Publicity and Propaganda paid particular attention to the competition between the Bureau of Reclamation and the Corps of Engineers, particularly in the developing conflict over who should build Folsom Dam on the American River in central California. Its report charged that Reclamation, through its propaganda, was trying to influence citizens against the Corps, which Congress had authorized to build the dam. This effort "to inflame the citizens of the American River district against the Army engineers," the report declared, was contrary to sound government administration. The problem, concluded the subcommittee, rested with the key administrators of the Bureau of Reclamation in Washington and in the regions whose backgrounds and training were primarily in the field of publicity and public relations. The result, in the view of the subcommittee, was the "selling of the public social theories and ideologies rather than construction of great engineering projects."[147] The congressional message rang loud and clear: the Bureau placed propaganda for social planning and ideology ahead of its primary construction goals. The fault was in its leadership, a leadership that was no longer made up of solid engineering personnel but rather publicity agents.

A Republican-controlled Congress was in no mood to consider such matters and passed legislation that denied salary to a commissioner of Reclamation and any regional directors who lacked engineering qualifications. President Truman reluctantly signed the Interior Appropriation Act in order to forestall shutting down the Department of the Interior but called the actions of Congress "arbitrary" and "diametrically opposed to the principles on which the government is founded." Commissioner Straus remained on the job confident the law was either unconstitutional or would soon be reversed by Congress. A Democratic-controlled Congress restored the salaries of both Straus and Boke in 1949. By 1950 much of the rhetoric over the land limitation rule began to calm down. Though Straus remained a controversial figure, Central Valley water users accepted the idea of technical compliance. The Friant-Kern Canal was put into service in 1949, and two years later, large landowners in Kern County signed water delivery contracts, agreeing to dispose of lands in excess of 160 acres. *Reclamation Era* reported, "The contracts allow companies 10 years to conclude sales." In the meantime, those lands still received water.[148]

---

147 House Subcommittee on Publicity and Propaganda, *Investigation of the Bureau of Reclamation*, 14.

148 Truman's letter to Congress voicing his opposition to the engineering requirement for Reclamation managers in the Interior Appropriation Act of 1949, January 6, 1949, *Reclamation Era*, 35 (March 1949): 49; see also "Senate Refuses to Reinstate Straus," *New York Times*, March

Begun in 1937, initial phases of the Central Valley Project became fully operational in 1951. The delay allowed opponents of public power in California and of the 160 acre limitation on farms in the Central Valley to gain momentum. While business interests cheered the trend in Reclamation's interpretation, the Central Valley Projects Conference deplored it. Composed of the State Grange, organized labor, veterans' organizations, consumer cooperatives, and church groups, the Conference viewed "technical compliance" as little more than a cover for the Bureau of Reclamation's surrender on the issue. And while Commissioner Straus used tough rhetoric in support of the 160 acre rule, the Bureau of Reclamation's actions increasingly identified it with the interests of commercial agriculture, not the family farm, and with the interests of private power, not public power. This alliance became so pronounced by the late 1950s and into the 1960s that "public interest groups," including the rising environmental movement, started to regard the Bureau of Reclamation as the enemy rather than a voice for public power and small farmer democracy in the West.[149]

The movement of the Bureau of Reclamation away from what one author called "the redistributive" policies of the New Deal, which sought to reinvigorate the 160 acre limitation in its application to the CVP, represented a dampening down of those sentiments in the postwar Democratic Party. After all, a Republican Congress had been elected in the fall of 1946, President Roosevelt had died in April of 1945, and the new Truman administration struggled to make its way through the maze of postwar anti-New Dealism fanned by partisan politics. But the lagging enthusiasm for the 160 acre rule and the requirement that farm owners holding over a quarter section of land must divest themselves of excess land in order to be entitled to receive Reclamation water cannot be entirely attributed to the end of the New Deal and party partisans. From its early years on, the Bureau of Reclamation only erratically enforced the 160 acre rule on the projects. Realistically it was an uphill battle from the beginning to enforce the rule in California's Central Valley. Though few individual landholdings in the Central Valley exceeded 160 acres, these

---

23, 1949; Truman Veto Is Seen For Federal Pay Bill," *New York Times*, May 10, 1949; "Truman Regrets, But Signs Pay Bill; Deplores the Rider Withholding Straus and Boke's Salaries—Calls It 'Arbitrary,'" *New York Times*, May 13, 1949; for information on the signing of water delivery contracts see "Kern County Officials Sign Recordable Contracts," *Reclamation Era*, 37 (August 1951): 163; Richard Boke, "Taking Inventory," *Reclamation Era*, 37 (August 1951): 179-80, 83.

[149] Richard S. Kirkendall, "Social Science in the Central Valley of California: An Episode," *California Historical Quarterly* 43:3 (September 1964): 195-218; Lee, "California Water Politics," 405, 410-11; David Carle, *Water and the California Dream: Choices for the New Millennium* (San Francisco, California: Sierra Club Books, 2000), 151-2.

10.7. The U.S. Army Corps of Engineers' Lake Wallula behind McNary Dam, Lock, and Powerhouse on the Columbia River. Courtesy of the History Office, U.S. Army Corps of Engineers.

10.8. The McNary Dam Powerplant. Courtesy of the History Office, U.S. Army Corps of Engineers.

large landowners and, more importantly, corporate farms still expected water from the CVP, and the smaller farmers needed their large landholding neighbors to help pay construction costs. Enforcement would have made Reclamation law an instrument of land redistribution in the Central Valley rather than a law designed simply to prevent government subsidized water from aiding and abetting the growth of concentrated land monopolies and ownership in the West. In the end, however, there was only a fine line of difference. On new projects the law served its intended purpose as a preventative measure to quell speculation and land monopoly, though some charged that it stifled ambition and enterprise. But on projects that enhanced water supplies to well-established agricultural enterprises, strict enforcement of the 160 acre rule continued, for many, to appear as confiscatory.[150]

Hells Canyon and Public Power

Even though a Columbia River Basin Authority, similar to the TVA, remained a political possibility, both the Bureau of Reclamation and the Army Corps of Engineers chose to move on with their own plans for additional dams on the Columbia River and its major tributary, the Snake River. In 1947 the Corps began building McNary Dam on the lower Columbia, and in the following year Reclamation presented its plan for development of the Snake River with a high dam in Hells Canyon as the centerpiece. Since 1944 Reclamation tried to head off creation of TVA-like river basin authorities on major rivers, and part of that effort went to the newly empowered regional offices. Planning responsibility for the Columbia River basin fell to Regional Director R. J. Newell, in Boise, who in May 1948 presented a report to Commissioner Straus titled "Development of the Hell's Canyon Project, Idaho-Oregon." Coincidently, Newell transmitted the report just one month before the Vanport Flood devastated the lower Columbia River area. That flood destroyed a city of 19,000 and killed fifty-two people. The report recommended a major construction project with a dam approximately 607 feet above river level in Hells Canyon on the Snake River, 198 miles upstream from Lewiston, Idaho, on the Oregon-Idaho state line. Newell's report estimated power production from the dam at 900,000 kilowatts, creating a 93-mile long reservoir covering 24,800 acres.

Newell also implied that "prospective developments on the lower Snake River" might impede consideration for construction, and the report

---

[150] Clayton R. Koppes, "Public Water, Private Land: Origins of the Acreage Limitation Controversy, 1933-1953," *Pacific Historical Review* 47:4 (November 1978): 607-8.

called for quick action for the immediate production of electricity. It was a simple argument maintaining that unless Reclamation began constructing Hells Canyon Dam soon other entities, both private and public, would take the initiative, making the Bureau's proposal irrelevant. As the Corps's construction of McNary Dam indicated, the Pacific Northwest presented a wide-open market for electricity to supply an exploding urban and industrial market.[151] Moreover a decision in favor of a large dam on the upper Snake River, Newell contended, offered protection for salmon by delaying dams on the lower Snake providing "a longer opportunity for solution of migratory fish problems which construction on the lower river dams will create." The reference to fish protection underlines growing concerns in the Pacific Northwest over the impact of dams on fish populations which slowly curbed much of the enthusiasm for dams. Later in the century, the question resolved itself to one of "fish versus dams" in the popular press. In the late 1940s, however, dam builders, both federal and private, expressing confidence that technical and hatchery solutions were the best means to address the fish problem, chose dams over fish with the support of most agencies created for fish protection. These included the Washington and Oregon Fish Commissions as well as the U.S. Fish and Wildlife Service.[152]

In addition to providing for continued growth of the urban centers of Puget Sound and Portland, the massive power production from Hells Canyon Dam held out possibilities for new developments in the Inland Empire of the Pacific Northwest. The report noted that the upper Snake River Basin housed the greatest phosphate resources in the United States, including an estimated five billion tons in Idaho alone. A growing market for phosphate fertilizers, to replace natural phosphates used up by crops, resulted in projections of a prosperous future for a western phosphate industry. It was estimated that it would take two billion kilowatt-hours of power to develop and maintain the industry. On the subject of irrigation, the report acknowledged that drawing water directly from the Hells Canyon Reservoir to irrigate nearby lands was impossible. More promising and the most ambitious was the prospect of diverting waters from a new reservoir on the Payette River in central Idaho to

---

151 United States Department of the Interior, Bureau of Reclamation, *The Columbia River: Supplemental Reports on the Bitterroot Valley, Cambridge Bench, Canby, Council, Crooked River, The Dalles (West Unit), Hells Canyon, Mountain Home (Payette Unit), Upper Start Valley, and Vale (Bully Creek Extension) Projects*, Vol. 2, H. Doc. 473, 81st Cong., 2nd sess., (Washington, D.C.: U.S. Department of Interior, 1950); Roger Johnson, "City of 19,000 Wiped Out in Dike Break," *Salt Lake Tribune*, May 31, 1948.

152 USDOI, BR, *The Columbia River*, 117-9; see also Keith C. Peterson, *River of Life, Channel of Death* (Corvallis: Oregon State University Press, 1995), 115.

another reservoir on the Boise River to the south via a Payette-Boise Aqueduct. Another aqueduct would then transport water nearly one hundred miles farther south to the Mountain Home Desert Project, creating twenty-five hundred new farms on one hundred thousand irrigated acres. Covering much of the costs of this scheme, the report proposed, were revenues from the dam's power production. This idea of supporting irrigation projects with power revenues was of, course, nothing new. Beginning with the Boulder Canyon Project Act in 1928 and later institutionalized in the 1939 Reclamation Projects Act, both Congress and Reclamation had already admitted that most irrigation projects required power revenues in order to repay construction costs.[153]

In 1949, with an eye toward this ambitious irrigation undertaking in Idaho, the Bureau of Reclamation struck an agreement with the Corps of Engineers, known as the Newell-Weaver Accord. It included a free hand to build a high dam in Hells Canyon in return for the Corps' control of the lower Snake and Columbia rivers. The pact allowed Reclamation to take control of the upper Snake Basin along the Oregon-Idaho state line and everything from there toward the Continental Divide as part of an expanded irrigation frontier in Idaho. While the two dam building arms of government (Reclamation and the Corps) attempted to split their chores on these two rivers of the Pacific Northwest, there continued to loom in the background the possibility of an overarching Columbia Valley Authority. That possibility of a new river basin authority may have spurred both Reclamation and the Corps to greater cooperation.[154]

Their concerns were needless. With the political winds blowing strongly in favor of Republicans in the region's state houses, local interests and private power groups gained the upper hand against anything resembling New Deal style river authorities or federal direction of resource policies in land, water, minerals, or grazing issues. This, in turn, led to revival of arguments in favor of private power development so long eclipsed in the region by the imposing success of the Bureau of Reclamation's giant Grand Coulee Dam and the numerous municipally owned utilities in the Pacific Northwest. Nevertheless, the public versus private power debate continued to rage throughout the region. Public power advocates still championed the social and economic

---

[153] USDOI, BR, *The Columbia River*; 141-2 see also Jim Lichatowich, *Salmon without Rivers: A History of the Pacific Salmon Crisis* (Washington, D.C.: Island Press, 1999), 185; Karl Boyd Brooks, *Public Power, Private Dams: The Hells Canyon High Dam Controversy* (Seattle: University of Washington Press, 2006), 50.

[154] Brooks, *Public Power, Private Dams,* 133-4.

Investigating HELLS CANYON

by
RAYMOND G. LEONARD, Information Specialist
Region 1, Boise, Idaho

THIS TREACHEROUS GORGE is one of the last remaining great damsites in the United States. Today, the vacuumlike gap, deeper than the Grand Canyon, is a menace to mankind, but in spite of the hazards, Bureau of Reclamation engineers calmly take topography for a proposed project. *Photo bv Phil Merritt, Region 1 photographer.*

WHEN CONGRESS APPROPRIATES MONEY FOR "INVESTIGA-ɪONS," the word itself may sound very prosaic and common-ɫace. The actual work, however, is often far from that. 'ake the Hells Canyon investigations, for example.

If the early pioneers who came West and saw the spectacu-ɑr Snake River Gorge were on earth today and heard about ɪe Bureau of Reclamation's plan to build Hells Canyon)am, no doubt they would sadly shake their heads and ıutter, "Fools rush in where wise men fear to tread."

For the deepest chasm in America gave them a rough time. t thwarted an attempt by Lewis and Clark to pass through :s turbulent and churning water. The John Jacob Astor xpedition lost three men, and a fourth went mad trying) ride the rough Snake through this gorge. The rest of ıe party were at death's door before they managed to tagger out of the abyss.

No wonder even our hardy forefathers, upon seeing this layground of Satan for the first time, decided it was a ne-way ticket to the hereafter instead of a passage to the reen valleys of Oregon. They avoided it like a plague.

The Bureau of Reclamation, however, in spite of the past ecord of the canyon, is investigating the possibilities of uilding a dam that is worthy of all the Hollywood-type djectives imaginable. The height of this concrete giant would be a lofty 742 feet, or the equivalent of a 70-story building. It would be situated in one of the few remaining primitive areas in the United States—a region of breath-taking ruggedness.

Each abutment would touch the base of part of a mountain range in a different State—the Seven Devils of Idaho and the Wallowas in Oregon. It is one of the greatest remaining dam sites on the North American continent. To top it all off, the undertaking would rank as one of the greatest engineering feats since the Bureau entered the dam-building business in 1902.

The huge development would make available 900,000 kilowatts of hydroelectric energy to a power-hungry Pacific Northwest, where industrial expansion is stifled by lack of generating capacity. Besides power, the dam would provide flood control and aid navigation of the lower Columbia and Snake Rivers. It would also provide financial assistance to future irrigation developments in the region.

Last summer, the Bureau sent a seven-man survey party into the Hells Canyon area to survey possible routes for access roads, and to make a reconnaissance study necessary for the preparation of a planning report in connection with legislation to authorize construction of the project.

At present, the only way into this region of primitive

30 THE RECLAMATION ERA

9. A three page article in the December 1950 *Reclamation Era* demonstrated Reclamation's interest in the ject.

beauty and desolation, unsullied by bill boards, hamburger stands, and filling stations, is over a so-called one-way road to Eagle Bar on the Idaho side of the canyon, sixteen miles downstream from Homestead, Oreg. This road, strewn with boulders, basalt fragments and full of chuck holes, is a rugged route to travel even for the most experienced driver, and few cars escape damage.

Building access roads down precipitous grades and along the sheer canyon sides would be the first big job the Bureau would have to tackle before starting actual construction on the dam.

The Hells Canyon surveyors suffered all the discomforts not found at home as they worked along the slopes or on the rim of the canyon, thousands of feet above the river. Mother Nature keeps the gorge mighty warm during the summer months, the average peak temperature being 102 degrees. Some days it is not uncommon for the mercury to soar to a blistering 117°.

Transportation on the job was by horseback or foot. Both men and animals found climbing along the rocky and narrow ledges on the canyon wall a gruelling ordeal. The horses developed ugly, running saddle sores, because saddles constantly shift back and forth as the animals make their way over the rocks without trails.

Mountain goats had nothing on the surveyors when it came to scaling the sides of the canyon. In places the rim of the chasm is several thousand feet above the river. A fall is certain death. When working along these precipitous slopes, the men often raised and lowered their instruments by rope.

Not only is surveying in Hells Canyon dangerous to life and limb, but it is equally severe on clothing and shoes. Most footwear must be replaced every 10 days.

Lurking rattlesnakes were an ever-present danger. Nineteen have been killed to date and the crews saw dozens of others. Then there was always the poisoned oak which is unusually prevalent throughout the area.

During off-hours the surveyors fished for sturgeon in the Snake River. The largest catch was a 200-pounder. On re hot days the men took to the river to keep cool. The part slept in sleeping bags and cooked its own meals. All suppli were brought in by horse or boat.

The first casualty has already been chalked up to Hel Canyon. On July 31, Photographer Phil Merritt of th Bureau's Regional office in Boise was pinned to the canyo wall by a one-half yard rock while on a photo assignmen Merritt would have been the first fatality had he been alon but fortunately the surveyors were on the scene to lift th rock off him. Luckily, there were no bones broken. Ph escaped with a few bruised muscles in his right knee.

If built, Hells Canyon Dam would resemble Hoover Dar A concrete arch-gravity type structure is contemplated, wit a maximum base thickness of 674 feet, tapering to 45 feet a

NO WORK FOR A SISSY. At right, Ted Sather lowers himself down the precipitous canyon wall by rope, as Dean Ellis, with survey instruments, looks on calmly. Below, a slip for these two men, at the axis of the proposed dam, would mean a plunge of several thousand feet to the floor of the canyon. Below, at left, Lynn Brown and Dean Ellis try to figure out a good place for a road. Scaling and descending the heights, plus the excruciating heat, really make this a man's job. *Photo directly below by Phil Merritt, others by Stanley Rasmussen, both Region 1 photographers.*

ON TOP OF THE WORLD! At extreme left, Regional Director H. T. Nelson, Ivan Tuescher, survey party chief, and Francis Hart, planning engineer, ponder possible locations for access roads.

AT THE BOTTOM OF HELLS CANYON surveyors make their weary way by boat. *Photo at left by Stanley Rasmussen, at right by Phil Merritt, both Region 1 photographers.*

the crest. The roadway across the top would be 1,740 feet long and the elevation of the crest would be 2,082 feet above sea level. The structure, third largest in the world, would contain 6,200,000 cubic yards of concrete.

Construction would take 6 to 7 years. At the peak of activity 5,000 to 6,000 men would be employed.

The multiple-purpose project would cost the Government approximately $333,000,000. This entire amount, including interest, would be paid back to the Federal Treasury through the sale of power.

The dam would also save thousands of dollars in flood damages to the lower Snake River and Columbia River areas by impounding spring run-off in its 4,400,000 acre-feet reservoir. Furthermore, the structure would also be a boon to navigation. It would increase channel depths in the lower Snake and lower Columbia Rivers during the low flood periods.

Irrigation would not be one of the benefits from the dam, but revenues would help make economically feasible future irrigation developments in the Upper Snake Basin and other parts of the Pacific Northwest which otherwise could not be constructed.

Situated as it is in one of the most isolated spots in the country, which never before has been available to the general public, the dam should prove to be a tourist attraction capable of luring half a million visitors a year to the canyon rim, with resultant benefits to business in the neighboring Idaho and Oregon towns and in the Pacific Northwest.

Congress has not yet authorized the project, but if and when it does, many similar investigations by reclamation engineers will provide the necessary data to help pave the way for the Bureau of Reclamation to add still another masterpiece to its long list of spectacular engineering achievements.

THE END

POUNDS OF COMPENSATION—100 pounds, to be exact, in this sturgeon caught by Bob Bond. For recreation at day's end, Bureau engineers swim in the Snake River or fish for wild denizens of the deep like this one. LUCKY FIND was a pioneer miner's cabin (below) handy to water—a "must" for the roving surveyors. SURE-FOOTED STEEDS are essential in Hells Canyon. At right, Ivan Tuescher talks over the day's work with Lynn Brown. *Photos by Stanley Rasmussen, Region 1 photographer.*

growth cheap electricity provided, while private power interests derided the inefficiency, extravagance, and waste of public facilities. These sentiments influenced later generations of environmentalists in reference to Hells Canyon. The prospect of converting the central Snake River into a vast elongated lake to produce power and extravagant irrigation projects for the sake of growth might seem the height of hubris and extravagant folly.[155]

As arguments grew fierce over Hells Canyon, a looming power shortage in the region exacerbated the controversy. Pro-dam forces embraced a growth vision for the region and an economy of abundance in which the Bureau of Reclamation emerged as a chief enabler. When Reclamation first undertook "public investment" on a large scale during construction of Hoover Dam on the Colorado River, the Bureau became an instrument of growth far larger than its original emphasis on irrigation. With that project, Reclamation graduated to multipurpose growth projects that provided an infrastructure suitable for the coming wartime economy and postwar development of the West. Setting aside agency competition on the Columbia River watershed, the big dam in Hells Canyon was a doorway to opportunity for boosters of Pacific Northwest regional development. Just as in the case of the Colorado River development, the Bureau of Reclamation stood ready to make the opportunity occur. Through its agreement with the Corps, Reclamation hoped to sweep away upstream competition from private and public utilities for multipurpose development of the Snake.[156] It was a broadly conceived plan designed to continue the New Deal's initiatives in dam construction and provide cheap power for development of this large river valley or as one source noted, the "federalization of Hells Canyon."

Federalization in postwar America faced daunting obstacles. In Idaho the advantages of large projects under a federal paternalism reminiscent of the New Deal proved a hard sell among already established irrigation communities. They feared that a new project threatened their traditional water supplies and would develop additional competition from a new farming population in the state. Idaho Power Company and other corporate utilities played upon these fears. They argued that costly large federal projects threatened local control over resource use that brought questionable multiple-use benefits in

---

155 Brooks' *Public Power, Private Dams* is an excellent analysis that reflects the values of late twentieth-century environmentalists who celebrate the failure of Reclamation to build the high Hells Canyon Dam and divert water to Idaho for new irrigation.

156 Robert M. Collins, *More: The Politics of Economic Growth in Postwar America* (New York: Oxford University Press, 2000), 8.

terms of navigation, flood control, and hydroelectric production. Private power interests portrayed smaller dams as more cost effective and maintained that they posed less danger to the region's valuable salmon industry. The latter appeal attracted the support of fish and wildlife organizations, many of which started petition drives against more dams on more rivers. In the process, Hells Canyon, in the words of historian Karl Brooks, became "a national controversy" in the politics of the 1950s that "symbolized deep postwar political differences over electricity's ownership and water's social purposes."[157]

As the Bureau of Reclamation planned for new projects in the 1950s, these "political differences" intensified. The struggle over the high dam in the Snake River Canyon in some ways portended the Bureau's path into the future. Its grand plans for Hells Canyon faced growing political strength of private power forces recovering from their defeats in the New Deal Era, fear among local communities of greater federal control over water resources in Idaho, and budding environmental concerns over the fate of salmon in the Pacific Northwest. Nationally the Republican presidential and congressional victory in the elections of 1952 brought together a combination of forces opposed to the expansion of federal influence in the Columbia and Snake river basins.

Since and during the Roosevelt/Truman administrations, standard Republican Party critiques of the ruling Democratic Party majority (i.e., the New Deal) usually began with charges of fiscal irresponsibility and "over spending." The rhetoric, and the concerns behind it, did not change in the postwar era when the Truman administration and even Congress became committed to policies of economic growth. A Democratic Congress passed the Employment Act of 1946 that committed the government "to promote maximum employment, production, and purchasing power," lest the economy fall back into prewar depression. According to the Keynesian economic model of the New Deal, economic growth might require periodic infusion of government stimulus money into the economy. Under these policies, Reclamation stood to benefit, as it had benefitted from Depression-era public works projects that translated into big dams, hydroelectric development, and other assorted water projects.

A partial reversal of this view of the relation of government to the economy began as Republican strategists planned to retake the White House

---

157 Brooks, *Public Power, Private Dams*, 47-51,134, 118.

10.10. Secretary of the Interior Douglas McKay. January 21, 1953, to April 15, 1956.

in the 1952 election. Their strategies had a direct bearing upon the fate of the Hells Canyon high dam and the expansionist plans of the Bureau of Reclamation for the Snake River basin and irrigation development in southern Idaho. After the Republican National Convention in July, candidate Dwight D. Eisenhower launched his campaign before a crowd at the state capitol building in Boise, Idaho. The choice of this location for the kick-off of the campaign was careful and deliberate. Although the electoral votes of the four northwestern states were not significant in the larger picture, all four—Washington, Oregon, Idaho, and Montana—had Republican governors and there was the already-brewing opposition to the high dam project in Hells Canyon. These conditions provided an excellent backdrop for the Republican candidate's positions which denounced the expansion of federal authority, praised local control over resources, and emphasized small federal government coupled with fiscal responsibility. None of these boded well for Reclamation's plans in the Snake River basin. What followed was a Republican victory in the fall 1952 presidential election and control of the House of Representatives. The election marked the beginning of what has been termed a "Republican Interlude" in the 1950s during Eisenhower's two-term presidency.[158]

In 1953 Eisenhower's appointment of Oregon governor Douglas McKay as secretary of the interior to replace Oscar Chapman, soon followed by the resignation of Reclamation Commissioner Straus, marked the end of lingering New Deal influences in the Department of the Interior. Certainly McKay's presence as secretary of the interior signaled the application of his pro-business policies as governor of Oregon and his friendship with private power interests. The new commissioner of Reclamation, Wilbur A. Dexheimer, did not harbor the activist administrative style of Straus who championed the dynamic role Reclamation had played in the expansion of public dams and public power during the New Deal. Reading the political tea leaves, the

[158] *Ibid.*, 47-51, 42.

10.11. Commissioner Wilbur A. Dexheimer served from 1953 to 1959.

Federal Power Commission eventually looked with favor upon Idaho Power's application for construction of a series of three small dams in Hells Canyon with only mild and barely audible objections from Reclamation under this new administration.[159]

Finally, in 1956, the Federal Power Commission approved the first of those small dams. The Bureau of Reclamation and the champions of public power in the Pacific Northwest resigned themselves to defeat in this long struggle. It had been a battle pitting the well entrenched forces of public power in the region against the renewed energies of private power interests allied with those who criticized the expansion of federal power and the federalization of water resources and hydroelectric power. The outcome represented the resurgence of Republican strength in the region in this decade of Eisenhower Republicanism. In reference to the development of a broad scale Columbia Valley Authority, that possibility faded in the final year of the Truman administration. One historian of public power concluded: "As twenty years of Democratic rule ended, it seemed that complete Federal domination

[159] "Hell's Canyon Dam Stirs Bitter Fight," *New York Times*, July 6, 1953.

of the electric power system of any single area would not spread beyond the Tennessee Valley."[160]

Columbia Basin Project: The Irrigation Phase

The failure of the Bureau of Reclamation's comprehensive plan for the Snake River had little bearing on the forward momentum of the Columbia Basin Project. While World War II stalled the irrigation phase of the Columbia Basin Project for almost a decade, the delay provided the opportunity for extensive studies, probably more so than on any other project. Initially, New Deal planners of the late-1930s viewed the project as a refuge for down and out Dust Bowl farmers who might prosper on small homesteads (80 acres for a husband and wife). But when war industries sprang up in West Coast shipyards and airplane factories, migrants from the interior of the country turned away from the land to seek those jobs. At a loss to find a new pool of likely settlers, Under Secretary of the Interior William Warne nevertheless cited figures indicating that thousands of returning veterans looked to opportunities to develop farms on new irrigation projects.[161]

Warne's pronouncement echoed the optimism many in the Bureau of Reclamation had concerning the Columbia Basin Project. Since the inception of Grand Coulee Dam, Reclamation's philosophy was: "If you build it, they will come." In the summer of 1946 *Reclamation Era* asserted that the Columbia Basin Project would have ready settlement opportunities for 5,000 families by 1950-1951. It noted that the project was one of the world's largest single irrigation enterprises that would provide jobs for veterans and, of course, irrigated farms, "to enable the Pacific Northwest's agricultural expansion to keep pace with its industrial development and greatly increased population." Later that year, a *Reclamation Era* article made the point that industrial development was greatly dependent upon power supplies to cities, and power revenues were also necessary to repay the high costs of construction on irrigation projects. The two developments complemented one another, and both were vitally necessary for the continued growth of the region. All these claims recognized a renewed emphasis and vigor on the part of Reclamation to continue with its mission of developing the waters of the West as the new postwar era began.[162]

---

160 John Richard Waltrip, *Public Power During the Truman Administration* (New York: Arno Press, 1979), 129.

161 Cannon, "Creating a 'New Frontier' Opportunity," 11.

162 "Columbia Basin Program Speeded," *Reclamation Era*, 32 (July 1946): 148; Frank A. Banks, Supervising Engineer, Columbia Basin Project, "New Empire in the Northwest,"

10.12. When completed this receiving basin delivered water pumped from behind Grand Coulee Dam to the feeder canal which empties into Banks Lake. January 27, 1947.

10.13. Pouring the cut-off wall of the South Dam of Banks Lake in the Grand Coulee. April 24, 1947.

10.14. Cut-off wall along the axis of the South Dam, after removal of concrete forms. May 20, 1947.

Easily the most ambitious of the Bureau's postwar projects, the estimated cost to irrigate almost 1.1 million acres on the Columbia Basin Project in 1940 was $487,030,228. Begun in the midst of the Depression with detractors asserting that its hydroelectric output far exceeded the present and even the future needs of the Pacific Northwest, the first priority was completion of Grand Coulee Dam with development of its grand irrigation goals put on hold until after the war. Yet even the projected development of new irrigation lands with thousands of new farms did not escape criticism. The perennial questions arose: Why the need for more farms when excessive agricultural production drove prices down? Did America need new farms at a time when cities offered greater opportunities as demonstrated by the dominant demographic trends during the first half of the twentieth century? Nevertheless, the war highlighted the crucial role that American agriculture played in providing for the nation's security. Congress recognized the importance of a healthy farm

---

Reclamation Era, 32 (September 1946): 203-4; the Bureau of Reclamation suspended publication of *Reclamation Era* from April 1942 to May 1946 due to World War II labor and material shortages.

economy when it gave draft exemptions to young farmers whose work served to provide food and fiber for the war effort. And while the war effort could probably have succeeded without the power from Grand Coulee Dam, products from American farms were probably more critical. Still, Grand Coulee's electricity guaranteed that both civilian and war needs were met in the Pacific Northwest. The power generated at Grand Coulee allowed incredible growth in the region's industrial and urban centers that had far-reaching effects. Electricity produced at Grand Coulee meant that the civilian population could work in war industries and, when they left the factories, avoid any inconveniences that power shortages might cause. The availability of abundant power reinforced a prevailing belief that Grand Coulee Dam was instrumental both to winning the war and the success of the Manhattan Project.[163]

Originally the planners of the Columbia Basin Project saw it in far different terms than the pattern along which it developed in the postwar years. In 1939 Commissioner John C. Page engaged Harlan Barrows, a planner in the Tennessee Valley Authority, to devise plans for the project's future. Historian Paul C. Pitzer relates that Page saw the project as "a blank sheet of paper where the government could create a project as nearly perfect as planning could devise." Under Barrows's direction at least 300 people worked on a series of reports titled the *Columbia Basin Joint Investigations*. Published between 1943 and 1945, plans and recommendations from the investigations reflected a rural agricultural world far different from the one emerging in eastern Washington during the war.[164]

By 1943 Congress, paying little attention to the ongoing planning efforts, replaced the Anti-Speculation Act of 1937 for the project with the Columbia Basin Project Act. The 1943 Act authorized the Bureau of Reclamation to buy private land to add to the project, affirmed the utilization of power revenues to subsidize irrigation, and identified the 160 acre land unit as the ideal size of farms but allowed for retention of additional acreage for landowners present before 1937. The vagueness of these provisions, as on other

---

163 Pitzer, *Grand Coulee*, 249; see also James Shepherd, "The Benefits and Costs of the Columbia Basin Project: Earlier Perspectives and Changing Perceptions," *Agricultural History* 76:2 (2002): 476-7; Gina Bloodworth and James White, "The Columbia Basin Project: Seventy-Five Years Later," in *Yearbook of the Association of Pacific Coast Geographers*, Vol. 70, 2008, 97.

164 United States Bureau of Reclamation, *Columbia Basin Joint Investigations; Character and Scope United States Department of the Interior, Harold L. Ickes, Secretary, Bureau of Reclamation, John C. Page, Commissioner* (Washington, D.C.: United States Government Printing Office, 1941); see also Pitzer, *Grand Coulee*, 271; "Joint Investigation, Columbia Basin Project," *Reclamation Era*, 30 (August 1940): 219.

projects, opened the door for a continuing trend toward larger farm units. In addition, farmers could withdraw their land from the project under condition that the land would receive no water from the project. This was significant for the east side of the project. The high price of wheat in the postwar period enabled farmers to keep land out of the project and farm it under dry land conditions without any need to sell the excess lands. Moreover, some of the land had been farmed since the 1880s without irrigation and now there was little incentive for the farmers to go into the irrigation project where increased debt was required simply to develop and farm 160 acres. By 1946 farmers on the east side withdrew over 300,000 acres—almost a third of the project. Publicly, Reclamation planners remained calm, saying that it eventually planned to bring most of the land under irrigation. Their actions, however, belied their optimism and Bureau planners delayed construction of eastside canals. In reality Bureau of Reclamation officials were "traumatized" by the withdrawals which imposed a "piecemeal construction" pace on the project that would now extend development over many years.

Local farming wisdom and experience saw greater efficiency and staple production on large farm units using machinery and fewer farm laborers. Farmers, working successfully under these conditions, resisted attempts to impose a landholding system based upon small proprietorships. Events also dictated that the Columbia Basin Project would not be the home of Dust Bowl refugees, contrary to the vision of earlier New Deal planners. Other possible sources of settlers included returning veterans who saw an opportunity in farming and the lurking possibility that a depressed industrial economy would force families back to the land. Indeed the ominous possibility of a postwar depression rivaling that of the 1930s haunted many Americans. In that event, construction of the Columbia Basin Project would stand as a much needed and gigantic public works project to shore up the Pacific Northwest that stood to suffer severely in a postwar economic contraction and depression. Also, when the War Department withdrew land for construction of an air base at Moses Lake and land for the atomic facilities at Hanford, it introduced an urban population that altered the economy, society, and character of the Columbia Basin, making the Bureau's ambitious plans for settlement of the Columbia Basin Project somewhat obsolete. As historian Pitzer writes of the early studies of the project, "The Joint Investigations were largely outdated, inappropriate, and flawed before the government even printed them."[165]

[165] Pitzer, *Grand Coulee*, 274, 272-3.

As a result of the war, economic and demographic shifts in the Pacific Northwest changed Reclamation's views on project settlement. In the shuffle to move along the irrigation phase of the project, earlier thoughts of installing poor veterans returning from the war or impoverished refugees from worn out farms on the Great Plains disappeared. In 1948 Reclamation announced that a family must have $7,500 in cash to be eligible for a farm unit and access to over $20,000 credit to develop it. Already-present-landowners with experience in eastern Washington farming seemed more eager than anyone else to step forward and farm on the project. Congressional legislation helped spur interest. The Columbia Basin Project also fell under the new terms of the 1939 Reclamation Project Act that offered up to fifty years of repayment time and consideration of land quality that took into account the economic value of crops that the land could reasonably produce. As noted earlier, the 1943 Columbia Basin Project Act increased the acreage to the traditional 160 acre parcel with wives, children, and relatives each able to claim 160 acres of land entitled to water. Ultimately, the result was larger land units operated under one farm manager or owner which reflected the mechanization occurring in agriculture under wartime labor shortages; not to say anything of increased sympathy in Congress to demands for larger landholdings on other Reclamation projects. Larger capital outlays necessary for emerging mechanization meant a demand for economies of scale even on American irrigated farms. The originally slated 40 to 80 acre land units reverted back to the traditional 160 acre homesteads, with the longstanding practice that married couples could double that amount to say nothing of other relatives combining land with them—all of which would essentially be under one operation. The result, as many sources point out, was a far different farm population on the project than the one Reclamation planners envisioned during the Depression.[166]

Most startlingly, of the project area's two million acres, only one million could be irrigated. Surveys divided the lands into six classes. One large class should receive no water, and the remaining classifications indicated the lands that might benefit from varying degrees of water service according to the types of crops deemed fit for the soil conditions. "The surveys," according to historian Pitzer, "were the most comprehensive preliminary investigations undertaken by the government on any reclamation project." Within this vast area at the beginning of the project, banks owned 85 percent of the land with

---

[166] Pitzer, *Grand Coulee*, 176-87, 280; Donald J. Pisani, "Federal Reclamation and the American West in the Twentieth Century," *Agricultural History* 77:3 (Summer 2003): 401-2; "Prevention of Land Speculation, Columbia Basin Project," in USDOI, BR, *Federal Reclamation and Related Laws Annotated*, Volume I, 556-7.

the federal government holding only 5 percent; the state of Washington and the Northern Pacific Railroad each controlled 5 percent of the land. The immediate area around the project was sparsely populated, containing only 10,000 people, and most of them lived in the southern area near Pasco, Washington. Unlike other Reclamation projects, Congress authorized the Bureau of Reclamation to purchase the great bulk of the lands for the project from private land holders to ease the ever-present issue of land speculation.[167]

However, before government land acquisitions, a wave of speculation swept the intended project lands prompting the 1937 Anti-Speculation Act. It called for an appraisal of the lands at their value before irrigation with the requirement that owners sell their excess lands (over 80 acres in the case of a husband and wife) at only the appraised price or below. While curbing runaway speculation, the provision was instrumental in convincing some landowners to avoid joining the project and selling their lands to the government, especially when the price of dry-land-cultivation wheat rose to a profitable point. The law also required formation of an irrigation district to assume the responsibilities of establishing a repayment schedule to cover the government's cost for the project. After a good deal of local controversy, Bureau of Reclamation officials accepted formation of three irrigation districts on the project, generally positioned east, west, and south on the project.[168]

The Bureau of Reclamation demonstrated remarkable commitment to carrying out the Columbia Basin Project irrigation phase as well as flexibility as it adjusted its visions of the project to the new realities of postwar agricultural economics. In 1946 Reclamation hurriedly dammed the "ancient Grand Coulee" at both ends to form Banks Lake. This became the equalizing reservoir filled with water from Franklin D. Roosevelt Lake pumped by the power generated by Grand Coulee Dam's turbines. On May 15, 1948, while awaiting completion of the CBP's major works, Reclamation put the Pasco Pumping Unit into operation by pumping water directly from the Columbia River for delivery to eighty farms in the Pasco area. As water rose between Dry Falls Dam on the south and North Dam on the north, the filling of Banks Lake in the Grand Coulee began. A main line canal led south to Billy Clapp Lake initializing this elaborate irrigation scheme which would eventually alter the environment of about 600,000 acres of central Washington dry lands. The completion of canals and secondary reservoirs resulted in the official opening

---

167 Pitzer, *Grand Coulee*, 268.

168 Pitzer, *Grand Coulee*, 269; Shepherd, "The Benefits and Costs of the Columbia Basin Project," 478.

10.15. Site of the Farm in a Day, viewed from the east, on May 28, 1952.

10.16. Clearing sagebrush on the Farm in a Day. May 29, 1952

of the irrigation phase of the project in 1951. On June 14, 1951, Secretary of the Interior Oscar L. Chapman sent a signal from the nation's capital that started the first pump sending water into the canal feeding into Banks Lake. In an event dubbed the Water-of-All-States Ceremony, young ladies from throughout the state of Washington lined the canal and poured in water from all the states and territories of the union to represent the project's opening as a national accomplishment.[169]

Washington State celebrated completion of the irrigation works in the following year. According to an enthusiastic article in *Reclamation Era*, the state announced that the Columbia Basin Project's irrigation works would eventually serve an estimated 87,000 farms. From May 22 to June 1, 1952, communities throughout the basin hosted an eleven-day celebration featuring the opening of canals carrying water pumped from the gigantic reservoir behind Grand Coulee Dam. Each community conducted its own festivities, which included the Cavalcade at Soap Lake, Pioneer Day at Quincy, at Ephrata a Little World's Fair, and an Aqua-Rama at the Tri-Cities. Commissioner Straus began the celebration by personally opening the valve that began the flow of water. To top off festivities, a widely publicized volunteer work day went forward on May 29 to construct "A Farm in a Day" for a new owner chosen by the Veterans of Foreign Wars. The major construction centered on building a home designed by the American Institute of Architects in association with the Extension Service of Washington State College in Pullman. A sketch of the home and the layout of its interior rooms appeared in *Reclamation Era* along with the information that plans and specifications would soon be available upon request from the Extension Service and the Farmers' Home Administration. Contributed materials and labor for the buildings cost $19,000 and the total value of the farm home and land was estimated at $50,000.[170]

Beginning in the early hours of May 29, volunteers arrived to begin framing a house upon an already-poured cement foundation. By early light others began leveling, plowing, and planting fields—some in the quick cash crops of potatoes and beans and a larger number of acres in alfalfa. At the end of the day some 80 acres stood ready to receive water. Also standing ready to receive the new homestead was a young 30-year-old farmer, Donald D. Dunn, his wife, and two daughters. Mr. Dunn was a World War II veteran who had served in a tank battalion in France. In addition, he was a Kansas farmer who recently lost his farm to floods in 1951 near Marion, Kansas, and ended up as a farm equipment salesman in Yakima. Dunn's farm experience, his service

---

169 Pitzer, *Grand Coulee*, 277-9; Water and Power Resources Service, *Project Data*, 383; Paul C. Pitzer, "A 'Farm-in-a-Day': The Publicity Stunt and Celebrations That Initiated the Columbia Basin Project," *Pacific Northwest Quarterly* 82:1 (January 1991): 3-4.

170 The entire issue of *Reclamation Era*, 38 (April 1952) was dedicated to development of the Columbia Basin Project; see also "Design for Modern Farm Living," *Reclamation Era*, 38 (May 1952): 103, 112.

10.17. Leveling equipment on the Farm in a Day at dawn on Thursday, May 29, 1952.

10.18. Work on leveling the Farm in a Day for irrigation. May 29, 1952.

in the war, and his recent move to Washington made him a prime candidate to win the contest.

Stories and pictures of the Dunn family taking possession of the farm appeared in newspapers coast to coast with *Reclamation Era* quoting Dunn when he took possession of his new irrigated farm, "It's a far cry from hoeing sunflowers in Kansas." The item captured human interest and dramatized the story of the Columbia Basin Project as serving all of America. Dunn himself appeared before the House Interior and Insular Affairs Committee in Washington, D.C., to receive a Concurrent Resolution from Congress congratulating Mr. and Mrs. Dunn "on the record of heroism and fortitude in the face of misfortune which has won for them this first family to receive irrigation water from the pumps at Grand Coulee Dam on the Columbia River." The National Reclamation Association presented Dunn with a gold-plated irrigation shovel in honor of Reclamation's Golden Jubilee dating from the June 1902 Reclamation Act.

While a great success in both its execution and its publicity, the outcome of the Dunn family's attempt "to make a go of it" on the project was troubling. After three years Dunn sold out and took his family to Colorado complaining that the small land unit of 80 acres was uneconomical. Others attributed his exit to poor management on his own land and on the lands he rented to expand his acreage. Beyond the disadvantages of small unit agriculture, Columbia Basin farmers increasingly faced a cost-price squeeze as the wartime and postwar demand for agricultural products fell. Dunn was caught in this trap although receiving the boost of a free farm and many of the initial tools and buildings required to develop it.[171]

By 1967 as landowners reduced their holdings to conform to Bureau of Reclamation acreage limitations, there were 5,463 irrigated farms on the Columbia Basin Project. The number fell far short of the 1946 projection of between ten and fifteen thousand farms for the project. In 1968 the Bureau of Reclamation turned over operations to the three irrigation districts with the "tacit acknowledgement" that further expansion of the project would not occur. While some may claim disappointment at this modest achievement in the number of irrigated farms compared to the ambitious numbers projected at the beginning, historians generally applaud the abridgement of the project because

---

[171] "New Life Pumped to Columbia Basin," *New York Times*, May 30, 1952; Pitzer, *Grand Coulee*, 295-6; Hugh H. Moncrieff, "Columbia Basin's Showplace," *Reclamation Era*, 38 (August 1952): 214-7, 225.

10.19. Carpenters work on the home on the Farm in a Day. May 29, 1952.

10.20. Donald D. Dunn, the recipient of the Farm in a Day near Moses Lake on the Columbia Basin Project, at the turnout which symbolically delivered the first water on the Columbia Basin Project.

to continue an emphasis upon the creation of small farms would have been unfortunate. Of any effort to force small holdings upon the vast acreages on the Columbia Basin Project, historian Pitzer observes, "It would be a collection of family farms ranging from forty to eighty acres, none of them capable of supplying their owners with a satisfactory living. The area would be a rural slum. It is for the best that this aspect of the project failed."[172]

The Upper Colorado River Storage Project: To Be or Not to Be

While the Columbia Basin Project moved forward, Congress committed Reclamation to development of the upper Colorado River basin. Prior to World War II Hoover and Parker dams and the All-American Canal were approved for lower basin development. Now it was the upper basin's turn. The 1922 Colorado River Compact made it clear that the states of the upper Colorado River basin had a share in the river's water and power development. In 1946 the Bureau's Region IV office in Salt Lake City played the lead role in putting forth Reclamation's plan for a comprehensive Colorado River Storage Project in the upper basin. The report, subtitled *The Colorado River: "A Natural Menace Becomes a Natural Resource,"* captured the Bureau's dam-building ethos that rivers left in their natural state were both destructive and wasteful. Reclamation saw the Colorado River as a leading example of a river in its natural state continuing to be "a natural menace," at the time a commonly held viewpoint in the West. The report described the river's history:

> Yesterday the Colorado River was a natural menace … unharnessed it tore through deserted flooded fields, and ravaged villages. It drained the water from the mountains and plains, rushed it through sun-baked thirsty lands, and dumped it into the Pacific Ocean—a treasure lost forever.[173]

States on the headwaters of the Colorado River stood eager for the Bureau of Reclamation to move ahead with its plans for the upper basin. During the war, the industrial base of the upper Colorado River states grew, and that growth made it abundantly clear that future expansion required more water for cities and power for industry. Many saw development of the water

---

[172] Brian Q. Cannon, *Reopening the Frontier: Homesteading in the Modern West* (Lawrence: University Press of Kansas, 2009), 32; Pitzer, *Grand Coulee*, 367; Pisani, "Federal Reclamation and the American West in the Twentieth Century," 401-2.

[173] USDOI, BR, *The Colorado River*, 25; see also Harvey, *A Symbol of Wilderness*, 42-3.

resources on the upper Colorado River as a key to growth. Utah with its deep canyon lands, which were ideal for reservoirs, went to the top of the list for dam construction. Although the war laid the foundations for manufacturing growth in these states, many pro-dam supporters saw water and power revenues as the gateway to an even more prosperous future. As railroads served to bring population and establish communities during the nineteenth century, dam advocates believed dams and hydroelectricity would finally bring a permanent basis for economic growth to the Intermountain West.[174]

The Bureau of Reclamation's projections for development of Colorado River resources sparked both enthusiasm and concerns. Its report outlined 134 potential projects: "100 in the upper basin and 34 in the lower basin." It estimated 1,533,960 acres of new lands brought into cultivation, along with 38 hydroelectric powerplants providing 9.2 billion kilowatt-hours annually. Costs for all the projects were as impressive as the report's scope, because Bureau planners estimated that construction costs would exceed $2 billion in 1944 dollars. At a benefit/cost ratio of 1.3:1 the benefits to the nation would create "a stable agriculture and abundant low-cost power [to] provide a sound basis for industrial and commercial expansion and thus are important to the region and the Nation but are not readily susceptible of dollar evaluation."[175]

Yet for all the possibilities the report highlighted, it also spoke of limitations. Bureau of Reclamation planners and engineers forthrightly explained that it was impossible to implement all the projects the report identified. They looked at this text as a list of possibilities that Congress and all the Colorado River basin states must deliberate upon and prioritize. Most critically, there was just not enough water in the Colorado River to construct all of the projects outlined. Difficult decisions lay ahead. The upper basin states of Colorado, Utah, New Mexico, and Wyoming had yet to sign a subordinate compact to divide their share of the river. In the lower basin, California and Arizona bitterly argued the terms of their subordinate compact and how much water each state was to receive. Concern about California was not Arizona's alone, all of the Colorado River basin states kept a careful eye on California: the only state that had actually developed its compact share of the Colorado River and was always looking for more.[176]

---

[174] Harvey, *A Symbol of Wilderness*, 40.

[175] USDOI, BR, *The Colorado River*, 14-9.

[176] USDOI, BR, *The Colorado River*, 5; for insights into the wariness of the Colorado River Basin States toward California see Daniel C. McCool, "Politics, Water, and Utah," in Daniel C.

The 1922 Colorado River Compact negotiations relied on a limited history of stream gauging which resulted in overestimation of the average annual flow. That error threatened each state's expectations for water resource development. This historic agreement divided the waters of the Colorado between the upper and lower basins, allotting each basin roughly 7.5 million acre-feet per year. Each sub-basin was to then agree on the amount of water their respective states received. By 1946 only the lower basin states of California, Arizona, and Nevada had water allotments assigned by the Boulder Canyon Project Act, though Arizona and California remained in a bitter conflict over interpretation of the allotments. To make matters worse, the water treaty with Mexico committed 1,500,000 acre-feet of water to Mexico thus further limiting water available for development.[177]

The Mexican Water Treaty and Protocol, signed in 1944, was the culmination of diplomatic efforts that went back to the turn of the twentieth century and included agreements on water allocations for the Colorado River, the Rio Grande, and the Tijuana River. Most representatives of the Colorado River basin states supported Congress's and the State Department's efforts to reach an accord with Mexico. They foresaw this day coming because the Colorado River Compact recognized Mexico's right to water allocation from the Colorado River. The general sense among basin states was to establish a limit for Mexico before it could expand its already impressive 200,000 acres of developed agricultural lands near Mexicali—an expansion that would entitle it to a larger share of the river. The urgency of the situation was clear to the upper basin states if they intended to develop their shares of the river's water resources. Only California launched a determined effort to defeat the treaty, but largely failed because of lack of support from the other basin states.[178]

The report, *The Colorado River: "A Natural Menace Becomes a Natural Resource"* released to the public in 1946, created more widespread

---

McCool, editor, *Water of Zion: The Politics of Water in Utah* (Salt Lake City: University of Utah Press, 1995), 2.

[177] For information on the 1922 Colorado River Compact see Norris Hundley Jr., *Water and the West: The Colorado River Compact and the Politics of Water in the West* (Berkeley: University of California Press, 1975); Worster, *Rivers of Empire*, 209-10.

[178] The story of the 1944 Mexico/United States Water Treaty has been expertly described in Norris Hundley, Jr., *Dividing the Waters: A Century of Controversy Between the United States and Mexico* (Berkeley and Los Angeles: University of California Press, 1966); for a detailed description of California's efforts to block the treaty see Norris Hundley, Jr., "The Politics of Water and Geography: California and the Mexican-American Treaty of 1944," *Pacific Historical Review* 36:2 (May 1967): 209-26.

concerns over the Water Treaty with Mexico. In February 1946, at the second annual conference of the Colorado River Water Users Association, members expressed outrage over what one representative from Arizona called "an act of piracy." Sydney Kartus, president of the Highline Reclamation Association in Arizona, was especially vehement in his denunciation of the treaty, claiming that it "would destroy the State of Arizona and irreparably injure all the Basin States within the Colorado River System." Joining Kartus was Chauncey Sandberg, Mayor of Hurricane, Utah, who maintained that "the Mexican Treaty betrays the best interest of the water users in the Colorado River Basin." Despite the bitterness expressed by some members of the association, they reluctantly recognized the futility of any effort to revoke the treaty. Instead, the association voted to accept a resolution calling for completion of works on the lower Colorado River to protect American interests before "construction of the proposed Mexican diversion dam."[179] Underlying Kartus's warnings about Mexico lay the long-term struggle of the state of Arizona to receive what it perceived as its fair share of the Colorado River, especially in regards to California's claims. Like other basin states, Arizona equally feared the strength of California and its burgeoning population. Moreover, Arizona historically had resisted the entire Colorado River Compact, because of its disagreement with California over interpretation of the provisions of the Compact. Unable to win concessions through various maneuvers, Arizona finally ratified the Compact in February 1944 and moved quickly to secure its rights to the waters of the Colorado.[180]

The Gila River was of particular concern to Arizonans. They believed their state had exclusive rights to the Gila River and that it should not be counted as part of the allocation of the Colorado River. Indeed, at the 1946 annual meeting of the Colorado River Water Users Association, Kartus went so far as to claim that the upper basin had no business with the Gila River or how Arizona developed its resources, asserting "With the exception of a minor area in New Mexico, the upper basin states have no more interest in the Gila River than they have in the Atlantic Seaboard." He defended Arizona's rights, and decried upper basin proposals to take transfer water out of the basin. He called the plans "unnatural" and contrary to the interests of the basin and Arizona. Kartus's declarations represented the determination of Arizona to receive and maximize its fair share of the Colorado River.[181]

---

179 *Proceedings, Second Annual Conference, Colorado River Water Users Association*, Salt Lake City, Utah, February 11, 12, and 13, 1946, 56, 104, 171.

180 Hundley, *Dividing the Waters*, 135.

181 *Proceedings, Second Annual Conference, Colorado River Water Users Association*, 82, 87.

By 1947 Arizona's congressional delegation, with avid support from the state's water users, introduced legislation for the Central Arizona Project. The project called for a large dam at Bridge Canyon just below Grand Canyon National Park on the mainstem of the Colorado River. Earlier proposals contained plans for a 78-mile long tunnel to bring water to central Arizona, but the cost of the tunnel proved prohibitive. By 1949, legislation for the proposed Central Arizona Project kept the dam at Bridge Canyon and included a series of aqueducts and a pumping plant at Lake Havasu to bring water to central Arizona. Critics of the project argued that the only reason Bridge Canyon Dam remained in the proposal was to produce hydroelectric power to help pay construction costs. Proponents claimed that the dam was necessary to provide electricity to the pumping station at Lake Havasu. The estimated cost of the project was $750,000,000.[182]

Arizona's venerable Senator Carl Hayden led the fight for the Central Arizona Project and successfully guided various bills in the Senate through sessions from 1947 to 1951. The entire project, however, ran into the determined effort of California's large congressional delegation in the House of Representatives. According to historian Charles Coate, California's success stemmed from its argument "that Arizona did not have sufficient water rights to run the project."[183] The two states finally took their dispute over how to divide the water of the Colorado River to the courts. This continuous fighting temporarily halted any development of the lower Colorado River by the Bureau of Reclamation.

Amid the interstate struggle, the Bureau of Reclamation found itself in a state of limbo regarding the Central Arizona Project. With many studies on the feasibility of the project stacked on office shelves, Reclamation waited for congressional approval to begin work. The delay allowed Reclamation and the states of the upper Colorado River basin time to explore various suggestions for resource development. The Bureau's 1946 *The Colorado River: "A Natural Menace Becomes a Natural Resource"* report was especially expansive in proposing a great number of projects in the upper basin. It identified up to 100 projects throughout the states in the upper basin that proposed to bring 1,230,810 acres of new lands into cultivation. Along with the added growth in agricultural

---

182 United States Senate, Committee on Interior and Insular Affairs, *Central Arizona Project and Colorado River Rights,* Hearings on S. 75 and S. J. Res. 4, 81st Cong., 1st sess. (Washington, D.C.: U.S. Government Printing Office, 1949), 48-51, 55.

183 Charles Coate, "The Biggest Water Fight in American History: Stewart Udall and the Central Arizona Project," *Journal of the Southwest* 37:1 (Spring 1995): 82; Worster, *Rivers of Empire*, 275.

production, the report identified sites for construction of 29 hydroelectric plants capable of producing 9.2 billion kilowatt-hours annually. Estimated cost for full development of the upper basin was $930,142,000.[184]

Before any work could begin in the upper basin, the states of Colorado, New Mexico, Utah, and New Mexico had to reach an agreement on allocating the upper Colorado River basin's apportionment under the Colorado River Compact. In October 1948 the upper basin states signed the Upper Colorado River Basin Compact which allotted 51.75 percent of the water to Colorado, 11.25 percent to New Mexico, 23 percent to Utah, and 14 percent to Wyoming.[185] Similar to many areas of the American West, World War II transformed the upper Colorado River region by promoting urbanization and industrialization. The Bureau of Reclamation's report saw the potential for hydroelectric power to broaden the area's economic foundation.

Signing the Upper Colorado River Basin Compact allowed each basin state to actively work on and lobby for projects that had been on the back burner for years. In particular, the compact permitted Utah leaders to move ahead with plans for the Central Utah Project. For years Utah's political and business leaders looked for avenues to bring more water into the growing communities and industries on the front range of the Wasatch Mountains. Their idea consisted of taking water from the rivers on the southern slope of the Uinta Mountains, transporting it to Bonneville Basin via a series of gravity-fed aqueducts and tunnels. In addition, Utahans also saw a series of hydroelectric powerplants along the aqueduct to produce much needed electricity for emerging industrial centers. By the time the upper basin states signed the Upper Colorado River Basin Compact, the idea for the Central Utah Project "was already in place."[186]

---

[184] USDOI, BR, *The Colorado River*, 14-5; see also Russell Martin, *A Story that Stands Like a Dam: Glen Canyon and the Struggle for the Soul of the West* (Salt Lake City: University of Utah Press, 1999), 48; Harvey, *A Symbol of Wilderness*, 42-4; Jared Farmer, *Glen Canyon Dammed: Inventing Lake Powell and the Canyon Country* (Tucson: University of Arizona Press, 1999), 134.

[185] Upper Colorado River Compact," in United States Department of the Interior, Bureau of Reclamation, *Federal Reclamation and Related Laws Annotated*, Volume II, Richard K. Pelz, editor (Washington, D.C.: United States Government Printing Office, 1972), 909-23; "Upper Colorado Compact Ok'd," *Reclamation Era*, 36 (June 1949): 140; see also Martin, *A Story that Stands Like a Dam*, 49; Harvey, *A Symbol of Wilderness*, 69; Farmer, *Glen Canyon Dammed*, 133.

[186] United States Senate, Committee on Interior and Insular Affairs, *Colorado River Storage Project: Hearings before the Subcommittee on Irrigation and Reclamation,* June 28, 29, 30, July 1, 2, 3, 1954, 83rd Cong., 2nd sess. (Washington, D.C.: United States Government Printing Office,

10.21. Flaming Gorge Dam, one of the features of the Colorado River Storage Project.

Other upper basin states also harbored similar ambitions for water resource development and looked to the Bureau of Reclamation for guidance. As noted earlier, the Bureau's 1946 report, *The Colorado River: "A Natural Menace Becomes a Natural Resource,"* was largely a proposal packet; not formulated plans ready for presentation to Congress. By 1950 the Bureau of Reclamation submitted a comprehensive plan to Congress for development of the waters of the Colorado River entitled the "Colorado River Storage Project" (CRSP). Reclamation's proposal reflected its interest in large-scale projects that included an entire river basin. It was a mammoth undertaking. The Colorado River Storage Project proposed construction of ten storage dams throughout the upper basin, with the two most important being the Echo Park Dam on the Green River and Glen Canyon Dam on the mainstem of the Colorado River. Echo Park and Glen Canyon dams were important for two reasons: the first was storage to ensure water deliveries to the lower basin states as part of fulfilling the upper basin's 1922 Compact obligation; the second was production of hydroelectricity to help pay for the irrigation phase

1954), 101-5; see also Kurt Vedder, "Water Development in Salt Lake Valley, 1847-1985," in McCool, *Waters of Zion*, 34.

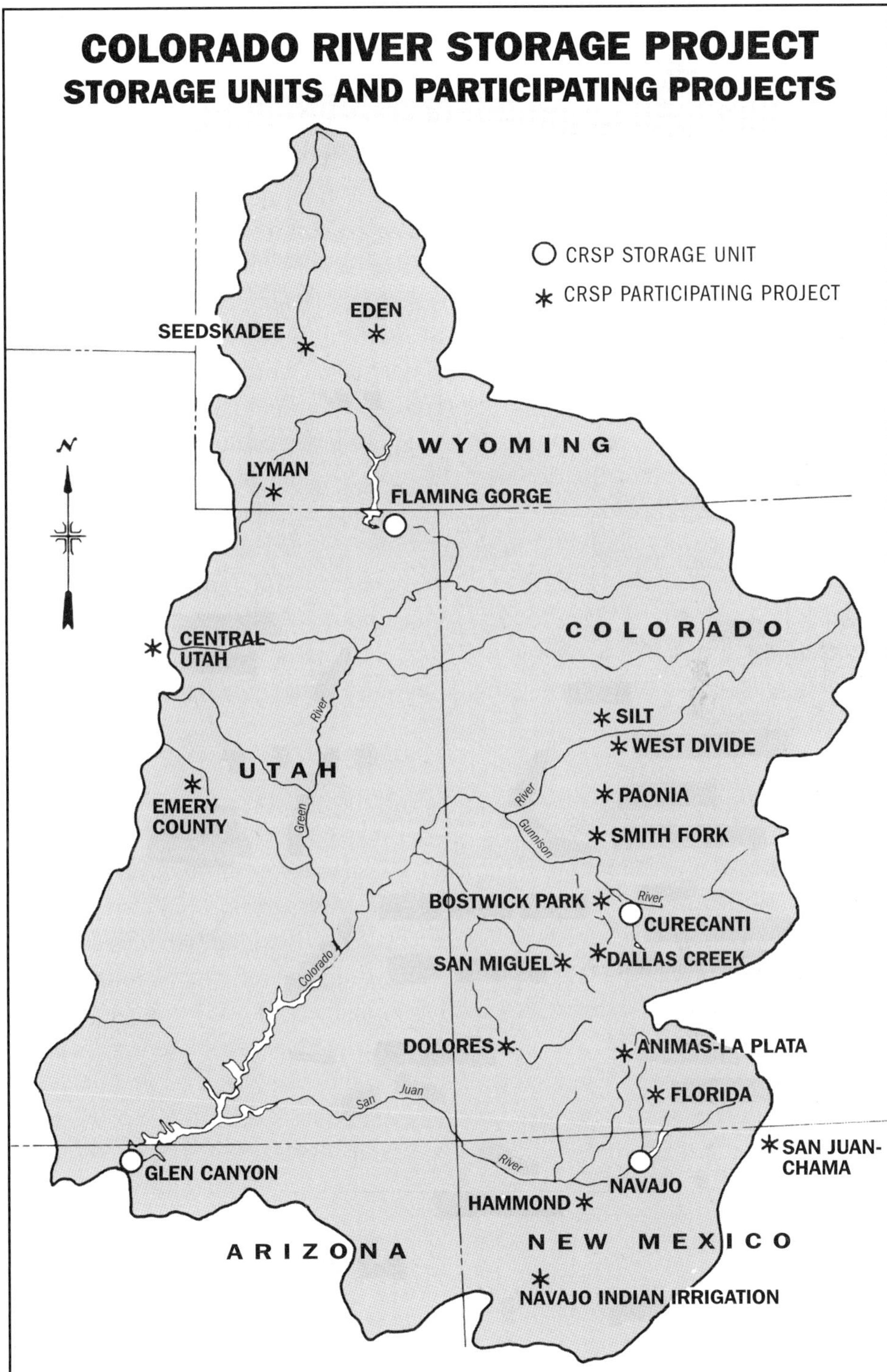

10.22. Congress authorized Reclamation's Colorado River Storage Project in 1965.

of the project and become a source of revenue for the upper basin's future development.[187]

Other dams proposed in what became known as "participating projects" included: Kendall Reservoir in Wyoming, Flaming Gorge and Split Mountain in Utah, Cross Mountain, De Beque, Whitewater, Crystal, Curecanti, in Colorado, and Navajo in New Mexico. These dams along with the ones in Echo Park and Glen Canyon would produce 1,622,000 "megawatts worth of power," and "store 48.5 million acre-feet of water." Along with a large expansion in hydroelectric power coming from CRSP, the Bureau of Reclamation also reported that almost 300,000 acres of new lands would be opened to irrigation. These various undertakings represented a long-term investment in the future of the upper basin including the initial phase of the Central Utah Project and also involved the Bureau in the construction of irrigation works on Indian reservation lands. The cost of this impressive project was (for its day) an astounding $1.5 billion. In its depth and brashness, the proposal was a testimony to the optimism about the future of water development in the West. Despite all the proposed benefits and regional political support, the Colorado River Storage Project faced skeptics in Congress, motivated not only by the frightening costs, but also a new generation of critics who feared destruction of the West's scenic beauty.[188]

The Changing Critique

For over five decades, the Bureau of Reclamation survived withering economic critiques. Critics targeted project cost overruns and farm failures, despite zero interest loans for the cost of delivering water, combined with incessant requests for extension of loan periods, all of which required Reclamation to develop a constant stream of explanations and justifications. Defenders of Reclamation inevitably reverted back to the larger social benefits conferred upon the nation by the development of projects that made homes for small farmers possible, created new wealth, and provided markets for industrial goods. Included in this new wealth, Reclamation hastened to

[187] "Echo Park—Split Mountain Dams Approved," *Reclamation Era*, (August 1950): 148; Farmer, *Glen Canyon Dammed*, 134-5; Harvey, *A Symbol of Wilderness*, 134-5.

[188] Susan Rhodes Neel, "Newton Drury and the Echo Park Controversy," *Forest and Conservation History* 38:2 (April 1994): 57; Martin, *A Story that Stands Like a Dam*, 54-6; Committee on Interior and Insular Affairs, *Colorado River Storage Project*, 76; "Storage Project Held Key to Water Pact," *Salt Lake City Tribune*, February 22, 1950; "Bureau Urges $1.1 Billion Basin Project," *Salt Lake City Tribune*, October 29, 1950.

point out, was hydroelectricity production that not only paid for cost overruns but acted as a local and regional economic multiplier. With these arguments, Commissioner Mead artfully defended Reclamation in its battle for survival in the 1920s. In the next decade, the Great Depression salvaged Reclamation by presenting it with opportunities to build great public works to improve the infrastructure of a faltering western economy.

The decade of the 1950s marked a turning point for critics of the Bureau of Reclamation. Some harkened back to controversial issues of protecting National Park Service units, the classic example having been the Yosemite National Park Hetch Hetchy crisis at the beginning of the twentieth century. Other criticisms even questioned Reclamation's engineering expertise and science. These concerns coalesced around Reclamation's proposed dams in Dinosaur National Monument on the upper Colorado River system. While Reclamation was not involved in the earlier crisis, the confrontation between the city of San Francisco and the defenders of Yosemite pitted preservationists against utilitarian conservationists. Though San Francisco was determined to build a dam in Yosemite National Park's Hetch Hetchy Valley, John Muir and other defenders of the natural aesthetic beauty of the park fought back against what they considered the philistinism of utilitarian commercialism that threatened nature's handicraft in the park. The defeat of the preservationists imposed a long silence that yielded to the prevailing arguments of the utilitarian conservation movement that rivers flowing freely to the sea represented wasted water power and water resources for farms and cities. Now in the post-World War II period, arguments flared anew as the value, vitality, and excitement of free flowing rivers and the inspiring presence of beautiful canyons in a uniquely American wilderness became more revered and valued by the public. Amidst the newfound interest in the esthetics of wild and untamed nature in an increasingly urbanizing America, opponents criticized Bureau of Reclamation plans to flood a portion of Dinosaur National Monument for aesthetic reasons and because of flaws in its scientific water studies. Reclamation's announced plans to push ahead with damming the upper Colorado River basin revived old arguments about dams in national parks and raised new ones about the science of reservoirs as well as fundamental questions about the wisdom of continued dam building in the American West.

Critics alerted the American public to a number of questions aimed at the core mission of the Bureau of Reclamation. Should dam-created reservoirs inundate places of beauty and great natural phenomena? Does the public's desire to preserve these places take precedence over utilitarian goals of water storage, irrigation, and power production that primarily served only local areas? While the utilitarian arguments of western members of Congress like Senator Arthur Watkins of Utah and the leadership of the Bureau of Reclamation still generally carried the day, the vastness of the American West with all its sites of grandeur and curiosity held a special appeal to the American public. Postwar prosperity, increased leisure time, good roads, and automobile ownership created a constituency reluctant to see the destruction of prized domains of natural beauty. Suddenly the old preservationist argument took on greater cogency against the utilitarian conservationists. The Bureau of Reclamation and its supporters faced new challenges that questioned their motives and their ideas of progress. The result was a consortium of conservationists-preservationists groups aligning to protect natural park and monument lands from dam construction.

In 1915 President Woodrow Wilson by executive order established Dinosaur National Monument. The purpose in setting aside 80 acres in a remote corner of northeast Utah was to protect a fabulous collection of well-preserved dinosaur bones and fossils before they fell into the hands of private collectors. In 1916 the newly created National Park Service assumed management of Dinosaur National Monument. Until the 1930s, the monument had few visitors and remained practically unknown to the general public. By the late 1930s, newspaper reports on the scenic wonders found in the nearby canyons of the Green and Yampa rivers piqued interest in Dinosaur National Monument. John Wesley Powell's earlier *History of the Exploration of the Canyons of the Colorado* vividly described scenic wonders in the canyons of the Colorado River and its tributaries and assigned imaginative names such as Steamboat Rock, Canyon of Lodore, and Echo Park. Renewed fascination with these handiworks of nature prompted a drive, spearheaded by the National Park Service with the full support of Secretary of the Interior Harold Ickes, to preserve the Green and Yampa river canyons. In July 1938 President Franklin D. Roosevelt signed a proclamation vastly expanding the size of Dinosaur National Monument to over 200,000 acres. According to historian Mark Harvey, "Literally overnight, a tiny national monument in northeast Utah became one of the biggest areas in the national park system, with more than 360 square miles."[189]

The canyons, especially the waters of the Green and Yampa rivers, also drew the attention of other federal agencies. Both the Bureau of Reclamation

---

[189] Harvey, *A Symbol of Wilderness*, 6-8, 9-14; see also Martin, *A Story that Stands Like a Dam*, 45.

and the Federal Power Commission recognized the potential in these sites for development of hydroelectric power. While they raised no objections to the enlargement of Dinosaur National Monument, both agencies protected their rights to any future power development in the monument. The 1938 Presidential Proclamation expanding the monument contained provisions protecting future power development rights of the Bureau of Reclamation and the Federal Power Commission. In the 1930s, of course, the power demands of the upper basin states of Utah, Colorado, New Mexico, and Wyoming were not pressing. Planning for development proceeded at the same time as World War II prompted growth in the region.

In the planning stages, the National Park Service (NPS) had no objections to the future possibility of dams within Dinosaur National Monument. In fact, the Bureau of Reclamation and the Park Service cooperated in investigating potential water development of the Green and Yampa rivers. Moreover, the Park Service was already cooperating with Reclamation managing recreation facilities at a number of Reclamation reservoirs best represented by the Boulder Dam National Recreation Area. The presumption prevailed on the part of Reclamation officials that a willing partnership existed between the Bureau and the NPS. Indications of trouble came in June 1943 when the Bureau "filed a 'reclamation withdrawal' with the Department of the Interior to protect the Echo Park damsite and other damsites within the monument." Reclamation's failure to inform the Park Service of its intentions to file a withdrawal was met by dismay in the NPS. Much more disconcerting to the Park Service, however, was the discovery that Reclamation personnel had conducted surveys and field studies within the monument without Park Service knowledge or permission.[190]

As a result of the Bureau of Reclamation's forays into Dinosaur National Monument, the National Park Service became the first organization to oppose the Echo Park damsite. Part of this opposition stemmed from Reclamation's failure to keep the Park Service informed of its activities and plans. In general, however, Park Service officials, under the leadership of Director Newton Drury, sought to fulfill its assigned mission to protect the sanctity of national parks and monuments and keep them in their natural and pristine state. Initially, the disagreement between the two agencies remained out of the public spotlight; confined to the halls and conference rooms of the Department of the Interior. The first public hint of the Park Service's opposi-

---

[190] Harvey, *A Symbol of Wilderness*, 27-9, 30-5.

tion to Reclamation's plans for the Colorado River appeared in the Bureau of Reclamation's 1946 report *The Colorado River: "A Natural Menace Becomes a Natural Resource."* As a "cooperating interest," the Park Service submitted a review of Reclamation's Colorado River development plan. The Park Service report emphasized the unique geological and scenic qualities of these relatively unknown river canyons. Although not unprecedented, it expressed concerns about dams within national monument lands:

> The policy of the National Park Service … has been and is to make the protection of the natural and archeological values of the area the controlling factor in administering it. The question of whether this policy is to be changed to permit development for water-control would require for its solution a review of all probable advantages and disadvantages by authorities superior to either the National Park Service or the Bureau of Reclamation. Before changes in the canyon unit are authorized … it should be shown that it would be in the greater national interest to develop the area for such use than to retain it in its natural state … for the enjoyment of them by the nation.

The statement was cautious and measured but nonetheless highlighted an issue between the two bureaus in the Department of the Interior that refused to go away.[191]

Between 1946 and 1951, an open and often highly public break developed between these two Department of the Interior bureaus. Park Service Director Drury and Reclamation Commissioner Straus came to personify the differences. Both men grounded their arguments in ideologies of Progressive Era conservationism, which is to say that it played out as an example of the conflicted views Americans often brought to resource issues. Commissioner Straus, backed by congressional, business, and civic leadership from the upper basin states, saw the Echo Park Dam not only as a crucial feature for economic and social growth, but also representing a fulfillment of allocations to the states of the waters of the Colorado River under the 1922 Colorado River Compact. Director Drury found support from the Wilderness Society, National Park Association, and the Izaak Walton League all of which sought to defend against congressional approval of the development plans of the Bureau

191 USDOI, BR, *The Colorado River*, 244-5.

of Reclamation. Eventually, Secretary Oscar Chapman pushed aside the arguments of the preservationists and backed the Bureau of Reclamation's plans for the Colorado River forcing Drury to resign as director in April 1951.[192]

Drury's dismissal from the Park Service alarmed conservation groups around the country about the dangers facing not only Dinosaur National Monument, but also the entire national park system in the form of dams in the parks. The opening salvo in the public debate over the Echo Park damsite began in 1950. Renowned author Bernard DeVoto wrote a scathing article for the *Saturday Evening Post* condemning the federal government's plans to construct dams within the boundaries of national parks and monuments. DeVoto's "Shall We Let Them Ruin Our National Parks" alerted the public to the activities of the Bureau of Reclamation and the Army Corps of Engineers. DeVoto noted the tremendous power both agencies possessed and the influence created from ties with political and business leaders. He wrote,

> Each of them has about a third of a billion dollars of public funds to spend every year, and so can exert incomparably more pressure than any corporation that ever cast a covetous eye on the wilderness beauties which were set aside for prosperity to enjoy.

DeVoto characterized Reclamation and the Corps as power-hungry federal entities willing to destroy pristine wilderness landscapes set aside for the enjoyment of the people to feed agency ambitions to control and regulate nature.[193]

In the article DeVoto initiated a two-pronged attack on water development policies of the Bureau of Reclamation and the Corps of Engineers. First and most persistently, he cried foul over the intrusion of these engineering agencies into the national parks. DeVoto reminded readers that national parks, monuments, and sites were important features of the nation's heritage whose preservation was for the benefit of future generations. He spoke of the aesthetic qualities national parks possessed and the immeasurable values and pleasures the public gained by visiting these national treasures. Sec-

---

[192] Martin, *A Story that Stands Like a Dam*, 52; Farmer, *Glen Canyon Dammed*, 135-6; Harvey, *A Symbol of Wilderness*, 30-5.

[193] Bernard DeVoto, "Shall We Let Them Ruin Our National Parks," *Saturday Evening Post*, 223 (July 22, 1950): 18, see also Mark Harvey, *Wilderness Forever: Howard Zahniser and the Path to the Wilderness Act* (Seattle: University of Washington Press, 2005), 95-6; Martin, *A Story that Stands Like a Dam*, 53; Harvey, *Wilderness Forever*, 133; Farmer, *Glen Canyon Dammed*, 136.

ondly, DeVoto questioned the overall necessity for the dams, maintaining that Reclamation's proposed dams offered dubious benefits. He asserted that the Intermountain West did not have the economic infrastructure to accommodate, at present, the expansive development the Bureau of Reclamation proposed. DeVoto saw these money-laden agencies running roughshod over the National Park Service's effort to defend the less tangible benefits of esthetic beauty found in remote canyons and waterways. The economic side of DeVoto's arguments actually preceded those used by Bureau of Reclamation critic Raymond Moley, especially the $207 million dollar price tag for the two dams inside the Dinosaur National Monument. Finally, the article included spectacular photographs of Echo Park, which were perhaps the first images of these hidden canyons that most Americans had ever seen.[194]

DeVoto's protests and the controversy with the National Park Service were not the only problems encountered by the Bureau of Reclamation and its Colorado River Storage Project plans. Questions arose over the viability and feasibility of the project within the Army Corps of Engineers, but the most vocal opposition came from General U. S. Grant III. Grant was president of the American Planning and Civic Association and a former staff member of the Corps of Engineers. He argued that there was no need for Reclamation to construct dams within Dinosaur National Monument and suggested that other damsites existed that better suited the Bureau's requirements. Grant's opinion and his former position in the Corps weighed heavily on the mind of Secretary Oscar Chapman. As a result, Chapman delayed giving his final approval for the project, which was necessary in order for CRSP supporters to introduce legislation in Congress. Chapman's uncertainty caused concern within the congressional delegations from the upper basin. In December 1952 Chapman reversed his original support for the Echo Park and Split Mountain dams. The secretary maintained that any legislation regarding the CRSP faced difficulty without approval from the Corps of Engineers. At the same time, he emphasized that his actions were in no way influenced by the protests of conservation groups. This hesitation gave conservationists time to regroup, knowing full well that the congressional representatives from upper basin states would reintroduce the CRSP bill in the following 1953 legislative session.[195]

Concerns were well founded given the support for the Colorado River Storage Project in the upper basin states. Congressional representatives from this region held powerful positions on nearly all the congressional committees

---

[194] DeVoto, "Shall We Let Them Ruin Our National Parks," 18-9, 48.

[195] Harvey, *A Symbol of Wilderness*, 133-7, 144-7; Farmer, *Glen Canyon Dammed*, 137.

reviewing the CRSP bill. Their influence and placement was such as to assure a final vote in the House of Representatives and the Senate. Attached to their support was a determination to protect the rights of their states to the waters of the Colorado River. For many of these congressional representatives, their fears rested upon the proposition that without the CRSP the lower basin states, especially California, would eventually claim a greater share of the Colorado River earlier guaranteed to the upper basin states under the 1922 Colorado River Compact. In their minds, the future growth and prosperity of the region depended upon full development of the Colorado River, and this included the dams at Echo Park and Split Mountain.

10.23. Bernard DeVoto. Courtesy of Professor Mark DeVoto.

Regional supporters of CRSP did not appreciate criticism of the project, especially from conservationists. Newspaper articles and public statements from officials in the upper basin labeled the conservationists elitists whose only desires were to keep Echo Park and Split Mountain as private playgrounds. They argued that the dams would "democratize" the monument making it more accessible to most Americans. Finally, the local press claimed that conservationists were ignorant of the importance of water to the development of western states. As much as the leaders of the upper basin states condemned the protests of conservationists, they kept an ever-wary eye on the actions of water and power interests within the lower basin states. Throughout the entire controversy over dams in Dinosaur National Monument, dam proponents continually portrayed conservationists as pawns of the special interests in California.[196]

In 1953 Dwight D. Eisenhower became the first Republican president to lead the nation in twenty years; his administration also inherited the Colorado River Storage Project controversy. Both sides of the debate anxiously wondered where the economy-conscious president stood on this issue. In the beginning, it did not look good for CRSP proponents when the budget-minded administration announced its "no new starts" policy. Secretary of the Interior

[196] For information on other groups wary of CRSP see, Upper Colorado River Commission, *Seventh Annual Report* (Grand Junction, Colorado: April 1, 1956), 8-9; Harvey, *A Symbol of Wilderness*, 264-5.

Douglas McKay vetoed any possibility for the Bureau of Reclamation's big dam project in Hells Canyon, opening the way for private interests to build smaller dams on the Snake River. McKay's action confirmed Eisenhower's campaign promise to pull back from the New Deal's funding of projects that could be better accomplished by private enterprise, or in this case, favoring support of private power interests against public ones. According to the administration, government must never compete with private interests, but rather develop cooperative efforts between government and private enterprise for natural resource management. In American politics, however, ideology often gives way to pragmatic political considerations. When western Republicans, especially those from the upper basin states, began clamoring for the administration's support for CRSP, prospects for the project's future improved. In December 1953 McKay and Bureau of Reclamation Commissioner Wilbur A. Dexheimer renewed the Department of the Interior's commitment to the Colorado River Storage Project and Echo Park Dam.[197]

McKay's announcement reignited conservationists' efforts to stop the construction of any dams in Dinosaur National Monument. In his announcement, the secretary maintained that the Echo Park damsite was essential for the project. Most critically, McKay argued that a dam at Echo Park would be less subject to evaporation losses, and referred to the dam as the "turbine" that ran the entire CRSP. Opposing conservationist voices argued that cheaper power resources existed in local coal deposits, oil burning turbines, or even atomic energy. The evaporation issue, however, provided conservationists with their best opportunity to challenge the Bureau of Reclamation's science that defended the Echo Park Dam site. David Brower, executive secretary of the Sierra Club, led the effort to discredit the Bureau's evaporation studies. In the now famous evaporation controversy, Brower found errors in Reclamation's evaporation estimates, creating doubt about the Bureau of Reclamation's contention that the Echo Park damsite was the best location.

Much has been written about Brower's testimony before Congress in June 1954 that revealed discrepancies in the Bureau of Reclamation's evaporation claims. Historian Mark Harvey writes,

> His appearance before the subcommittee has since become a famous episode in the annals of American environmental history, taking up almost legendary proportions. The story has been told with overtones of David and Goliath, and generally

[197] Harvey, *A Symbol of Wilderness*, 179,183-4; Harvey, *Wilderness Forever*, 135-58.

it has been assumed that Brower brought an end to Echo Park Dam almost single handedly.[198]

Though significant, the evaporation controversy did not mark the end of the fight over dams in Dinosaur National Monument. Of greater importance is how Brower's testimony challenged the expertise of the Bureau of Reclamation, and in turn opened questions about overall feasibility of the Colorado River Storage Project.

Miscalculations, embarrassingly brought forward in congressional hearings, ballooned into a larger issue that mirrored a transition taking place in American thought and culture by the 1950s. The growing popularity of national parks and wilderness experiences registered the importance that the American public assigned to protecting natural settings and scenery. Conservationists and opponents of the dams recognized, modified, and played to these sentiments. Especially appealing was the wilderness aesthetic that stood for peace and serenity in a natural setting to offset the artificiality and stress of urban life.[199]

In contrast, the pro-development factions inside and outside the Bureau of Reclamation failed to grasp how forcefully the wilderness argument resonated throughout American society. Instead, they relentlessly painted conservationists as elites who only desired to save the beauties of nature at the expense of many. Dam supporters argued that the dam, impounded reservoir waters, and extensive waterways would allow greater access to beautiful canyons of the area. National parks and monuments were playgrounds for a larger American public rather than exclusive aesthetics preserved for those who could afford to access them through expensive and strenuous means of travel. CRSP supporters continually claimed that opponents acted as spokespersons for Southern California interests. The arguments of Brower and DeVoto seemed farfetched to commercial interests of the upper basin and to engineers within the Bureau of Reclamation. As Marc Reisner argued in *Cadillac Desert*, "They couldn't fathom that a sea of change in public feeling toward the natural world was taking place, one of epochal shifts that guaranteed things will never be the same again."[200]

---

198 Harvey, *A Symbol of Wilderness*, 191; Martin, *A Story that Stands Like a Dam*, 62; Farmer, *Glen Canyon Dammed*, 142.

199 Harvey, *Wilderness Forever*, 66; Harvey, *A Symbol of Wilderness*, 239.

200 Marc Reisner, *Cadillac Desert: The American West and Its Disappearing Water*, revised and updated (New York: Penguin Books, 1993), 287; Martin, *A Story that Stands Like a Dam*, 63.

Between 1953 and 1956 debate over the Colorado River Storage Project continued. Each side actively lobbied to influence public opinion. Dam opponents used multiple media formats to make their case: films, brochures, articles, and full-length books. One of the more notable was David Brower's film production *Two Yosemites*, wherein he compared the plight of Dinosaur National Monument with the iconic desecration of Hetch Hetchy that occurred almost a half century earlier in Yosemite National Park. Another was *This is Dinosaur*, a book collection of essays and photographs edited by Wallace Stegner and published by Alfred A. Knopf, highlighting the scenic wonder of the monument. Dam proponents responded by advertising the benefits of the project. Upper basin political and commercial interests also insisted that the CRSP ranked as an entitlement under the 1922 Colorado River Compact. To press their claims, like-minded upper basin residents formed Upper Colorado Grass Roots, Inc. Its purpose was to "disseminate information and to counteract the opposition of Southern California interests who stood to benefit if the enactment of the authorizing legislation were prevented."[201]

By the spring of 1956 dam opponents, with a continual outpouring of public support for their position, still blocked passage of the Colorado River Storage Project in Congress. In an effort to break the impasse, the Department of the Interior and the Bureau of Reclamation gave up the Echo Park and Split Mountain dams after some hard negotiations with Sierra Club director Brower. This action effectively removed any opposition to the project on the part of the conservationists. On April 11, 1956, the Colorado River Storage Project became law with the stipulation "that no dam or reservoir constructed under the authorization of the Act shall be within any National Park or monument." The conservationists preserved the sanctity of the National Park system. In the end, project supporters came to the realization that continued debate might ruin any chance for attaining legislation for water development in the upper basin and struck a compromise. In the place of two dams in Dinosaur National Monument, the Act authorized the Bureau of Reclamation to build a high dam in a remote area along the border of Arizona and Utah. Glen Canyon Dam eventually became the centerpiece of the Colorado River Storage Project.[202]

---

[201] Harvey, *A Symbol of Wilderness*, 238, 271-2; Wallace Stegner, editor, *This is Dinosaur: Echo Park Country and Its Magic Rivers* (New York: Alfred A. Knopf, 1955); Farmer, *Glen Canyon Dammed*, 148-9; Martin, *A Story that Stands Like a Dam*, 65-6; for information on the Upper Colorado River Grass Roots, Inc. see, Upper Colorado River Commission, *Seventh Annual Report*, 6-7; Harvey, *A Symbol of Wilderness*, 233-4, 265-6.

[202] Roderick Nash, *Wilderness and the American Mind*, third edition (New Haven: Yale University Press, 1982), 219; Farmer, *Glen Canyon Dammed*, 144; Harvey, *A Symbol of Wilderness*, 284-5; Harvey, *Wilderness Forever*, 183-185; Martin, *A Story that Stands Like a Dam*, 71.

Conclusion

Both the Bureau of Reclamation and upper basin supporters of CRSP confronted an unexpected shift in public attitudes and values regarding the scenic landscape of the American West. The growing strength of the preservationist wing of the conservationist movement reflected the interests of Americans who now enjoyed more leisure time in the first glow of postwar prosperity. In the 1950s, record visitor numbers to national parks testified to the public's awe and appreciation of natural wonders entrusted to the park system and also to the affordability of travel to points of incredible scenic beauty. When leading conservation publicists and organizations drummed up wide support in the public, and thence in Congress, against the desecration of a site of great natural beauty within a protected national monument, astonished backers of river and power development, including the Bureau of Reclamation, recognized a powerful counter force to their objectives. So powerful that spokespeople for various organizations forced retreat and compromise on the part of those supporting such traditional values of industrial, commercial, and agricultural growth. It was confirmation by the public and more sensitive political leaders that American society demanded not only "guns" in the form of an expanding economy to support a military establishment, but also "butter" in the form of consumer goods and amenities of life embodied in the beauties of the American landscape. The story of the defeat of the Echo Park Dam and its partner project the Split Mountain Dam offered clear signs that challenges to future dam building by the Bureau of Reclamation, even in remote canyons of the West, faced stiff challenges from new, largely urban voices demanding a stake in the future of the American West's landscapes and waterscapes. In spite of the loss of Echo Park and Split Mountain dams, the Bureau of Reclamation looked forward to constructing the still impressive Colorado River Storage Project.

CHAPTER 11:

END OF AN ERA AND NEW BEGINNINGS, 1956 TO 1968

Introduction

Conservationists' determined opposition and subsequent public outcry against the Echo Park Dam caught the Bureau of Reclamation off guard. Its partial defeat, however, did little to slow down the incredible pace of construction activities spurred on by the Colorado River Storage Project Act of 1956. In the upper basin, Reclamation immediately went to work on four major storage projects: Glen Canyon Dam in Arizona, Flaming Gorge Dam in Utah, Navajo Dam in New Mexico, and the Curecanti Unit in Colorado. These long-awaited projects marked the first steps in meeting the aspirations of upper basin water development boosters. In construction of these dams and powerplants, Reclamation continued its legacy of underwriting the growth and economic expansion of the American West. Remote locations meant the building of new roads and bridges to bring material and personnel along with the construction of communities to house them, interconnecting these so-called empty places into the mainstream of American life.

Scattered Indian reservations throughout the American West drew the attention of the Bureau of Reclamation. During the late 1950s and into the 1960s, as interest and concern regarding the plight of Native Americans gained momentum, Reclamation began construction on one of the few projects in its long history designed solely to benefit Native peoples—Navajo Indian Irrigation Project. Authorized in 1962, the project promised not only economic prosperity to the Navajo people in northwestern New Mexico, but also brought into sharp focus the issue of Indian water rights, a matter generally side-stepped by Reclamation and water development boosters in the West.

11.1. Commissioner Floyd E. Dominy served from 1959 to 1969 under four presidents.

Floyd Dominy, one of the most forceful and colorful commissioners in Reclamation history, headed the Bureau of Reclamation from 1959 to1969. Dominy led Reclamation with a brashness unmatched in any other commissioner. Seen by many as a ruthless technocrat, Dominy was intelligent, forceful, and politically astute, using all of his dynamic personality to push the Reclamation agenda forward. He was not alone in this endeavor. Stewart L. Udall, secretary of the interior from 1961 to 1968, had his own vision of the role of Reclamation in the development of the American West, and that vision especially embraced his native state of Arizona. Paradoxically, Udall brought to the Department of the Interior a conservationist credo that saw the need for continued water resource development in the West, which often ran squarely into a more nuanced, though less-developed, personal attitude toward preservation and protection of natural resources.

11.2. Secretary of the Interior Stewart Udall served from January 21, 1961, until January 20, 1969. Courtesy of the Office of the Secretary of the Interior.

Udall and Dominy worked together under the Kennedy and Johnson administrations, both representing in their own ways the ambitions and energies of the New Frontier and the Great Society. For the first time in many years, Reclamation encountered tightening budget constraints brought on by increased expenditures for military and social programs. At times the secretary and commissioner worked in unison as when they strove to put together the Pacific Northwest/Pacific Southwest Intertie that integrated federal, municipal, and private power producers to supply electricity throughout the Far West. At other times Udall and Dominy butted heads. Efforts to implement the Pacific Southwest Water Plan (PSWP) largely failed because they did not present a unified front.

Construction on the CRSP

On July 2, 1956, President Dwight D. Eisenhower signed the Colorado River Storage Project Act. With the Echo Park controversy resolved, the Bureau of Reclamation pushed ahead with three large dam construction projects simultaneously: Glen Canyon Dam on the main stem of the Colorado River; Flaming Gorge Dam on the Green River; and Navajo Dam on the San

11.3. Glen Canyon Dam during construction. April 9, 1963.

Juan River. The Curecanti Unit, renamed the Wayne Aspinall Unit in 1980, consisted of Blue Mesa, Morrow Point, and Crystal dams on the Gunnison River in west-central Colorado. Though authorized at the same time as Glen Canyon, Flaming Gorge, and Navajo dams, construction of the Curecanti units

11.4. The generator floor at Glen Canyon Powerplant, Colorado River Storage Project, Page, Arizona. July 1966.

did not begin until 1960. The Colorado River Storage Project initiated one of the most active construction periods in Reclamation's history. In total, CRSP called for ten reservoirs capable of storing 48,555,000 acre-feet of water, along with associated powerhouses with a generating capacity of 1,622,000 kilowatts.[203]

Glen Canyon was by far the largest and most important of the CRSP units. This huge dam on the Utah/Arizona border rose from the canyon floor and topped off at 710 feet when the concrete stopped pouring. Its reservoir backed water for 186 miles to the mouth of the Green River and 71 miles up the San Juan River, covered 176,620 acres, and had a capacity of 27,000,000 acre-feet of water. Glen Canyon's powerhouse contained eight generators "each with a total of capacity of 900,000 kilowatts." The dam was unique in its purpose and pivotal to the success of the Colorado River Storage Project. Glen Canyon Dam and Lake Powell were not built to directly supply irrigation, municipal, or industrial water. The overriding purpose of this 27 million acre-

---

[203] United States Department of the Interior, Bureau of Reclamation, Region 4, "Annual Project History, Flaming Gorge Unit, Colorado River Storage Project, Vol. I," 1956, RG 115, Entry 10, Project Histories, Feature Histories, and Reports, 1902-1932, Box 124, Colorado River Storage Project, Flaming Gorge (UT-WY); see also C. B. Jacobson, "Largest Reclamation Development Underway," *Reclamation Era*, 43 (February 1957): 2.

11.5. Glen Canyon Dam in 2001. Photograph courtesy of Carol D. Storey.

feet of storage was to guarantee the upper basin's ability to meet its obligation under the Colorado River Compact to deliver water to the lower basin. Importantly, too, the power generated at the dam was crucial to repaying CRSP to the U.S. Treasury. According to one source,

> Glen Canyon Unit was one of four major projects, but it was by far the largest and most expensive—a 'cash register' that would finance the other three [Flaming Gorge Dam, Navajo Dam, and the Curecanti Unit] and several smaller participating projects.[204]

From the early 1900s, Glen Canyon drew the attention of water resource developers. While the canyon contained breathtakingly beautiful scenery, its deep and steep sides offered multiple natural damsites. In 1922 Southern California Edison Company researched the possibilities for a power

---

[204] United States Department of the Interior, Bureau of Reclamation, "Glen Canyon Unit, Colorado River Storage Project, Annual Project History, Vol. I," 1956, RG 115, Entry 10, Project Histories, Feature Histories, and Reports, 1902-1932, Box 126, Colorado River Storage Project, Flaming Gorge (UT-WY) and Glen Canyon; see also Jedediah Rogers, "Glen Canyon Unit," Denver, Colorado: Bureau of Reclamation Project History Program, 2006, 2; Steven W. Carothers and Bryan T. Brown, *The Colorado River through the Grand Canyon: Natural History and Human Change* (Tucson: University of Arizona Press, 1991), 181.

production dam in Glen Canyon. During that same year Arizona water boosters proposed plans to make the canyon the centerpiece of a state-financed Colorado River storage project. Long-time irrigation advocate George Maxwell, founder of the National Irrigation Association and a booster of federal reclamation, pushed hard for Arizona to claim the Glen Canyon damsite for itself. He was joined by Arizona State Senator Fred Colter of the Highline Reclamation Association which looked to Glen Canyon as a major water source for central Arizona. W. S. Norviel, Arizona State Water Commissioner, also proposed a similar idea but in response to growing support for the Swing-Johnson Bill that Arizona desperately hoped to defeat. Arizona failed in the face of the 1922 Colorado River Compact, the Boulder Canyon Project Act, and the lack of resources to fund such a state-based Colorado River project. All the while Glen Canyon remained a remote, seldom visited location, which was never far from the engineers' drawing boards.[205]

Construction on Glen Canyon Dam officially began on October 16, 1956. Ceremonies surrounded the occasion with invited officials and dignitaries from the states of Utah and Arizona. President Eisenhower telegraphed a signal from Washington that triggered an explosion to mark the start of construction. When Eisenhower had to push the button twice before the switch was connected, the delay caused a momentary crisis. Afterwards, the president humorously remarked, "I guess it takes electricity a long time to travel out West." Much work remained ahead before Reclamation poured any concrete into the canyon. Under the direction of construction engineer L. F. "Lem" Wylie, Reclamation first needed to construct facilities to house workers, upgrade roads to the remote location, and construct a bridge to join both sides of the chasm.[206]

---

[205] For information on Fred T. Colter's High-Line Canal plan see, United States Senate, *Glen Canyon-Bridge Canyon Dam Project and Arizona High-Line Canal: Summary of the Arguments Made by Senator Fred T. Colter, President of the Arizona Highline Association, on S. 3414, A Bill Providing for the Building of the Glen-Bridge Dams and Arizona High-Line Canal, Introduced in the United States Senate March 1, 1926, as a Substitute for the Swing-Johnson Boulder-Black Dam and Compact Bill*, S. Doc. No. 113, 69th Cong., 1st sess. (Washington, D.C.: United States Government Printing Office, 1926); see also Hundley, *Water and the West:* 123, 161, 171, 247-8; Rich Johnson, *The Central Arizona Project, 1918-1968* (Tucson: University of Arizona Press, 1977), 15; Donald J. Pisani, *Water and American Government: The Bureau of Reclamation, National Water Policy, and the West, 1902-1935* (Berkeley: University of California Press, 2001), 7; Worster, *Rivers of Empire*, 160.

[206] Martin, *A Story that Stands Like a Dam*, 87; see also United States Department of the Interior, Bureau of Reclamation, "Glen Canyon Unit, Colorado River Storage Project, Annual Project History, Vol. I," 1956, RG 115, Entry 10, Box 126, Colorado River Storage Project, Flaming Gorge (UT-WY) and Glen Canyon; Jacobson, "Largest Reclamation Development Underway," 2.

Named construction engineer for the Glen Canyon Unit in 1956, Wylie began his career with the Bureau of Reclamation during the construction of Boulder/Hoover Dam in 1932. An engineering graduate of the University of New Mexico, Wylie worked his way up the ladder reaching the position of assistant engineer on Hoover. He worked on numerous projects after Hoover Dam including the All-American Canal. By the time he received the call to Glen Canyon, Wylie was the assistant regional director in Region V at Amarillo, Texas. Glen Canyon Dam became the apex of Wylie's career with Reclamation, and he retired just before Reclamation put the dam into operation in 1964. L. F. Wylie died in 1984 and, according to Russell Martin in *A Story that Stands Like a Dam*, had his ashes distributed "over Antelope Island—around a bend of the blue lake from the dam he had built back in the day when you could build things."[207]

To get construction underway, Wylie had substantial hurdles to overcome. The damsite was in the middle of nowhere 15 miles north of Lee's Ferry, Arizona. Roads to the site were rough and primitive, and it was a 190-mile drive just to get from one side of the canyon to the other—a distance of 1,200 feet. Good road construction was essential to the success of the project because Glen Canyon Dam was the first major Bureau of Reclamation dam "without the service of a railroad." Cognizant of the importance of the dam to their economies, the states of Utah and Arizona raced to construct roads to handle the enormous amount of supplies, personnel, and equipment required for the project. According to Wylie,

> The logistics involved in the construction of Glen Canyon Dam are comparable to those of any army fighting on two fronts—these fronts being separated by only 1,200 feet of Colorado Gorge. Men and material must move to the damsite from two locations—Flagstaff, Ariz., 135 miles to the south, and Kanab, Utah, 76 miles to the west.

Consequently Wylie's first task was to connect the two sides of the canyon. Plans for the bridge included 30-foot-wide roadway supported by towers on each end. Standing 700 feet above the canyon floor and spanning 1,028 feet, by August 1957 Reclamation completed the "world's highest steel-arch bridge."[208]

---

207 "L. F. Wylie Heads Glen Canyon Job," *Reclamation Era*, 42 (August 1956): 77; see also Martin, *A Story that Stands Like a Dam*, 75-6, 326.

208 L. F. Wylie, "Roads to Scenic Treasures," *Reclamation Era*, 44 (August 1958): 58; Francis J. Murphy, Project Manager, Kiewit-Judson Pacific-Murphy, "Highway in the Sky," *Reclama-*

In March 1957 Commissioner Wilber Dexheimer named the construction town "Page" on the east side of the river, in honor of former Commissioner of Reclamation John C. Page (1936-1943). To obtain the land on which to build Page, Reclamation negotiated a land deal with the Navajo Tribe. Successful negotiations with the Navajo Tribal Council resulted in acquisition of 53,000 acres from the tribe that included land not only for the town of Page, but also the damsite and the reservoir area. On September 2, 1958, the president signed Land Exchange, Navajo Tribe (Public Law 85-868), which transferred Indian lands to the United States. In return, the exchange allowed the Navajo to select an equal amount of land near McCracken Mesa, Utah, and provided "monetary compensation" for Navajo possessing grazing permits or leases on the land transferred to the government. While negotiations with the Navajo progressed and after Reclamation addressed the transportation issue, the next step was building facilities to house Reclamation and construction personnel. On April 11, 1957, in a school house in Kanab, Utah, Reclamation awarded Merritt-Chapman and Scott Corporation the $107,955,552 general construction contract for Glen Canyon Dam and powerhouse, at the time the single most expensive contract in Reclamation history.[209]

By 1958 Reclamation activities at the Glen Canyon Dam site increased at a frantic pace. Construction of the upstream and downstream cofferdams commenced in October, while the town of Page, Arizona, grew to approximately 4,500 residents by the end of the year. Plans for the new town included 300 permanent homes for Reclamation personnel and a trailer park for 1,000 trailers to house construction workers. Early on, services for residents of Page were sparse, consisting of a grocery store, a service station, a beauty shop, and a bank. According to Jared Farmer, in *Glen Canyon Dammed*, Page resembled on old-West boomtown that had a frontier atmosphere and construction workers compared themselves to early-American pioneers who saw their task as "transforming the wilderness rather than exploring it connected them to a vital, ephemeral part of America." By 1963 Page was a vibrant community with a population of over 6,000 and, according to Russell Martin, was "the biggest

tion Era, 44 (November 1958): 102-4; Farmer, *Glen Canyon Dammed*, 34; Martin, *A Story that Stands Like a Dam*, 13; Rogers, "Glen Canyon Unit," 17-20.

209 United States Department of the Interior, Bureau of Reclamation, "Glen Canyon Unit, Colorado River Storage Project, Annual Project History, Vol. III," 1958, RG 115, Entry 10, Project Histories, Feature Histories, and Reports, 1902-1932, Box 127, Colorado River Storage Project, Glen Canyon, Navajo, and Hammond Units; see also "Glen Canyon Unit, Colorado River Storage Project, Annual Project History, Vol. III," 1958, RG 115, Entry 10, Box 126, Colorado River Storage Project, Flaming Gorge (UT-WY) and Glen Canyon; Rogers, "Glen Canyon Unit," 23.

community alongside the Colorado River for 350 miles upstream and for 390 miles downstream."[210]

In February 1959 Glen Canyon Bridge opened to traffic with completion of the highway between Bitter Springs, Arizona, and Kanab, Utah, and air and bus service to Page began later in the year. Throughout 1959 construction moved at a steady pace as the waters of the Colorado River now ran into the diversion tunnels. The steady progress construction engineer Wylie and Reclamation associates envisioned came to halt in July 1959 when the "five Basic crafts" unions went on strike and remained off the job until December. For six months, the damsite lay quiet and Page became virtually a ghost town, as workers left to wait out the strike. In January 1960 Merritt-Chapman and Scott Corporation finally settled with the striking unions. Page came back to life and action down on the canyon floor resumed. By June 1960 Wylie's contractors began pouring concrete for both the dam and the powerhouse; Glen Canyon Dam began to rise.

The same telegraph signal that initiated the explosion that started Glen Canyon Dam construction in 1956 also triggered another explosion in a remote canyon of Utah on the Green River initiating construction of the Flaming Gorge Unit. Just 32 miles downstream from the Utah/Wyoming border, Flaming Gorge Dam rose in Red Canyon just below the mouth of Cart Creek. An integral part of the Colorado River Storage Project, the dam design included a thin-arch concrete dam, 490 feet high with a reservoir capacity of 3,788,900 acre-feet. Its adjoining powerplant would contain three 36,000 kilowatt generating units capable of producing 1,622,000 kilowatts. According to the project history, "This unit will help to provide the long-term regulatory storage needed to permit the States of the Upper Basin to meet their flow obligations at Lee Ferry, Arizona, under various compacts and laws concerning the Colorado River and still utilize their apportioned share of the river."[211] Thus, similar

---

210 Martin, *A Story that Stands Like a Dam*, 13, 100, 137; Farmer, *Glen Canyon Dammed*, 110-11.

211 United States Department of the Interior, Bureau of Reclamation, "Annual Project History, Flaming Gorge Unit, Colorado River Storage Project, Vol. II, 1957; "Annual Project History, Flaming Gorge Unit, Colorado River Storage Project, Vol. I," 1956; Vol. III , 1958, RG 115, Entry 10 Box 124; see also Toni Rae Linenberger, "Flaming Gorge Unit: Colorado River Storage Project," Denver: Bureau of Reclamation History Program, 1998, 11; United States Department of the Interior, Water and Power Resources Service, *Project Data* (Washington, D.C.: United States Government Printing Office, 1991), 357; Dick and Vivian Dunham, *Flaming Gorge Country: The Story of Daggett County, Utah* (Denver: Richard R. Dunham; printed by Eastwood Printing and Publishing Company, 1997), 333.

to Glen Canyon Dam, the Flaming Gorge Dam included no significant water delivery works and was to act as a water storage and power producing unit.

The appointment of construction engineer for Flaming Gorge Dam went to Lem Wylie's fellow University of New Mexico alumnus Jean R. Walton. Similar to Wylie, Walton had a long and distinguished career with the Bureau of Reclamation. He joined Reclamation in 1936 as an assistant engineer on the Pine River Project in Colorado. After he concluded his military service in World War II, Walton became a field engineer at Davis Dam in 1946, and in 1950, he was appointed construction engineer at Davis Dam. After finishing Davis, Walton became construction engineer "on the second barrel of the San Diego Aqueduct Project" and served as a construction advisor on the Snowy Mountain Project in Australia. Like Wylie, Walton retired from the Bureau of Reclamation upon the completion of his unit of CRSP.[212]

Flaming Gorge Dam never quite attracted either the praise or the controversy Glen Canyon Dam garnered, despite the similar circumstances and challenges. The damsite was nearly as remote as the Glen Canyon site sitting amidst the red sandstone canyons of the Green River and required considerable logistical efforts before construction began. Located in Daggett County, with a population of approximately 350 people, Flaming Gorge Dam sits in the extreme northeast corner of the State of Utah. Construction engineer Walton wrote,

> Prior to the construction of a bridge across the river by the Bureau of Reclamation the inhabitants of the eastern part of the county were obliged to travel some 100 to 120 miles through Rock Springs and Green River, Wyo., to reach Manila, Utah, the county seat.

As with Glen Canyon Dam, the building of a bridge was a step preliminary to dam construction. Along with the bridge, Reclamation built a community to house construction and bureau employees and named it Dutch John, Utah. Dutch John was named after John Honselena (or Henselena) from Schleswig, Germany, who according to legend was a horse trader. According to Dick and Vivian Dunham in *Flaming Gorge Country*, little is known of Dutch John, but one legend claims he was shot for horse stealing. By 1959 the population of Dutch John alone nearly tripled the entire population of Daggett County. As

---

212 "Jean R. Walton Named Construction Engineer on Flaming Gorge Dam," *Reclamation Era*, 42 (November 1956): 92; see also United States Senate, *Colorado River Storage Project: Letter from the Assistant Secretary of the Interior Transmitting the Seventh Annual Report*, VII-VIII.

11.6. Flaming Gorge Dam and Powerplant in August of 1984. Photographer: Tom Fridmann.

with many large-scale Reclamation projects, the roads and bridges needed to construct large dams and powerhouses connected sparsely settled and relatively remote areas of the West with the rest of the nation.[213]

After completion of Dutch John, access roads to the damsite, and the bridge across the Green River, construction of Flaming Gorge Dam began in earnest. In 1958 Reclamation awarded the dam and powerhouse construction contract to Arch Constructors of Omaha, Nebraska. In March 1959 the contractor completed the diversion tunnel and, by November, began diverting the river away from the damsite. As with most construction sites, the Flaming Gorge damsite presented its own particular challenges. A year later Jean Walton reported that the Green River Bridge was partially destroyed by an ice jam, where "large sections of ice were observed to be about 12 feet square and 1½ to 2 feet thick." Though the ice field apparently did no damage to the damsite itself, the bridge repairs were a slight setback to the construction schedule. Despite the delay both upstream and downstream coffer dams were soon in place, and by September Arch Constructors began pouring concrete for the dam and powerplant.[214]

---

213 "Annual Project History, Flaming Gorge Unit, Colorado River Storage Project, Vol. III," 1958, RG 115, Entry 10, Box 124; Jean R. Walton, "Flaming Gorge," *Reclamation Era*, 45 (August 1959): 69, 80; Dunham, *Flaming Gorge Country*, 92-3.

214 United States Department of the Interior, Bureau of Reclamation, "Annual Project History, Flaming Gorge Unit, Colorado River Storage Project, Vol. IV," 1959, in RG 115, Entry 10, Box

The third large dam authorized by the Colorado River Storage Project was Navajo Dam in New Mexico. Navajo Dam originated from a 1955 Bureau of Indian Affairs report that proposed a "distribution system for irrigation of 137,250 acres of new land within and adjacent to the Navajo Indian Reservation, all in New Mexico." While the 1956 Colorado River Storage Project Act authorized construction of Navajo Dam, it did not authorize irrigation funds for reservation lands. Reclamation engineers designed an earth and rock-filled dam about 370 feet high and 6,100 feet long with a total reservoir capacity of 1,450,000 acre-feet. According to one report, Navajo Dam "will be the second largest earth dam built by the Bureau of Reclamation.... It will be 2½ times the size of Grand Coulee Dam." Reclamation touted the dam as the centerpiece of water resource development in New Mexico and "the key to agricultural and industrial growth in the area." Future proposed projects resulting from Navajo Dam included not only the Navajo Indian Irrigation Project, but also the San Juan-Chama Project, which later diverted water to the Rio Grande basin.[215]

Located on the San Juan River in northwestern New Mexico 34 miles east of Farmington, New Mexico, construction began on Navajo Dam in 1958. Reclamation awarded the construction contract to Morrison-Knudsen, Henry J. Kaiser Co., and F&S Contracting Company in June. Throughout 1959 work on Navajo Dam proceeded steadily on the embankments, outlet works tunnel, and the intake and outlet portal structures. By January 1960 the contractor had diverted the San Juan River. While authorization for the Navajo Indian Irrigation Project and the San Juan-Chama Project languished in Congress, Reclamation proceeded on the construction of the Hammond Project. This was a relatively small project to provide irrigation water for 3,900 acres in a narrow valley between Farmington and Blanco, New Mexico. Project works included the Hammond Diversion dam located 19 miles downstream from Navajo Dam, the 27-mile-long Hammond Main Canal, and the Hammond Pumping Plant to lift water to the East and West Highline laterals.[216]

124; see also Vol. V, Box 126; Linenberger, "Flaming Gorge Unit: Colorado River Storage Project," 12.

215 United States Department of the Interior, Bureau of Reclamation, "Annual Project History, Navajo Unit, Colorado River Storage Project, Vol. I," Farmington, New Mexico, 1956, RG 115, Entry 10, Box 128, Colorado River Storage Project-Navajo Unit (NM), Transmission Division (UT); United States Department of the Interior. Bureau of Indian Affairs, *Navajo Project New Mexico: Feasibility Report, Appendix D, Agricultural Economic Financial Analysis* (Window Rock, Arizona: Navajo Agency, January 1955); see also Herbert E. Simison, "Action on the San Juan River," *Reclamation Era*, 46 (February 1960): 8, 10.

216 United States Department of the Interior, Bureau of Reclamation, "Annual Project History, Navajo Unit, Colorado River Storage Project, Vol. III," Farmington, New Mexico, 1958, RG 115, Entry 10, Box 128; Vol. IV, Farmington, New Mexico, 1959, Box 128; Vol. V, Farmington,

The three large dams of the Colorado River Storage Project testified to the Bureau of Reclamation's ability to orchestrate multiple, highly technical projects simultaneously. Each dam marked a new beginning for water resource development in the upper basin of the Colorado River. In September 1962 construction on Navajo Dam was essentially complete, and the gigantic earth-filled structure was ready to serve irrigation needs in New Mexico. According to one *Reclamation Era* article, Navajo Dam made possible the gravity diversion of water directly from the reservoir to the 110,000 acre Navajo Indian Irrigation Project, and from an upstream point to the San Juan-Chama Project, which would take supplemental irrigation, municipal, and industrial water from the San Juan River into the Rio Grande drainage.

In Utah minor construction remained, but in the fall of 1963 Flaming Gorge Dam became operational. In a ceremony similar to the one that initiated its construction, President John F. Kennedy sent the signal that started power generation on September 27, 1963. The following year, First Lady Lady Bird Johnson, who saw the dam as embodying many of the visions of early American explorers of the West, officially dedicated Flaming Gorge Dam. In her speech, the First Lady recalled the legacy of John Wesley Powell, who not only named the canyon Flaming Gorge, but also envisioned a modern-day reclamation program. She declared, "About a hundred years ago, he made his daring journey down these rapids while studying the water system of the mountain region. And he dreamed dreams about a huge network of dams which would cause this arid land to flower."[217]

Perhaps because of the magnitude of the project, Glen Canyon Dam lagged behind the other Colorado River basin dams. The labor troubles in 1959 probably contributed to this slower schedule, but did not stop the Bureau of Reclamation from planning to put the dam to work as soon as possible. In 1963, however, Mother Nature did not cooperate as a western drought affected the rate at which Lake Powell filled. This circumstance had a domino effect downstream on the Colorado River. A crisis began to emerge because of concerns about the lack of water in Lake Mead to run the generators at Hoover Dam. In

---

New Mexico, 1960, Box 128; United States Department of the Interior, Bureau of Reclamation, "Annual Project History, Hammond Participating Project, Colorado River Storage Project, Vol. I," Farmington, New Mexico, 1960, RG 115, Entry 10, Box 128; USDOI, Water and Power Resources Service, *Project Data*, 358.

[217] "1963—First Payoff on the CRSP," *Reclamation Era*, 49 (August 1963): 57-61, 70-3; for the start of power generation at Flaming Gorge Dam and the First Lady's remarks see "Flaming Gorge Power is Signaled by President Kennedy," *Reclamation Era*, (November 1963): 106; "A First for the First Lady," *Reclamation Era*, 50 (November 1964): 77-8.

January 1964 a memorandum to Bureau of Reclamation Commissioner Floyd Dominy noted, "The filling criteria provided the storage at lake [*sic*] Mead will not be drawn down below the rated head at the powerplant while storage of the minimum pool is being attained at Lake Powell." As a result, Secretary of the Interior Stewart Udall ordered, in March 1964, the outlet gates of Glen Canyon Dam opened to maintain a proper operating water level in Lake Mead. It took seventeen years for Lake Powell to fill. Nevertheless, on September 22, 1966, Glen Canyon Dam was officially dedicated, once again by Lady Bird Johnson, before 3,000 people that a *Reclamation Era* article noted included "many officials and dignitaries of the Federal, state, and local level."[218]

Although the final construction phase of the upper Colorado River projects went well into the 1960s, the beginning of that decade closed a chapter on Republican Party rule and conservative economic policy. Yet despite Eisenhower's pronouncement of "no new starts," appropriations for the Bureau of Reclamation's programs continued to grow under his administration. Indeed, CRSP led the way by authorizing funds of over three-quarters of a billion dollars. During Eisenhower's presidency, significant appropriations kept coming for the Columbia Basin and Missouri Basin programs. Under Reclamation Commissioner Wilbur Dexheimer, 53 projects or units were authorized. The estimated total cost of these units was $1,434,151,773. All in all, the storage capacity for these units was 41,533,400 acre-feet. They were intended to irrigate 855,875 acres, provide 200,970 acre-feet of municipal and industrial water, and have a generating capacity of 1,380,500 kilowatts. In total, there were thirty-seven "new starts" during the Eisenhower administration. Still, Eisenhower tried to rein in spending on Reclamation projects. The president attempted to institute reforms on how projects were authorized that "emphasized comprehensive planning." Eisenhower also sought to establish tougher reviews under criteria established by the Bureau of the Budget, along with seeking cost-sharing and project reimbursement by states and project

---

[218] For information on the problems regarding filling Lake Powell see, Memorandum, J. R. Riter to Chief Engineer, Subject: Chances of attaining minimum power pool at Lake Powell in 1964, January 13, 1964, in Dominy Papers, Box 4, Folder, 1964 Blue Envelope Letter; United States Department of the Interior, News Release, March 26, 1964, in Stewart L. Udall Papers, Western Waters Digital Library, University of Arizona, Tucson, Arizona, http://content.library.arizona.edu/cdm/ref/collection/udallCAP/id/3382, hereafter cited as the Udall Papers; Martin, *A Story that Stands Like a Dam*, 253; for information on Glen Canyon dedication see Project Construction Engineer to Office of the Chief Engineer, Denver, September 23, 1966, Subject: Weekly Progress Report, September 17 to September 23, 1966—Glen Canyon Unit—Colorado River Storage Project, in RG 115, Office of the Chief Engineer, Project Histories, 1958-1971, ACC# 8NN-115-85-008, Box 9, General Reports and Histories, Colorado River Storage Project, Aug 1966 through Navajo Irrigation Project, 1966.

beneficiaries. Finally Eisenhower hoped to take away some of the control over project planning by Reclamation and the Corps of Engineers through creation of an independent review agency for water projects. According to political scientist Daniel McCool, Eisenhower's "impact on federal water policy was temporary at best. He managed to slow the authorization of new projects, and spending increased at a lower rate for a few years, but the decision-making process remained the same."[219] Few attempts to control development of water resources in the West came from the subsequent Democratic administrations whose main interests focused on foreign policy and social issues.

Nevertheless, for the Bureau of Reclamation, the ending of construction for Glen Canyon, Flaming Gorge, and Navajo dams marked the beginning of new era. A 1963 *Reclamation Era* article reveled in the "harvest of benefits" from new dams proclaiming that "in the coming years, the benefits will grow in magnitude to equal or exceed the hopes and dreams of the most optimistic Upper Basin proponents." Concluding construction activities on these particular upper Colorado River Storage Project structures coincided with the arrival of a new decade and new administrations in the White House that brought the Bureau of Reclamation budget constraints and competition stemming from Cold War expenses, the Vietnam War, and the social agendas of the New Frontier and especially the Great Society. The "guns and butter" economic strategy of the previous decade under which Reclamation prospered was now threatened as the domestic purse strings tightened. Reclamation encountered challenges to its Progressive Era conservation and utilitarian ideals and witnessed the re-emergence of old rivalries among its constituencies in the western states over the diminishing availability of new water resource development opportunities.

Kennedy Administration and a "New Frontier" for Reclamation

In one of the hardest fought and contested presidential campaigns, Massachusetts Senator John F. Kennedy narrowly defeated Vice President Richard Nixon in 1960. While Kennedy campaigned against Eisenhower's lack of leadership on conservation issues, his own thoughts on conservation remained a mystery. According to historian Thomas G. Smith, Kennedy "criticized the no-new start policy of the Eisenhower Administration which

---

219 "Colorado River Storage Project Heads New Reclamation Authorization," *Reclamation Era*, 42 (May 1956): 54; "New Commissioner of Reclamation," *Reclamation Era*, 45 (May 1959): 29, 38; Daniel McCool, *Command of the Waters: Iron Triangles, Federal Water Development and Indian Water* (Tucson: University of Arizona Press, 1994), 104-5.

left water and power development to private initiative. In eight 'arid' years, JFK reminded voters, austerity-prone Republicans had failed to generate new programs for resource development." Yet Smith also pointed out that Kennedy's time in Congress provides little evidence that he would become a proponent for large-scale water projects. Smith maintains that "Kennedy only gave sporadic support to the construction of hydroelectric power projects" and "opposed both the Echo Park and Glen Canyon Projects."[220]

The Eisenhower years left the Kennedy administration a rich legacy of water projects including the Colorado River Storage Project and multiple other projects throughout the West. At the beginning of the Kennedy administration, the Bureau of Reclamation made steady progress on Glen Canyon, Flaming Gorge, and Navajo dams. At Glen Canyon, there were reasons to celebrate when "the three millionth cubic yard (of concrete) was placed on the dam," on March 15, 1962. A year later Glen Canyon began backing up the water of the Colorado River when two of the three outlet gates were closed. Gate 2 was slowly lowered throughout the month of March to allow for the "slow rise of the Lake."[221] In 1960, the Bureau of Reclamation was also making significant progress on other projects outside CRSP. On the Pick-Sloan Missouri Basin Program, Reclamation completed construction on Anchor and Granite Reef dams in Wyoming along with the Helena Valley Unit in Montana. In addition, one report stated, "Significant starts were made on Yellowtail Dam on the Big Horn River in Wyoming, key feature to further downstream development; Merritt Dam, key structure of the Ainsworth Unit in Nebraska; and on Arcadia Diversion Dam, a key feature of Nebraska's Farwell Unit."[222] Water resource development projects appeared hugely popular.

Though the president himself showed little personal interest in conservation matters, the new Kennedy administration entered the White House

[220] Thomas G. Smith, "John Kennedy, Stewart Udall, and the New Frontier Conservation," *Pacific Historical Review* 64:3 (August 1995): 332-3.

[221] Speedletter, Project Construction Engineer to Assistant Commissioner and Chief Engineer, Denver, Colorado, May 18, 1962, Subject: Weekly Progress Report for May 12, to 18, 1962—Specifications No. DC-4825-Glen Canyon Unit-Colorado River Storage Project, RG 115, Office of the Chief Engineer, Project Histories, 1958-1971, ACC# 8NN-115-85-008, Box 2, General Reports and Histories Central Utah, 1962 through Hardin, 1962; Glen Canyon Unit—Construction Progress Reports (L-29) Technical Reports, Colorado River Storage Project, 1963, RG 115, Office of the Chief Engineer, Construction Progress Reports for the Upper Colorado Region, 1961-1969, ACC# NRG-115-97-582, Box 6.

[222] Interior Missouri Basin Field Committee, *Missouri Basin Project Report*, Billings, Montana, June 1961, RG 115, Office of the Chief Engineer, Project Histories, 1958-1971, ACC# 8NN-115-85-008, Box 1, General Reports and Histories, Utah 1961 through Glendo 1961.

with a less restricted vision of water resource development in the West than its predecessor. Kennedy's Secretary of the Interior Stewart L. Udall became not only the voice of the administration's conservation policy but also its conscience. Udall brought to the Department of the Interior a wealth of experience in reclamation matters. A true native of the arid West, the Arizona native grew up learning and understanding the importance of water in an arid environment. Udall's six years in Congress sitting on the House Committee on Interior and Insular Affairs and its Subcommittee on Irrigation and Reclamation gave him beneficial knowledge of the politics of water in the West. While maturing over time, Udall's conservation ethic echoed that of Progressive conservationists Theodore Roosevelt and Gifford Pinchot. Steeped in New Deal ideology, Udall saw government playing a large role in natural resource development and sincerely believed that technology was the ally of conservation efforts. In what amounted to a restatement of early twentieth-century conservation essentials, his 1965 book, *The Quiet Crisis*, asserted, "Full-fledged collaboration of science and industry and government, quickened by the spur of business competition, will enable us to write bright new chapters in the conservation of some natural resources."[223]

To a certain degree, Udall was a transitional figure in the movement from conservation thinking to late twentieth-century environmentalism. As a congressman during the 1950s, Udall supported and forcefully defended the Echo Park Dam, reproving the "rising tide of antireclamation sentiment ... which threatens western development." Yet, Udall also had a sincere affection towards the natural environment. According to one source,

> Increasingly he was attracted to the preservationist pleas of Brower and the Sierra Club, Howard Zahniser of the Wilderness Society, and Ira Gabrielson of the Wildlife Management Institute. He also enjoyed the works of Henry Thoreau and Robert Frost which examined man's relationship with nature.

While secretary of the interior, Udall attempted to reconcile these competing ideals by trying to balance the ever-increasing demands for more water resource development projects with rising environmental concerns and esthetics in American society.[224]

---

[223] Stewart L. Udall, *The Quiet Crisis*, introduction by John F. Kennedy (New York: Holt, Rinehart and Winston, 1965), 179; see also "Stewart L. Udall—New Secretary," *Reclamation Era*, 47 (February 1961): 1.

[224] Smith, "John Kennedy, Stewart Udall and the New Frontier Conservation," 335, 357.

11.7. Granite Reef Diversion Dam on the Salt River Project.

11.8. Yellowtail Dam in Montana is part of the Yellowtail Unit of the Pick-Sloan Missouri Basin Program. July 12, 1967. Photographer: Lyle C. Axthelm.

Though Udall often battled inner conflicts, his commissioner of the Bureau of Reclamation, Floyd E. Dominy, had no such issues. Dominy was not just the face of Reclamation during the 1960s, he was the force behind the policies, and he relished the role. Dominy joined the Bureau of Reclamation in 1946 and began a meteoric rise to the position of commissioner. In 1946, he served as Reclamation's Chief of the Allocation and Repayment Branch of the Operation and Maintenance Division, eventually becoming director of the division in 1953. By 1957 Dominy had moved up to assistant commissioner for legislative liaison and in 1958 was named associate commissioner and, was "second in authority to W. A. Dexheimer." According to one *Reclamation Era* article, Dominy had "overall charge of policy and program execution pertaining to the functions of irrigation, power, project investigation, and budget considerations, in addition to his present responsibility for Bureau legislative affairs." By comparison, Commissioner Dexheimer was a caretaker administrator. In May 1959 Dominy's drive and visibility led to his appointment as commissioner of the Bureau of Reclamation.[225]

Clearly Dominy was a motivated and highly ambitious individual, but his background also provided him with an intimate understanding of the problems agriculture faced in the arid West. Born in Hastings, Nebraska in 1909, Dominy earned a B.A. degree in agriculture at the University of Wyoming in 1932, after studying civil engineering at Georgia Tech. Between 1934 and 1938, he worked as an agricultural agent for Campbell County, Wyoming, where he assisted farmers through the early years of the Great Depression, and later worked for the Agricultural Adjustment Agency's Western Division. Even in this position, Dominy revealed a knack for self-promotion. In a 1942 addendum to an application to the War Department, Dominy asserted, "As a county director of the Drought Cattle and Sheep-Purchasing Program, … I acted with energy and dispatch and established a reputation for getting the work done and done right." By 1942 Dominy had landed himself a position in Washington, D.C., working for Nelson Rockefeller as assistant director of the Food Supply Division of the Coordinator of Inter-American Affairs. After a short stint in the Navy during World War II, Dominy set his sights on the Bureau of Reclamation.[226]

---

225 "Dominy Named Reclamation Assistant Commissioner for Legislative Liaison," *Reclamation Era*, 43 (November 1957): 84; "Dominy Named Associate Commissioner," *Reclamation Era*, 44 (May 1958): 32; "New Commissioner of Reclamation," *Reclamation Era*, 45 (May 1959): 29, 38; Reisner, *Cadillac Desert*, 219-25.

226 "New Commissioner of Reclamation," *Reclamation Era*, 45 (May 1959): 38; Civil Service, Addendum to application to the War Department: "Statement of Experience," Dominy Papers,

11.9. Interior of the Yellowtail Powerplant, soon after the units all went on line. January 25, 1967. Photographer: Lyle C. Axthelm.

Dominy's energy, ability, and, most importantly, ambition propelled his rapid rise through the ranks of the Bureau of Reclamation. In his early years with Reclamation, there is evidence that Dominy at times disapproved of the direction the bureau was heading. In a 1952 private letter to acquaintances in Gillette, Wyoming, Dominy professed his belief that Reclamation leadership should "be recruited from technically trained persons with wide knowledge, experience, and training in dealing with engineering and agricultural problems in Western states." Of course the then commissioner, Michael Straus, with his

Box 2, Folder, Biographical Information, c. 1932-1942; see also Dominy, *Oral History Interviews*, 8; Reisner, *Cadillac Desert*, 218-9.

eastern newspaper background did not fit Dominy's thoughts on the proper leadership for Reclamation. But these 1951/1952 letters also reveal another trait of Dominy that was to serve him well in the future. He was beginning to develop meaningful working relationships with members of Congress.[227]

Dominy's rapport with Congress was legendary making him the most successful commissioner since Elwood Mead. His goals were clear. Dominy desired to ensure that the West received the greatest benefits out of water and power development, and that the taxpaying public got the most from its investment in western water. At the same time his independence and his close ties to Congress caused tension between the commissioner and his boss the secretary of the interior. Stewart Udall once commented,

> Well Dominy, like all reclamation commissioners or at least like many of them, he was a skillful politician. He had his own ties on the Hill. He played that game much too much.... In other words, without talking to me go to and talk to senators or congressmen ... and get things set up that I had reservations about. And then confronted with a fait accompli [*sic*], in essence, where I was forced to fight my own bureau in order to get changes made.

Dominy effectively used his congressional relationships, gaining easy access to some of the most powerful members of Congress involved with Reclamation policy, most notably Senator Carl Hayden from Arizona and Congressman Wayne Aspinall from Colorado. In many ways, Dominy was an example of the bureaucratic side of Daniel McCool's fabled "iron triangle," wherein "administrative agencies expand their budgets, personnel, and turf." Dominy worked the other two sides of the triangle that included Congress and the Bureau of Reclamation's water users in the West. His aggressive and pro-development style earned him both praise from his supporters and severe condemnation from his critics. Historian Jared Farmer writes, "Floyd Dominy looms larger than life in the history of water development in the twentieth-century American West. The commissioner is remembered as one of the most able and ruthless bureaucrats ever to serve Washington." Dominy pursued Reclamation's

---

227 Floyd E. Dominy to Mr. and Mrs. Jesse Spielman, Gillette, Wyoming, November 18, 1952, Folder Biographical Information, 1951-1969; for information on Dominy courting members of Congress see Floyd E. Dominy to Hon. Wesley A. D´Ewart, Member of Congress, November 18, 1952, Folder Biographical Information, 1951-1969, Dominy Family; Wayne Aspinall to Floyd E. Dominy, December 24, 1951, Folder, Assoc. Commr., 4/25/59, Dominy Papers, Box 2.

agenda with unabashed determination in the 1960s—a decade of domestic upheaval, foreign policy defeats, social reforms, and the emergence of environmentalism in legislative form with, among others, the National Historic Preservation Act of 1966 and the National Environmental Policy Act of 1969 (NEPA).[228]

Indian Water Rights and the Navajo Indian Irrigation Project

Amongst rapid social changes, America's dominant society began to reconfigure attitudes toward Native Americans or the Indian peoples of the American West, who, in turn, gained greater recognition of their rights and benefits as citizens. The Bureau of Reclamation began to take more interest in Indian irrigation projects and their inclusion in larger multiple-use river basin development. Most prominent among these were the 1956 Colorado River Storage Project and the Central Arizona Project in 1968. Reclamation, however, initiated only one project aimed primarily to benefit Indians: the Navajo Indian Irrigation Project authorized in 1962. Even that project had to be tied to a non-Indian project—the San Juan-Chama Project, designed to divert water from the San Juan River to serve the Albuquerque area—to garner congressional support.

Though the welfare of Native American peoples was gaining greater sympathy in American society, Indian irrigation projects remained highly contested issues in the West. Controversy was closely associated with the meaning of Indian water rights and impacts on non-Indian irrigation projects. Historically, the Bureau of Reclamation's record regarding Indian irrigation was spotty at best. The general consensus appears to be that Reclamation normally favored non-Indian projects over those benefiting Native Americans and did little to protect or develop Indian water resources—a common approach within U.S. Government policies.[229]

---

[228] Floyd E. Dominy to W. R. Whitman, Director, Yuma Irrigation District, December 28, 1960, Dominy Papers, Box 2, folder, Biographical Information, Employment; Stewart L. Udall, *Oral History Interview: Stewart L. Udall*, William M. Moss, Interviewer, comprised of eight interviews between January and September 1970 (Washington, D.C.: John F. Kennedy Library, 1981), 151-2; see also McCool, *Command of the Waters*, 5; Farmer, *Glen Canyon Dammed*, 147.

[229] There is substantial literature on the Bureau of Reclamation's half-hearted efforts toward developing Indian irrigation projects, see Donald J. Pisani, "Water Rights and the Betrayal of Indian Allotment," *Environmental History Review* 10:3 (Autumn 1986): 157-76; Pisani, *Water and American Government*, 181-201; McCool, *Command of the Waters*; Ann Caylor, "'A Prom-

The overriding question hinged upon the degree to which Indians owned or held water rights. In 1908 the U.S. Supreme Court ruled in *Winters v. United States* that Indians had a "reserved right" to water based upon treaty agreements that established Indian reservations. For Reclamation and most westerners, this "reserved right" challenged the prior appropriation doctrine instituted throughout the American West. Reclamation law stipulated that the Bureau of Reclamation observe state laws, yet the Winters Doctrine "implied water rights for Indians that trumped non-Indians' state water law rights." In 1963 the Supreme Court ruling in *Arizona v. California* further strengthened so-called "Winters rights." The Court's decision gave Indian reservations "enough water to irrigate 'practically all irrigable acres.'"[230]

Both rulings profoundly affected western water issues in the late twentieth century. Though little was done to protect Indian water rights as stipulated in the *Winters* case, its implications caused concern and revised plans in water resource development proposals. With the Supreme Court further asserting the Winters Doctrine in *Arizona v. California*, Native American tribes began to demand greater control of all their natural resources, not only water. Both rulings helped reshape many of the Bureau of Reclamation's subsequent water resource management policies. Developing controversies led to a string of water settlements that quantified Indian water rights in an era of decreasing water supply.

Attempts to develop irrigation projects on Indian reservations began in the late nineteenth century. The federal government saw projects as part of a larger goal to assimilate Indians into the dominant society by transforming them into individual landowners. The 1887 Dawes General Allotment Act primarily was the instrument of this policy that sought to establish individual Indians on small landholdings. For arid reservation lands, irrigation would seal the bond between the allotee and his/her land. According to historian Donald Pisani, "Irrigation promised the same benefits to whites and Indians—smaller more compact farms, immunity from drought, higher value crops, and larger yields."[231]

---

ise Long Deferred': Federal Reclamation on the Colorado River Indian Reservation," *Pacific Historical Review* 69:2 (May 2000): 193-215.

230 Robert T. Anderson, "Indian Water Rights, Practical Reasoning, and Negotiated Settlements," *California Law Review* 98:4 (August 2010): 1139-40; Judith Eva Jacobsen, "A Promise Made: The Navajo Indian Irrigation Project and Water politics in the American West," PhD Thesis, University of Colorado, 1989, 93.

231 Pisani, "Water Rights and the Betrayal of Indian Allotment," 158.

Irrigation projects on Indian reservations were the responsibility of the Office of Indian Affairs (later renamed the Bureau of Indian Affairs). Throughout the late nineteenth century, Indian Affairs actively tried to construct irrigation works on many reservations, including the Colorado River Indian Reservation in 1868, the Navajo Reservation in 1886, and the Fort Hall Reservation in 1894. Most of these efforts, however, failed to live up to expectations because Indian agents lacked engineering experience and were unable to coordinate and implement funding. In regards to the Colorado River Indian Reservation, historian Ann Caylor observed that "this first irrigation plan failed because planners underestimated the technical difficulties and the high cost of arid land redemption." Nevertheless, efforts to construct and maintain irrigation projects continued and by 1905 the foundation of the Indian Irrigation Service working in tandem with the U.S. Reclamation Service was in place.[232]

The nineteenth-century irrigation initiatives by the Indian Office's irrigation projects implied that Indian reservations possessed water rights. In addition, an 1894 report from the U.S. House of Representatives "recognized the need to reserve and protect a *future* water supply for the Indians." Evidence of this intent by Congress surfaces in various Acts concerning Indians and Indian reservations: 1887 General Allotment Act; 1891 on the Umatilla Reservation in Oregon; 1894 on the Yakima Reservation in Washington; 1899 on the Uintah Reservation in Utah. In 1903 even some Reclamation Service officials noted that there might be a need to protect Indian water rights lest the Indians lose them. Daniel McCool notes that Chief Engineer Frederick Newell "warned that there was 'great danger' that the Indians would lose their water because it was not being put to beneficial use and would eventually be appropriated by others."[233] Recognition of Indian water rights, however, did not translate into a concerted effort by Congress and the Reclamation Service to aid tribes in developing their water resources to the fullest extent. However, in 1907, as part of an agreement between the Reclamation Service and Indian Affairs, Reclamation

---

[232] Caylor, "'A Promise Long Deferred,'" 193, 197-201; Pisani, "Water Rights and the Betrayal of Indian Allotment," 158-60; Andrew Denny Rogers, "Indian Irrigation: Its Organization and Achievements," an unpublished manuscript, n.d., 7-9 in Andrew Denny Rogers Papers, American Heritage Center, University of Wyoming, Laramie.

[233] United States House of Representatives, *Umatilla Irrigation Company: Report*, Report No. 3379, 51st Cong., 2nd sess., January 6, 1891; United States House of Representatives, *Right of Way through Yakima Indian Reservation: Report*, Report No. 714, 53rd Cong., 2nd sess., April 17, 1894, italic in the original; see also Pisani, *Water and American Government*, 162; Lloyd Burton, *American Indian Water Rights and the Limits of the Law* (Lawrence: University Press of Kansas, 1991), 2; Caylor, "A Promise Long Deferred," 205; McCool, *Command of the Waters*, 112.

took over the planning and construction of irrigation projects on Indian reservations. The three largest "Indian" projects taken on by the Reclamation Service were located on the Fort Peck, Blackfeet, and Flathead reservations. According to Donald Pisani in *Water and American Government*, "the Reclamation Service promised to reclaim more than four hundred thousand acres." These projects, however, would serve both Indian and non-Indian land, and any success these projects gained stemmed from their emphasis on irrigating non-Indian property. Indeed, this characteristic became so pronounced that by the 1930s many came to view these projects as "white" projects.[234]

There were many reasons for the Bureau of Reclamation's failure to provide irrigation works solely for the benefit of Indian tribes. The larger social and cultural aspects of the federal government's goal of assimilation and allotment of Indian lands to individuals played an important part in setting priorities. Once lands were allotted, the remaining reservation lands were opened to non-Indian settlers. In addition, Reclamation focused on water delivery systems to non-Indians who were more likely able to pay for the services. As a rule, Indians had much smaller landholdings and often sold or rented their allotments to whites. Historian Pisani argues that when Reclamation did pay attention to Indian reclamation development, its efforts served to keep the Reclamation program going. Indian lands and water sometimes provided the last untapped resources in the West for Reclamation to develop. Pisani notes, "The Reclamation Service had little interest in the welfare of Native Americans, but it needed Indian land and money." And one might add water to the list.[235]

Still, Indian apathy and Reclamation self-interest only supply partial answers. Progressive Era conservation ideals played an important role in determining how successful the Reclamation Service could be in irrigating Indian lands. The perception, among most westerners who adhered to the prior appropriation doctrine, was that Indians were not utilizing their resources, and unused Indian water was going to waste. In addition, Indian tribes faced the paternalistic and, at times, patronizing qualities inherent in federal Indian

---

234 Pisani, *Water and American Government*, 161; Garrit Voggesser, "The Indian Projects: The Crow Project," Denver, Colorado: Bureau of Reclamation History Program, 2001, 3; Garrit Voggesser, "The Indian Projects: Blackfeet Project," Denver, Colorado: Bureau of Reclamation History Program, 2001, 1; McCool, *Command of the Waters*, 163-5.

235 Pisani, *Water and American Government*, 159; Garrit Voggesser, "The Flathead Project: The Indian Projects," Denver, Colorado: Bureau of Reclamation History Program, 2001, 22; Voggesser, "The Indian Projects: Blackfeet Project," 22-3; Harold Shepherd, "Conflict Comes to Roost! The Bureau of Reclamation and the Federal Indian Trust Responsibility," *Environmental Law* 31:901 (2001): 926.

policy that subscribed to a popular notion that non-Indian settlers were better suited to put both reservation lands and water to beneficial use, and through that example, Indians would learn the value of private property. Absent from this discussion was recognition that despite the apparent lack of Indian interest in farming, most tribes understood the need to protect their water rights. Furthermore, most Indian irrigation projects were authorized without tribal consent, and tribal members protested the appropriation of reservation resources to fund projects that primarily benefited non-Indian farmers. In sum, Indian concerns, traditions, and cultures were virtually ignored in pursuit of federal and Reclamation goals to "make the desert bloom."

Indians were not alone in recognizing the unbalanced benefits provided by the Reclamation Service's Indian irrigation projects. Over time, the Office of Indian Affairs expressed concern about the Reclamation Service's preference for non-Indian irrigators. By 1913 Indian Affairs developed its own engineering division, and in 1924 severed its relationship with Reclamation and "reclaimed" control over Indian irrigation projects. Congress, exemplifying the national trends, continually underfunded Indian irrigation projects in comparison to the appropriations given to the Bureau of Reclamation. Work on Indian irrigation remained a constant struggle for the Bureau of Indian Affairs throughout the twentieth century. The final failure of Indian irrigation stems from competition for scarce resources between Indians and non-Indians. Ann Caylor suggests that Indian Affairs and the Reclamation Service simply "could not aid local western interests and simultaneously assist Indian development …." Pisani puts the argument more succinctly, "In a contest between whites and Indians, political expediency, if nothing else, dictated that the Reclamation Service would support white farmers. The 1902 act, after all, had been written for those homesteaders."[236]

Despite the mixed results of the Bureau of Reclamation's efforts to develop irrigation projects on Indian land, the issue of Indian water rights remained a constant concern regarding the future of water development in the West. The hallmark event for Indian water rights occurred with the 1908 Supreme Court's ruling in *Winters v. United States* decision. The federal government on behalf of the Indians on the Fort Belknap reservation in Montana sued upstream non-Indian irrigators in the Milk River Valley on the grounds that the tribe had a superior water right based on the treaty that created the reservation. According to legal scholar A. Dan Tarlock,

---

[236] Shepherd, "Conflict Comes to Roost!," 915; Caylor, "'A Promise Long Deferred'," 195; Pisani, "Water Rights and the Betrayal of Indian Allotment," 160-1.

> *Winters* held that the reservation had an implied water right with a priority date of the 1888 Fort Belknap Agreement, the treaty that modified the reservation. Thus, by looking backward, Indians reserved rights, as they came to be called, are superior to all pre-reservation state appropriation rights.

In the view of the Bureau of Reclamation and proponents of the Reclamation program, the "Winters Doctrine" threatened the doctrine of prior appropriation that had become the foundation of irrigation development. Daniel McCool explains, "The very invocation of implied Treaty rights directly contradicts existing prior appropriation water law … *Winters* rights exist *irrelevant of usage*; …" perhaps more importantly, the "Winters Doctrine is the principal source of water rights for Indians whereas Prior Appropriation Doctrine is the principal source of water rights for non-Indians."[237]

Throughout the first half of the twentieth century, Indian "reserved rights" had little bearing on Bureau of Reclamation projects. In reality, non-Indian groups benefitted. The Supreme Court must shoulder some of the burden for the shortcomings in Indian irrigation development. *Winters* left unsaid the amount of water due reservations. This ambiguity posed a theoretical threat to water development in the West, but theory was a long way from reality in the western landscape. On the ground, the Bureau of Reclamation projects had the backing of statutory law, while *Winters* was a case decided by the high court. With no one willing or able to defend or protect Indian water rights, authorized Reclamation projects routinely infringed upon Indian claims. Congress continued to approve projects, and Reclamation faithfully performed its duty by fulfilling the intent of Congress. On those rare occasions when criticism arose over the pork-barrel spending on water projects, Reclamation and its allies promoted some projects as aiding Indian reservations. Critics have come to term this development as the "Indian blanket," where, according to Lloyd Burton in *American Indian Water Rights and the Limits of the Law*, "Proponents would add to the disputed proposal a feature calling for the deliv-

---

[237] A. Dan Tarlock, "Tribal Justice and Property Rights: The Evolution of *Winters v. United States*," *Natural Resources Journal* 50:2 (Spring 2010): 478; for more information on the *Winters* case see Norris Hundley, Jr., "The 'Winters' Decision and Indian Water Rights: A Mystery Reexamined," *Western Historical Quarterly* 13:1 (January 1982): 17-42; Norris Hundley, Jr., "The Dark Bloody Ground of Indian Water Rights: Confusion Elevated to Principle," *Western Historical Quarterly* 9:4 (October 1978): 454-82; John Shurts, *Indian Reserved Water Rights: The Winters Doctrine in Its Social and Legal Context* (Norman: University of Oklahoma Press, 2000); see also McCool, *Command of the Waters*, 44, italics in the original.

ery of water to an Indian reservation in the vicinity of the desired reclamation facility."[238]

11.10. Senator Thomas H. Kuchel Republican from California from 1951 to 1968. Courtesy of the U.S. Senate History Program.

From 1924 to 1946, the Bureau of Reclamation limited its involvement in Indian irrigation projects. It deferred to the Indian Irrigation Service under the auspices of the Office of Indian Affairs, which took the lead in developing water resources on reservations. By 1956, however, when Congress passed the Colorado River Storage Project Act, the Bureau of Reclamation showed a newfound interest in Indian irrigation projects. In its 1946 report on the Colorado River subtitled "*A Natural Menace Becomes a Natural Resource*," the Bureau of Reclamation discussed new opportunities for Indian irrigation projects throughout the river basin. The Office of Indian Affairs advocated new plans and contributed proposals to Reclamation's 1946 report as a "cooperating interest." In its recommendations, the Indian Office sought to more than double the amount of irrigable acres on Indian reservations from 262,290 to 566,440. This total included Indian lands then irrigated by non-Indians. To accomplish this, the Indian Affairs Office estimated that it would require almost tripling the amount of water already delivered to Indian reservations from 1,034,308 acre-feet to 2,845,420 acre-feet. The Office of Indian Affairs argued that the need to develop Indian land and water resources was critical to the well-being of Indian tribes. It stressed the federal government's responsibility "to protect Indian rights and to provide resources sufficient to enable Indian people to attain economic independence at a level comparable to other citizens."[239]

In 1955 the Bureau of Indian Affairs pushed for irrigation development on the Navajo Indian Reservation in New Mexico as part of the Colorado

[238] Burton, *American Indian Water Rights and the Limits of the Law*, 84; McCool, *Command of the Waters*, 174.

[239] United States Department of the Interior, Bureau of Reclamation, *The Colorado River*, 261-8.

River Storage Project. A report from Indian Commissioner Glen L. Emmons proposed that a dam and reservoir combined with a distribution system for irrigation could serve 137,250 acres of new land within and adjacent to the Navajo Indian Reservation. That same year proponents of the Colorado River Storage Project touted the benefits of this project for Indians and used it as a rallying cry to win support for the legislation. Eventually they were successful. Congress named the Navajo Indian Irrigation Project a participating project of the Colorado River Storage Act of 1956. Still, appropriations in the original legislation only provided funding for the construction of Navajo Dam on the San Juan River in New Mexico. The larger Navajo Indian Irrigation Project had to wait for later authorization.[240]

In debates over the Colorado River Storage Project, opponents brought up the issue of Indian water rights. They portrayed the unstated and unknown quantitative water rights Indians possessed as a threat to the feasibility of the project. California Senator Thomas Kuchel claimed that the Senate Committee on Interior and Insular Affairs was

> unable to obtain a clear statement of the attitude which the United States takes in regards to the quantities of water to which the Indian tribes may eventually be entitled. Officials of the Department of the Interior, while maintaining the Indian rights issue is immaterial to S.500, could not state what priorities are claimed on behalf of the tribes in the basin.

As a Californian and U.S. Senator, Kuchel opposed the overall legislation and there is little question that he sought to raise the Indian water issue to provoke antagonism among the basin states concerned about Colorado River flow rates. The question was important because any water diverted for Indian purposes affected everyone else. More importantly, his question was never answered. Congress ignored Indian water rights in the legislation but did endorse the principle that Indian projects were non-reimbursable, which meant Indians were freed from paying construction costs.[241]

---

240 Bureau of Indian Affairs, *Navajo Project New Mexico: Feasibility Report*, 2; "Annual Project History, Navajo Unit, Colorado River Storage Project, Vol. I," 1956, RG 115, Entry 10, Box 128; see also United States Senate, Committee on Interior and Insular Affairs, *Report: Authorizing the Secretary of the Interior to Construct, Operate, and Maintain the Colorado River Storage Project and Participating Projects*. S. Rpt. No. 1893, 83rd Cong., 2nd sess., July 26, 1954 (Washington, D.C.: United States Government Printing Office, 1954), 5; Jacobsen, "A Promise Made," 126.

241 United States Senate, Committee on Interior and Insular Affairs, *Colorado River Storage Project: Report, A Bill to Authorize the Secretary of the Interior to Construct, Operate,*

11.11. Clinton P. Anderson, Democrat from New Mexico, served in the U.S. House of Representatives from 1941 to 1946 and in the U.S. Senate from 1949 until 1972. Courtesy of the U.S. Senate History Program.

In 1958, New Mexico senators Clinton Anderson and Dennis Chaves introduced legislation to authorize and fund the Navajo Indian Irrigation Project (NIIP). As part of the bill, the senators bound the Navajo Indian Irrigation Project to the San Juan-Chama Project: a trans-basin diversion plan to take water from the San Juan River to supply irrigation, municipal, and industrial water to central New Mexico and the Albuquerque area. Some scholars have suggested that connecting NIIP to the San Juan-Chama Project presents an example of lawmakers and water resource development boosters embracing the so-called "Indian blanket" to drum up support for water projects. According to historian Steven Schulte, "In the early 1960s, Native American water rights were beginning to be considered in larger western water-allocation formulas." He further states that New Mexico's "political leadership realized that it might be possible to achieve larger reclamation goals by using the cover of a 'Navajo blanket.'" Clearly irrigation projects for Indians were becoming more politically popular as awareness permeated American society about the poor living conditions found on Indian reservations.[242]

In contrast to earlier efforts to bring irrigation to Indian reservations, the Navajo actively lobbied for the Navajo Indian Irrigation Project between

and Maintain the Colorado River Storage and Participating Projects in the States of Arizona, Colorado, New Mexico, Utah, and Wyoming, S. Rpt. No. 128, 84th Cong., 1st sess., March 30, 1955 (Washington, D.C.: United States Government Printing Office, 1955), 43; see also United States House of Representatives, *Committee Conference Report: Authorizing the Secretary of the Interior to Construct, Operate, and Maintain the Colorado River Storage Project and Participating Projects*, S. Rpt. No. 1950, 84th Cong., 2nd sess., March 27, 1956 (Washington, D.C.: United States Government Printing Office, 1956), 9.

[242] Steven C. Schulte, *Wayne Aspinall and the Shaping of the American West* (Boulder: University Press of Colorado, 2002), 106; Burton, *American Indian Water Rights and the Limits of the Law*, 84; see also Leah S. Glaser, "The Navajo Indian Irrigation Project," Denver: Bureau of Reclamation History Program, 1998, 15.

1958 and 1962. Throughout the congressional life of the bill, Navajo leaders travelled to Washington, D.C., to testify in favor of the legislation. Their testimony continually stressed the importance of the project to the welfare of the tribe. Department of the Interior officials also emphasized the tremendous benefits of the proposed project to the Navajo people. In 1961 Secretary of the Interior Udall specifically outlined the economic benefits. Udall claimed, "The Navajo Indian Irrigation Project by providing irrigation for 110,000 acres of land within and adjacent to the Navajo Indian Reservation would give a powerful shot in the arm to this area." According to the secretary, the project meant 1,120 new family farms and the promotion of "allied industries," to create 2,240 jobs for Navajo families.[243]

11.12. Dennis Chavez, Democrat from New Mexico, served in the U.S. House of Representatives from 1931 to 1934 and in the U.S. Senate from 1935 to 1962. Courtesy of the U.S. Senate History Program.

The Bureau of Reclamation indeed had grand plans for the Navajo Indian Irrigation Project. Provisions within the legislation called for an enlarged canal to supply industrial and municipal water to help spur economic diversification on the reservation. While a new day seemed to have arrived for Indian reclamation, the Navajo were obliged to negotiate for the project and make compromises. Most importantly, tribal leaders waived their "reserved rights" under the Winters Doctrine and agreed to a guaranteed or limited delivery of 508,000 acre-feet of water. The tribe recognized the significance of the concessions. Tribal Executive Secretary J. Maurice McCabe spoke for all the Navajo when he stated:

> The Navajo Tribe consented to this, and relinquished its rights under the Winters doctrine for the water necessary to irrigate

243 United States House of Representatives, Hearings before the Subcommittee on Irrigation and Reclamation of the Committee on Interior and Insular Affairs, *San Juan-Chama Reclamation Project and Navajo Indian Irrigation Project*, 87th Cong., 1st sess., April 24, 25, 26, and June 1, 1961 (Washington, D.C.: United States Government Printing Office, 1961), 91-2.

> the Navajo Indian irrigation project, in order to provide a practical plan for comprehensive development of the resources and industrial potential of the San Juan basin of New Mexico. We have taken this important and far-reaching step because such development is necessary for our very survival.

There were further concessions. Congress deemed that the "primary purpose" of the Navajo Indian Irrigation Project was to irrigate Navajo lands. Congress did recognize that some water would go to municipal and industrial uses. However, it held that before Congress would authorize funds to expand the main canal from Navajo Reservoir to meet municipal and industrial demands repayment contracts for M&I water must be signed and approved by Congress before enlargement of the canal began.[244]

Observers have paid special attention to this particular aspect of the project and its relationship to the Indian water rights issue. The limited flow of the San Juan River and the two projects that vied for the same resource created an intense negotiating atmosphere. The San Juan-Chama project proposed to divert water from the San Juan River to the Rio Grande for supplemental irrigation water in the Rio Grande Valley and to supply municipal and industrial water for Albuquerque. Historian Leah Glaser writes, "Because they competed for the same water source, final approval of the two projects followed complicated and often heated negotiations balancing Indian rights against non-Indian claims to the water of the San Juan River." Others suggest that the tribal concession was necessary to attain congressional approval, and they note House Interior and Insular Committee chairman Wayne Aspinall's demand that the Navajo place some limit on their water rights. Indeed this was Aspinall's *modus operandi* as chairman. Steven Schulte maintains, "No new federal policy would take effect until it had passed through Wayne Aspinall's Interior Committee. Aspinall would … need to be satisfied that any new law would be a law the West—or western Colorado—could benefit from." Moreover, the Navajo quantification of their water rights shaped the scope of the San Juan-Chama Project. Some arguments claim that for Reclamation and its allies the trans-basin diversion project commanded greater interest than the overall San

---

[244] *Ibid.*, 33; see also McCool, *Command of the Waters*, 183-4; Schulte, *Wayne Aspinall and the Shaping of the American West*, 105; for the congressional conditions on the main canal see United States House of Representatives, Committee on Interior and Insular Affairs, *Authorizing the Construction of the Navajo Indian Irrigation Project and the Initial Stage of the San Juan-Chama Project as Participating Projects of the Colorado River Storage Project*, H. Rpt. No. 685, 87th Cong., 1st sess., July 10, 1960 (Washington, D.C.: United States Government Printing Office, 1960), 12.

Juan River basin development. If so, the effect was to place limits on Indian water projects in favor of non-Indian projects. The casual observer needs only to compare and contrast the San Juan-Chama project with NIIP. According to one source, "Work began immediately on the San Juan-Chama Project and was sustained for the fourteen years required to complete project works," while work on NIIP lagged. By 1976, the San Juan-Chama Project was complete, but only 10,000 acres of the Navajo Indian Irrigation Project had received water. However, it is also important to note that the Indian Irrigation Service and the Bureau of Indian Affairs had to balance budget requests for NIIP with their other budget priorities.[245]

Moreover, the Navajo Indian Irrigation Project was the only Bureau of Reclamation project of the Colorado River Storage Project primarily dedicated to Indian lands. Navajo acceptance of a quantitative water right suggested that the Navajo saw ahead important strides in the improvement of living conditions on the reservation. They no doubt perceived NIIP as more than just an irrigation project to improve agricultural development, but also envisioned industrial benefits through the promise of a stable industrial and municipal water supply. Equally significant, the Navajo maintained that their concession on water rights depended upon completion of the project, and did not affect Winters Doctrine rights on other water sources throughout the reservation. In a written statement to the Subcommittee on Irrigation and Reclamation, Maurice McCabe stressed the tribe's attitude on this issue. He wrote,

> It should be known to the committee and other interested parties that the Navajo Tribe will not consider itself bound by this agreement unless the irrigation project is in fact established. It is clearly understood by all parties, I believe, that the tribe's concession to the Winters doctrine applies to no other situation other than this one.[246]

On June 3, 1962, Congress authorized initial construction of the Navajo Indian Irrigation Project and the San Juan-Chama Project. The Bureau of Reclamation figured NIIP to be a long term project taking fourteen years to complete and costing an estimated $135,000,000. Reclamation's plans called for a 600 mile delivery system originating from Navajo Dam on the San Juan River. In its

---

[245] Glaser, "The Navajo Indian Irrigation Project," 15-6; Schulte, *Wayne Aspinall and the Shaping of the American West*, 70; Jacobsen, "A Promise Made," 132-3.

[246] Subcommittee on Irrigation and Reclamation, *San Juan-Chama Reclamation Project and Navajo Indian Irrigation Project*, 46.

entirety, the project consisted of a main canal, two significant tunnels, multiple siphons, laterals, and several smaller tunnels, all to service eleven 10,000 acre blocks on the Navajo Reservation. After four years of intense negotiations, the Navajo seemed to have a legitimate Reclamation project.[247]

Almost immediately, delays hampered construction. Initially, the Bureau of Reclamation contended that the original Bureau of Indian Affairs plans were incomplete and required further investigations. Reclamation engineers found so many faults in the BIA's initial proposal that they proposed a new investigation. Geographer Judith Jacobsen reported that Reclamation project director Bert Levine found that "the preplanning wasn't worth the paper it was written on," noting that the "BIA didn't have the funds to do a real plan … we threw the whole thing out."[248] Levine's comment reflected more than Reclamation's professional hubris. As the organization responsible for the construction, Reclamation engineers needed to be satisfied with the plans. As a result, construction on the project did not begin until 1964.

There were other factors that contributed to snail-paced construction activities on the Navajo Indian Irrigation Project. NIIP legislation stipulated that appropriations come through the Bureau of Indian Affairs' budget. Congress required this condition to ensure "that the Navajo Indian Irrigation project would be funded and operated as an Indian project under Indian laws." Once again critics note Wayne Aspinall's role in requiring this particular stipulation. In the scathing Nader Task Force report on the Bureau of Reclamation, *Damming the West*, authors Richard Berkman and Kip Viscusi claim, "Aspinall insisted on the unusual procedure because the Bureau of Indian Affairs has considerably less influence in Congress than does the Bureau of Reclamation." This meant that Reclamation's only responsibility was to the project's engineering and construction aspects. It had little motivation to actively pursue funding. Over the course of construction, Congress regularly underfunded NIIP which led to delays. On the other hand, some critics suggest that lack of progress was representative of Reclamation's historic disinterest in Indian projects, or its preference for non-Indian projects. As noted earlier, Reclamation detractors look at the on-time progress made on the San Juan-Chama Project as evidence of the Bureau of Reclamation's lack of commitment to NIIP.[249]

---

[247] Glaser, "The Navajo Indian Irrigation Project" 16-7; Jacobsen, "A Promise Made," 131-2.

[248] The Levine quote comes from an interview with Judith Jacobsen, see Jacobsen, "A Promise Made," 138.

[249] Jacobsen, "A Promise Made," 156; Richard L. Berkman and Kip Viscusi, *Damming the West: The Nader Task Force Report on the Bureau of Reclamation* (Washington, D.C.: Center for

The Bureau of Indian Affairs continued to have difficulty with the congressional funding process. As mentioned above, the historic competition for funding found the Indians consistently holding the short straw when it came to a choice between Indian and non-Indian irrigation. Historian David Dejong notes, "As was frequently the case in the West, funding for Indian projects lagged, as they competed directly with settlers for federal funding." This condition carried well into the mid- to late twentieth century demonstrating that old habits are hard to break. In his study on "iron triangles," Daniel McCool notes that the BIA was never able to accumulate the same cohesive power block that the Bureau of Reclamation attained and that "the Indian triangle has long suffered from a low level of political effectiveness." This ineffectiveness very much affected construction progress on the Navajo Indian Irrigation Project.[250]

Indeed, many interested parties voiced dismay at the slow progress on the NIIP; none more so than the Navajo. In May 1967 the Navajo Tribal Council issued a resolution expressing dissatisfaction with the ability of the Bureau of Indian Affairs to obtain adequate project funding by noting that "as a result the construction of said project is at a stage that the delivery of water to project lands will be three years later than originally scheduled." A growing sense of frustration pervaded the Navajo community. The resolution went further to state, "In order to make up for lost time in the construction it would be more appropriate to have funds appropriated through the Bureau of Reclamation."[251] For many within the Tribal Council, concern grew that the Navajo Indian Irrigation Project was another episode in the federal government's long history of failing its obligations to Native Americans.

The Navajo were not alone in their concerns as others voiced similar dissatisfaction with the allocations of funds for NIIP. In 1969 Senator Joseph Montoya from New Mexico expressed his dismay at the slow progress and the meager appropriations. He claimed that "through the fiscal year 1969, a total of $75,800,000 has been programmed for this project, however, only

---

the Study of Responsive Law, November 1971), 228.

250 David H. Dejong, "'Abandoned Little by Little:' The 1914 Pima Adjudication Survey, Water Deprivation, and Farming on the Pima Project," *Agricultural History* 81:1 (Winter 2007): 61; McCool, *Command of the Waters*, 5.

251 Resolution of the Navajo Tribal Council, May 4, 1967, in United States Senate, Committee on Interior and Insular Affairs, Subcommittee of Water and Power Resources, *Navajo Indian Irrigation Project*, Hearing on S. 203, 91st Cong., 1st sess., July 15, 1969 (Washington, D.C.: United States Government Printing Office, 1969), 28; see also Jacobsen, "A Promise Made," 156.

a total of $28,300,000 has been requested by the Bureau of the Budget and appropriated by Congress." A year later another New Mexico representative echoed similar sentiments. Testifying before the House Subcommittee on Irrigation and Reclamation, Congressman Ed Foreman stated, "The Navajo Indians are not looking for a welfare handout. They are expecting from the Government of the United States fulfillment of the promises made by this Government which date from the 1858 treaty." Wayne Aspinall took particular umbrage at Forman's remarks, reminding the congressman that it was the state of New Mexico that pushed the San Juan-Chama Project instead of NIIP. He scolded his colleague stating, "So far as any delay is concerned in this project, I doubt very much if my colleague can point his finger to either Congress or the administration."[252]

Aspinall did acknowledge that there were problems funding the Navajo Indian Irrigation Project through the Bureau of Indian Affairs, "so the Bureau of Reclamation had never been in position to go ahead full speed with this construction." His acknowledgement appeared disingenuous. After all, Aspinall ensured that appropriations for NIIP would have to go through the BIA and that his committee oversaw both Reclamation and Indian policy. Moreover, the Colorado congressman was more than likely well aware of the BIA's poor track record in Indian irrigation. In 1962, testifying before Aspinall and the Subcommittee on Indian Affairs, Associate Commissioner of the Bureau of Indian Affairs Martin P. Mangan bluntly stated, "I might point out to the committee that the Bureau of Indian Affairs has been working on Indian Irrigation projects since before the turn of the century. To my knowledge we have never completed any of them in place." By 1977 nothing much had changed for the BIA in obtaining funding for Indian irrigation projects. The commissioner of Indian Affairs commented on the serious problem of just maintaining what facilities the Indian Irrigation Service had constructed. "Very candidly, a lot of existing projects that we have were authorized and were never completed. So what we do is we try to keep the project operational with what we have, and it is kind of giving enough money so the project doesn't completely collapse."[253]

---

[252] For Senator Montoya's quote see Berkman and Viscusi, *Damming the West*, 227; for Foreman's quote and Aspinall's response see United States House of Representatives, Subcommittee on Irrigation and Reclamation of the Committee on Interior and Insular Affairs, *Navajo Indian Irrigation Project: Hearings on S.203 and H.R.13001 to Amend the Act of June 13, 1962, with Respect to the Navajo Indian Irrigation Project*, 91st Cong., 2nd sess., June 11, 1970 (Washington, D.C.: United States Government Printing Office, 1970), 5, 7.

[253] United States House of Representative, Subcommittee on Indian Affairs, Committee on Interior and Insular Affairs, *Increasing the Appropriation for the Completion of the Construction*

While the Bureau of Reclamation did not have control over the funding of the Navajo Indian Irrigation Project, it did not mean that Reclamation was silent regarding the slow pace of construction. Reclamation reports to Congress consistently pointed out that the slow progress on NIIP was related directly to the meager congressional appropriations. For example in one report a Reclamation official stated, "Under the rate of funding of the construction of the Navajo irrigation project, the first water will not be delivered to project lands until 1975 … A much longer period is possible if funds are not forthcoming to maintain the revised schedule of construction." In addition, it did not help matters that the scope of the project continually changed. The first alterations came in the form of Reclamation's revision of the original BIA plans when it found lands originally destined for irrigation unsuitable and suggested substitution of better lands outside of the reservation. Though not an insurmountable obstacle, revisions called for land transfers and congressional approval. More delay came in the decision making on the type of irrigation farming to undertake. The original legislation called for creation of individual family-type farms. Over time that plan morphed to a corporate agribusiness model.[254] These alterations plus Congress's reluctance to fund NIIP made for slow progress.

The question of Indian "reserved rights" under the Winters Doctrine was never far from the mind of Congressman Wayne Aspinall. While conducting hearings on amending NIIP legislation in 1970, he questioned Assistant Secretary of the Interior Harrison Loesch and Raymond Nakia, chairman of the Navajo Tribal Council, as to the future Navajo water claims. Beyond the congressional concerns, Indian water rights continued to trouble the Bureau of Reclamation. Historically, Reclamation officials often chose to ignore or dismiss any recognition of Indian "Winters" rights because, according to Daniel McCool in *Native Waters*, "any water allocated to Indians could not be used to fuel the growth of the reclamation program." In

of the Irrigation and Power Systems of the Flathead Indian Irrigation Project, Montana, Hearing, 87th Cong., 2nd sess., August 27, 1962, unpublished hearing, 5; quote of the commissioner of Indian Affairs taken from McCool, *Command of the Waters*, 125.

254 United States Senate, Committee on Interior and Insular Affairs, *Amending the Act of June 13, 1962, with Respect to the Navajo Indian Irrigation Project*, S. Rpt. No. 91-363, 91st Cong., 1st sess., August 7, 1969, 6; United States House of Representatives, Committee on Interior and Insular Affairs, *Amending the Act of June 12, 1962 (76 Stat. 96), with Respect to the Navajo Indian Irrigation Project*, H. Rpt. No. 91-1331, 91st Cong., 2nd sess., July 27, 1970, 2-3; for information on the change from family farms to corporate farming see Water and Power Resources Service, "Navajo Indian Irrigation Project," May 1980, 40; Glaser, "The Navajo Indian Irrigation Project," 19.

connection with NIIP, there is evidence to suggest that Reclamation sought to limit water deliveries to the reservation by employing new water saving technologies. A 1974 Reclamation report, titled "Navajo Indian Irrigation Project, New Mexico, All-Sprinkler System," claimed that a sprinkler system could irrigate the entire 105,000 acres using only 330,000 acre-feet of water "with a depletion of 236,000 acre-feet" in annual deliveries. On the surface, this observation appears as a reasonable and highly effective water conservation solution. Yet others contend that it was an effort to take water away from the Navajo. According to Lloyd Burton in *American Indian Water Rights and the Limits of the Law*, "This in turn stimulated a legal opinion within the Interior Department to the effect that the Navajo were entitled to only enough water to irrigate the amount of land mentioned in NIIP legislation …"[255]

By 1988 the federal government had spent $450 million on the Navajo Indian Irrigation Project with a little more than half the project completed. Although inflation increased costs, only six of the eleven 10,000 acre plots received irrigation water. In spite of slow development, the Navajo pushed ahead. In 1986, according to Judith Jacobsen, the Navajo agricultural corporation, Navajo Agricultural Products Industry, had gross receipts exceeding $20 million, with profits averaging $2 million.[256] Beginning with NIIP, the Bureau of Reclamation's engagements with Indian water rights grew more complicated. Many of those complications, both legislative and legal, arose from the increasingly limited available water resources of the Colorado River. Indian reserved rights under the Winters Doctrine gained new recognition with the Supreme Court's 1963 ruling in *Arizona v. California*. Furthermore, the acrimonious legislative battles over the Central Arizona Project (CAP) and the Pacific Southwest Water Plan kept the issue of Indian water rights in sharp focus. Indian water rights threatened the security of the alliances among the so-called "iron triangle" members—Congress, water users, and federal bureaucracies.

---

[255] For Aspinall's concerns over the Navajo asserting "reserved rights" under Winters see Subcommittee on Irrigation and Reclamation of the Committee on Interior and Insular Affairs, *Navajo Indian Irrigation Project: Hearings on S.203 and H.R. 13001 to Amend the Act of June 12, 1962, with Respect to the Navajo Indian Irrigation Project*, 30-1; see also Daniel McCool, *Native Waters: Contemporary Indian Water Settlements and the Second Treaty Era* (Tucson: University of Arizona Press, 2002), 22; for information on the All-Sprinkler System see, Water and Power Resources Service, "Navajo Indian Irrigation Project," 2; Burton, *American Indian Water Rights and the Limits of the Law*, 31; Glaser, *The Navajo Indian Irrigation Project*, 22-3.

[256] Jacobsen, "A Promise Made," 134.

Central Arizona Project and the Pacific Southwest Water Plan

For many in Arizona, passage of the Colorado River Storage Project in 1956 was bittersweet. Beginning in the 1920s with Fred Colter and the Arizona Highline Reclamation Association, Arizonans had longingly dreamt of bringing the waters of the Colorado River to central Arizona. Colter went so far as to file a water right for the entire flow of the river through the state. Although well beyond the state's resources at the time, Colter's proposal rallied those who believed that the water of the Colorado River belonged to Arizona, because "it crossed their lands." Arizona was further stymied by its adamant refusal to sign the 1922 Colorado River Compact and its dogged determination to prevent passage of the Boulder Canyon Project Act, which spurred worries that California would take Arizona's share of the river.[257] New hopes arose in 1944 when Arizona finally ratified the Compact, opening the door to Arizona development of Colorado River water. In the late 1940s and early 1950s, Arizona's political and civic leaders fought a heroic fight to obtain passage of the Central Arizona Project (CAP); only to see those hopes dashed by a bitter suit with California in the Supreme Court (*Arizona v. California*). According to historian Donald Worster Arizonans worried, "that a bigger more advanced neighbor like California, a state which contributed virtually no runoff to the river, would get a larger mouthful to drink."[258]

The postwar debates over the Central Arizona Project resurrected decades-old animosities between the two states. The 1963 *Arizona v. California* ruling offered a partial respite. Although the decision appeared to recognize many of Arizona's long-sought rights to the Colorado River, the state still faced an uphill battle to use those rights. Arizona's determined push to obtain legislation for the Central Arizona Project, from 1963 to 1968, faced fierce opposition from not only California but also the states of the upper basin. Clearly CAP opened up old wounds and renewed interstate rivalries.

For the Bureau of Reclamation, the emerging controversies brought forth strident criticism which at times questioned Reclamation's very purpose. Pragmatically, Reclamation saw the project as a marvelous challenge in basin-

---

[257] Johnson, *The Central Arizona Project, 1918-1968*, 15; Hundley, *Water and the West*, 247-8; Byron E. Pearson, "'We Have Almost Forgotten How to Hope': The Hualapai, the Navajo, and the Fight for the Central Arizona Project, 1944-68," *Western Historical Quarterly* 31:3 (Autumn 2000): 297.

[258] Worster, *Rivers of Empire*, 209.

United States
Department of the Interior
Stewart L. Udall, Secretary

Pacific Southwest WATER PLAN

REPORT
January 1964

BUREAU OF RECLAMATION
Floyd E. Dominy, Commissioner

11.13. The Pacific Southwest Water Plan, published in January 1964, included several maps of proposed project works.

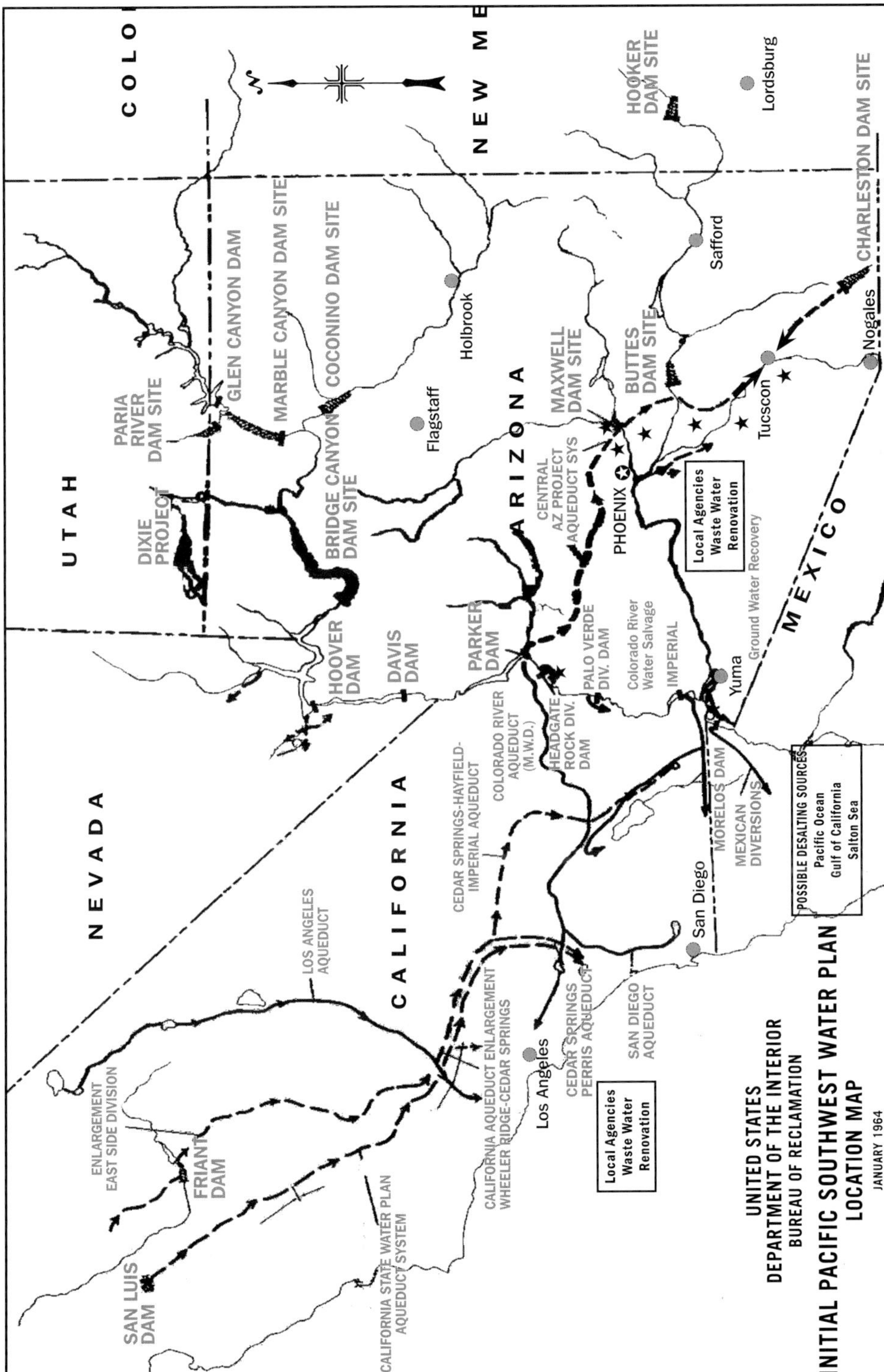

11.14. An early concept map for the entire Pacific Southwest Water Plan.

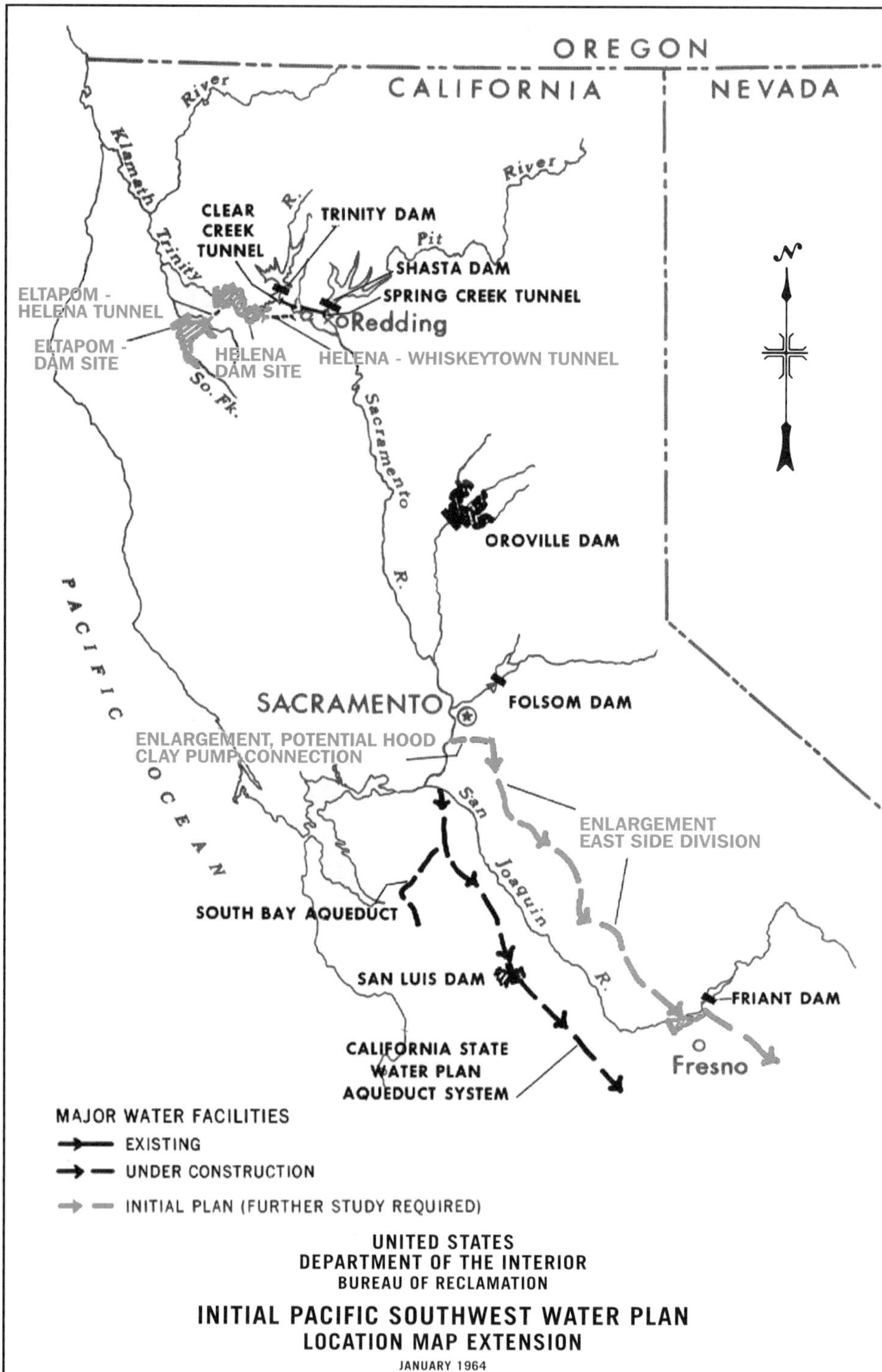

11.15. An early concept plan for the Pacific Southwest Water Plan in California..

wide multipurpose development. The Pacific Southwest Water Plan, which included the Central Arizona Project, outlined an idea that proposed construction of two large dams on the Colorado River along with a grandiose plan to augment the Colorado River from other sources. But Reclamation also found itself caught in the middle of political maneuverings. Traditional Reclamation political allies—Senators Carl Hayden and Henry Jackson, Congressman Wayne Aspinall, and Secretary of the Interior Stewart Udall—all championed their individual views and protected their turfs. The dispute that arose along the path toward authorization of the Central Arizona Project shook the foundations of Reclamation and marked the beginning of the end for the most dramatic and energetic dam-building era in western America's Reclamation history.

The history of the Central Arizona Project idea was almost as long as that of the congressional career of Arizona Senator Carl T. Hayden. Hayden's career mirrors the political, social, and economic history of the state of Arizona. He became Arizona's first and only congressman after statehood in 1912; eventually elected to the Senate in 1924 where he served until 1968. The Arizona senator was one of the Bureau of Reclamation's most avid supporters, seeing water development projects as instrumental to the growth and prosperity of not only his state, but the West in general. In *Vision in the Desert*, Jack August Jr. notes Reclamation Commissioner Floyd Dominy's observation that "Arizona and reclamation grew up together, and Carl Hayden nurtured them both." According to August, Hayden

> played a key role in periodically finding new sources to increase the reclamation fund to finance new and expanded water projects. He familiarized himself with reclamation techniques such as multiple-purpose projects, basin funds, reimbursable benefits, treasury repayments, and annual appropriations.

In 1922 Hayden was one of the few Arizona politicians who supported the Colorado River Compact. He argued "that the upper basin would never consume its share." Notwithstanding this observation, he joined the majority in Arizona to fight against the Boulder Canyon Act for the protection of Arizona's rights to the waters of the Colorado River.[259]

---

259 Jack L. August, Jr., *Vision in the Desert: Carl Hayden and Hydropolitics in the American Southwest* (Fort Worth: Texas Christian University Press, 1999), 43, 92; Ross R. Rice, *Carl Hayden: Builder of the American West* (Latham, Maryland: University Press of America, 1994),

In 1944 when Arizona eventually signed the Colorado River Compact, Hayden instantly saw his chance and began pushing for the Central Arizona Project. He sought legislation that would allow Arizona to develop 2.8 million acre-feet of Colorado River water allotted to it in the Boulder Canyon Project Act. Throughout the late 1940s and early 1950s, the Arizona senator doggedly pursued his agenda only to be continually thwarted by California's legislative power in the House of Representatives. The primary issue remained water rights and consumption. According to Rich Johnson, former director of the Arizona Interstate Stream Commission in *The Central Arizona Project, 1918-1968*,

> The California position on this was that Arizona was consumptively using more than 2 million acre-feet of water from the Gila River, and since the Gila was part of the Colorado River system Arizona's use from the Gila should be deducted from her allotment of 2.8 million acre-feet of Colorado River water.

This impasse stalled Hayden's every effort, forcing the state of Arizona to seek relief from the Supreme Court. On January 19, 1953, the Supreme Court granted Arizona's original motion, and the bill of complaint was filed to hear the case of *Arizona v. California*.[260]

Taking over twelve years, $5 million, 50 lawyers, 340 witnesses, 43 volumes of testimony, and more than 4,000 exhibits, the case between Arizona and California was one of the longest and most expensive court proceedings ever brought before the Supreme Court. Relying on the Colorado River Compact, the California Self-Limitation Act (a California law that imposed a limit of 4.4 million acre-feet of Colorado River water to the state), the Boulder Canyon Act, and Arizona's water delivery contracts with the secretary of the interior, "Arizona sought to obtain a Court Decree which would legalize its claim to 3.8 million acre-feet of Colorado River system water, including 2.8 million acre-feet from the mainstream and one million from the Gila River system." California attorneys argued that because Arizona did not ratify the Colorado River Compact until 1944, it was not a party to the Compact. They further claimed, "California had been required to adopt her Self-Limitation

---

4; for information on Hayden's support of the Colorado River Compact see Hundley, *Water and the West*, 241-2.

[260] For information on Hayden's and state officials' activities after Arizona signed the Colorado River Compact see August, *Vision in the Desert*, 155; Rice, *Carl Hayden*, 122; see also Johnson, *The Central Arizona Project, 1918-1968*, 50, 89.

Act only if Arizona failed to ratify the Compact within six months after enactment of the Boulder Canyon Act." Finally California maintained its rights by virtue of prior appropriation. In June 1963 the Supreme Court found in favor of Arizona, recognizing Arizona's claim to 2.8 million acre-feet of water per year from the mainstream of the Colorado River. In its ruling, the Supreme Court included an opinion regarding Indian water rights. The impact of that opinion was not fully recognized at the time, but it had a ripple effect on water resource development throughout the remainder of the century. In his article "The Effects of the Central Arizona Project on the Fort McDowell Indian Community" William Coffeen noted, "To the surprise of everyone concerned, the court declared that Indian Reservations situated on the rivers of the Colorado were to have prior rights to approximately one million acre-feet of Arizona's entitlement."[261]

11.16. Carl T. Hayden, Democrat from Arizona, served both the territory and state in the U.S. House of Representatives from 1911 until 1926 and served the state in the U.S. Senate from 1927 until 1968. Courtesy of the U.S. Senate History Program.

Upon the Supreme Court's ruling assuring Arizona's rights, Carl Hayden once again introduced Central Arizona Project legislation. Hayden's bill was essentially the same he brought to Congress in 1947. It proposed a large dam on the mainstream of the Colorado River at Bridge Canyon between Lake Powell and Lake Mead. Hayden's legislation also included a pumping plant at Lake Havasu with canals and laterals to transport water to central

[261] Philip L. Fradkin, *A River No More: The Colorado River and the West* (New York: Alfred A. Knopf, 1981), 250; Johnson, *The Central Arizona Project, 1918-1968*, 111; Robert Dean, "'Dam Building Still had Some Magic Then': Stewart Udall, the Central Arizona Project, and the Evolution of the Pacific Southwest Water Plan," *Pacific Historical Review* 66:1 (February 1997): 84; Coate, "'The Biggest Water Fight in American History'," 80; for more information on the Court's ruling regarding Indian water rights see William R. Coffeen, "The Effects of the Central Arizona Project on the Fort McDowell Indian Community," *Ethnohistory* 19:4 (Fall 1972): 355; Jacobsen, "A Promise Made," 93; McCool, *Command of the Waters*, 129; McCool, *Native Waters*, 15; Burton, *American Indian Water Rights and the Limits of the Law*, 29.

Arizona. The only material difference from the 1947 bill was a greater emphasis on providing water for municipal and industrial water users recognizing fast-paced urbanization taking place in the Phoenix and Tucson areas. The legislation reflected Hayden's desire for a simple bill that dealt only with CAP. He also expected Arizona's political and civic leadership to provide a united front to show that all Arizonans favored the project. That unity included the support of Arizona native and Secretary of the Interior Stewart L. Udall. In a 1963 strategy meeting with Arizona's congressional delegation that included Secretary Udall, Hayden stressed the need to work doubly hard to line up support. Perhaps fearing a repeat of the Echo Park controversy because of the dam's proximity to Grand Canyon National Park, Hayden stressed to Udall the importance of getting "the birdwatchers in line, by the way, the very birdwatcher we want to get in line is your National Park Service Director [Conrad Wirth]."[262]

Senator Hayden's trust in Udall's unqualified support was over-optimistic. Udall saw larger hurdles ahead for CAP than just "birdwatchers." Secretary Udall understood that Arizona's largest threat came from the powerful water interests of California. Despite what the Supreme Court had awarded Arizona, he knew full well that California's much larger congressional delegation stood ready to hold up or even defeat a "simple" Central Arizona Project bill. Instead Udall envisioned a basin-wide project that gave something to everyone. Political expediency was in the air, and he and Arizona very much needed to placate California concerns and gain its support for CAP. According to Ross Rice, in *Carl Hayden: Builder of the American West*,

> Stewart Udall … had come to the conclusion that opposition to an Arizona project was so strong that central Arizona would only get the water if it was included in a broad regional water project, taking in both lower and upper reaches of the river …

Udall's assessment was prescient. He also knew that any CAP legislation must go through Wayne Aspinall's committee and deal with the congressman's deep concern, in spite of the Colorado River Storage Project, regarding what he

[262] Memorandum, Commissioner of Reclamation to Assistant Secretary, Water and Power, Subject: Central Arizona Project, February 23, 1962 in Udall Papers, http://content.library.arizona.edu/cdm/compoundobject/collection/udallcoloradoAZU/id/8235/rec/9; for Hayden's "birdwatcher" quote see Memorandum to the Files, Subject: Arizona Authority and Power Program Meeting, Draft 2/7/63, Udall Papers, http://content.library.arizona.edu/cdm/compoundobject/collection/udallcoloradoAZU/id/8105/rec/1.

considered over-development of the lower basin to the detriment of the upper basin.[263]

In 1963 Udall came up with a plan known as the Pacific Southwest Water Plan (PSWP). The Bureau of Reclamation was very involved in preparing Udall's initial proposal sent to Congress, and its reports echoed the Secretary's belief in a basin-wide project. One report stated, "As water becomes more critical in the West, river basin boundaries will become even less rigid in water and land resource development." Over the course of five years, the proposal took on many different forms, but the Central Arizona Project remained the centerpiece. PSWP proposed another dam on the Colorado River upstream from Grand Canyon National Park at Marble Canyon. In addition, the plan proposed to augment water to southern California and Arizona from storage reservoirs in northern California. This design called for enlarging the California Aqueduct. Also there was a farsighted proposition to provide an additional one million acre-feet of water through conservation methods such as canal lining and piping systems; along with desalinization plants and improved effluent treatment facilities.[264]

Udall's Pacific Southwest Water Plan contained other enticements that included water projects in New Mexico and Nevada. It also provided a method for continuous funding of Colorado River basin projects through a Pacific Southwest Development Fund that would derive revenues from not only the proposed Bridge Canyon and Marble Canyon dams but also from the Hoover and Parker-Davis powerplants, "after those projects have met their financial obligation." According to one report, "These revenues would assure repayment of 95 percent of the cost of the projects proposed for construction under the Initial Plan." This proposal was not only ambitious in its construction scheme but was laid out in a manner designed to please all interested parties

---

263 Rice, *Carl Hayden*, 135; Dean, "'Dam Building Still had Some Magic Then'," 84; Pearson, "'We Have Almost Forgotten How to Hope'," 304; Schulte, *Wayne Aspinall and the Shaping of the American West*, 178; Coate, "'The Biggest Water Fight in American History'," 80.

264 United States Department of the Interior, Bureau of Reclamation, "Appendix to the Pacific Southwest Water Plan," 1, Udall Papers, http://content.library.arizona.edu/cdm/compoundobject/collection/udallcoloradoAZU/id/13231/rec/1; United States Department of the Interior, "Pacific Southwest Water Plan Supplemental Information on Bridge Canyon Project," December 1963, 1, RG 115, Office of the Chief Engineer, Bureau of Reclamation Project Reports, ACC# 8NS-115-95-083, Box 60; see also *A Plan to Transport Surplus Columbia River Water to the Arid Pacific Southwest: Snake-Colorado Project*, Samuel B. Nelson, General Manger and Chief Engineer, Department of Water and Power, City of Los Angeles, October 1963, 3, Udall Papers, http://content.library.arizona.edu/cdm/compoundobject/collection/udallcoloradoAZU/id/11060/rec/1; Coate, "'The Biggest Water Fight in American History'," 81.

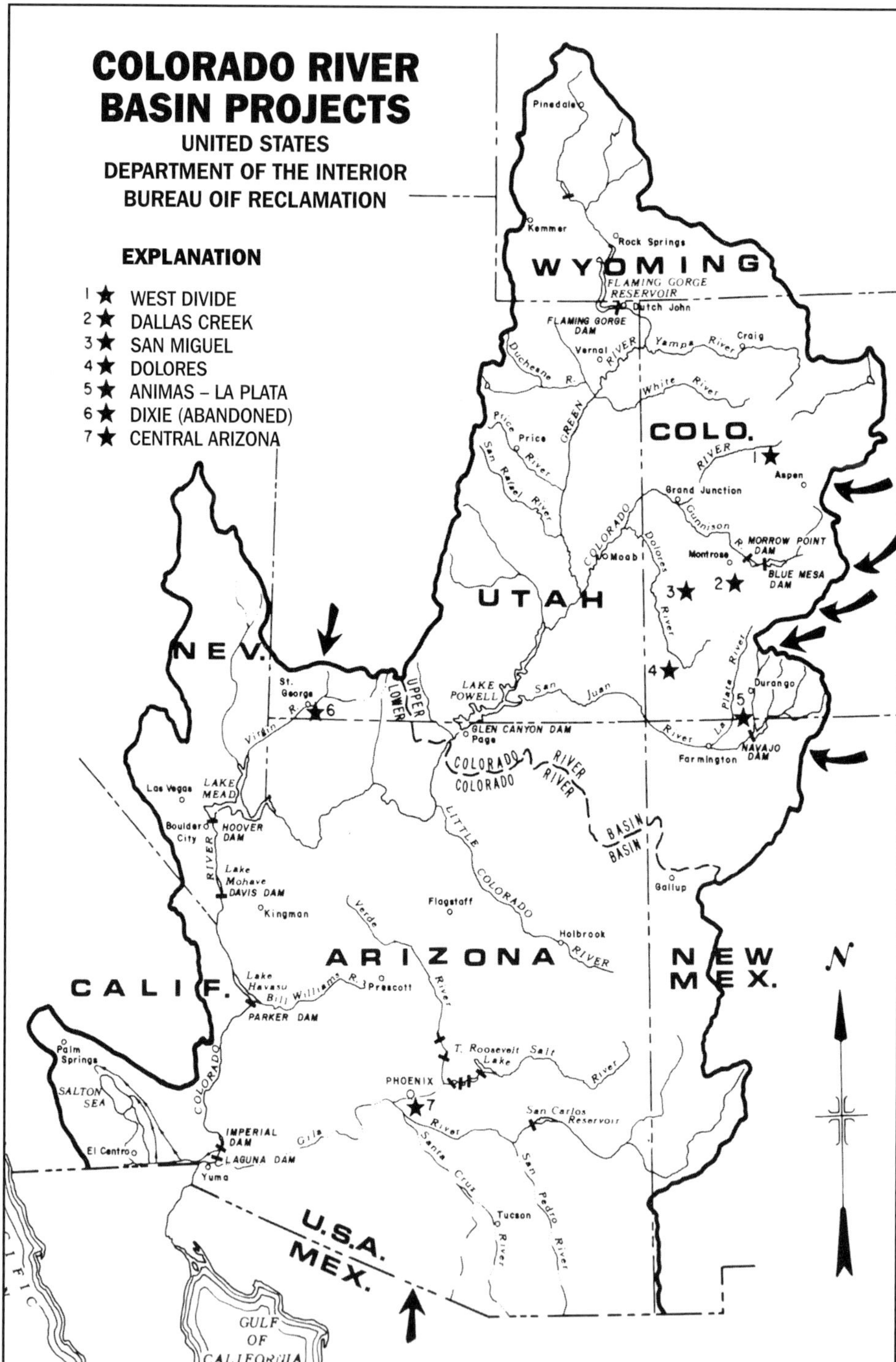

11.17. Map of authorizations under the Colorado River Basin Project Act of 1968, the last really large Reclamation authorization approved by Congress.

and calm any fears of over-appropriation of the Colorado River. Its augmentation design

> would guarantee the availability for consumptive use of 7.5 million acre-feet of water a year for the lower Colorado River, or its equivalent, regardless of any anticipated future shortages, and would protect 'areas of origin'—areas from which water may be exported—from any future damage.

Udall's far-reaching program was estimated to cost over $2 billion. Paul Dean, in "Dam Building Still Had Some Magic Then," noted in retrospect, however, that the Central Arizona Project alone eventually took fifteen years to construct at a hefty price tag of $4.4 billion.[265]

For the Bureau of Reclamation, it was an exciting and nerve-racking time requiring engineers and officials to constantly readjust plans to suit the ever-changing political climate. Through this flux, the Bridge Canyon and Marble Canyon dams remained key features of the project. The location of Marble Canyon Dam was downstream of Glen Canyon just upstream from the entrance to Grand Canyon National Park. Marble Canyon was to be a 310-foot high thin arch concrete dam, creating a reservoir with a capacity of 386,000 acre-feet of water. As one of the primary so-called "cash register" dams of the PSWP, designers conceived a powerhouse containing four 150,000-kilowatt generating units with a total capacity of 600,000 kilowatts, producing an annual average of 2,473,000,000 kilowatt-hours. Project proposals included recreation developments and facilities and wildlife enhancements. Estimated cost for Marble Canyon Dam and appurtenances was $238,654,000, which also included a small silt control dam on the Paria River. Plans for the Bridge Canyon Dam, located 117.5 miles upstream from Hoover Dam and just outside the boundary of Grand Canyon National Monument, were equally impressive. According to one report, "The Bridge Canyon designed dam is a conventional, variable-radius concrete dam 736 feet high above the foundation, with a crest length of approximately 1,650 feet." Bridge Canyon's powerhouse was designed to hold six 250,000-kilowatt generators able to produce 1,500,000

---

265 Fact Sheet on Interior's Pacific Southwest Water Plan, no date, Udall Papers, http://content.library.arizona.edu/cdm/compoundobject/collection/udallcoloradoAZU/id/11689/rec/7; see also "Resume—Pacific Southwest Water Plan, with Particular Reference to the Bridge Canyon and Marble Canyon Units, no date, Dominy Papers, Box 16, Folder, Professional Files, Pacific Southwest Water Plan; Dean, "'Dam Building Still had Some Magic Then'," 81-4.

kilowatts. Reclamation estimated the price tag for the dam and powerhouse at $511,320,000.[266]

Bureau of Reclamation reports substantiated many of Secretary Udall's public claims regarding the PSWP by presenting technical data relating to the project's feasibility and emphasizing the dire need to improve water resource development in the lower basin of the Colorado River. As early as January 1963, before the Supreme Court had ruled on *Arizona v. California*, Reclamation released a supplemental report on the Central Arizona Project discussing elements of the PSWP. The report openly admitted that CAP would not end water shortages in the Pacific Southwest. "It is … clear that water supply problems of the Central Arizona Project must be considered as but a part of the general water supply problem of the Lower Colorado River Basin and southern California as a whole." To help alleviate some of the concerns, Reclamation argued that it was "essential that all water supplies be distributed in lined canals or underground piping systems as soon as possible." The report also recommended the development of desalinization units to increase water surpluses but noted that this would require "huge amounts of nuclear and hydro energy for operation." In short, this particular report, emphatically called for an all-encompassing approach to water resource development in order "to permit the growth and development of the entire region."[267]

Bureau of Reclamation reports continually emphasized a basin-wide approach to solve the problems facing the water-starved Pacific Southwest. In these reports, there were also statements that at the time may have appeared as objective observations that would later come back to haunt both the Central Arizona Project and the Pacific Southwest Water Plan. Perhaps the most technically challenging was the issue of augmenting the waters of the Colorado River. In Secretary Udall's initial PSWP proposal, he mentioned the possibility of diverting water from rivers in northern California

---

[266] United States Department of the Interior, "Pacific Southwest Water Plan—Supplemental Information on Marble Canyon Project," December 1963, 2-3,7, Udall Papers, http://content.library.arizona.edu/cdm/compoundobject/collection/udallcoloradoAZU/id/2231/rec/34; United States Department of the Interior, "Pacific Southwest Water Plan—Supplemental Information on Bridge Canyon Project," December 1963, 2-4, Udall Papers, http://content.library.arizona.edu/cdm/compoundobject/collection/udallcoloradoAZU/id/2264/rec/67; see also Dean, "'Dam Building Still had Some Magic Then'," 88; Coate, "'The Biggest Water Fight in American History'," 81; Martin, *A Story that Stands Like a Dam*, 9-10.

[267] United States Department of the Interior, Stewart L Udall, Bureau of Reclamation, Floyd E. Dominy, Region 3, A. B. West, Regional Director, *Supplemental Report Central Arizona Project: Project Development Report*, January 1963, RG 115, Phoenix Office Reports, ACC# NRG-115-98-007, Box 68.

that flowed into the sea; projects that the Bureau of Reclamation had been looking into for some time. In 1953 Reclamation sponsored the United Western Investigations that

> suggested the possibility of bringing water by exchanges into the Colorado River watershed by diversion from streams with surplus flows which after providing for all foreseeable needs, might still empty over 200 million acre-feet of water annually into the Pacific.

There was little doubt that plans to import water from sources outside the Colorado River basin were under discussion in the highest circles of the Bureau of Reclamation.[268]

Almost instantly after the release of the Pacific Southwest Water Plan interested parties began to express comments and concerns. Interior Secretary Udall correctly assessed that California representatives would have issues with the plan. Overall, California water interests favored the basin-wide approach proposed in PSWP and approved greater federal support for the state's water plan. Californians, however, vehemently objected to any transfer of water between California and Arizona to augment the Colorado River. The Los Angeles Chamber of Commerce stated that "the plan would endanger water supply … now existing or planned in the future of California." California Senator Kuchel claimed that

> for every drop Southern California loses from her historic Colorado River supply, there is created thereby a corresponding hazard to Northern California. This comes from the danger of the Hayden-Goldwater bill or the Udall proposal which might try to turn the northern regions of our State into a potential waterhole for a vast part of the whole southwest…

Similar concerns occurred throughout California, prompting Department of the Interior and Reclamation officials to concede that other sources of water to supplement the Colorado River needed to be found.[269]

---

268 *Ibid.*, 87.

269 Los Angeles Chamber of Commerce, Committee Report to the Board of Directors, "Pacific Southwest Water Plan," October 23, 1963; Udall Papers, http://content.library.arizona.edu/cdm/

The question of water transfers into the Colorado River faced difficult technical problems, yet the idea attracted attention and was given serious consideration throughout the Colorado River basin. Of course, the Central Arizona Project prompted these concerns, with all interested parties understanding that the Colorado River was close to if not already over-appropriated. Many believed that if CAP legislation passed and the project was built, someone somewhere faced water losses. In 1963 the Los Angeles Department of Water and Power (DWP) presented a plan to transfer water from the Snake River downstream from Twin Falls, Idaho. This proposal, suggested pumping water to the high plateau of eastern Nevada then dropping it into Lake Mead to flow to Lake Havasu. Department of Water and Power engineers estimated the plan would cost $1.4 billion and maintained that the "project would generate 10% more energy than it would consume." It is uncertain whether or not the so-called "Snake-Colorado Project" ever left DWP board rooms, but it does show that some water interests in the Colorado River's lower basin looked north for more water.[270]

Transferring water from the Columbia River or its tributaries faced intense opposition in the Pacific Northwest, primarily in the person of Senator Henry "Scoop" Jackson of Washington, the powerful chairman of the Senate Committee on Interior and Insular Affairs. To succeed, the idea required a complete transformation in water resource development culture throughout the American West. Urban and agricultural centers throughout the West perceived water projects as a local or regional activity founded on the belief that those regions and or states controlled the water. The Colorado-Big Thompson and the Fryingpan-Arkansas Project revealed the controversial nature of water transfers even between communities within the same state. To achieve a water transfer of the magnitude some lower basin interests proposed, although technically feasible and incredibly expensive, required a complete overhaul regarding water rights and water issues by all concerned parties. Prospects for change were unlikely because all states and regions jealously guarded their inherent and legal water rights.

---

compoundobject/collection/udallcoloradoAZU/id/9308/rec/1; Press Release from the Office of U.S. Senator Thomas H. Kuchel, April 8, 1964, Udall Papers, http://content.library.arizona.edu/cdm/compoundobject/collection/udallcoloradoAZU/id/8896/rec/1; see also Coate, "'The Biggest Water Fight in American History'," 81-2.

[270] *A Plan to Transport Surplus Columbia River Water to the Arid Pacific Southwest: Snake-Colorado Project*, 1-5, Udall Papers, http://content.library.arizona.edu/cdm/compoundobject/collection/udallcoloradoAZU/id/11060/rec/1.

The Pacific Southwest Water Plan and the Central Arizona Project raised concerns and fears in the Colorado River's upper basin. Upper basin states believed that any additional water projects on the Colorado River for the lower basin threatened water allocated to them. Colorado's Wayne Aspinall guarded upper basin interests. As chairman of the House Committee on Interior and Insular Affairs, Aspinall controlled when, and if, and in what form CAP or PSWP legislation appeared. According to Aspinall biographer Steven Schulte, Aspinall's "fundamental principle and the basis for his reclamation philosophy" was that the "Upper Colorado River states needed to devise 'some way to keep water from running down hill' to California and the lower Colorado River states." This overriding concern meant that there would be no CAP or PSWP without some measure of protection for the water rights of the upper basin.[271]

In Arizona, there were mixed feelings regarding Secretary Udall's Pacific Southwest Water Plan. Though many of the state's leaders looked favorably on many of the proposed benefits and the fact that the Central Arizona Project remained the centerpiece, they worried that this new broad proposal would hamper the legislative process. Most believed, like Senator Carl Hayden, that with the state's stunning victory in *Arizona v. California* a clear path to CAP was at hand. Hayden wrote to Udall expressing his misgivings, "I am prepared to support the Lower Basin Account principle and to approach the water problem of the entire area served by the Colorado River, but I must insist that the proper vehicle for undertaking this must be legislation authorizing the Central Arizona Project." Udall made extensive efforts to ease the concerns of the venerable Arizona senator, arguing that the basin-wide approach was the best solution. The secretary believed that a single CAP bill would only "stir up a major controversy in the House that might defeat it." In conclusion, Udall asked Hayden to leave it to the experts and allow "the Bureau of Reclamation experts a full opportunity to develop the best plan available to maximize the water supplies for all the people in the Pacific Southwest."[272]

There is no doubt that both men saw the Central Arizona Project as crucial to the continued growth and prosperity of Arizona. For Senator

---

271 Schulte, *Wayne Aspinall and the Shaping of the American West*, 14; Stephen C. Sturgeon, *The Politics of Water: The Congressional Career of Wayne Aspinall* (Tucson: University of Arizona Press, 2002), 81.

272 Senator Carl Hayden to Stewart Udall, Secretary of the Interior, January 22, 1963, Udall Papers, http://content.library.arizona.edu/cdm/compoundobject/collection/udallcoloradoAZU/id/756/rec/20; Stewart L. Udall to Sen. Carl Hayden, Personal and Confidential, June 12, 1963, Udall Papers, http://content.library.arizona.edu/cdm/compoundobject/collection/udallcoloradoAZU/id/7716/rec/10.

Hayden, CAP represented over forty years of frustration and now, with the benefit of the Supreme Court's ruling, the goal was in sight. Naturally, he expressed his frustration over a quickly conceived plan that might not lead to a Central Arizona Project. Hayden believed that "the overdrawn PSWP a method for delaying consideration of CAP that played into the hands of California and upper basin opponents." Udall also believed that Arizona desperately needed CAP, but he maintained that a basin-wide project was the best approach. Ultimately, the two men disagreed on tactics rather than the goal. Udall saw PSWP as the only means to assure California and basin-wide support for CAP, while Hayden held to the conviction that "simple" CAP legislation could easily get through Congress.[273]

The "experts" at the Bureau of Reclamation, and especially its charismatic Commissioner Floyd Dominy, embraced the idea of the Pacific Southwest Water Plan and the technical challenges it contained. The Central Arizona Project alone was an engineering dream. Not only did the original proposal include the two dams on the Colorado River, the project proposed the Havasu and Hassayampa pumping plants, the Granite Reef Aqueduct, the Salt-Gila Aqueduct and Pumping Plant, Maxwell Dam (later renamed Orme Dam), the Tucson Aqueduct and Pumping Plant and Buttes Dam and reservoir. Facilities for the upper tributaries zone consisted of Hooker Dam and reservoir, Charleston Dam and reservoir, and a San Pedro River source of the Tucson Aqueduct. The PSWP expanded these units to include water conservation efforts such as canal lining along the All-American Canal and desalinization units, which were still in the experimental stage. Perhaps the most challenging proposal was transferring water into the Colorado River basin from a source as yet unnamed. In *A Story That Stands Like a Dam*, Russell Martin noted that "to Dominy's mind, this was the kind of grand and complex scheme that made it worthwhile to head up Reclamation, to endure the million headaches, the piss-ant politicians, and the crybaby conservationists."[274]

Martin's comment fairly describes the willingness of Dominy to accept the "headaches" of his job between 1963 and 1968. Dominy adjusted to the ever-changing views of the politicians, all the while protecting the Bureau of Reclamation's grandiose plans for the Colorado River's lower basin. He matched the enthusiasm of leading political figures and water boosters within

---

273 August, *Vision in the Desert*, 189; Rice, *Carl Hayden*, 135; Coate, "'The Biggest Water Fight in American History'," 82-3.

274 West, *Supplemental Report Central Arizona Project: Project Development Report*, January 1963, 28, 48, RG 115, Phoenix Office Reports, ACC# NRG-115-98-007, Box 68; see also Martin, *A Story that Stands Like A Dam*, 251.

the Colorado River basin, and likewise he underestimated the tenacity of the conservationists. Suffice it to say at this point that the actions and rhetoric of conservationists during the legislative effort to pass a CAP bill or a PSWP bill significantly altered the outcome. Dominy and his allies faced determined opposition and misread the strength of an emerging environmental consciousness spreading throughout American society. Nevertheless, it was not just a vocal environmental movement that undermined Secretary Udall's far-reaching Pacific Southwest Water Plan. Rather, it was also disunity and regional concerns within Reclamation's constituency in the Colorado River basin.

Water, or, perhaps, the lack thereof, was the issue in the Colorado River basin. California water interests were not about to sacrifice any part of the water they took from the Colorado River in order for Arizona to have its Central Arizona Project. Its concern was water California routinely appropriated, while Arizona refused to recognize the 1922 Colorado River Compact, that was over the 4.4 million acre feet apportioned to the state by the Boulder Canyon Project Act. Similarly, upper basin states, under the ever-watchful eye of Congressman Aspinall, jealously guarded their own water rights. In 1967 according to a Bureau of Reclamation interoffice correspondence, Aspinall had

> concluded that not more than 50,000 acre-feet of water from the Colorado will be available for the Central Arizona Project after the upper basin accomplishes full development … Our figures show water availability in the range of 600,000 acre-feet under similar conditions.

The irony in all this was that all Colorado River basin states still clung to the division of water established in the 1922 Colorado River Compact, 7.5 million acre-feet per basin, while knowing full well that the river's flow was rarely that high and the division would at some point likely prove unrealistic. These were the difficult forces Secretary Udall and Commissioner Dominy faced as they struggled to construct a plan to appease everyone.[275]

For Secretary Udall, the most important task in securing legislation was the full support of California congressional delegation and water interests.

---

[275] Notes—Asst. Commissioner Stamm, October 5, 1967, Dominy Papers, Box 4, Folder, Correspondence, 1967 Blue Envelope Letter; see also Wayne Aspinall to the Secretary of the Interior, March 8, 1968, Udall Papers, http://content.library.arizona.edu/cdm/compoundobject/collection/udallcoloradoAZU/id /13900/rec/2; Schulte, *Wayne Aspinall and the Shaping of the American West*, 81.

California representatives voiced critical concerns about earlier versions of the Pacific Southwest Water Plan, especially the provision to move water from northern California. They were, however, open to the idea of water augmentation. During a 1964 meeting between Udall and the directors of the Metropolitan Water District, the secretary discussed an upcoming meeting with Senator Henry Jackson of Washington on "a feasibility study and the eventual taking of ten to fifteen million acre-feet of water from the Columbia River." Perhaps sensing Jackson's distaste for such a proposal, Udall stressed to the directors that it was "essential to prove to the northwestern states that there is no harm in such a project and that there would be some benefits." Udall's role at this particular meeting with the staff of one of California's most powerful water brokers revealed the secretary of the interior's skillfully lobbying for acceptance of PSWP.[276]

California's primary interest was and remained holding on to its historic appropriation to waters from the Colorado River. State water interests remained adamantly against sacrificing any water to supply the Central Arizona Project. What California really demanded was a reversal of Arizona's hard won rights obtained by the Supreme Court's ruling in *Arizona v. California*. Historian Norris Hundley maintains,

> As a price for dropping their opposition to CAP, they [Californians] demanded a first priority for California's apportionment of 4.4 million acre-feet. In effect, Arizonans would have to promise to regulate CAP diversions so that Californians never received less than that amount.

This condition was perhaps the most controversial concession Arizona made in its effort to pass Central Arizona Project legislation. Arizona stood virtually alone. Even with its tepid support of Secretary Udall's regional plan, Arizona's congressional delegation simply did not match California legislative strength in Congress.[277]

---

[276] Confidential Report to M.W.D. Directors, "Conference with Secretary of the Interior Udall, December 15, 1964; Udall Papers, http://content.library.arizona.edu/cdm/compoundobject/collection/udallcoloradoAZU/id/10989/rec/1.

[277] Norris Hundley, Jr., "The West Against Itself: The Colorado River—An Institutional History," in Gary D. Weatherford and F. Lee Brown, editors, *New Courses for the Colorado River: Major Issues for the Next Century*, foreword by Governor Bruce Babbitt (Albuquerque: University of New Mexico Press, 1986), 35; see also Confidential Report to M.W.D. Directors, "Conference with Secretary of the Interior Udall," December 15, 1964, 4, Udall Papers, http://content.library.arizona.edu/cdm/compoundobject/collection/udallcoloradoAZU/id/10989/rec/1;

By 1965 Arizona reluctantly agreed to the California water allocation condition, and Secretary Udall's Pacific Southwest Water Plan was no more. For Arizona's political leadership this was a bitter pill to swallow. It was a necessary political expedient to end over 40 years of acrimony and obtain the Central Arizona Project. The far-reaching and ambitious PSWP never passed the watchful eye of the Bureau of the Budget, which demanded major revisions. To appease the Bureau of the Budget, the Department of the Interior proposed a new plan that included nearly all the elements of the old PSWP except for Bridge Canyon Dam. Nevertheless, Arizona and California, indeed the entire lower basin, were unified in their efforts to pass new legislation entitled the Lower Colorado River Basin Project Act. Despite the administration's budgetary concerns, the new legislation reinserted the Bridge Canyon unit. Much of the debate concerning this bill revolved around the proposal to study water "importation," based on an article in the bill that made the provisions of the 1944 Mexican Water Treaty a national obligation.

Still the idea of water importation was not dead. Conceivably, it could solve many of the longstanding issues and inconsistencies that lay within the so-called "law of the river," softening some of the provisions of the 1922 Colorado River Compact. All parties understood that the inflated Colorado River flow rates that established the criteria for dividing the river between the basins did not conform to the river's actual flow rates nor were the parties willing or able to adjust their demands. On behalf of the upper basin states, Wayne Aspinall contended, "The value to the upper basin States was the stoppage of the operation of the laws of appropriation with respect to the waters of the Colorado River." In other words, the Compact was their insurance policy that protected upper basin water rights against its perception of the insatiable thirst of the lower basin. For the lower basin, especially California, the Compact guaranteed their rights to water already put to "beneficial" use. On the other hand, some perceived water importation as a panacea for the deficiencies of the Colorado River. As Arizona Congressman Morris Udall declared, importation was "essential to the adequate development of both the upper and lower basins."

With insertion of the Mexican Water Treaty stipulation, Colorado River basin states argued that importation was all the more necessary. It was no longer the sole responsibility of one region or section of the country to

---

From: M. Weinberg, February 14, 1964, Dominy Papers, Box 16, Professional Files 1960-1968, Folder, Professional Files, Pacific Southwest Water Plan; Coate, "'The Biggest Water Fight in American History'," 86.

ensure the delivery of 1.5 million acre feet of water to Mexico. The entire nation was obligated and therefore other regions might be expected to contribute to the water deficiencies of the Colorado River basin. Senator Thomas H. Kuchel stressed to his colleagues:

> This is a national problem. We are all ... representatives of the national interest and of the American people, and if surplus unnecessary to the future growth in one area may be utilized to stave off stagnation in another, to that extent I think the equities and justice demand that type of use.

Carl Hayden held similar sentiments but was much more direct. Testifying before the House Subcommittee on Irrigation and Reclamation in 1964, he stated, "If enough water cannot be found there [Colorado River] he [secretary of the interior] should go to the Columbia or even the Yukon to get water that we in the Southwest must have." Though most members of Congress were less straightforward than the Arizona senator in designating an importation source, there is little doubt that most were looking to the Pacific Northwest.[278]

Advocates of huge water transfers realized their proposal entered new territory in terms of Reclamation projects and moved beyond traditional thinking regarding river basin development. They were quick to point out that this provision of the Lower Colorado River Basin Project Act only provided for studying an importation plan and in no way opened the door to such a project. Language in the bill clearly guaranteed the protection of all water rights and future water needs of this unnamed source. These assurances, however, did little to ease the trepidation of congressional representatives from the Pacific Northwest. Indeed, some members of the House Subcommittee on Reclamation and Irrigation believed that the importation scheme was the only thing holding up the alliance between not only California and Arizona, but the entire Colorado River basin states. Idaho Congressman Compton I. White went to the heart of the matter when he claimed, "It seems to me we are talking mainly about the transfer from the Pacific Northwest and I think again that this is the glue that is sticking California together with Arizona and that is sticking the

---

[278] United States House of Representatives, Subcommittee on Irrigation and Reclamation of the Committee on Interior and Insular Affairs, *Lower Colorado River Basin Project*, Hearing, 89th Cong., 1st sess. (Washington, D.C.: United States Government Printing Office, 1965), 48, 245, 93; United States House of Representatives, Subcommittee on Irrigation and Reclamation of the Committee on Interior and Insular Affairs, *Central Arizona Project*, Hearings, 88th Cong., 2nd sess., November 9, 1964, Phoenix, Arizona (Washington, D.C.: United States Government Printing Office, 1965), 5; see also Sturgeon, *The Politics of Western Water*, 74.

upper basin with the lower basin."[279] White's statement represents not only wariness from the Pacific Northwest, but also evidence of the difficulty of turning the Mexican Treaty burden borne by Colorado River basin states into a national obligation.

Supporters of water importation fashioned their arguments in a number of ways to garner support. First and foremost, they stressed the urgent water needs of the Pacific Southwest in order to maintain growth and prosperity. This argument relied on the arid West tradition and ideal that any water flowing unused to the sea was a wasted resource. In 1965 Arizona Congressman John Rhodes claimed, "As long as there is water running into the Pacific Ocean unused, as long as there are water needs unsatisfied in the West, there will be a job for reclamation." Another line of argument spoke to the economically minded. The importation plan called for diverting no less than 2.5 million acre-feet of water into the Colorado River below Lee Ferry, allocating 1.5 million acre-feet to settle the Mexican Water Treaty obligation. Because the treaty was a national burden, costs for construction of the diversion system were non-reimbursable. With provisions in place to protect current and future water rights for "area of surplus" [that is, the area from which water would be imported] supporters of the augmentation plan felt that they had a well-developed proposal.[280]

Central Arizona Project supporters in California and Arizona believed that they finally had the necessary protections and incentives in place for quick congressional action on the Lower Colorado River Basin Project Act. Despite these hopes, the bill moved slowly. The cause for some of this delay was the determined and strenuous opposition to the so-called Grand Canyon dams by the conservationists. By 1966, their opposition eventually forced Secretary Udall to agree to remove the two Colorado River dams from the legislative proposal, and in their place, he substituted a coal-fired generating unit near Page, Arizona. Even with this concession, movement for legislation on CAP stalled. Behind the scenes, concern persisted over whether or not there was enough water in the Colorado River to satisfy both CAP and the needs of the upper basin. Spearheading the stalling tactics was Colorado's Wayne Aspinall who had yet to be convinced that CAP was not taking water that rightly

---

279 Subcommittee on Irrigation and Reclamation, *Lower Colorado River Basin Project*, 2, 3, 11, for White's statement, 274; see also Dean, "'Dam Building Still had Some Magic Then'," 91; Rice, *Carl Hayden*, 154.

280 Subcommittee on Irrigation and Reclamation, *Lower Colorado River Basin Project*, 51, 40-1.

belonged to the upper basin. A 1967 Bureau of Reclamation interoffice memo warned, "Aspinall expects to go out and kill the Central Arizona Project." Aspinall's position as chairman of the House Interior and Insular Affairs Committee guaranteed that any form the CAP legislation took must pass under his critical gaze before reaching the full House. He intended to use his power to ensure the protection of upper basin water rights to the Colorado River.[281]

CAP supporters in Arizona knew that the Colorado congressman was delaying the Lower Colorado River Basin Project Act jeopardizing the entire project. From his position as chairman of the Senate Appropriations Committee, a frustrated Carl Hayden threatened to pull funding for the Fryingpan-Arkansas Project in Colorado. According to sources, Reclamation Commissioner Floyd Dominy convinced the Arizona senator not to resort to such a drastic measure in reaction to Aspinall's "misgivings regarding the water supply of the Colorado River Basin." Political leaders in Arizona made their irritation known by discussing plans for the state financing its own project. Richard Johnson, executive director of the Arizona Interstate Stream Commission, wrote to Secretary Udall expressing the frustration of the entire state. "Because the urgency of Arizona's need for Colorado River water and the failure of the 89th Congress," he stated, "the State of Arizona has now initiated a study of and planning for a State financed and constructed diversion system." These were idle threats. Arizona did not have the resources to finance the project. Nevertheless, Secretary Udall took no chances on the Arizona "go it alone" proposal and went before the Federal Power Commission to prevent the Arizona Power Authority from receiving a license "to develop hydropower potential of the Marble Canyon site on the Colorado River."[282]

---

281 Notes—Asst. Commissioner Stamm, October 5, 1967, Dominy Papers, Box 4 Folder, 1967 Blue Envelope Letter; Sturgeon, *The Politics of Western Water*, 76.

282 For information on Hayden's threat to pull funding for the Fryingpan-Arkansas Project see Sen. Carl Hayden to Floyd Dominy, June 10, 1965, Udall Papers, http://content.library.arizona.edu/cdm/compoundobject/collection/udallcoloradoAZU/id/10462/rec/27; Sturgeon, *The Politics of Western Water*, 78; Rice, *Carl Hayden*, 151; for information on the Arizona "go it alone" plan see Richard Johnson, Executive Director, Arizona Interstate Stream Commission to Stewart Udall, October 14, 1966, Udall Papers, http://content.library.arizona.edu/cdm/compoundobject/collection/udallcoloradoAZU/id/14601/rec/9; Sturgeon, *The Politics of Western Water*, 88; Coate, "'The Biggest Water Fight in American History'," 89; Schulte, *Wayne Aspinall and the Shaping of the American West*, 199; for information on Udall's response to the Arizona threat see United States Department of the Interior, Office of the Secretary, Press Release, "Interior Files Petition of Intervention in Marble Canyon Power License Application," October 1, 1962, Udall Papers, http://content.library.arizona.edu/cdm/compoundobject/collection/udallcoloradoAZU/id/14368/rec/2.

Aspinall's support for CAP or a larger Colorado River basin-wide project remained a requirement for successful passage of the bill. Aspinall wrote to one colleague, "I think I can say in all modesty that the Central Arizona Project legislation cannot pass without my support." Still believing that there was not enough water in the Colorado River to sustain both CAP and future development of the upper basin, Aspinall introduced new amendments to the Lower Colorado River Basin Project Act that included five upper basin projects all in Colorado. Central Arizona Project supporters had little choice but to acquiesce in the chairman's proposal and renamed the new legislation the Colorado River Basin Project Act. Aspinall's projects quickly came under the scrutiny of the Bureau of the Budget when the Colorado projects "added $360 million to the bill's cost," bringing "the total 1966 authorization proposal to $1.7 billion, compared to Stewart Udall's original proposal for $1.1 billion." The Bureau of the Budget questioned whether the five projects were economically feasible and could meet the benefit/cost ratio criteria. Nevertheless Aspinall insisted that the five Colorado projects remain part of the legislation, and that he would not hold hearings on the bill until his projects received "favorable reports."[283]

Aspinall's move forced the Bureau of Reclamation to undertake intense studies on his newly proposed projects. Wishing to ensure passage of the legislation and conscious of Aspinall's ability to hold up progress, Secretary of the Interior Udall ordered Reclamation to work overtime to complete the reports on the five Colorado projects. They included the Animas-La Plata, Dallas Creek, Dolores, West Divide, and San Miguel, which had been in the works for years but had yet to receive authorization. Reclamation personnel completed a Herculean task by producing and distributing the reports in near record time. The results, however, were hardly a cause for celebration. According to Steven Schulte, "When the reports finally arrived, only two Colorado projects received unconditional executive branch clearance." Indeed, Reclamation's analysis indicated that only four out of the five projects "would barely break even," and one project, the San Miguel, "would actually lose money." This information did little to dampen Aspinall's commitment, and he used his formidable influence to pressure the Bureau of the Budget and the administration into accepting the reports. With favorable reports in hand, Aspinall gave his support to the Central Arizona Project and was ready to move forward on the Colorado River Basin Project legislation.[284]

---

283 Coate, "'The Biggest Water Fight in American History'," 89; Aspinall quote found in Sturgeon, *The Politics of Western Water*, 89, 87.

284 Sturgeon, *The Politics of Western Water*, 86-7, 89; see also Schulte, *Wayne Aspinall and the Shaping of the American West*, 194-5.

There is little doubt that Wayne Aspinall used his position as chairman to construct the legislation to his liking and forced other members of Congress to accede to his demands. According to Stephen Sturgeon, in *The Politics of Western Water*,

> While Aspinall justified his decision as an attempt to protect Colorado's Upper Basin rights, the contemporary perception of his action was that the Chairman had committed a blatant act of extortion, demanding his 'pound of flesh' in return for not killing the Central Arizona Project.

Arizona interests had long expected as much and were disappointed that CAP did not have an easier route to passage. Their expectations rested on the concept of reciprocity from colleagues who in the past received their support for Reclamation projects not directly benefitting Arizona. In a 1964 hearing before the House Subcommittee on Irrigation and Reclamation, Carl Hayden reminded the committee of his long-time support for "reclamation developments … in the districts of most members of your committee." In a sense, Hayden was asking his congressional brethren to return his past favors. What Arizonans found instead in Chairman Aspinall was a person who, according Stephan Sturgeon, "felt no obligation to return past favors and instead exploited the situation to further benefit his district."[285]

Meanwhile, in 1967 Secretary of the Interior Stewart Udall had misgivings over successful passage of the CAP or the Colorado River Basin Project Act. Moved by the vociferous opposition of the conservationists, Udall presented a new plan to Congress that eliminated the controversial Grand Canyon dams and replaced them with a coal fired generating plant. The secretary's proposal erased some of the most important and hard-fought agreements made among Colorado River basin states. His proposition kept the Central Arizona Project but said nothing regarding Aspinall's Colorado projects. The plan removed the feasibility study for river augmentation from the Columbia River, leaving any study of "shortage problems" to the National Water Commission. Finally, Udall asked that the issue of California's guaranteed 4.4

[285] Stephen C. Sturgeon, "Operation Blackmail: Upper Basin Reaction to the Central Arizona Project," an unabridged excerpt from the doctoral dissertation "Wayne Aspinall and the Politics of Western Water," a paper presented at the American Society of Environmental History Conference, Tucson, Arizona, April 15, 1999, 5-6; for a fuller rendition of the quote and more information on contemporary views of Aspinall's actions see Sturgeon, *The Politics of Western Water*, 84, 121-4; for information on Hayden and reciprocity see Subcommittee on Irrigation and Reclamation, *Central Arizona Project*, 4.

million acre-feet be left to Congress. Udall's plan gained support within the Johnson administration, and in one single act removed all opposition of the conservationists.[286]

The secretary's proposal upset basin representatives. Congressman Aspinall vehemently opposed elimination of the Grand Canyon dams and his Colorado projects, while California representatives feared losing their priority water right in a floor fight on the bill. Arizona's congressional representatives were not only surprised by the secretary's actions but concerned about whether a coal-fired powerplant was the best source of electricity for CAP pumping plants. When Udall's proposal entered into serious discussion in Congress each interest introduced amendments that guaranteed their hard fought gains in previous bills. The legislation that began to take shape appeared to placate all interested parties containing nearly all the items in the original Colorado River Basin Project Act except any large dams on the mainstem of the Colorado River.[287]

Despite the apparent consensus among Colorado River basin states, Aspinall's fears of the CAP taking water that rightfully belonged to the upper basin reemerged. The final climax of the tortuous road to CAP came when the chairman refused to hold hearings on the latest version of the Colorado River Basin Project Act. For Senator Carl Hayden, Aspinall's decision was the last straw, and taking advice from Reclamation Commissioner Floyd Dominy, Hayden threatened to add the Central Arizona Project to a public works bill, bypassing Aspinall's committee. One Bureau of Reclamation interoffice memo noted, "Senator Hayden finally got his dander up sufficiently to adopt the 'Dominy Formula' and … filed notice to move for a suspension of the Senate rules for the purpose of offering S.1004 as an amendment to the Public Works Appropriation Bill." Furious with the Arizona senator's action, Aspinall quickly called his committee together and in a short while submitted the Colorado River Basin Project Act to the full House.[288]

---

286 "Elements of the Administration Plan," January 28, 1967, in Udall Papers, http://content.library.arizona.edu/cdm/compoundobject/collection/udallcoloradoAZU/id/14252/rec/85; see also Coate, "'The Biggest Water Fight in American History'," 91, 96; Sturgeon, *The Politics of Western Water*, 108; Schulte, *Wayne Aspinall and the Shaping of the American West*, 203.

287 Coate, "'The Biggest Water Fight in American History'," 91.

288 Memorandum, Ted Riggins to Commissioner Dominy, Subject: Status Report—Central Arizona Project Legislation, October 5, 1967, Dominy Papers, Box 4, folder, Correspondence, 1967 Blue Envelope Letter; see also Sturgeon, *The Politics of Western Water*, 115; Schulte, *Wayne Aspinall and the Shaping of the American West*, 211-2; Rice, *Carl Hayden*, 151.

By the fall of 1968 Congress passed the Colorado River Basin Project Act, and after over twenty years of frustration, court cases, and strenuous effort, Arizona finally achieved authorization for the Central Arizona Project. On September 30, 1968, President Johnson signed the bill into law in front of a host of Colorado River basin state representatives, none more pleased than ailing octogenarian Carl Hayden. The law contained something for everyone. It authorized the Central Arizona Project, five projects in Colorado, and water projects in Utah and New Mexico. California and Nevada retained their priority rights over CAP water during periods of drought or water shortages, and the Mexican Water Treaty became a national obligation. In deference to Washington Senator Henry M. Jackson, the legislation contained a ten-year prohibition on studies to import water to the Colorado River basin.[289] It also marked the end of an era in water resource development in the arid West, though most attending the bill's signing were unaware of that fact. The legislation that Johnson signed gave Arizonans their long sought-after Central Arizona Project and proved to be the last Reclamation basin-wide authorization.

Conclusion

In terms of public works legislation, one might argue that the Colorado River Basin Project Act was the final chapter of FDR's New Deal. Democratic presidents from Truman to Johnson attempted to keep alive the promise of prosperity first enunciated by Franklin Roosevelt in 1932; even Republican Eisenhower did little to upset the New Deal momentum. While primarily public works projects under FDR, water resources developments after World War II became symbols of American strength and ideals. Bureau of Reclamation commissioners from Michael Straus to Floyd Dominy continued to promote Reclamation activities as extensions of American greatness, arguing they were necessary to maintain the United States's postwar status that helped to fuel the economic juggernaut. In testimony before the House Committee on Interior and Insular Affairs in 1967, Commissioner Dominy boasted, "Our Federal reclamation program, through development and stabilization of the basic economy of the West, plays a very effective and significant role in achieving national objectives. Economic development from our water resources projects is generative and long lasting, not a one-shot or aid type program." Though serving principally the western states, the idea behind advocating Reclamation works remained very much that it was and continued to be a national benefit.

[289] Rice, *Carl Hayden*, 154; Coate, "'The Biggest Water Fight in American History'," 93, 97.

By 1968, however, that message became ever-more blurred as costs rose and new environmental values began to envelop American culture. Moreover, the race for water projects became even more intense. The easy dam sites had been taken, and it was becoming more difficult to justify projects whose benefits hardly measured up to costs. In addition, there was a growing sense of distrust in the general public regarding the activities of the federal government, brought on by the growing strains of the Vietnam War. By mid-1970s, the era of consensus that marked most the 1950s and early 1960s gave way to an era of cynicism and skepticism in the wake of the end the war in Vietnam and the Watergate scandal. For the Bureau of Reclamation, this was a period of painful transition as appropriations for water projects became mired in economic stagnation.

CHAPTER 12:

MEETING NEW CHALLENGES, 1956-1980

Introduction

From 1956 to 1980, the Bureau of Reclamation went about its business guided by policies developed during the previous fifty years. It generally enjoyed cordial relationships with its western constituents and their political representatives. With regional and even national support, Reclamation burgeoned into one of the most powerful bureaus within the Department of the Interior. Work continued and even expanded in all the West's major river basins as challenges arose. On the Columbia River, Grand Coulee Dam received a makeover with the addition of a new powerplant alongside its already formidable hydropower facilities. Also the construction of the Pacific Northwest/Pacific Southwest Intertie connected Grand Coulee and Hoover dams with the major public and private power producers on the West Coast. In Utah the Bureau of Reclamation began construction on the long-promised Central Utah Project, designed to allow that state to use its share of the Colorado River. Despite these accomplishments, tightening budgets as a result of competing demands from the Vietnam War, the social programs of the 1960s and early 1970s, and rising inflation placed severe strains on Reclamation plans.

As Reclamation construction proceeded throughout the American West, subtle changes in the region and the nation were afoot with long-range implications for the Bureau of Reclamation's future. In the isolation of Glen Canyon, teams of archaeologists, biologists, geologists, and historians quickly surveyed Glen Canyon Dam's reservoir site to document the area's cultural and physical history. Somewhat routine at the time, these efforts presaged policies and programs that moved beyond salvage work at damsite locations to statutes requiring the federal government to protect cultural resources on federal government property. At the same time, in the wake of Rachel Carson's *Silent Spring* (1962), an environmental awareness emerged in American culture and society. In a remake of the Echo Park controversy, the Bureau of Reclamation encountered changing attitudes about the relationship between humans and the natural environment when it suggested dam building in the vicinity of the Grand Canyon. Once again, in the 1960s, conservationists, soon to be termed environmentalists, successfully forced Reclamation to alter plans and remove two proposed dams from the Central Arizona Project. No short-lived victory

12.1. Don Fowler, on the right, and Richard Ross chalking petroglyphs in order to take photographs at Smith Fork Bar rock art panel. July 1959. Gift of photographer W. J. Owen to Don Fowler. Courtesy of Don Fowler.

for defenders of the natural world, soon Congress approved far-reaching environmental legislation that forever altered the way the Bureau of Reclamation could conduct business.

Two events forced a fundamental transformation of the Bureau of Reclamation. The first came in 1976 when disaster struck in southeastern Idaho with the collapse of Reclamation's Teton Dam. Teton's failure struck at the heart of Reclamation's well-earned engineering reputation in dam construction. Less than a year later, President Jimmy Carter announced a hit list, proposing to eliminate many western water projects and challenging how Reclamation determined the economic feasibility of water resource development projects. Determined to rein in federal water project spending, Carter tried to reveal the extent of subsidies in Bureau of Reclamation projects. His economy-minded administration saw economically unjustified projects as items to be cut from the budget. Taken together, these events signaled the winding down of Reclamation's active dam-building era. Shrinking budgets, environmental concerns and regulations, along with concerns over dam safety compelled the American public and their political representatives to rethink the management of water in the West.

12.2. About 1946 Harry Aleson (left) and Ralph Badger visited "Fort Moki" or "The Watch Tower" in upper Glen Canyon. This is an example of one of the more spectacular archaeological sites inundated by Lake Powell behind Glen Canyon Dam. Courtesy of the Utah State Historical Society.

Finding the Past: Archaeology and Cultural Resources on Reclamation Projects

While the Bureau of Reclamation's Lem Wylie and his associates assembled at the Glen Canyon construction site, another group of adventurers also moved into this isolated canyon of the Colorado River. Not attached to the dam building team, these intrepid individuals sought to document past human activity in Glen Canyon before the rising waters of the reservoir buried it forever. Under the sponsorship of the National Park Service, the objective of the Upper Colorado River Basin Archaeological Salvage Program was "to locate and record historical sites that would be lost" because of completion of Glen Canyon Dam. Between 1957 and 1963 teams of archaeologists and historians endeavored to locate and record all remnants of human occupation in Glen and San Juan canyons while biologists and geologists studied the canyons' natural history. Faculty members, students, and volunteers under the direction of the University of Utah and the Museum of Northern Arizona were responsible for carrying out this survey.[290]

[290] C. Gregory Crampton, *Ghosts of Glen Canyon: History Beneath Lake Powell* (St. George, Utah: Publishers Place, Inc, 1986), 16; Jesse D. Jennings, *Glen Canyon: An Archaeological Summary*, foreword by Don D. Fowler (Salt Lake City: University of Utah Press, 1998), 1, 3.

12.3. This photo from Glen Canyon archaeological work titled "Fowler's Folly" is indicative of the issues field personnel in remote Reclamation project locations faced continually regardless of whether they were doing cultural resources work or technical field studies. Harris Wash, June 1961. Gift of photographer Thomas Oliver to Don Fowler. Courtesy of Don Fowler.

Authorization for the "salvage project" came from the 1935 Historic Sites Act. According to Russell Martin, in *A Story that Stands Like a Dam*, the Act "charged the Department of the Interior, through the National Park Service, with the preservation and scientific dissemination of the nation's antiquities, and it had called for their 'emergency salvage' in situations where major development threatened to destroy them." But the Act alone was not enough to justify expending resources to rescue the past. It took the determined effort of archaeologists and other scholars to convince the federal government of the need to embark on such a program. By 1944 a group of serious academics argued that the federal government's water resource development policy "would eventually destroy many archeological sites." They were able to formulate an Inter-Agency Agreement among the Bureau of Reclamation, the National Park Service, and the Army Corps of Engineers to conduct archaeological surveys in proposed reservoir sites. In 1945 a Memorandum of Understanding between the Park Service and the Smithsonian led to creation of the Inter-Agency Archeological and Paleontological Salvage Program.[291]

[291] Martin, *A Story that Stands Like A Dam*, 105; Frank H. H. Roberts, Jr., "River Basin

12.4. The isolation of parts of Glen Canyon is emphasized by this October 1960 picture of a Museum of Northern Arizona pack train on the Oak Canyon Trail during salvage archaeology work. Photographer: Christy Turner. Courtesy of the Museum of Northern Arizona: photograph S.GCP 611 B/W.

Initial efforts toward salvaging the past on proposed reservoir sites on both Bureau of Reclamation and Army Corps of Engineer projects began in 1946. Teams of archaeologists and associated researchers fanned out throughout much of the nation to start primary investigations. According to Frank Roberts from the Bureau of American Ethnology and director of the River Basin Surveys, these teams first set out to become familiar with the "potential reservoir areas" then conduct intensive surveys, followed by excavation efforts. There was an understanding among all involved that time constraints placed on the salvage program meant that only cursory excavations could occur. Depending on the size of the site, time limitations might allow a thorough investigation of smaller sites, but on larger finds researchers had little choice but to limit their time at any location. By 1948 the surveys identified archaeological material on fifty-seven Reclamation projects and nine Corps projects.[292]

Surveys: The First Five Years of the Inter-Agency Archeological and Paleontological Salvage Program," from the Smithsonian Report for 1951, 351-83 (Washington, D.C.: United States Government Printing Office, 1952), 351-2.

292 Smithsonian Institute, Bureau of American Ethnology, Bulletin 154, *River Basin Survey Papers: Inter-Agency Archeology Salvage Program, Numbers 1-6* (Washington, D.C.: United

For the most part the Bureau of Reclamation had little direct involvement in the salvage program because of the Park Service's lead. Its primary responsibility came in funding the operations, sharing the cost of the surveys with the Park Service. By 1947 the Bureau of the Budget had determined that any request for funds to conduct surveys "should appear in the Interior Department budget." Yet Reclamation projects held the lion's share of attention from survey teams. In 1948 Reclamation "supplied additional funds ... for surveys at its projects in the Columbia-Snake Basin." Salvage projects on the Bureau of Reclamation's Pick-Sloan Missouri Basin Program occurred at the Angostura Unit in South Dakota, the Medicine Creek Dam in Nebraska, and the Boysen Dam in Wyoming.[293]

In 1957 when salvage efforts began in Glen Canyon, preparations for cultural resource retrieval were firmly in place. The University of Utah's Archaeology and History Departments along with the Museum of Northern Arizona took the lead in the archaeological and historical investigations. According to one of the survey's participants, Don D. Fowler, "The project lasted six years. It cost about one million dollars ... At least 205 people were on the payroll of the two institutions for varying periods of time, including twenty-six members of the Navajo Nation." In the course of the salvage program in Glen Canyon researchers located and identified over 2,000 archaeological sites and 250 historical sites.[294]

Overall researchers unearthed a plethora of archaeological evidence suggesting a vast amount of human activity in Glen Canyon spanning thousands of years. For example, historian C. Gregory Crampton noted that

> Wright Bar was distinguished in Glen Canyon history as the site of a magnificent 50-foot panel of ancient petroglyphs pecked into the smooth face of a cliff overlooking the Colorado River. The panel contains dozens of figures and designs making up a style which may date anywhere from 100 B.C. to 1050 A.D., the oldest rock art in Glen Canyon.

States Government Printing Office, 1953), 3-4, RG 115, Binder: Special Lists, Vol. II, An Index of Cultural Resource Reports Located at the National Archives, ACC# 8NS-115-95-122, Box 1; see also Roberts, "River Basin Surveys."

293 Roberts, "River Basin Surveys," 353, 361; see also Don D. Fowler, *The Glen Canyon Country: A Personal Memoir* (Salt Lake City: University of Utah Press, 2001), 244-6.

294 Don D. Fowler, foreword, in Jennings, *Glen Canyon*, xi, xv-xvi; Paul V. Long, Jr., *Archaeological Excavations in Lower Glen Canyon, Utah, 1959-1960*, Museum of Northern Arizona Bulletin no. 42–Glen Canyon Series no. 7 (Flagstaff, Arizona: The Northern Arizona Society of Science and Art, Inc., 1966), 60-1; Martin, *A Story that Stands Like A Dam*, 106.

Most of the research indicated that, despite the long period of human activity in the canyon, Native American societies never built lasting communities. Evidence found at archaeological sites suggested that Native American peoples over time utilized Glen Canyon intermittently primarily for agricultural purposes. Researchers generally concluded that "in the Glen Canyon area ... repeated short term occupation, possibly on a seasonal basis at least over a number of seasons or alternating years, was the habitation pattern for the riverine environment."[295]

Investigations into historical sites within Glen Canyon and San Juan Canyon yielded similar results. Led by University of Utah history professor C. Gregory Crampton, a former student of historian Herbert Bolton at the University of California, he followed Bolton's suggestions to explore the locations and environments of historical subjects. Crampton enthusiastically approached the investigation of Glen Canyon for it allowed him the opportunity to explore the route of the 1776 Dominguez and Escalante expedition, peer into the side canyons described by John Wesley Powell, follow the trails of Mormon pioneers, and view the remnants of Anglo farming and mining enterprises.[296] What Crampton and his fellow researchers found was a rich tapestry of Euro-American activity in Glen Canyon. The salvage program opened new vistas on the history of the upper Colorado River basin. Crampton wrote:

> When it comes to history most writers have fallen victims to the charms of the Grand Canyon and the lower basin. The upper basin as a unity scarcely exists in the historiography; the Glen Canyon country has no entity at all. Even river runners from Powell onward have generally rested their pens as they have their oars in Glen Canyon ...

Yet despite the dearth of archival records regarding Glen Canyon, Crampton and his colleagues found physical evidence left by people who entered the canyon. Much of what researchers discovered centered on enterprises that sought to extract some measure of wealth from the canyon. These included abandoned mines and mining equipment "ruins of buildings, inscriptions on cliff walls, trails and roads, corrals, reservoirs, farming areas, and graves."[297]

---

[295] Crampton, *Ghosts of Glen Canyon*, 26; Long, *Archaeological Excavations in Lower Glen Canyon, Utah, 1959-1960*, 65-6.

[296] Crampton, *Ghosts of Glen Canyon*, 7; Martin, *A Story that Stands Like A Dam*, 103.

[297] C. George Crampton, *Anthropological Papers: Outline History of the Glen Canyon Region, 1776-1922*, Charles E. Dibble, editor, Number 42 (Glen Canyon Series Number 9) University

It is uncertain whether or not members of the salvage program in Glen Canyon had any opinions regarding the merit of the dam that Project Construction Engineer Lem Wylie and the Bureau of Reclamation were constructing. Most were probably aware of the Echo Park controversy and associated events that pushed Glen Canyon Dam to the top of Reclamation's to-do list. After almost six years of roaming the Glen Canyon, they were most certainly conscious of the canyon's natural beauty and the breadth of cultural resources soon to be lost when waters behind the dam filled the canyon. Crampton pragmatically noted that by 1922 both Glen Canyon's "scenic beauties" and "its potentialities for commercial utilization" had been recognized, and Crampton realized that this dichotomy simply stood as part of the history of Glen Canyon. In 1960 he remarked that one of the canyon's famous landmarks, named by John Wesley Powell, would fall victim to the reservoir. "When Lake Powell is at its maximum level the top of Sentinel Rock, less than two miles above Glen Canyon Dam, will be approximately 300 ft. below the surface."[298]

By 1963 the salvage program was running out of time and money, and program participants at Glen Canyon no doubt wished for more time to develop research and further investigate findings. Yet, their efforts produced an impressive amount of scholarly work. Edward B. Danson, director of the Museum of Northern Arizona, wrote that the salvage effort yielded "twenty-six monographs in the Glen Canyon Series of the University of Utah and [that] thirty-three other contributions have rendered Glen Canyon one of the most thoroughly studied areas in America." The Glen Canyon Salvage program stood as the precursor to other endeavors in the upper basin of the Colorado River. At the Navajo Dam site in New Mexico researchers from the New Mexico State Museum of Natural History uncovered and studied Native American sites with assistance from the Bureau of Reclamation and the National Park Service. Most important, "salvage archaeology" paved the way for the institutionalization of "cultural resource management" within the Bureau of Reclamation and elsewhere in the federal government.[299]

---

of Utah, Department of Anthropology, September 1959, 2; Udall Papers, http//content.library.arizona.edu/cdm/compoundobject/collection/udallcoloradoAZU/id/11998/rec160; Fowler, *The Glen Canyon Country*, 255.

298 Crampton, *Anthropological Papers*, 59, Udall Papers, http//content.library.arizona.edu/cdm/compounobject/collection/udallcoloradoAZU/id/14189/rec/160; see also C. Gregory Crampton, *Historic Sites in Glen Canyon Mouth of San Juan River to Lee's Ferry*, Charles E. Dibble, editor, Number 46 (Glen Canyon Series Number 12) University of Utah, Department of Anthropology, June 1960, 68; Udall Papers, http//content.library.arizona.edu/cdm/compoundobject/collection/udallcoloradoAZU/id/12147/rec/1.

299 Edward B. Danson, Director, Museum of Northern Arizona, foreword, in Long, *Archaeo-*

While the salvage operations in Glen Canyon drew to a close, the Sierra Club published Eliot Porter's *The Place No One Knew*, a pictorial treatise on the natural beauty lost because of Glen Canyon Dam. The book showed the world the natural beauty within Glen Canyon that the closing of Glen Canyon Dam's diversion gates were about to destroy. Eliot's images in *The Place No One Knew* suggested that the place never saw human activity and to dam the canyon or even think about building dams close to the Grand Canyon was an affront to nature. Of course, the salvage program offered irrefutable evidence of the canyon's long history of human activity. Porter's book and photographs, however, persuasively portrayed Glen Canyon as an isolated and lonely wilderness. Even if human activity were present in the canyon for thousands of years, the salvage program study revealed sporadic occupation and minimal environmental impact.[300]

John Wesley Powell's expeditions down the Colorado River in 1869 and 1870 introduced most Americans to the wonders of the river's canyons and mesas. Powell and his men surveyed the last unknown lands on the map of the continental United States. As such, this blank space on the map stood virtually untouched by the progressive forces that changed the face of the nineteenth-century American West. Jesse D. Jennings, of the University of Utah and a leading member of the salvage program, wrote enthusiastically about the new studies underway and this new chapter in the history of the American West. He also noted that "the Glen and San Juan canyons are paradoxically the focus of modern history less because of their own worth than because they impeded the march of progress." Glen Canyon Dam and Lake Powell inevitably broke that pattern, ensuring that these isolated canyon lands no longer "impeded" progress, but were in fact firmly connected to the economic development of the arid West.[301]

---

logical Excavations in Lower Glen Canyon, Utah, 1959-1960, iii; for information on salvage archaeology at Navajo Dam see United States Department of the Interior, Bureau of Reclamation, Annual Project History, Navajo Unit, Colorado River Storage Project, Volume II, Farmington, New Mexico, 1957, in RG 115, Records of the Bureau of Reclamation, Entry 10, Project Histories, Feature Histories, and Reports 1902-1932, Box 128; for reference to the transition from salvage archaeology to cultural resource management see Don D. Fowler, foreword, in Jennings, *Glen Canyon*, xiv.

300 Eliot Porter, *The Place No One Knew: Glen Canyon on the Colorado* (San Francisco: Sierra Club, 1963).

301 Jennings, *Glen Canyon*, 66, 70.

Getting Our Share: The Central Utah Project

Historians of the irrigation movement in the American West often look first to Native American systems and to the early Mormon experience in Utah as a starting point in the evolution of federal reclamation policy. For early reclamation advocates, irrigation in Utah provided an instructive pattern for government water resource management policies. As Reclamation historian William Rowley notes, "While not exactly free enterprise, industrious enterprise and community cohesiveness inspired by religious commitment in Mormon Utah set an example for how lands might be brought under cultivation by irrigation." Utah Mormon settlements stood as an example for the American West demonstrating the value of irrigation agriculture in the second driest state in the Union. By the time Congress passed the 1902 Reclamation Act, even Utah realized the limitations of "community cohesiveness" and "industrious enterprise." If it were to continue to grow and prosper, Utah needed the assistance of the federal government to expand its irrigation empire. When World War II ended, Utah, similar to other upper basin states of the Colorado River, experienced tremendous industrial and population growth. To maintain that economic development, the people of Utah dreamt of grander water resource development and looked expectantly to the waters of the Colorado River.[302]

A proposed Central Utah Project (CUP) was a bold trans-mountain plan to divert water from the south side of the Uinta mountain range, in the Colorado River drainage, northward into the Bonneville Basin. Utah water interests and their political allies conceived of the idea prior to the state ratification of the Upper Colorado River Compact in 1948. The Central Utah Project became one of the most time consuming and complicated construction projects undertaken by Reclamation during the second half of the twentieth century. Consisting of multiple projects, all designed to utilize Utah's allotment of the Colorado River, many units (such as the Vernal and Jensen projects in the Uintah Basin) were largely "stand alone" projects, meaning that they were independent of the grander goals of CUP but still instrumental in the state's overall water development plans. Yet they were crucial to successful completion of the larger Bonneville Project whose purpose was to increase the water supply to the more populated areas of central Utah. During CUP's seemingly never-ending construction period, both Reclamation and project supporters faced delays and controversies. The Central Utah Project struggled

---

302 Rowley, *The Bureau of Reclamation*, 56.

through continuous funding woes from a hesitant Congress, encountered opposition from environmentalists, and attempted to resolve disputes with Native Americans showing greater determination to assert their water rights. In the end, CUP never accomplished the grandiose plans envisioned by Utah water users and Reclamation officials and was eventually scaled down to meet the economic and social realities of the late 1980s and early 1990s.

In 1938 the Bureau of Reclamation began investigating projects in northeastern Utah that eventually became part of the Central Utah Project. From its office in Vernal, Utah, Reclamation engineers studied irrigation possibilities of the Ashley Valley, named after General W. M. Ashley one of the principals in the early nineteenth-century Rocky Mountain Fur Company. The valley saw earlier Mormon irrigation efforts primarily associated with cattle ranching. By May 1943 Reclamation completed its study of Ashley Valley as part of a larger investigation of the Uintah Basin. By September 1943 the survey of Stanaker (later renamed Steinaker) Dam and reservoir site was complete. World War II halted work because the project was not deemed crucial to the war effort. Reclamation, in February 1949, incorporated this planning, generally known as the Vernal Project, into the larger Central Utah Project.[303]

The Bureau of Reclamation's 1946 report on the development of the water resources of the Colorado River, *The Colorado River*, made only vague references to the Central Utah Project. This ambiguity stemmed from the uncertainty of future development of the Colorado River's upper basin because of the lack of an accord among the upper basin states regarding allocation of Colorado River water. In addition, Reclamation warned that the river did not contain enough water to fulfill the requirements of all the proposed projects found in the report. Moreover, the authors noted, "If some States elect to use part of the water to which they are entitled in out-basin or export diversion projects, a corresponding elimination of within-basin projects will be necessary." In the view of the Bureau of Reclamation, according to the report, states must determine how to best utilize water.

---

303 United States Department of the Interior, Bureau of Reclamation, "Annual Project History, Vernal Unit, Central Utah Project," Volume I, Calendar Year 1919-1959, 3-4, RG 115, Office of the Chief Engineer, Project Histories, Alphabetically Arranged by Projects, ACC# 8NN-115-88-053, Box 68, Cedar Bluff, 1963 through Vernal, 1963, for more information on the early investigations of the Vernal Project see United States Department of the Interior, Bureau of Reclamation, *Beyond the Wasatch: The History of Irrigation in the Uinta Basin and the Upper Provo River Area of Utah*, edited by Gregory D. Kendrick, introduction by Charles S. Peterson (Washington, D.C.: United States Government Printing Office, 1989), 25; Adam R. Eastman, "The Central Utah Project, Vernal Unit," Denver: Bureau of Reclamation History Program, n.d.

Regardless, the Bureau of Reclamation continued to outline grand proposals for water resource development in Utah. In the case of Vernal, the report claimed that water storage on Ashley Creek could irrigate 1,900 acres of "new land and furnish a supplemental supply to 22,300 acres of cultivated land near Vernal." Reclamation offered similar projections for the proposed Jensen Unit in Uintah County in northeastern Utah. The report made its most ambitious estimates about the Central Utah Project indicating that 625,000 acre feet annually was available "from the streams in the Uinta Basin to the Bonneville Basin in Utah." It continued,

> Water could be pumped from the potential Echo Park Reservoir on the Green River to replace irrigation supplies not used on lands in the Uintah Basin which would be diverted to the Bonneville Basin under this project and to permit expansion of irrigation to the Uintah Basin.

Ambitious planning, however, remained just that. While detailed plans excited the imaginations of local water development proponents, construction faced many hurdles.[304]

A prerequisite to move plans off the drawing board required an agreement on dividing the waters of the Colorado River among the upper basin states. Prompted by the successful construction of water projects in the lower basin, and fear of losing water rights because of those and future developments, the states of Colorado, New Mexico, Utah, and Wyoming reached an agreement. Contrasted to the lower basin, the upper basin based its division of the Colorado River waters on percentages rather than a hard acre foot allotment, with Utah granted 23 percent of the water. The Upper Colorado River Compact, signed by the upper basin states and approved by Congress in 1948, opened the door for the introduction of the Colorado River Storage Project legislation with the Central Utah Project named a "participating project."[305]

Legislation for the Colorado River Storage Project went through many variations before finally passing Congress in 1956. The controversial dams proposed for Dinosaur National Monument presented the primary hurdle,

---

304 USDOI, BR, *The Colorado River*, 17-8, 117.

305 "Upper Colorado River Compact," in USDOI, BR, *Federal Reclamation and Related Laws Annotated*, Volume II, 909-23; "Upper Colorado Compact Ok'd," *Reclamation Era*, 36 (June 1949): 140; see also Martin, *A Story that Stands Like a Dam*, 49; Harvey, *A Symbol of Wilderness*, 69; Farmer, *Glen Canyon Dammed*, 133; "Annual Project History, Vernal Unit, Central Utah Project," Volume I, 4, RG 115, Office of the Chief Engineer, ACC# 8NN-115-88-053, Box 68, Cedar Bluff, 1963 through Vernal, 1963.

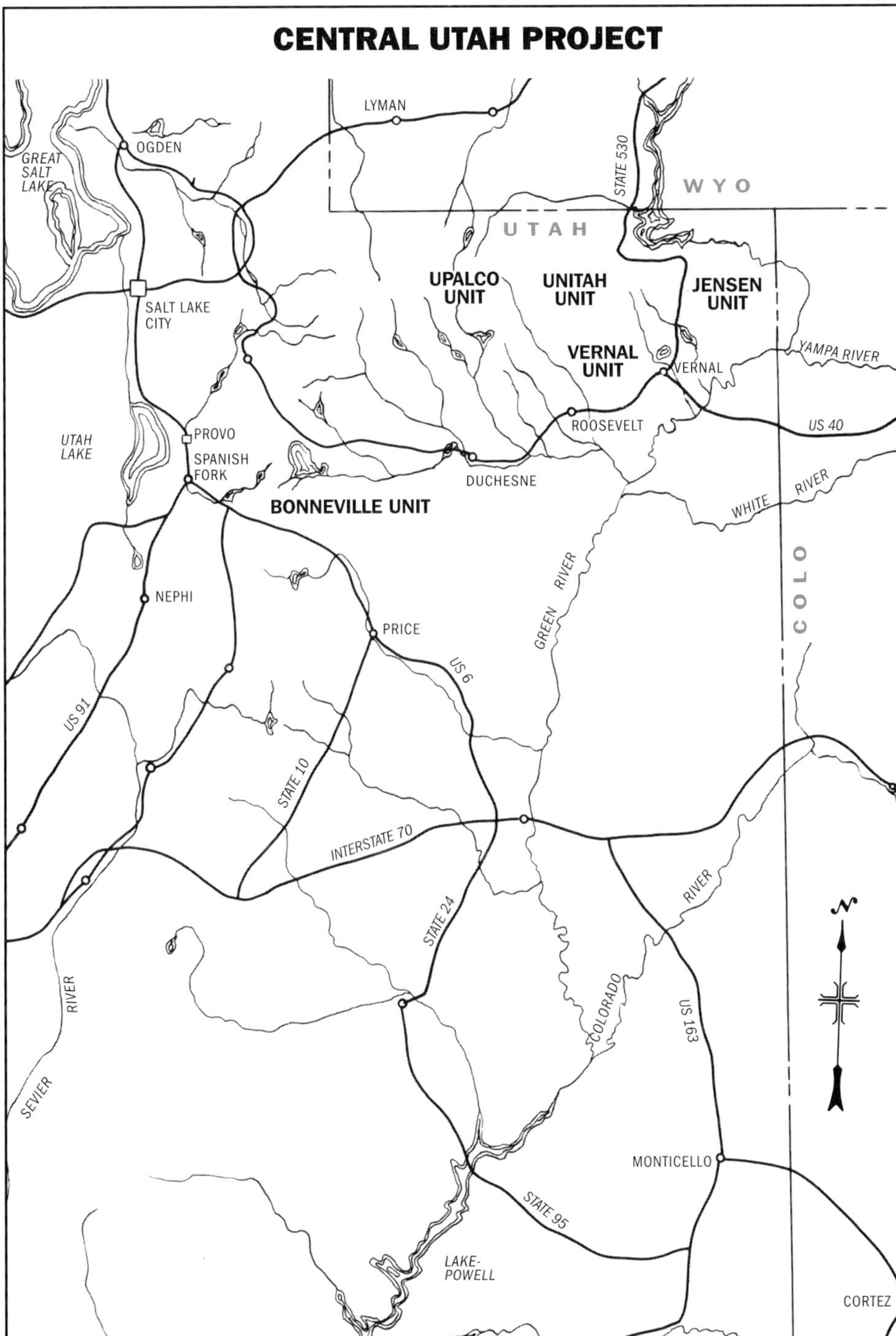

12.5. The Central Utah Project as envisioned by Reclamation in 1978.

and many deemed them crucial to development of the Central Utah Project. According to CUP proponents, Echo Park Dam, at the confluence of the Yampa and Green rivers, assured a stable water supply for CUP. According to one source:

> Under the ultimate phase of the Central Utah Project, Reclamation planned on diverting into Strawberry Reservoir the entire flows of all the major streams and rivers draining off the southern face of the Uinta Mountains. This water would then be diverted into the Bonneville Basin for use along the Wasatch Front and in central Utah.

Echo Park Dam was the safety valve that insured the availability of water for the Uintah Basin and northeastern Utah. Cancellation of Echo Park and Split Mountain dams in Dinosaur National Monument forced the Bureau of Reclamation and Utah water interests back to the drawing board to devise new ways to tap into the waters of the Green River.[306]

Reclamation planning for the Central Utah Project envisioned six separate units: Vernal, Bonneville, Jensen, Upalco, Uintah, and the Ute Indian. Each unit was to receive authorization and appropriations individually, which resulted in requiring decades to complete various phases of the project, and some units still have not been completed.[307] For example, the Upalco unit sought to develop the waters of the Yellowstone and Lake Fork Rivers, providing supplemental irrigation water for 42,520 acres of Indian and non-Indian land. Similarly, Reclamation designed the Uintah unit, located in Duchesne and Uintah counties, to serve over 67,000 acres of Indian and non-Indian lands with supplemental irrigation water from the Uinta and Whiterocks rivers.[308] Few of these projects ever reached the level of development that boosters envisioned when the Colorado River Storage Project Act passed in 1956. Progress on these projects represented the low level of planning each received and the time it took for these projects to receive the full attention of Reclamation planners. Though perhaps not recognized at the time, these reports foreshadowed the veiled requirement that Utah water users would have to embrace development of Indian water projects in order to fulfil their goals.

---

[306] Adam R. Eastman, "The Central Utah Project, Jensen Unit," Denver: Bureau of Reclamation History Program, 2006; see also, Harvey, *A Symbol of Wilderness*, 38-9.

[307] The Vernal and Jensen Units are completed. The Bonneville Unit is still under construction. The Uintah and Upalco Units were deauthorized by Congress in 2002. The Ute Indian Unit—also known as the Ultimate Phase—was only authorized for study. It was never approved for construction.

[308] Water and Power Resource Services, *Project Data*, 121-5.

In 1958 Congress appropriated $1,035,000 for "initial construction of the Vernal Unit" to begin in the spring of 1959. As planned the unit contained seven features that included: Steinaker Dam located 3.5 miles north of Vernal; Fort Thornburgh Diversion Dam on Ashley Creek 4 miles northeast of Vernal; a 3.1-mile-long feeder canal from Thornburgh to Steinaker; an 11.8 mile canal from Steinaker to the main irrigation system. According to the first Reclamation project history, "the Vernal Unit will provide an average of 18,000 acre-feet of supplemental irrigation water annually for 14,781 acres of irrigable land in Ashley Valley." The project also added supplemental municipal and industrial water to the communities of Vernal, Maesser, and Naples, Utah. News of the project appropriation was greeted with enthusiasm and elation, and in August 1958 the community of Vernal held an elaborate celebration that included a parade featuring Utah Governor George D. Clyde. This was followed in May 1959 with a groundbreaking ceremony to mark the beginning of construction of Steinaker Dam.[309]

Compared with other CRSP projects, the Vernal Unit was relatively small and straightforward. Construction progress on the project moved along at a steady and uneventful pace. As planned, the earthen Steinaker Dam rose 162-feet with a crest length of 1,997 feet, and a reservoir capable of holding 38,173 acre-feet of water that had a surface area of 820 acres. While work moved forward on Steinaker Dam, in July 1960 construction started on the Fort Thornburgh Diversion Dam, Steinaker Feeder Canal, and the Rock Point Canal Extension. By January 1961 the Bureau of Reclamation accepted as completed Steinaker Dam, and during 1962 "there were 18,007 acre-feet of water diverted from Ashley Creek through Steinaker Feeder Canal to Steinaker Reservoir."[310]

While work on the Vernal Unit went forward, other events outside the northeastern Utah community surfaced to concern project supporters. With

---

[309] Eastman, "The Central Utah Project, Vernal Unit," 2; "Annual Project History, Vernal Unit, Central Utah Project," Volume I, 17, RG 115, Office of the Chief Engineer, ACC# 8NN-115-88-053, Box 68, Cedar Bluff, 1963 through Vernal, 1963.

[310] "Annual Project History, Vernal Unit, Central Utah Project," Volume I, 86; Water and Power Resources Service, *Project Data*, 159; United States Department of the Interior, Bureau of Reclamation, "Annual Project History, Vernal Unit, Central Utah Project," Volume II, Calendar Year 1960, 65; United States Department of the Interior, Bureau of Reclamation, "Annual Project History, Vernal Unit, Central Utah Project," Volume III, Calendar Year 1961, 1; United States Department of the Interior, Bureau of Reclamation, "Annual Project History, Vernal Unit, Central Utah Project," Volume IV, Calendar Year 1962, 11, RG 115, Office of the Chief Engineer, ACC# 8NN-115-88-053, Box 68, Cedar Bluff, 1963 through Vernal, 1963.

the successful Soviet launch of Sputnik in 1957, the Cold War loomed behind the scenes. The Uintah Water Conservancy District worried that Congress and the Eisenhower administration might curtail funding for the Vernal Unit and the entire Colorado River Storage Project. After intense lobbying by the district and Utah's congressional delegation, Congress appropriated $1 million to continue construction. Two years later, the ghost of the Echo Park controversy reappeared. Under the leadership of Howard Zahniser of the Wilderness Society, environmentalists began efforts toward securing legislation designating wilderness areas. The conservancy district took a dim view of actions taken by conservationists. It warned,

> The interests striving to eliminate any possible water development in Dinosaur National Monument will not stop their efforts to make Dinosaur a National Park. They will continue their battle to change the present status, to construct facilities in the Echo Park Reservoir Area, to alter boundaries, or they will pass Wilderness Legislation to accomplish their aims in Dinosaur.

These two events, though having minimal affect on the Vernal Unit or CUP, represented important factors that would have long-term implications not only for water resource development in Utah, but also for the overall goals and ambitions of the Bureau of Reclamation during the 1960s. Sputnik spurred the Eisenhower administration into increased military spending on missile defense and led to the spiraling costs of the space race, leading to fewer funds available for water projects. The threat of "wilderness legislation" was another major concern for Reclamation policy, because it could limit sites for future Reclamation projects.[311]

By 1963 the Vernal Unit was essentially complete. Water deliveries began under Bureau of Reclamation control for the next three years until relinquished to local administration. While the Vernal Project represented the first stage of the Central Utah Project, it was also the only unit of CUP that the Bureau of Reclamation completed as originally planned. CUP's remaining units faced appropriation and environmental difficulties that the Vernal Unit

---

[311] "Second Annual Report of the Uintah Water Conservancy District," December 15, 1958, in "Annual Project History, Vernal Unit, Central Utah Project," Volume I, 162; "Fourth Annual Report of the Uintah Water Conservancy District," December 19, 1960, in "Annual Project History, Vernal Unit, Central Utah Project," Volume II, 127, RG 115, Office of the Chief Engineer, ACC# 8NN-115-88-053, Box 68, Cedar Bluff, 1963 through Vernal, 1963; for more information on Howard Zahniser and the wilderness movement see, Harvey, *Wilderness Forever*.

avoided. A more immediate threat to a prompt construction schedule for the Central Utah Project also appeared in 1963 just as the Vernal Unit was beginning to service water users. That concern came from the Supreme Court's decision in *Arizona v. California*, which many believed would open the door for a competing Central Arizona Project (CAP).

Water interests from the Colorado River's upper basin feared the possibility of the Central Arizona Project (CAP) taking water that rightfully belonged to them. Debates throughout the Colorado River watershed about the Mexican Water Treaty of 1944 dealt with the understanding that the river was over-appropriated. Arizona's 2.8 million acre-feet, granted by the Supreme Court in 1963, raised concerns about where that water was to be found. The Uintah Water Conservancy District was representative of Utah apprehensions regarding the impact of CAP on the Central Utah Project. It pushed hard for creation of a seven-county water district to reaffirm state unity for the project and prove Utah's commitment to the Bureau of Reclamation. The district argued that "we must recognize that the Central Utah Project offers the State of Utah its only opportunity to use its share of the Upper Colorado River water." A display of unity and commitment might provide the impetus for faster completion of CUP before authorization of the Central Arizona Project occurred, thus preserving Utah's Colorado River water allocation. The district argued,

> We have fought long and hard for authorization of the Colorado River Storage Project and the Central Utah Project.... The people of Utah must stand together to assure that this project will become a reality so that the benefits can be fully realized.

Such proclamations as this one echoed throughout the upper basin, reaffirming the basic "first in time, first in right" principle of the prior appropriation doctrine.[312]

The centerpiece of the Central Utah Project was the Bonneville Unit: a transbasin diversion from the headwaters of the Duchesne River in the Colorado River basin to the Bonneville Basin of central and western Utah. In addition to increasing irrigation water supplies, the project included elements to

---

312 "Sixth Annual Report of the Uintah Water Conservancy District," December 1962 , "Annual Project History, Vernal Unit, Central Utah Project," Volume IV, Calendar Year 1962, 11, RG 115, Office of the Chief Engineer, ACC# 8NN-115-88-053, Box 68, Cedar Bluff, 1963 through Vernal, 1963.

supply municipal and industrial water and power production. Planning for the Bonneville Unit began in 1956 with Reclamation finally completing its report in August 1964. Based on 1963 prices, the project's price tag was an estimated $324 million to provide 315,000 acre-feet of water annually to the Bonneville Basin. Reclamation reported, "Operation of the Bonneville Unit will deplete the flow of the Duchesne River to the Colorado River by an estimated 165,900 acre-feet annually. This is well within Utah's allotment of Colorado River water." In 1965 Congress approved a $3.5 million appropriation to begin construction of the Bonneville Unit.[313]

The Bonneville Unit was one of the most complicated engineering challenges Bureau of Reclamation engineers faced at the time—the type of project that the bureau relished. In sum, the unit included "ten new reservoirs and enlargement of two reservoirs; 140 miles of aqueducts, tunnels and channels; three powerplants; 13 miles of dike; and 200 miles of drains." It was designed to provide irrigation water for 156,530 acres of land that included 43,740 acres of "full service land." In 1967 work began in earnest when Reclamation awarded contracts for construction of Starvation Dam, Starvation feeder canal, Knight Diversion Dam, Water Hollow Tunnel and Channel No.2, along with road improvements, field stations, "and improvements of the Government Community of Duchesne, Utah." Also, in 1967, Congress appropriated another $4,285,000, with the promise of $11,145,000 indicated in President Johnson's 1968 budget request. To celebrate the initial construction, May 31, 1967, Utah officials hosted "Bonneville Days" to mark the ground breaking ceremonies at the Starvation Damsite. An estimated 1,500 people attended and heard the keynote address delivered by Colorado Congressman Wayne Aspinall.[314]

Construction of Starvation Dam marked an important event in the progress of the Central Utah Project. This earth-rolled structure was 155 feet high, 2,920 feet long with a 167,300 acre-foot storage capacity. The reservoir provided water for the Strawberry and Duchesne rivers as it began the long

313 United States Department of the Interior, Bureau of Reclamation, "Project History, Bonneville Unit, Central Utah Project," Volume I, 1966, RG 115, Project Histories, ACC# 8NN-115-92-130, Box 29, Central Utah (Vernal) through Central Valley.

314 United States Department of the Interior, Bureau of Reclamation, "Project History, Central Utah Project, Bonneville Unit, Utah," 1974, Volume 9, 73, RG 115, Project Histories, Box 229, Central Utah (Bonneville), 1974 Through Cedar Bluff, 1975; United States Department of the Interior, Bureau of Reclamation, "Project History, Bonneville Unit, Central Utah Project," Volume II, 1967, 8-9, 77, RG 115, Project Histories, ACC# 8NN-115-92-130, Box 41, Wichita, 1966 through Central Utah, (Bonneville), 1967.

journey from the Uinta Basin to the Bonneville Basin. The Bureau of Reclamation quickly completed it in March 1970. This accomplishment along with completion of Knight Diversion Dam, which diverts water from the Duchesne River into Starvation Reservoir, and the Starvation Feeder Conduit, completed in 1968, allowed for the next phase of the Central Utah Project. This structure was Soldier Creek Dam, seven miles downstream from Strawberry Dam, which increased "the present Strawberry Reservoir capacity from 283,000 to 1,106,500 acre-feet.... This reservoir will be the prime storage facility for the Bonneville Unit providing the necessary carry over storage for use in years of short supply." Reclamation completed construction of the dam in 1974.[315]

By 1970 the Bureau of Reclamation boasted that it had completed 16 percent of the Central Utah Project, but that accomplishment belied persistent issues that slowed further construction. In 1967 Utah water users, under the auspices of the Central Utah Water Conservancy, anxiously watched developments in Washington, D.C., that might impact continued development of CUP. In July the district passed a resolution urging delay in any legislative action on the proposed High Uintah Wilderness Area. The district was "vitally concerned about the timing of the establishment of the Wilderness Area in relationship to the development of the Central Utah Project." It argued that because of Utah's limited water supply the development of the waters of the Colorado River were essential "to the State's future." Any designated wilderness area might limit Utah's ability to gain the greatest benefits from its water resources. Equally disturbing was the pending Central Arizona Project legislation. The prospect of the CAP threatened Utah's rightful allotment from the Colorado River. Utah water users, like others in the upper basin, pinned their future hopes on some sort of augmentation scheme to offset the impact of the CAP. They looked with favor on various proposals that sought to take water from the Columbia River watershed and transport it to the Southwest. In Vernal the Uintah Water Conservancy District asserted, "Utah's compact water allocation could be drastically curtailed if the Central Arizona Project is approved and constructed without provisions for a meaningful augmentation of water supply in the Colorado River."[316]

---

315 "Project History, Central Utah Project, Bonneville Unit, Utah," 1974, Volume 9, 76, 79, RG 115, ACC# 8NN-115-92-130, Box 229, Central Utah (Bonneville), 1974 through Cedar Bluff, 1975; for more information on Starvation Dam and Soldier Creek Dam see also Water and Power Resources Service, *Project Data*, 135, 1192.

316 United States Department of the Interior, Bureau of Reclamation, Upper Colorado Region, *Central Utah Project, Bonneville Unit: Municipal and Industrial System, Summary Statement*, August 1979; Central Utah Water Conservancy District, *4th Annual Report*, December 31, 1967, in "Project History, Bonneville Unit, Central Utah Project," Volume II, 1967, RG 115, Project

Concurrently, Central Utah Project advocates encountered another threat: Indian water rights. At one time much of the Uinta Basin had been an Indian reservation. In 1864 the federal government created the 2,039,040 acre Uintah Indian Reservation as the new home for one band of the Ute Tribe. According to one source, "The newly created reservation … stretched from the crest of the Wasatch Mountains on the west, to the Sand Ridge on the east, and from the summit of the Uinta Mountains on the north, to the top of Tavaputs Plateau on the south." Over the intervening years, allotments, leases, and land sales depleted Indian land ownership. Nevertheless, Central Utah Project boosters and the Bureau of Reclamation understood that any project success must consider and include Native American interests. This meant stepping into the complex legal, social, and cultural issues of Indian water rights.[317]

In 1964 the Bureau of Reclamation's Region IV in Salt Lake City released the *Definite Plan Report* on the Bonneville Unit of the Central Utah Project. It noted that although the Ute people had appropriated water under Utah law "they still claimed an inherent water right prior to all non-Indian rights under … Winters v. United States [*sic*]." While no formal agreement existed between Indian and non-Indian water users, informal agreements allowed cooperation. Earlier investigations came to the conclusion that Ute water rights remained a powerful element in the ultimate success of CUP. To clear up ambiguities, all parties needed to reach an agreement to initially limit irrigation water to Indian land with the promise of expanding irrigation services to the Ute at some future date. In 1965 the Ute Indian Tribe, the Central Utah Water Conservancy, and the United States reached an accord that "recognized the first priority of Indian water rights and provided for the deferment of certain of these rights." In short, the Ute agreed to postpone irrigation services to undeveloped Indian land until completion of the ultimate phase of the Central Utah Project, or until 2005 whichever came first. In return, the federal government would institute measures to mitigate fish and wildlife losses on the reservation.[318]

Histories, ACC# 8NN-115-92-130, Box 41, Wichita, 1966 through Central Utah, (Bonneville), 1967; "Uintah Water Conservancy District Eleventh Annual Report," December 18, 1967, in United States Department of the Interior, Bureau of Reclamation, "Project History, Vernal Unit, Central Utah Project, Volume IX, Calendar Year 1967, RG 115, Project Histories, ACC# 8NN-115-92-130, Box 42, Vernal through Central Utah, 1967.

[317] Bureau of Reclamation, *Beyond the Wasatch*, 16.

[318] United States Department of the Interior, Bureau of Reclamation, Region 4, *Central Utah Project, Initial Phase: Bonneville Unit, Definite Plan Report* (Salt Lake City: Region 4, August 1964), 72-3; see also "Project History, Central Utah Project, Bonneville Unit, Utah," 1974, Volume 9, 73, RG 115, Project Histories, ACC# 8NN-115-92-130, Box 229, Central Utah

Initially the Bureau of Reclamation attempted to bring some benefits to Indian lands but was hamstrung by the 1965 deferral agreement in making a greater effort. By 1974 it had finished construction of Bottle Hollow Dam and Reservoir that received water diverted from the Uinta River. Bottle Hollow was part of a two unit plan to compensate the Ute Tribe for "economic losses associated with stream fishing in Rock Creek within the Uintah and Ouray Indian Reservations." However, the second unit, Lower Stillwater Reservoir, still awaited construction. By 1977 grumblings over the lack of progress from the Ute Tribe became more intense.[319] According to Lloyd Burton, in *American Indian Water Rights and the Limits of the Law*, "tribal leadership had changed, and the terms of the original deferral agreement came to be seen in a much more unfavorable light." Evidence for Burton's claim can be found in the actions of some tribal members a year earlier. Discontented members of the Ute Tribe filed suit against the Bureau of Reclamation seeking guarantees "that the Ute Tribe would receive benefits from the Bonneville Unit in the same time period as other project water users." The case was eventually dismissed because the court found "the plaintiffs did not have standing" within the tribal hierarchy.[320]

By the 1980s the issues of Indian water rights and Ute Tribe demands for equal development of its water resources were contributing factors in the relatively slow construction progress of the Central Utah Project. The Utah State legislature attempted to remedy the situation by offering a proposed version of a compact between the state and the Ute Tribe "to resolve present and future controversies over quantification, distribution, and the use of all water claimed by or through the Ute Indian Tribe." Though the legislature ratified the compact, neither the Ute Tribe nor the federal government took any action on the proposed compact. For the Bureau of Reclamation, the tribe's intransigence blocked congressional appropriations for construction. According to Daniel McCool's study *Native Waters*, the Central Utah Project was suddenly

---

(Bonneville), 1974 Through Cedar Bluff, 1975; for more information on the 1965 Deferral Agreement see McCool, *Command of the Waters*, 227; McCool, *Native Waters*, 28; Burton, *American Indian Water Rights and the Limits of the Law*, 67-8.

319 "Project History, Central Utah Project, Bonneville Unit, Utah," 1974, Volume 9, 5, RG 115, Project Histories, ACC# 8NN-115-92-130, Box 229, Central Utah (Bonneville), 1974 Through Cedar Bluff, 1975; for more information on Bottle Hollow Dam and Reservoir see Water and Power Resources Service, *Project Data*, 142.

320 Burton, *American Indian Water Rights and the Limits of the Law*, 68; McCool, *Native Waters*, 179; for information on the 1976 lawsuit brought against the Bureau of Reclamation, see Upper Colorado Region, *Central Utah Project, Bonneville Unit: Municipal and Industrial System, Summary Statement*, 28-9.

"without a legal title to the water it was diverting." Without some sort of resolution, CUP was essentially in limbo, and Reclamation had little choice but to wait for Utah water users to reach an accord with the Indians.[321]

Along with the growing controversy about Indian water rights, the Central Utah Project encountered obstacles in the form of environmental regulations growing out of the National Environmental Policy Act of 1969. Reclamation began to feel the effects of this law in relationship to CUP in 1970 when "new construction contracts on project features were delayed pending submittal of an environmental impact statement in compliance with the National Environmental Policy Act of 1969." Because the law required a period for public review and comment, Reclamation encountered legal challenges to its environmental evaluation further delaying construction progress. The Sierra Club sued the Department of the Interior and "challenged the adequacy" of the Bonneville Unit's Final Environmental Statement because it did not address the entire Bonneville Unit. Department of the Interior lawyers argued "that the statement was intended as a final only for the Strawberry Collection System, and that detailed environmental statements would be prepared for each of the other systems before reaching decisions on construction." Both the U.S. District Court and the Tenth Court of Appeals accepted the argument and ruled in favor of the Department of the Interior.[322]

It was not until August 1975 that the Bureau of Reclamation cleared the environmental hurdles to begin issuance of construction contracts on the Bonneville Unit. Nevertheless Reclamation's environmental reports revealed some disturbing aspects of the Central Utah Project. Because the plan called for the diversion of nearly every stream and river on the Uinta Range, the impact to stream fisheries appeared to be a problem. On the Jensen Unit, Reclamation reported, "Approximately 520 acres of farm and range lands and 2½ miles of Brush Creek fishery classified by the State as Class III (significantly important) will be inundated by Tyzack Reservoir." According to one source, the multiple diversions, canals, and aqueducts that made up a large portion of CUP "would dry up 245 miles of streams. Wildlife specialists estimated that 78 percent of the fish population in the streams would be lost." Even more

---

[321] United States Department of the Interior, Water and Power Resources Service, Upper Colorado Region, *Ute Indian Unit: Central Utah Project, Concluding Report*, May 1980, 3: see also McCool, *Native Waters*, 180; Burton, *American Indian Water Rights and the Limits of the Law*, 67-8.

[322] Upper Colorado Region, *Central Utah Project, Bonneville Unit: Municipal and Industrial System, Summary Statement*, 1-2, 28.

disconcerting were reports that diversions from the Colorado River system were showing signs of troubling downstream environmental effects. In 1975 engineers from the Bureau of Reclamation's Upper Colorado Region reported, "The flow of the Colorado River will be reduced by an average of 15,000 acre-feet annually. The salt-concentrating effect of the depletion will increase the salinity concentration of the river at Imperial Dam by an estimated 1.5mg/1." The report went on to conclude that at the moment Reclamation had no options available to alleviate the increasing salt content in the Colorado River. It concluded that "since it is considered that the right to divert stream flows in the Upper Colorado River Basin provided by the Colorado River Compact of 1922 are accompanied by a corresponding right to concentrate the salt load of the stream without penalty."[323]

By September 30, 1978, the Bonneville Unit of the Central Utah Project was only 19 percent complete, and the Bureau of Reclamation still planned to complete the project in 1995. Delays and inflation had affected project costs and Reclamation estimated the total to complete CUP to be $1,107,416,000. The Bonneville Unit missed becoming part of President Carter's hit list by meeting the criteria established by the White House "to ensure that only safe, economical, and environmentally sound projects would receive final approval and funding." This bit of good fortune, however, did little to insure that the Central Utah Project met all the expectations that planners and boosters envisioned. More importantly, the difficulties CUP encountered represented an evolving attitude in American society toward water resources development projects in general and the Bureau of Reclamation in particular. Environmental groups and Native American issues were beginning to garner greater attention and sympathy, thwarting Reclamation and its allies' efforts for continued water resource development. In addition, the failure of Teton Dam in 1976 created concerns regarding the safety of Reclamation structures. Equally significant were the growing costs basin-wide development incurred. Cost estimates for the Bonneville Unit of CUP were $324 million in 1966, by 1979 that figure had risen to over $1 billion, and as time went on that estimate continued to climb. Few administrations were able or willing to promote and defend large-scale Reclamation projects, as Americans generally questioned their value to the nation. The Bureau of Reclamation, born out of the Progres-

323 *Ibid.*, 1-2; United States Department of the Interior, Bureau of Reclamation, Upper Colorado Region, *Jensen Unit, Central Utah Project, Definite Plan Report*, December 1975, 72; Adam R. Eastman, "From Cadillac to Chevy: Environmental Concerns, Compromise, and the Central Utah Completion Act," Master's Thesis, Brigham Young University, August 2006, 85.

sive Era utilitarian conservation ideal, faced new challenges as it entered a new chapter in its history.

New Lessons: Bureau of Reclamation and the Environmental Movement

Much has been written regarding the conflict between the Bureau of Reclamation and the emerging environmental movement over the construction of the so-called Grand Canyon dams during the debate over the Central Arizona Project. The entire story need not be retold. Postwar America's changing views regarding human relationships with nature started to question the older idea of utilitarian conservation. Through the 1960s, Reclamation and its supporters clung to the conservation ethic proclaimed in the early twentieth-century Progressive Movement by Theodore Roosevelt and Gifford Pinchot. In reference to resource management it was a utilitarian doctrine that argued for proper scientific management of natural resources for the greatest good, for the greatest number, over the longest time. The environmental movement placed a new twist on the call and programs to protect natural resources, especially scenic sites. For an increasingly urbanized and industrialized society, that movement argued the natural state of nature and its scenic beauties must be preserved. In the early twentieth century John Muir and his disciples in the Sierra Club pioneered the arguments for defense and preservation of nature at the same time that the concept of utilitarian conservation developed. The shift toward preservationism also started to popularize an environmental ethic expressed in Aldo Leopold's *Sand Country Almanac* (1945). The appearance of Rachel Carson's *Silent Spring* in 1962 further defined threats to the environment and human health posed by the heavy use of pesticides. Environmental concerns in terms of aesthetics and safety assumed such broad-based appeal among the American public that Congress hastened to pass the Environmental Policy Act of 1969 that established the Environmental Protection Agency. The laws and regulations that flowed from this agency and Congress profoundly shaped and even limited the future of the Bureau of Reclamation.

In the aftermath of the Echo Park controversy, Reclamation plans and prospects faced sharp opposition from "conservationists" to any additional proposals for more dams on the Colorado River. Conclusions found in some Reclamation reports on the proposed Colorado River dams should have raised red flags. One report referred to the waters behind the proposed Bridge Canyon Dam backing up into Grand Canyon National Monument. It noted Bridge Canyon Dam raising the "water surface through Grand Canyon

National Monument for a distance of 39 miles, approximately 13 miles of which border the Grand Canyon National Park along the common boundary upstream from the mouth of Havasu Canyon." Incredibly, no one in Reclamation saw this development as a potential problem for the Bureau, and certainly not a serious detriment to the entire project. Indeed, the author of this report went on to claim that higher water levels would enhance recreational value by providing easier access for visitors, along with an "opportunity for the Havasupai Indians to increase their income from established tourist enterprises." One might conclude that Reclamation officials had short-term memory loss, but it also is important to point out that these reports discussed the project's overall engineering and economic feasibility not necessarily cultural or ideological issues that might arise. Of course, observations such as these may have been the cause for Senator Carl Hayden's request to Secretary of the Interior Stewart Udall to get his "birdwatchers in line."[324]

Another issue that plagued the Bureau of Reclamation was protection of Rainbow Bridge National Monument from the rising waters of Lake Powell. Provisions to protect the monument were written into the Colorado River Storage Project (1956) legislation, but by the early 1960s, little had been done. To fully isolate Rainbow Bridge from the rising waters required Reclamation to construct two dams on either side. In 1960 Glen Canyon Dam Chief Construction Engineer L. F. Wylie conducted some preliminary investigations for protection of the monument; he estimated the costs at $10 million. Eventually the responsibility to come up with funds fell on Congress. In his study on Glen Canyon Dam, Russell Martin claims that "Congress had refused to approve legislation that would have made the construction funds available, and President Kennedy carefully had steered clear of the controversy, saying repeatedly that Congress should decide the issue." With neither the administration nor the Congress willing to act, there was little recourse, and the monument remained unprotected.[325]

---

324 United States Department of the Interior, "Pacific Southwest Water Plan Supplemental Information on Bridge Canyon Project," December 1963, 23, RG 115, Office of the Chief Engineer, Bureau of Reclamation Project Reports, Box 60, Project Records 1910-1995, Bridge Canyon/Brookhurst/Buckeye/Buffalo Rapids/Buford-Trenton Brunt River/Cache Creek/Cachuma; see also *Supplemental Report Central Arizona Project: Project Development Report*, January 1963, 49, RG 115, Phoenix Office Reports, ACC# NRG-11598-007, Box 68.

325 Smith, "John Kennedy, Stewart Udall and the New Frontier Conservation," 338; L. F. Wylie to Commissioner, Subject: Proposed Financial Years 1961 and 1962 Programs, August 11, 1960, Dominy Papers, Box 3, folder, Correspondence 1960 Blue Envelope Letters; see also Martin, *A Story that Stands Like A Dam*, 9.

Commissioner Dominy, however, was not overly concerned with the Rainbow Bridge National Monument question. In 1960 the House Committee on Appropriations "deleted" $3.5 million for construction works at the monument. No evidence suggests that either the Bureau of Reclamation or the Department of the Interior made efforts to secure funding for Rainbow Bridge preservation. In 1962 Dominy told Congress to act on this question or risk delaying the entire Colorado River Storage Project. A penurious Congress chose to withhold funding and construction of Glen Canyon Dam proceeded.[326]

At the same time, Secretary of the Interior Stewart Udall continued to have misgivings over the apparent failure to protect Rainbow Bridge. The secretary was concerned with the aesthetic value of the monument. Nevertheless, Udall chose not to press Congress for funding to protect Rainbow Bridge. According to Thomas Smith, "Udall argued, probably correctly, that a diversion dam would mar the grandeur of 'the most magnificent piece of sculpture in the world.'" The argument for inaction did little to calm the consternation of environmental movement leaders such as the Sierra Club's David Brower. Some agreed that diversion dams would ruin the scenic splendor of Rainbow Bridge, and conservationists decided not to contest the secretary's decision. For others the Department of the Interior's judgment "to allow the waters to lap the monument's abutments" revealed unwillingness on the part of the federal government and Congress to keep promises.[327]

It is uncertain how much effect the failure of the Bureau of Reclamation and the Department of the Interior to protect Rainbow Bridge National Monument had on the intense efforts of conservationists in fighting the so-called Grand Canyon dams. At the same time that Udall decided not to push for diversion dams to protect Rainbow Bridge, the Department of the Interior announced its Pacific Southwest Water Plan. Concurrently, the Bureau of Reclamation completed much of Glen Canyon Dam and began filling Lake

---

[326] Martin, *A Story that Stands Like A Dam*, 217; United States House of Representatives, Subcommittee on Appropriations, *Public Works Appropriations for 1962*, Hearings, 87th Cong., 1st sess. (Washington, D.C.: United States Government Printing Office, 1961), 3-4; United States House of Representatives, Committee on Appropriations, *H.R. 9076: A Bill Making Appropriations for Civil Functions Administered by the Department of the Army, Certain Agencies of the Department of the Interior, Atomic Energy Commission, the Tennessee Valley Authority and Certain Study Commissions for the fiscal year ending June 30, 1962, and for other purposes*, H.R. Report No. 1125, 87th Cong., 1st sess., September 6, 1962, 11, Udall Papers, http//content.library.arizona.edu/cdm/compoundobject/collection/udallcoloradoAZU/id/6757/rec/1.

[327] Smith, "John Kennedy, Stewart Udall and the New Frontier Conservation," 338; Martin, *A Story that Stands Like A Dam*, 10.

Powell. Then came Eliot Porter's *The Place No One Knew* as a "eulogy" to the loss of Glen Canyon. Sierra Club Executive Director David Brower utilized the introduction to confess his own part in the inundation and destruction of Glen Canyon: "Glen Canyon died in 1963, and I was partly responsible for its needless death. So were you. Neither you nor I, nor anyone else, knew it well enough to insist that at all costs it should endure. When we began to find out it was too late." Deeply felt emotions about a beautiful place inspired the budding environmental movement to attack the host of proposals introduced in Congress for development of the lower Colorado River.[328]

The tactics used to oppose the Bureau of Reclamation's plans for the lower Colorado River resembled those utilized against the Echo Park proposal. Reclamation believed hard-headed utilitarianism would win against emotional appeals from aesthetic-minded conservationists. In February 1965 a Reclamation engineer participated in a debate with David Brower and found Brower's presentation

> delivered extemporaneously and in a rather emotional manner.... It is my opinion from this encounter that the Bureau should engage in face-to-face discussions with Mr. Brower before unbiased audiences because any Reclamationist, armed with the basic facts, could adequately defend the Bureau's position against his pure emotionalism.

This attitude reveals some arrogance and the reclamation mindset of this Reclamation official. More importantly it shows how dismissive the mindset of the Bureau of Reclamation was towards the arguments of the Brower and his associates.[329]

In the end, Brower's "emotionalism" defeated two dams on the Colorado River. Still the tactics used by conservationists did not sit well with Bureau of Reclamation Commissioner Floyd Dominy. Dominy tried to counter the tactics of his opponents by flying over the Grand Canyon in an open helicopter taking photographs of the contested damsites. He attempted to "show people that I wasn't going to do any damage, whatever, to Grand Canyon

---

328 David Brower, foreword, in Porter, *The Place No One Knew*, 7; see also Martin, *A Story that Stands Like A Dam*, 239-41; Farmer, *Glen Canyon Dammed*, 145-6.

329 Memorandum, Area Engineer, Phoenix, Arizona to the Regional Director, Boulder City, Nevada, Subject: Report on Debate with Mr. David Brower of the Sierra Club Relative to Bridge Canyon and Marble Canyon Dams, February 10, 1965, March 30, 1965, in Dominy Papers, Box 4, Folder, Correspondence, 1965 Blue Envelope Letters.

National Park." Dominy considered the arguments put forth by Brower and his associates "distortions, misrepresentations, innuendos, and errors of omission as well as commission." Among the many accusations Brower made against constructing the dams on the lower Colorado River, the ones that perhaps perturbed Dominy the most were those presented in a series of 1966 newspaper advertisements proclaiming the flooding of the Grand Canyon was equivalent to flooding the Sistine Chapel. Dominy accused Brower of deliberately creating "the false impression that Grand Canyon National Park will be totally flooded from rim to rim." In a draft letter to Brower, Dominy wrote:

> The picture used in the advertisement tends to support your erroneous portrayal for few people would realize that the area shown is 80 miles downstream from the Park boundary and that even in the immediate area of the damsite, the outer rim would tower 3,000 feet or more above the dam.

Almost twenty years later, Floyd Dominy retained an acrimonious opinion of David Brower. In an oral history he still referred to Brower as a "sanctimonious bastard" who "was guilty of misleading the public."[330]

In the final analysis Brower and his allies probably misrepresented the impact the dams would have on Grand Canyon National Park. The dams proposed by the Bureau of Reclamation posed few "visual" threats to the Park and may have made visitor access easier. But Dominy, Reclamation officials, and project supporters were on the wrong side of rising concern for environmental protection in American society. Brower and his associates fanned a passion most Americans had for their national parks and the idea that they be left in their natural condition. The conservationists were also careful not to condemn dam building in general in order to avoid assaulting the public's fascination with technological progress and achievement. Instead of hydroelectric power, they insisted that coal-fired powerplants or nuclear energy offered better sources of power. One study maintains:

> As far as Brower and his allied conservationists were concerned, the federal government had the blessing of every major conservation organization in the nation to build as many

---

[330] Dominy, *Oral History Interviews*, 76; see also George C. Googin, "Reclamation's Dominy—A Man Who Gets Things Done," *Constructors*, XLVII: No. 6 (June 1966): 26-27, Dominy Papers, Box 2, Folder, Biographical Information; for information on Dominy's opinions about David Brower see Floyd Dominy to Mr. Brower, no date, draft in Dominy Papers, Box 16, Professional Files; see also Dominy, *Oral History Interviews*, 153.

> coal-fired or nuclear plants as it needed to pump currently stored water wherever it was lacking, but those organizations would *never* … compromise on the issue of the destruction of the Grand Canyon.[331]

This steadfast determination to protect the Grand Canyon while welcoming the construction of other, later proven environmentally damaging, sources of power presents a confusing picture of the event. From this *milieu* of mixed messages, the beginning of an environmental movement appeared.

Attitudinal changes regarding American society's relationship with the environment were an emerging cultural transformation that arose long before environmentalists initiated their battle with the Bureau of Reclamation over the Grand Canyon Dams. On another front, beginning shortly after the Echo Park episode in the mid-1950s, Howard Zahniser and the Wilderness Society began lobbying members of Congress for legislation to protect America's rapidly dwindling wilderness country. In addition, the publication of Rachel Carson's *Silent Spring* (1962) explored technology's long-term environmental impacts by raising fears about the poisoning of the environment from pesticides and herbicides. According to environmental historian Ted Steinberg, Carson's book was "a stinging critique of America's chemical dependency. She embraced the idea that all of nature was bound up in an interdependent web of life, which humankind had the potential to destroy." The "web of life" meant an ecological approach to environmental issues pointing to the dangers that technological civilization posed to the environment's ecosphere.[332]

Much impressed with Carson's critique and a convert to the need to defend the ecological integrity and health of the environment, Secretary of the Interior Stewart Udall's *The Quiet Crisis* (1963) echoed Carson's warning. Yet Udall's conservationism did not totally discount the part technology might play in alleviating environmental degradation, through the development of environmentally friendly technologies such as harnessing the tides to produce electricity or creating new sources of fresh water by desalinization. Moreover, Udall saw the federal government as an active participant in not only assisting

---

[331] Dean, "'Dam Building Still had Some Magic Then'," 93-4; see also Martin, *A Story that Stands Like A Dam*, 265; italics in the original; Worster, *Rivers of Empire*, 276.

[332] Ted Steinberg, *Down to Earth: Nature's Role in American History* (New York: Oxford University Press, 2002), 242-3, 246-7; John Opie, *Nature's Nation: An Environmental History of the United States* (Fort Worth, Texas: Harcourt Brace College Publishers, 1998), 415; D. T. Kuzmiak, "The American Environmental Movement," *The Geographic Journal* 157:3 (November 1991): 270; see also Harvey, *Wilderness Forever*.

in technological advances, but also in protecting the nation's natural heritage. Indeed, during his tenure as secretary of the interior, Udall oversaw the expansion of the national park system and was a leading advocate for preservation of wilderness areas and passage of the 1964 Wilderness Act.[333]

For the most part, the Wilderness Act had little impact on the activities and policies of the Bureau of Reclamation. If anything, Reclamation officials welcomed the legislation to protect valuable watershed areas, in spite of some fears and misgivings about locking up of natural resources to prevent economic development. The Wilderness Act is important because it established and protected designated areas and "made the preservation of wilderness into national policy." Previously, the Forest Service initiated a policy of identifying primitive areas, but that designation had no legal standing. Officials could easily reverse the status. Although it faced stiff opposition from western congressmen, especially Colorado's Wayne Aspinall, the Wilderness Act reflected the views of a growing number of Americans who "believed that a portion" of public lands "should be set aside forever."[334]

A greater challenge to the Bureau of Reclamation activities, arising from the environmental movement, came in the form of the Wild and Scenic Rivers Act of 1968. Similar to the Wilderness Act, it sought the preservation of sections of the nation's river systems from development. Concern for river and stream protection to preserve the last vestiges of free-flowing rivers began in the late 1950s along with the rise of a wilderness protection movement. In 1960 the Senate Select Committee on National Water Resources asked the National Park Service to investigate the condition of America's steams. In its report to the committee, the Park Service pointed out the "rarity" of natural flowing streams in the United States and warned that because of continued pressure on the nation's water resources they may soon become "nonexistent." The report recommended, "That certain streams be preserved in their free-flowing condition because natural scenic, esthetic and recreational values outweigh their value for water development and control purposes." By 1963 the departments of Agriculture and Interior

---

333 Udall, *The Quiet Crisis*, 175-9; see also Smith, "John Kennedy, Stewart Udall, and the New Frontier," 360-1; for more information on Udall's conservation ideals see Stewart L. Udall, "Pausing at the Past: Reflections of a Native Son," in Stewart L. Udall, Patricia Nelson Limerick, Charles F. Wilkinson, John M. Volkman, William Kittredge, *Beyond the Mythic West* (Salt Lake City, Utah: Peregrine Smith Books in association with the Western Governors' Association, 1990), 1-34.

334 Steinberg, *Down to Earth*, 243; for information on the legislative history of the Wilderness Act and the role of Colorado Congressman Wayne Aspinall's role see Schulte, *Wayne Aspinall and the Shaping of the American West*, 115-138; Sturgeon, *The Politics of Water*, 63-5, 151.

had created a list of 650 rivers, or segments thereof, for possible wild and scenic river designation, and out of those 89 were chosen for study. In 1964 the first Wild Rivers Act was introduced in Congress, and a year later President Johnson's State of the Union address called for a "rivers bill."[335]

12.6. President Lyndon B. Johnson during his 1964 State of the Union Address. House Speaker John McCormack and Senator Carl Hayden, president pro tempore of the Senate, are seated behind him. January 8, 1964. Photographer: Cecil Stoughton. Courtesy of the Lyndon B. Johnson Library, National Archives and Records Administration.

In making a case for a wild and scenic river system, supporters in Congress employed many of the same verbal arguments expounded by members of the environmental movement. While emphasizing the aesthetic values inherent in a free-flowing river, some members argued that the loss of these river environments was tantamount to a national crisis. One Senate report noted that a wild and scenic river system would "preserve and protect some of America's unspoiled and free-flowing streams, or their segments, that symbolize this vanishing heritage of our national landscape." Such statements reveal the growing appeal of environmental preservation in American culture, and a new resonance for the writings and ideas of Henry David Thoreau, John Muir, and Aldo Leopold in an urban America increasingly denied access to nature. President Lyndon Johnson reflected the moment when he declared, "The wonder of Nature is the treasure of the United States."[336]

---

335 United States House of Representatives, Committee on Interior and Insular Affairs, *Providing for a National Scenic Rivers System, and for Other Purposes*, H. Rpt. No. 1623, 90th Cong., 2nd sess., July 3, 1968, 2, 7-8; Tim Palmer, *The Wild and Scenic Rivers of America* (Washington, D.C.: Island Press, 1993), 11, 22.

336 United States Senate, Committee on Interior and Insular Affairs, Senator Frank Church, *National Wild and Scenic Rivers System*, S. Rpt. No. 491, 90th Cong., 1st sess., August 4, 1967;

Wild and scenic river protection faced stiff opposition from federal water resource agencies—the Bureau of Reclamation, the Army Corps of Engineers, the Federal Power Commission, and even the Soil Conservation Service. Bureau of Reclamation Commissioner Floyd Dominy scoffed at the proposition asserting, "I've seen all the wild rivers I ever want to see." Congress responded to its larger constituency and took little notice of the concerns voiced by Dominy and other water agencies. Indeed, proponents of the measure formatted their arguments to attract the largest possible coalition of constituents. Wild and scenic rivers, supporters maintained, would provide "remarkable scenery, recreational, geological, fish and wildlife, historic, cultural, or other similar values." More importantly, to ease concerns of water interests in the American West, advocates said a designation of a wild and scenic river would have little to no alterable effect on the status quo. A 1967 Senate report on the bill stated that the proposed legislation was actually less restrictive than the 1964 Wilderness Act, and that establishing "a National and Scenic Rivers System is not intended to affect or impair any prior valid water right under State or Federal law."[337]

Despite such reassurances, opponents' rhetoric often stressed the long-held precepts of utilitarian conservation ethic that natural resources development was necessary for continued prosperity of the nation, and that locking up these resources, for whatever reason, stifled progress. One opponent of the bill proclaimed, "The conservation values sought by this legislation are meritorious. However, the preservation of these values should not be permitted to seriously jeopardize the future water resource development of this nation." In the end, "preservation values" prevailed. On October 2, 1968, Congress passed the Wild and Scenic Rivers Act, and within a few days, President Johnson signed the bill into law. The initial act designated parts of twelve rivers while "identifying 27 others for study." Ironically, Johnson signed the Act only two days after the Colorado River Basin Project Act that authorized, among other items, the Central Arizona Project. The juxtaposition of these two measures demonstrates a growing ambivalence about water resource development. The Colorado River Basin Project Act was the last major Reclamation project authorization by Congress, while the Wild and Scenic Rivers Act was perhaps the one piece of legislation that foretold, if not

---

United States House of Representatives, Committee on Interior and Insular Affairs, *National Wilderness Preservation System: Message from the President of the United States, Transmitting National Wilderness Preservation System*, H. Doc. 79, 89th Cong., 1st sess., February 8, 1965, 1.

337 Palmer, *The Wild and Scenic Rivers of America*, 26; Reisner, *Cadillac Desert*, 242; Committee on Interior and Insular Affairs, *National Wild and Scenic Rivers System*, 5.

the end, at least a drastic curtailment in water resource development in the American West.[338]

Perhaps the most significant environmental measure to impact the activities and policies of the Bureau of Reclamation was the National Environmental Policy Act of 1969 (NEPA) that created the national Environmental Protection Agency in 1970. The Act transformed the manner in which government agencies and even the private sector conducted business. For example, it forced water resource agencies such as Reclamation and the Corps of Engineers to consider myriad environmental consequences connected to dam building and other aspects of water projects. In considering these developments, the Act called for these agencies to investigate and write environmental statements (ES), outlining what environmental effects or benefits might arise as a result of proposed projects. More importantly, the ES was open to public review and comment. As one observer wrote, "Although the Act itself could not stop developments, it forced decision-making on a reluctant federal land bureaucracy that wanted as much autonomy as possible." In addition, NEPA suspended the federal government's right to sovereign immunity which meant that citizens could sue federal bureaus and departments regarding the adequacy of environmental statements.[339]

The required comment period allowed for environmental impact statements gave critics a platform from which to call attention to not only environmental degradation threatened by water projects, but also the questionable economics of water projects. As a result of this scrutiny, Reclamation, or any other water resource agency, often had to respond to and rewrite the ES, delaying authorization and appropriations. If this tactic was unsuccessful in halting the project entirely, opponents resorted to filing an injunction with the courts to force the Bureau of Reclamation to conform to NEPA. For example, lawsuits against Reclamation halted or delayed construction of Teton Dam and portions of the Centrtal Utah Project while the courts decided whether or not Reclamation had conformed to the Act. In both cases the courts ruled in favor of the Bureau of Reclamation, but clearly the legislation delayed the construction timetable and increased costs.[340]

---

338 Committee on Interior and Insular Affairs, *National Wild and Scenic Rivers System*, 55-6; Opie, *Nature's Nation*, 394; Palmer, *The Wild and Scenic Rivers of America*, 25.

339 Kuzmiak, "The American Environmental Movement," 271; see also Adam M. Sowards, "William O. Douglas's Wilderness Politics: Public Protest and Committees of Correspondence in the Pacific Northwest," *Western Historical Quarterly* 37:1 (Spring 2006): 41.

340 United States Department of the Interior, Bureau of Reclamation, Upper Colorado Region, *Central Utah Project, Bonneville: Municipal and Industrial System, Summary Statement*, August 1979, 1-2, 28; United States House of Representatives, Committee on Government Operations,

On the other hand, Bureau of Reclamation officials, aware of the rising apprehension in the public's mind over environmental questions, took steps to assure the public that their agency shared those concerns. Yet there was a wide gulf between what Reclamation and its supporters perceived as environmentally or ecologically sound practices and the views held by vocal environmentalists. Reclamation officials continued to argue that water resource projects enhanced and perhaps even helped nature. A strong proponent of this ideal was Bureau of Reclamation Commissioner Ellis L. Armstrong, who contended that Reclamation projects helped the environment by reducing silt in water, producing non-polluting hydroelectric energy, and enhancing fish and wildlife. In the early 1970s, on the heels of the Wilderness Act and the Wild and Scenic Rivers Act, however, there was a sense among Reclamation advocates that efforts to protect the environment threatened to close-off all future natural resource development. Armstrong wrote,

> Because the great movement is vital to the Nation and the world, environmental problems are being attacked by the Bureau of Reclamation with all-out concern. We feel the scope of the movement should combine all geophysical and manmade surroundings both for the present and the future. It should not be purely protective and preventive …

After passage of some of the most revolutionary environmental legislation in the nation's history, the long-standing debate between utilitarian and preservationist conservation persisted.[341]

A major environmental history text book of the United States, *Nature's Nation*, argues that because of the Wilderness Act and the Wild and Scenic Rivers Act the Bureau of Reclamation "was severely restricted by controls set on western waterways." Author John Opie writes, "Water resource projects now came under unprecedented scrutiny on the basis of water shortages, pollution, and environmental deterioration." No doubt Reclamation felt restricted under

---

Teton Dam Disaster: Thirtieth Report by the Committee on Government Operations, H. Rpt. No. 94-1667, 94th Cong., 2nd sess., September 27, 1976 (Washington, D.C.: United States Government Printing Office, 1976), 7; see also Marc Reisner and Sarah Bates, *Overtapped Oasis: Reform or Revolution for Western Water* (Washington, D.C.: Island Press, 1990), 22-3.

341 Ellis L. Armstrong, "Commissioner's Page: Big Non-Polluters," *Reclamation Era*, 56 (May 1970): back of front cover; Ellis L. Armstrong, "Commissioner's Page: What Can I Do?" *Reclamation Era*, 56 (August 1970): back of front cover; Ellis L. Armstrong, "Commissioner's Page: Striking A Balance," *Reclamation Era*, 57 (November 1971): back of front cover.

the mandates of the new environmental legislation. From a larger perspective, the National Environmental Policy Act perhaps democratized the authorization process of Reclamation projects. More to the point, the new environmental regulations began the restructuring of a cultural change within Reclamation. It was certainly becoming clearer that the hectic construction activity of the 1950s and 1960s was coming to a close, and that water resource development was taking on a new dimension compelling the Bureau of Reclamation to expand its outlook and consider other perspectives.[342]

Grand Coulee Dam's Third Powerplant

Grand Coulee Dam is a workhorse. No other Bureau of Reclamation dam does more in terms of harnessing the flows of one the nation's mightiest rivers to generate power and provide irrigation water. Located in the remote Pacific Northwest, its critics originally called it a "white elephant." But the crisis of World War II proved them wrong when its power became available for shipyards, aircraft factories, and the production of plutonium at Hanford, Washington, for the development of the atomic bomb. In the postwar years the power from the dam underwrote an industrial base in the Pacific Northwest and supplied water for agriculture in the Columbia Basin Project.

Still, the river and its vast watershed suggested that Grand Coulee Dam possessed additional potential for power production. The second half of the twentieth century saw tremendous change on the Columbia River as the Corps of Engineers remade the river into what historian Richard White called the "organic machine." Not only did hydroelectric production increase with new dams, but the Corps made the river a highway to transport goods from the hinterlands of the Pacific Northwest. Idaho wheat from the Palouse Country and other goods now made their way by river and sea to markets of the Pacific Rim. At Grand Coulee Reclamation officials sought to utilize more of the Columbia River's hydropower potential rather than allowing water to flow over the dam without generating electricity. Historian Paul Pitzer observed Reclamation's conundrum was "having water escape without producing power or allowing expensive machinery to sit unused reduced the efficiency of the Grand Coulee plant."[343]

---

342 Opie, *Nature's Nation*, 323.

343 Pitzer, *Grand Coulee*, 333.

Some of the problem lay in the lack of upstream flow control. According to one Bureau of Reclamation report, "Present hydroelectric power developments … utilize only a fraction of potential water power within the Columbia Basin." Columbia River power interests that included the Bureau of Reclamation, the Army Corps of Engineers, and the Bonneville Power Administration understood that more upstream storage with releases at optimal times would permit Grand Coulee Dam's powerhouses to run at a consistent pace. In 1950 Reclamation proposed a number of projects throughout the Columbia River watershed to help in regulating river flows. Some of the work began during the late 1940s with construction of the 564-feet high Hungry Horse Dam, completed in 1953, on the Flathead River in Montana. Even though Hungry Horse assisted with flow control by bringing upstream system storage to 15,000,000 acre feet, it was still not enough to permit Grand Coulee's powerhouses to run at maximum efficiency.[344]

Among sites considered for additional reservoir construction to regulate stream flows were Libby Dam on the Kootenai River in northwestern Montana and Glacier View Dam in western Montana. Each of these proposals surfaced issues that made their prospects dubious. For instance, the reservoir at Libby Dam would back water up into Canada and inundate two proposed Canadian hydroelectric sites. Glacier View faced stiff opposition from the National Park Service and "wildlife interests" because that reservoir "would be on lands in Glacier National Park." Development on the Kootenai River also faced long drawn out negotiations with the Canadian Government through the International Joint Committee. Though perhaps disappointed with the slow movement on better regulating the flow of the Columbia River, within the hydropower community there still remained a sense of urgency to increase power production within the basin. In 1954 M. E. Marts from the University of Washington noted, "The Columbia power system was planned as an extension of the existing federal system to meet a load estimate for the foreseeable future at 10,000,000 kilowatts." Thus there was the perception of a need that, according to some, would become dire in a relatively short time, along with the understanding that the Columbia River stood ready to meet those needs.[345]

---

[344] Pitzer, *Grand Coulee*, 333; Eric A. Stene, "Hungry Horse Project," Denver: Bureau of Reclamation History Program, 1994, 2-10; United States Department of the Interior, Bureau of Reclamation, *The Columbia River: Final Letters of Transmittal and Comments of the Affected States and of Federal Agencies*, Volume I, H. Doc. 473, 81st Cong., 2nd sess., February 1950, 60, 203-4.

[345] M. E. Marts, "Upstream Storage Problems in the Columbia River Development," *Annals of the Association of American Geographers* 44:1 (March 1954): 47-8; USDOI, BR, *The Columbia River*, 203.

During the 1950s Columbia River power interests rested their hopes on reaching an agreement with Canada for further development of the river basin. In 1960 the United States and Canada reached an accord in which the Canadian government agreed to construct three dams—Arrow, Duncan Lake, and Mica—bringing the total reservoir capacity in Canada to over 15.5 million acre-feet. According to Paul Pitzer, "The ten-year financial agreement … created a $485 million joint program and included the promise that Canada would not divert Columbia River water for at least sixty years." The agreement, however, faced difficulties in the Canadian Parliament that delayed the final treaty until September 1964. The renegotiated treaty stipulated that the U.S. would pay Canada $300 million for "downstream power, navigation, and flood control benefits." Half of the additional electricity generated downstream as a result of the treaty would revert to Canada by the year 2003. For the Bureau of Reclamation and other Columbia River water interests, the treaty marked the beginning of full power development at Grand Coulee Dam. It was not long after Congress ratified the Canadian treaty that legislation for a third powerhouse appeared in Congress.[346]

Not surprisingly Cold War competition with the Soviet Union spurred Congress to consider authorization for a third powerplant at Grand Coulee Dam. During the 1950s, the Soviets made dramatic gains in the production of hydroelectric power. One powerplant on the Volga River outstripped Grand Coulee's power production. By 1957 there were three dams on the Volga that exceeded Coulee's power production, and in 1958, a Soviet dam on the Angara River brought that number to four. With these developments, "The grand dam on the Columbia was dangerously close to losing all her first-place claims," according to many observers. In the highly competitive Cold War game the Soviet successes could not go unchallenged. There was more at stake than merely ensuring future power supplies for the Pacific Northwest; the reputation of the United States's technical prowess was at risk.[347]

This one-upmanship actually played into the hands of Third Powerplant proponents, presenting them with a nationalistic and patriotic theme to drum up support for the project. Pitzer notes that U.S. Senator Henry Jackson of Washington, chairman of the Senate Committee on Interior and Insular Affairs, utilized patriotic sentiments to push legislation authorizing the project through

---

346 Pitzer, *Grand Coulee*, 338-9; see also Peterson, *River of Life, Channel of Death*, 130; Wm. Joe Simonds, "The Columbia Basin Project," Denver: Bureau of Reclamation History Program, 1998, 49-66.

347 Pitzer, *Grand Coulee*, 341.

12.7. Exterior of the Third Powerhouse showing the design elements incorporated into the massive structure. Behind the powerhouse is the new dogleg in Grand Coulee Dam with the penstocks, which deliver water to the generating units, barely showing above the top of the powerhouse.

his committee. And he was not alone. Secretary of the Interior Stewart Udall, in a statement to Congress, noted that the United States lagged behind the Soviet Union in hydroelectric power development. He claimed:

> During the last decade, three hydroelectric plants in the Soviet Union surpassed Grand Coulee ... The third powerhouse will move Grand Coulee back into position of world leadership at 5.6 million kilowatts.... So I think this is something of special interest to the country and to the committee.

How much Cold War rhetoric moved Congress to support legislation for the third powerhouse is uncertain, but it clearly served to stifle criticism of the project.[348]

---

[348] *Ibid.*, 341; see also United States Senate, Committee on Interior and Insular Affairs, *Third Powerplant, Grand Coulee Dam*, Hearing, 89th Cong., 1st sess., April 13, 1965 (Washington, D.C.: United States Government Printing Office, 1965), 21-2.

12.8. Grand Coulee Dam after completion of the dogleg, penstocks, and Third Powerhouse. The Feeder Canal and Banks Lake, in the upper right corner of the picture, initiate the largest irrigation water delivery for the Columbia Basin Project.

Still, sponsors of the Third Powerplant, and in particular Senator Jackson, faced criticism by some members of the House Subcommittee on Irrigation and Reclamation in 1965. While there was little opposition to the project there was some expectation that Senator Jackson could be persuaded to consider the transfer of water into the upper Colorado River system from the Columbia River basin in return for support for his pet project. In the background, the Third Powerplant had the unwavering support of President Lyndon Johnson. In the foreground, Senator Jackson reassured the committee that the entire cost of $364,310,000 would be repaid in full and that annual revenues would exceed $4 million a year. "I submit," said the Senator to the subcommittee, "this is an extremely sound investment, from a dollar and cents position alone. The third powerplant will be a moneymaker." Most members of the subcommittee and the entire House Committee on Interior and Insular Affairs agreed, but some wanted something from Jackson, namely Columbia River water.[349]

Some committee members briefly turned Jackson's testimony to the Pacific Southwest Water Plan and the Central Arizona Project with an eye toward augmentation from the Pacific Northwest. California Congressman Craig Hosmer grilled him at length on the possibility of transferring surplus Columbia River water to areas of shortage. Jackson, however, remained steadfast in his opposition to transferring Columbia River water to augment the Colorado River basin. Hosmer hoped for more. He stated, "I concede that we do need some slight bit more reassurance that the Pacific Northwest is willing to cooperate with the rest of the country as much as it wants the rest of the country to cooperate with it."[350]

The jousting between Senator Jackson and Congressman Hosmer posed no threat to the smooth passage of the Third Powerplant legislation through Congress. It did reveal a chink in the water resource "iron triangle." Regional rivalries had always existed among representatives of the seventeen western

---

349 United States House of Representatives, Committee on Interior and Insular Affairs, Subcommittee on Irrigation and Reclamation, *Third Powerplant, Columbia Basin Project, Washington*, Hearings, 89th Cong., 1st sess., September 9 and 10, 1965 (Washington, D.C.: United States Government Printing Office, 1965), 7; see also United States House of Representatives, *Third Powerplant Columbia Basin Project, Washington, Communication from the President of the United States, the Report of the Secretary of the Interior Concerning Economic and Engineering Feasibility of a Third Powerplant at Grand Coulee Dam on the Columbia River, and a Draft Authorizing Legislation*, H. Doc. No. 142, 89th Cong., 1st sess. (Washington, D.C.: United States Government Printing Office, 1965), III-IV.

350 Subcommittee on Irrigation and Reclamation, *Third Powerplant, Columbia Basin Project, Washington*, 10-16.

states Reclamation served, but for the most part, reciprocity usually won out. As opportunities for water resource development lessened and costs for Reclamation projects rose, rivalries became more intense as each region struggled for a scarce resource. Despite the emergence of tensions among western water development supporters, Third Powerplant legislation easily passed.

On June 14, 1966, President Johnson signed the law authorizing construction of the Third Powerplant at Grand Coulee Dam. In his remarks at the signing, President Johnson lauded the Grand Coulee Dam's contribution to the nation:

> This authorization builds on a project which was begun more than thirty years ago. And not a year has passed which did not bring new benefits and greater prosperity to the people of the region it serves. The whole Nation, I think, has benefited greatly, for the development of the resources of any region always adds to the strength and prosperity of all the regions.

Initially the Bureau of Reclamation designed the new powerplant to contain 12 turbine generators with a capacity of 300,000 kilowatts each. By 1967 Reclamation altered its plans to double the power production of the Third Powerplant. Instead of the 300,000-kilowatt generators, Reclamation proposed to install six 600,000-kilowatt generators while constructing the powerhouse to have the capability to provide for another six units at some future date. According to one *Reclamation Era* article, "the proposed new third powerplant at Grand Coulee Dam … will ultimately bring the facility to 9.2 million kilowatts." The Columbia River would be put to work like never before.[351]

To house these gigantic floor generators, the Bureau of Reclamation designed a powerplant structure 1,128 ft. long, 150 ft. wide, and 200 ft. high to sit at an angle to Grand Coulee Dam. The installation of the new generators required Reclamation engineers and planners, along with its contractors to come up with new and innovative procedures. This sheer size alone dictated innovations. According to one report,

---

[351] "269 Remarks upon Signing Bill Authorizing a Powerplant at the Grand Coulee Dam, June 14, 1966," American Reference Library—Primary Source Documents, 2001, from the Public Papers of Lyndon B. Johnson, 1; see also Donald J. Duck, Deputy Director of Design and Construction, *Construction of Grand Coulee Third Powerplant* (Denver, Colorado: United States Department of the Interior, Bureau of Reclamation, 1975), 2; United States Department of the Interior, Bureau of Reclamation, Region 1, "Third Powerplant, Columbia Basin Project, Washington: Definite Plan Report," Boise, Idaho, September 1967, 4; "Plan Doubling Third Powerplant at Grand Coulee," *Reclamation Era*, 56 (May 1967): 52.

> The physical dimensions of the individual units in the Third Powerplant are of unprecedented record size. The generator rotor will be 68 feet in diameter and weigh 1,900 tons. The turbine runners will be 33 feet, 3 inches in diameter and weigh 550 tons. The combined turbine-generator shaft for each Third Powerplant will be 100 inches in diameter, 43 feet, 9 inches long, and weigh about 350 tons.

As a result of their incredible size, installation of the generators required the manufacturer to construct each generator on-site, as opposed to having completed units delivered and installed. When completed, these gigantic generators easily allowed Grand Coulee Dam "to exceed the total capacity of all 50 powerplants the Bureau has constructed," throughout the American West.[352]

Much to the chagrin of the Bureau of Reclamation and other Third Powerplant proponents an unexpected complication arose. The Department of the Interior received disconcerting news that the Soviet embassy had made inquiries asking permission for the Soviet Union to bid on Reclamation's contract for generator manufacturing. According to Paul Pitzer, the Soviets claimed that they already possessed expertise in the construction of large generators and that "they wanted to demonstrate their accomplishments by doing the same at Grand Coulee." This inquiry touched off a flurry of activity among the White House, the State Department, and the Department of the Interior. Secretary Udall argued that only United States companies have the opportunity to build the units, and he cited solid support in Congress. On the other hand, President Johnson, with the support of the State Department, maintained that allowing the Soviets to bid on the generators might lessen Cold War tensions.[353]

This situation presented a difficult dilemma for Third Powerplant supporters who had employed the full flourish of Cold War rhetoric to achieve congressional approval. Pitzer neatly summed up Udall's conundrum:

---

352 United States Department of the Interior, Bureau of Reclamation, "Project History, Grand Coulee, Third Powerplant, Columbia Basin Project—Washington," Volume IV, Calendar Year 1972 RG 115, Project Histories, ACC# 8NN-115-92-130, Box 201; see also B. P. Bellport, Bureau of Reclamation Chief Engineer, "12 600-Mw Units Planned for Grand Coulee No. 3," *Electric World*, (July 10, 1967): 144-6, in United States Department of the Interior, Bureau of Reclamation, "Project History, Grand Coulee, Third Powerplant, Columbia Basin Project—Washington," Volume I, Calendar Year 1967, 144, RG 115, Project Histories, ACC# 8NN-115-92-130, Box 262.

353 Pitzer, *Grand Coulee*, 343.

12.9. The huge turbine pits in the generating floor of the Third Powerhouse during construction. 1976.

> If the Russians won the bid and built the units, they could still claim at least a part in the world's largest powerplant. That would cloud Grand Coulee's glory when it reclaimed the title of the world's biggest. In propaganda terms, the Soviets seized a wonderful opportunity to embarrass the United States.

Without vigorous opposition from the State Department, Secretary Udall decided to take a made-in-America stance regarding the generators for the Third Powerplant. The State Department reported, "Interior has recommended exclusion of Soviet and presumably other foreign suppliers from bidding on Grand Coulee Generating Units because among other things, American industry should have the first opportunity on … world's largest generating units." Along with the Soviet Union, General Electric was one of the first casualties of this directive, taking itself out of consideration for generator manufacturing "after expressing concerns about government demands that they be made entirely in the United States." General Electric declared it impossible to build the units under the strict criteria requiring only American made parts. Ultimately the Department of the Interior decided that fifty-one percent American parts met the criteria of made-in-America.[354]

In 1967 installation of the generators was years in the future. Construction of all the Third Powerplant facilities required clearing out the entire area along the east bank of the Columbia River just below the dam. This meant removal and relocation of the Right 230-kV Switchyard facilities. For this segment of construction Reclamation planned to construct a new Consolidated 230-kV Switchyard that would not only replace the Right Switchyard but also the Left 230-kV Switchyard. This construction was probably the most intricate from a planning perspective, because it had to be done in a manner that would not drastically alter Grand Coulee Dam's power production. Reclamation Chief Engineer B. P. Bellport explained. "The existing Grand Coulee plant facilities are the key power producers for a far-flung area of the Pacific Northwest. Work Schedules therefore will require close integration with power production requirements to assure no major power outages occur throughout construction." Making room for the new powerhouse required that Reclamation take and remove portions of the city of Coulee Dam, and the project construction office reported "54 residences, 8 businesses, the Government-owned visitor center and some city-owned community facilities would have to be purchased, relocated, or destroyed."[355]

The Bureau of Reclamation's final structure for the Third Powerplant was the Forebay Dam located on the right abutment of Grand Coulee Dam. As

---

[354] Telegram, Department of State, May 23, 1967, in Dominy Papers, Box 18, folder, Professional Files, 1966-1967, Third Powerplant; see also Pitzer, *Grand Coulee*, 343, 349.

[355] Bureau of Reclamation, Region 1, "Third Powerplant,… Definite Plan Report," 5; Bellport, "12 600-Mw Units Planned for Grand Coulee No. 3," 144, RG 115, Project Histories, ACC# 8NN-115-92-130, Box 262; Bureau of Reclamation, "Project History, Grand Coulee, Third Powerplant, Columbia Basin Project—Washington," Volume IV, 5, RG 115, Project Histories, ACC# 8NN-115-92-130, Box 201; for information regarding the city of Grand Coulee see Samuel S. Rey, Administrative Officer Third Powerplant Construction Officer, "The Third Powerplant … Its Impact on People," *Reclamation Era*, 58 (May 1972): 20; Pitzer, *Grand Coulee*, 343.

one Reclamation report stated, "An open channel, or forebay, will be excavated upstream from the forebay dam and connected to the reservoir by removing a portion of [the right abutment of] Grand Coulee Dam." Reclamation designed the Forebay Dam as a gravity-type structure 200 feet high with a length of 1,125 feet and estimated that it would take 600,000 cubic yards of concrete. By January 1970 much of the preparatory work was complete, including removal of Right Switchyards, construction of the new Consolidated Switchyard, and modifications to the Left Powerplant, allowing work to commence on construction of all Third Powerplant facilities. A month later, the Bureau of Reclamation awarded the construction contract for the new powerplant and Forebay Dam to the construction firm of Vinnel-Dravo-Lockhead-Mannix. *Reclamation Era* reported that the $112 million contract was the largest in Reclamation history, and that it avoided the budget cutting measures imposed by the Nixon administration. The article explained, "The construction project was exempted from the President's reduction of spending for public works, because the of urgent schedule to meet power needs in the Pacific Northwest and to fulfill commitments of the Columbia River Treaty with Canada." With much fanfare, the first batch of concrete was placed in the dam and powerplant on October 21, 1970.[356]

Construction of the Third Powerplant and Forebay Dam used methods similar to those Reclamation employed on other construction projects. As mentioned earlier the first stage was modification of Grand Coulee's power transmission facilities to ensure that the dam's powerplants continued to function during construction. Reclamation excavated for the forebay, removed part of the right abutment of the original dam, erected the Forebay Dam, and then built the Third Powerplant structure. Once that was completed, installation of the generators and turbines began, and, then with the turbines and generators in place, work on the rest of the powerplant then proceeded. Reclamation contracted the architectural firm of Marcel Breuer and Associates to design the powerhouse to make the facility not only structurally sound, but also aesthetically pleasing. According to one report, "The concept includes 84-foot high walls cast in place with reinforced concrete.... The walls are of folded, faceted design and architectural finish provided by grooves in the

---

[356] For information on the Forebay Dam see Duck, *Construction of Grand Coulee Third Powerplant*, 8; Bureau of Reclamation, Region 1, "Third Powerplant,...Definite Plan Report," 6; for information on the Third Powerplant contract see Bureau of Reclamation, "Project History, Grand Coulee, Third Powerplant, Columbia Basin Project—Washington," Volume IV, 15-6, 62, RG 115, Project Histories, ACC# 8NN-115-92-130, Box 201; "$112 Million Contract for Third Powerplant: Reclamation's Largest—World's Largest," *Reclamation Era*, 56 (May 1970): 11-12; Pitzer, *Grand Coulee*, 347.

wood form materials." Former Bureau of Reclamation Chief Engineer Harold Arthur recalled that First Lady Lady Bird Johnson's "America the Beautiful" campaign was in full swing and "criticism that a lot of government works were not aesthetically pleasing" influenced Reclamation's decision to enhance the Third Powerplant's architecture. These plans also reflected an eye for detail that Reclamation planners and engineers had developed throughout its history.[357]

For the most part, construction of the Third Powerplant proceeded relatively smoothly with no significant delays or obstructions. There were, however, labor disputes, bickering among the main contractor firms, and the ever-troublesome problem of congressional funding. The most difficult aspect of construction involved alterations to Grand Coulee Dam. In order to make room for the new powerplant and the forebay that would supply water to its penstocks, Reclamation had to remove approximately 260 linear feet of concrete from the right end of Grand Coulee Dam. Interestingly, during periodic drawdowns of Franklin D. Roosevelt Lake for construction of the upstream and downstream cofferdams, the lowering lake level exposed Kettle Falls which had not been seen in twenty-eight years. Paul Pitzer notes, "For six weeks people came to see a sight hidden from view for nearly three decades."[358]

By the close of 1973, the Bureau of Reclamation's contractors had completed concrete work on the Forebay Dam, the penstock encasement, and most of the powerplant structure. In addition Westinghouse completed the fabrication and installation of the first three 600,000-kW units of the Third Powerplant. Westinghouse made good use of Secretary Udall's made-in-America policy by using 49 percent in foreign made components. Nevertheless, any sort of allegiance to American-made products came to an end in August of that year when Reclamation "awarded a contract to Canadian General Electric to furnish and install the larger 700,000 kW generator" for the remaining three units. In 1974 contractors removed the forebay cofferdam and the downstream dike, and by October the dam, powerplant, and waterway

[357] Bureau of Reclamation, "Project History, Grand Coulee, Third Powerplant, Columbia Basin Project—Washington," Volume I, 1, RG 115, Project Histories, ACC# 8NN-115-92-130, Box 262; Duck, *Construction of Grand Coulee Third Powerplant*, 9; Harold Arthur, *Oral History Interview*, Transcript of tape-recorded Bureau of Reclamation Oral History Interviews conducted by Brit Allan Storey, senior historian, Bureau of Reclamation, during 1995 and 1995, in Denver, Colorado, 212.

[358] Duck, *Construction of Grand Coulee Third Powerplant*, 6-7; Simonds, "The Columbia Basin Project," 51; see also Pitzer, *Grand Coulee*, 346; Bureau of Reclamation, Region 1, "Third Powerplant,...Definite Plan Report," 14.

were "substantially complete," and Reclamation had spent $322 million on construction.[359]

Pacific Northwest-Pacific Southwest Intertie

As construction on Grand Coulee Dam began in the 1930s, discussions regarding how best to utilize the dam's enormous hydroelectric power potential began. Conversations centered on connecting Grand Coulee's hydroelectric facilities to expanding markets in the Pacific Southwest, especially California. Beginning with the Pacific Northwest Regional Planning Commission's 1935 publication "The Columbia Basin," the idea of a Pacific Northwest-Pacific Southwest Intertie began to take root. By the late 1940s and early 1950s, the Bureau of Reclamation, the Corps of Engineers, and the Federal Power Commission weighed in supporting an intertie system. Yet, these early plans for a north-south intertie system became embroiled in regional controversies and the ever-present debate over public versus private power.[360]

It all began innocently enough in March 1948 when Assistant Secretary of the Interior William Warne sent a private letter to California Congressman Norris Poulson suggesting that a water transfer from the Columbia River to California was "practical." In an unrelated manner, this personal observation was soon followed by legislation introduced by California Congressman Clair Engle authorizing an intertie system. Engle's legislation found support from Bureau of Reclamation Power Manager Ben W. Creim of Region II in Sacramento, who advocated in intertie system to prevent energy shortages in California and inhibit peaking power shortages in the Northwest. Shortly thereafter, however, Warne's letter to Poulson discussing a water transfer appeared in the *Congressional Record* creating uproar in the Pacific Northwest. Because of the close timing between the introduction of Engle's legislation and the release of Warne's letter, many in the Pacific Northwest connected an intertie system with a California water grab. Political candidates throughout the state of Washington utilized this confusion to claim that "California intended to take Pacific Northwest water." In addition, private power advocates, who decried any government involvement with the production and marketing of electricity, used the issue to garner support for a Columbia Interstate

---

359 Bureau of Reclamation, "Project History, Grand Coulee, Third Powerplant, Columbia Basin Project—Washington," Volume IV, RG 115, Project Histories, ACC# 8NN-115-92-130, Box 201; Pitzer, *Grand Coulee*, 349-50; Duck, *Construction of Grand Coulee Third Powerplant*, 1.

360 Floyd E. Dominy, "A New Power Giant Materializes on the West Coast," *Reclamation Era*, 51 (August 1965): 64-5.

Compact, being discussed in the Pacific Northwest since 1950, claiming one objective of the compact was to prevent diversion of Columbia River water to California and Arizona. These disputes did not lessen discussion on or interest in the intertie, however, as both the Defense Electrical Power Administration and the Defense Power Administration advocated such a system in the name of national defense during the Korean War. Despite active support for the intertie, the idea never gained much interest in Congress throughout the rest of the 1950s.[361]

Momentum for the Pacific Northwest-Pacific Southwest Intertie began in earnest as negotiations between the United States and Canada over upstream storage on the Columbia River moved forward. One observer noted,

> Undoubtedly the greatest single development which evolved concurrently with the Columbia River Treaty was the Pacific Intertie. Indeed the Intertie may well have been the prime catalyst in getting the Treaty ratified and implemented in its present form.

The Treaty with Canada did much more than allow Grand Coulee Dam power production to increase with the addition of the Third Powerplant. Occurring concurrently with construction of the new powerhouse, the Bureau of Reclamation in partnership with the Bonneville Power Administration (BPA) began laying out plans for the Pacific Northwest-Pacific Southwest Intertie. Under the Columbia River Treaty, Canada was to receive fifty percent of the hydroelectric power produced by the new storage dams it was constructing. Yet, at the time, there was no market for that power in British Columbia. Treaty provisions permitted the Canadian government and power interests to sell the surplus electricity in the United States. Eager and willing markets for the Canadian entitlement existed in California, Arizona, and Nevada.[362]

---

361 Bonneville Power Administration, *Columbia River Power for the People: A History of Policies of the Bonneville Power Administration* (Portland, Oregon: United States Department of Energy, 1980), 240-2; Dominy, "A New Power Giant Materializes on the West Coast," 64-6; for information on the Columbia Interstate Compact see Harvey R. Doerksen, *Columbia River Interstate Compact, Politics of Negotiation* (Pullman, Washington: State of Washington Water Research Center, Washington State University, and the University of Washington, August 1972), 102-3.

362 Bonneville Power Administration, "Background: The Columbia River Treaty Revisited," March 1989, 4; United States Department of Energy, Bonneville Power Administration, *Colum-*

Nevertheless, regional concerns still controlled how legislation would move through Congress, and it was not until 1963 that Intertie supporters produced an acceptable bill. To protect Pacific Northwest interests, Washington Senator Henry Jackson inserted an amendment in the bill that guaranteed "electrical consumers in the Pacific Northwest first call on electrical energy generated at federal hydroelectric plants." With these regional concerns taken care of, Congress passed the 1964 Public Works Appropriations Bill, which allowed construction of the Pacific Northwest-Pacific Southwest Intertie.[363]

At an Intertie Victory Breakfast in Portland on September 17, 1964, President Lyndon B. Johnson proclaimed, "This intertie ... is the most exciting transmission system in history. It makes us the world leaders in direct current transmission." As a person who helped lead the effort in rural electrification in Texas, Johnson's comments were not as hyperbolic as they may seem. Indeed, Secretary of the Interior Stewart Udall attributed the success of the Intertie legislation to President Johnson's personal attention to the issue. For Udall the Pacific Northwest-Pacific Southwest Intertie was clearly the "truest" measure of conservation. "It will conserve energy, capital, manpower, and materials—the ingredients of a strong healthy economy." Bureau of Reclamation Commissioner Floyd Dominy also praised the benefits of the intertie, noting that it would reduce waste and "promote a maximum of electrical efficiency throughout the States."[364]

The Pacific Northwest-Pacific Southwest Intertie stood not only as a historic event in the history of the Bureau of Reclamation but also in the longstanding feud between public power and private power advocates. According to one proponent the intertie integrated "two of the largest hydro systems in the United States (BPA and the Bureau of Reclamation), the largest municipal-owned system (Los Angeles), and the largest single private

---

bia River Power for the People, 218; Dominy, "A New Power Giant Materializes on the West Coast," 64.

363 Bonneville Power Administration, "Background: The Columbia River Treaty Revisited," 6; Bonneville Power Administration, *Columbia River Power for the People*, 244; see also "Pacific Northwest Power Marketing," August 31, 1964, in United States Department of the Interior, Bureau of Reclamation, *Federal Reclamation and Related Laws Annotated*, Volume III, Richard K. Pelz, editor (Washington, D.C.: Bureau of Reclamation, 1972), 1760-4; "Public Works Appropriations Act, 1965," in *Ibid.*, 1757-8.

364 For President Johnson's Comments see *Reclamation Era*, 51 (August 1965); see also Bonneville Power Administration, *Columbia River Power for the People*, 211, 245; for Secretary Udall's and Commissioner Dominy's comments see *Reclamation Era*, 51 (August 1965); Dominy, "A New Power Giant Materializes on the West Coast," 63.

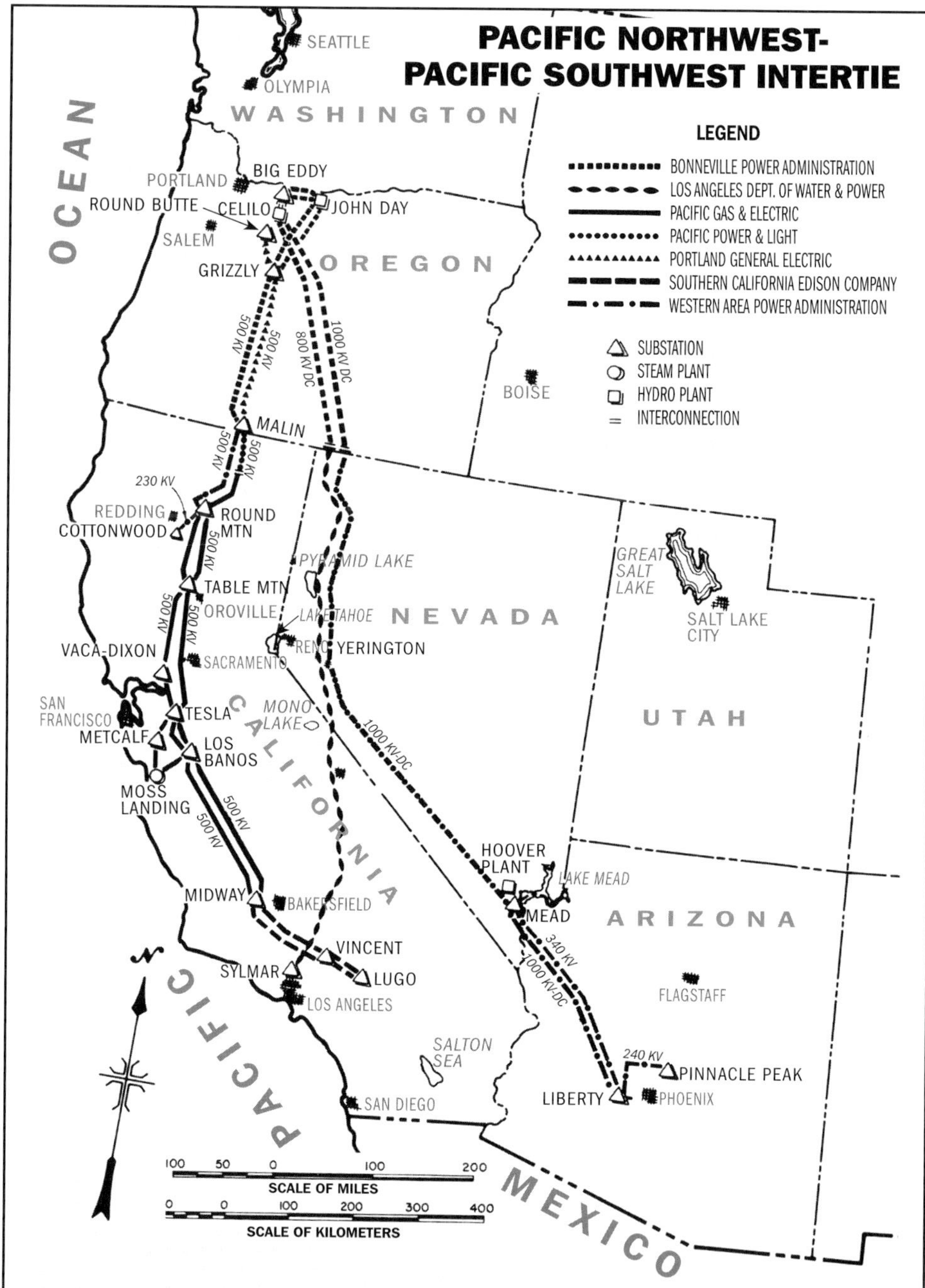

12.10. Reclamation participated with other public bureaus and private companies to create the Pacific Northwest-Pacific Southwest Intertie Project linking the power markets and generating facilities of the two regions to one another. The project allowed electricity to flow south for air conditioning in the summer and north for heating in the winter. Reclamation's role ended in 1977 when the Congress transferred power marketing and transmission responsibilities out of Reclamation in the Department of Energy Organization Act.

power system in the country (the California Power Pool)." And for the most part the Pacific Northwest-Pacific Southwest Intertie lived up to its hype. In 1980 BPA wrote, "The Pacific Southwest has enjoyed enormous benefits, mainly a net flow of 111 billion kWh of secondary energy from the Northwest. This energy from 1968 to 1976 equaled 186 million barrels of oil worth about $2 ½ billion at $12 per barrel." Later in their lives both Udall and Dominy remarked that the intertie was one of their greatest accomplishments in government service.[365]

On October 26, 1975, in front of an estimated 1,500 onlookers, Washington's Senator Jackson and Bureau of Reclamation Commissioner Gilbert Stamm simultaneously pushed the buttons that put Unit 19 of the Third Powerplant into operation. Among those watching was Hu Blonk, former managing editor of the *Wenatchee World* and the former chief of publications for the Bureau of Reclamation. At the time, Blonk was the chairman of the Third Powerplant Startup Committee, and he no doubt looked on with pride and satisfaction. He was a long-term resident of the area who had seen the dream of a large dam on the Columbia River come to fruition, and now that dam was beginning to fulfill its full potential. He looked back on all the naysayers who had called Grand Coulee Dam a "white elephant" and a waste of the taxpayers' money. He noted that the $490 million price tag for the Third Powerplant project would pay for itself in 36 years. "And," Blonk added, "by the end of the conventional payout period of 50 years, it will have earned a surplus of some $265 million to be applied to Reclamation's projects or to accumulate in the Federal Treasury." The powerplant that many ridiculed would "constitute 22 percent of the peaking capacity of the Federal hydropower system" when it was completed in 1979. The startup of Unit 19 was only the beginning; in 1976 Units 21 and 22 were online, and by 1978 all three of the 700,000-kw units went into production. For many like Hu Blonk, Grand Coulee Dam had reached its full potential.[366]

The Third Powerplant at Grand Coulee Dam and the Pacific Northwest-Pacific Southwest Intertie represented Bureau of Reclamation successes. Yet, controversies over the Central Arizona Project and the Central Utah Project and the environmental problems with dams in general cast a cloud

---

365 Udall, *Oral History Interview*, 151-2; Dominy, *Oral History Interviews*; see also T. M. Hollearin, "Construction Advances on the Pacific Northwest-Pacific Southwest Intertie," *Transmission & Description*, (July 1967): 66.

366 Hu Blonk, "A Braggart's Delight: The Startup of the Third Powerplant," *Reclamation Era*, 61 (Autumn 1975): 1-6; see also Pitzer, *Grand Coulee*, 350.

12.11. Teton Dam under construction, April 1, 1975.

over Reclamation's programs and activities. Unlike presidents Eisenhower and Kennedy, Lyndon B. Johnson actively pursued and promoted Reclamation projects. But even his support could not mask trends curbing the future programs of Reclamation. Johnson's own "guns and butter" policies strained the economy. The nation could no longer simultaneously afford new Reclamation projects whose prices now ran into the billions of dollars, fight a war in Vietnam, and implement the social reforms and programs of the Great Society. Throughout the 1970s Congress continued to authorize projects for the Bureau of Reclamation but appropriations for the construction money were hard to come by, and new environmental regulations strained Reclamation's ability to justify proposed projects. By mid-decade, even Reclamation's vaunted engineering reputation took a serious setback when one of its dams failed in Idaho.

Teton

By the end of November 1975 the Bureau of Reclamation had completed major construction on Teton Dam in southeastern Idaho and, on October 3, 1975, began filling the reservoir. Reclamation employees noticed small seeps downstream of the dam on June 3 and 4, 1976. On the morning of June 5, 1976, at around 7:00 a.m., the first signs that there was a problem with the dam appeared. Reclamation personnel noticed some seepage near the toe of the dam just above the embankment-abutment contact area. Two hours later,

12.12. Teton Dam during failure June 5, 1976.

they observed another area of seepage "just above the abutment-embankment contact and approximately 130 feet below the crest of the dam." At this time Reclamation supervisors initiated efforts to control the leaks by directing two bulldozers to channel the flow away from the powerhouse. According to one report on the disaster, "The project supervisor did not believe at this time the safety of the dam was jeopardized." By 11:30 a.m., that assessment had changed. Erosion created a large hole in the embankment that forced the dozer operators to flee their machines, which were eventually lost in the embankment hole. By noon the entire right embankment of the dam collapsed and the almost-full reservoir emptied into the Teton River heading for the communities and farms downstream. The devastation caused by the rampaging water was widespread, flooding parts of Rexburg and all of Sugar City, Idaho, and resulting in the deaths of 11 persons, 13,000 drowned cattle, 100,000 acres of prime farmland stripped clean, and an estimated $300 million in property damage and loss. If not for the efforts of Reclamation personnel alerting civic authorities in a timely manner and the cooperation of the communities downstream in evacuating the flood area the losses could have been much worse.[367]

[367] Independent Panel to Review Cause of Teton Dam Failure, "Report to the U.S. Department of the Interior and the State of Idaho on the Failure of Teton Dam," Idaho Falls, Idaho, December 1976, 1-1, RG 115, Office of the Chief Engineer, Teton Dam Records Related to Dam Failure, 1976-1981, ACC# 8NS-115-97-434, Box 8; Teton Dam Records; U.S. Department of the Interior, Teton Dam Failure Review Group, "Interim Report on Teton Dam Failure," July

Teton Dam was the main fixture in the Bureau of Reclamation's Teton Basin Project. Investigations to increase irrigation water to the basin began in the early 1930s, but it was not until September 1964 that Congress authorized the project. Reclamation's plans included not only the dam and reservoir, but also the Teton Powerplant and Teton Pumping Plant along with associated switchyard and irrigation canals. Construction began on the 305-foot high, 3,100-foot long earthfill structure in 1972, creating a reservoir with a capacity of 288,210 acre-feet of water. Teton Dam's primary purpose was to supply supplemental irrigation water to 111,210 acres of land in the Fremont-Madison Irrigation District of eastern Idaho. The project itself had the enthusiastic support of local irrigation communities and Idaho's congressional representatives. For many of these individuals, the filling of the reservoir that began in October 1975 marked a bright new era for the people of the Teton River basin.[368]

The catastrophic events in Idaho caused puzzlement, surprise, concern, and, even some defensiveness among Reclamation staff. On June 8, 1976, Bureau of Reclamation Commissioner Gilbert G. Stamm met with President Gerald Ford to brief him on the conditions and offer proposals to meet the crisis. Stamm's first concern was to reassure flood victims that help was on the way and "to demonstrate to the people that we mean business and are ready to go." Secondly, he and Secretary of the Interior Thomas Kleppe advised the president to utilize other federal agencies such as the Federal Disaster Assistance Administration (FDAA later renamed FEMA) and the Army Corps of Engineers to assist in salvaging irrigation works downstream of Idaho Falls to save "400,000 acres of highly productive lands … because if those lands go 10 days to 2 weeks without irrigation, we will have additional terrific crop loss from lands that are not affected at all by the flooding." Stamm also informed Ford that he was forming a "blue ribbon team" to determine what caused the failure of Teton Dam. For his part, President Ford wasted little time in calling on Congress to provide relief for flood victims. He not only declared the afflicted counties federal disaster areas, but five days after the dam collapsed he called on Congress to appropriate "$200 million for compensation of the victims of this terrible tragedy." Congress

---

14, 1976, RG 115, Office of the Chief Engineer, Teton Dam Records Related to Dam Failure, 1976-1981, ACC# 8NS-115-97-434, Box 9, Teton Dam Records; see also "583 Remarks at a Meeting to Discuss the Collapse of Teton Dam in Idaho, June 8, 1976," Public Papers of Gerald R. Ford, American Reference Library—Primary Source Documents, 2001, database: Master File Premier, 2.

[368] Eric A. Stene, "Teton Basin Project," Denver: Bureau of Reclamation History Program, 1996, 2-8; see also Water and Power Resources Service, *Project Data*, 1209-11.

approved the appropriation and President Ford signed the bill on September 7, 1976. In the end, the federal government paid more than $315,000,000 to over 7,500 claimants.[369]

In the immediate aftermath of the disaster, however, the Bureau of Reclamation wasted little time before beginning to assist flood victims. This was indeed a daunting task. Flood water had washed away the communities of Sugar City and much of Rexburg and devastated irrigation works on the Teton and Snake rivers all the way downstream to Idaho Falls. There was some concern about what would happen when flood waters reached American Falls Reservoir, and some questioned the dam's ability to withstand the sudden inflow. Reclamation officials quickly began releasing water from the reservoir to increase storage capacity, which raised the Snake River to near flood stage. While damage to the cities of Idaho Falls and American Falls was relatively minor, the rampaging water wreaked havoc on irrigation facilities. On June 8, 1976, Bureau of Reclamation Commissioner Gilbert Stamm recommended to President Gerald Ford that the first order of business was repairing the damaged canals and headworks.[370]

There were political considerations in Stamm's proposal meant to show the public that the Bureau of Reclamation was on the job mitigating the effects of Teton and bringing aid to those affected. But Reclamation's quick response to the disaster was much more than a public relations exercise; it achieved amazing results. Within a month of Teton's collapse, the Bureau of Reclamation oversaw reconstruction efforts that restored irrigation service to ninety-eight percent of the affected cropland. It was an incredibly rapid response to a dire emergency. In a relatively short period of time, "more than 90 contracts were awarded to 22 different contractors and approximately 2.5 million dollars were expended." While doing little to mitigate the condemnation Reclamation endured during the upcoming congressional hearings, these accomplishments did succeed in not allowing one part of the disaster to spiral out of control.[371]

---

369 "583 Remarks at a Meeting to Discuss the Collapse of Teton Dam in Idaho, June 8, 1976," 1-2; U.S. Department of the Interior, Teton Dam Failure Review Group, *Failure at Teton Dam: Final Report*, January 1980, 1-1; United States House of Representatives, *Supplemental Appropriations for Reimbursement for Damages Caused by Failure of Teton Dam: Communication from the President of the United States*, H.Doc. No. 94-523, 94th Cong., 2nd sess., June 11, 1976 (Washington, D.C.: United States Government Printing Office, 1976).

370 "583 Remarks at a Meeting to Discuss the Collapse of the Teton Dam in Idaho, June 8, 1976," American Reference Library—Primary Source Document, 2001, 1.

371 "583 Remarks at a Meeting to Discuss the Collapse of the Teton Dam in Idaho, June 8,

The much more difficult and long-lasting task for the Bureau of Reclamation came in aiding flood victims. Congress responded by authorizing $200 million of disaster relief on June 10, 1976, and Reclamation began the long process of settling property claims. Within a matter of days after the failure of Teton Dam, Reclamation developed a claims program and opened claims processing offices. For many Bureau employees who worked in the claims program, their participation turned out to be a gratifying experience. Neil Stessman recalled that his experience was "very positive," and that it brought together other federal agencies including the Federal Disaster Assistance Administration, the Farmers Home Administration, and the Small Business Administration. Reclamation opened claims offices in Rexford, Idaho Falls, and Blackfoot to expedite the relief effort. By fall 1976 a *Reclamation Era* article reported that "3,349 claims totaling over $116 million had been filed. This is over half the number of claims expected. The Bureau has already paid 1,962 claims totaling over $42 million." It took another two years for the Bureau of Reclamation to settle all the claims stemming from the failure of Teton Dam.[372]

Reclamation's speedy response to the water needs of Snake River basin irrigators to a certain extent demonstrated the Bureau's engineering and construction expertise. It showed to constituents and the general public alike Reclamation's ability to harness resources and effectively coordinate efforts in the face of calamity. Indeed even the quickness with which the Bureau of Reclamation developed and implemented the claims program testified to Reclamation's determination to serve its customers. Nevertheless, the fact remained that one of its dams had failed, resulting in hundreds of millions of dollars in property damages and eleven fatalities. Seeing the writing on the wall, the commissioner recognized the political backlash Teton created for the entire Reclamation program. Despite this "unprecedented" setback for the Bureau of Reclamation, Stamm wrote that "it would be an even greater tragedy … if those who have traditionally opposed water resource development programs should succeed in using the Teton disaster as a political weapon to stop all further development."[373]

1976;" Gilbert Stamm, "After Teton," *Reclamation Era*, 62 (Autumn 1976): 112; Stephan G. Wade, "The First Days," *Reclamation Era*, 62 (Autumn 1976): 13.

372 For a review of Reclamation's claims program see J. Neil Stessman, *Oral History Interview*, Transcript of tape-recorded Bureau of Reclamation Oral History Interviews conducted by Brit Allan Storey, senior historian, Bureau of Reclamation, from 1994-1996, in Billings, Montana, 90-106; see also Paul Winegar, "The First Months—Rebuilding Lives." *Reclamation Era*, 62 (Autumn 1976): 20-6.

373 Stamm, "After Teton," 3.

Shortly after relief work commenced in Idaho, both the executive and legislative branches of the federal government began investigations into the collapse of Teton Dam. Stamm, in association with Governor Cecil Andrus of Idaho, formed the Independent Panel to Review Cause of Teton Dam Failure, consisting of non-federal and highly respected engineers. The commissioner also established an in-house panel to investigate the cause of the dam's failure. By their very nature, both panels looked primarily at design, engineering, and construction aspects of the dam to discover probable explanations for Teton Dam's collapse. On the other hand, congressional investigations, while not overlooking engineering and construction factors, focused on the institutional and bureaucratic culture within the Bureau of Reclamation. These inquiries sought to determine whether or not there were issues within the Bureau of Reclamation that led to building an unsafe dam.

The first indication that the Bureau of Reclamation faced stern review of its practices from Congress actually came in late June 1976 during the brief congressional hearings held on President Ford's $200 million Teton relief measure. Idaho Senator Frank Church, chairman of the Subcommittee on Energy Research and Water Resources, set the tone. Church perceived the president's request for relief funding as a declaration of culpability on the part of the federal government to the disaster. The senator asserted, "It would appear that President Ford subscribes to my own belief that the Federal Government shares primary responsibility for this calamity and, as a result, must see that complete restitution is made to flood victims." The president's statement asking for additional relief funds neither inferred nor admitted negligence on the part of the federal government. When asking for additional appropriations for flood victims, President Ford stated, "These funds will compliment on-going Federal disaster assistance to provide further relief for injuries and damages inflicted by the flood." The carefully worded statement contrasted with statements coming from some members of Congress who wanted to know the causes of the disaster, and more importantly who to blame.[374]

Evidence emerged dating to 1973 that pinpointed problems with the Teton damsite. United States Geological Survey (USGS) memoranda expressing concerns about seismic activity in the vicinity of the dam surfaced. These interoffice memos found their way to the press, and in turn, piqued the interest of Congress. Officials at the USGS maintained, however, that the state-

---

[374] United States Senate, Subcommittee on Energy Research and Water Resources of the Committee on Interior and Insular Affairs, *Teton Dam*, Hearing, 94th Cong., 2nd sess., June 15, 1976 (Washington, D.C.: United States Government Printing Office, 1976), 2,11.

ments made by its geologists were initial observations that were incomplete and needed further study and not meant to question Reclamation's decision to construct a dam on this location. Indeed, the Bureau of Reclamation was well aware of the seismic activity in the area and installed seismographs throughout the construction area. As one observer wrote, the Teton damsite "became the project at which perhaps more seismographs were placed than anywhere else to date." Testimony further revealed that Dr. Robert Curry, professor of geology at the University of Montana, was the person responsible for leaking the memorandums to the press. Curry's place in this episode is interesting only because of his work with the Sierra Club in its fight against construction of Teton Dam. Between 1971 and 1974 the Sierra Club and other conservation organizations fought a number of court cases against the Bureau of Reclamation seeking an injunction to stop construction. In 1974 the U.S. District Court of Idaho denied the injunction and the Ninth Circuit Court of Appeals upheld that ruling.[375]

Bureau of Reclamation designers, engineers, and geologists were all aware that the dam's foundation sat upon a fractured and faulted rock base. Indeed, even before construction began, Reclamation contractors conducted a test to determine if grouting could seal the foundation. Former Bureau of Reclamation Director of Design and Construction Harold G. Arthur later recalled:

> The results of that program were positive. It appeared that the foundation at depth could be satisfactorily sealed by cement grouting, and that the near surface layers could not be grouted successfully because of the inability to exert much pressure on the ground without lifting the rock [it] could be treated by excavating a deep cutoff trench through the surface layers of rock, and this trench would be backfilled with impervious material. So we thought that we had a feasible site and we could build a dam there.

---

[375] Subcommittee on Energy Research and Water Resources, *Teton Dam*, 42; see also J. Laginha Serafim, "Disaster in 1976 (The Teton Dam Case)," (Desastres de Barragens em 1976 O Caso de Teton), Translated from Portuguese: Article in the engineering review *Tecnica*, reprint from issue No. 443, pp. 7-17, published in Lisbon by the Student Association of Higher Technical Institute, 1977. This article was based on a talk given at the National Civil Engineering Laboratory on December 30, 1976. Division of Design, Engineering Services Branch, Code 280, USBR Translation Group, Denver, Colorado, May 1978, 8; United States House of Representatives, Committee on Government Operations, *Teton Dam Disaster: Thirtieth Report by the Committee on Government Operations*, H. Rpt. No. 94-1667, 94th Cong., 2nd sess., September 27, 1976 (Washington, D.C.: United States Government Printing Office, 1976), 7.

Although these words came many years after the event, they represent a trait of Bureau of Reclamation culture that suggests technology can, in the words of Commissioner Gilbert Stamm, fix "deficiencies in nature." Members of Congress quickly detected that aspect of Reclamation culture as they pursued their investigations of the bureaucratic and procedural policies of the Bureau of Reclamation.[376]

In hindsight critics declared that the disaster could have been avoided. At the hearings in June 1976, Senator James McClure of Idaho exclaimed,

> I think it clear that no act of God created the destruction, no act of God took the toll of human lives, and no act of God caused the thousands of injuries suffered when the wall of water, which people of the area thought was peacefully stored behind this Government structure, suddenly swept down the valley.

McClure's statement surfaced the high degree of anger and frustration over a situation that many believed need not have happened. Two months later, during hearings conducted by the House Subcommittee on Government Operations, that anger focused squarely upon the Bureau of Reclamation. California Congressman Leo G. Ryan charged that the "root of the problem is the momentum to build." He explained, "Once a decision has been made to build, the force of that decision grows with every cost invested and every further decision concerning construction." Ryan saw a narrow-minded, goal-oriented Bureau of Reclamation hell bent upon construction projects that neglected dam safety.[377]

The so-called "momentum theory" dominated congressional investigations. It explained what many perceived as a series of gross misjudgments on the part of the Bureau of Reclamation during planning and construction of

---

[376] Arthur, *Oral History Interview*, 234-5; see also Committee on Government Operations, *Teton Dam Disaster*, 8.

[377] Statement of Warren Viessman, Jr., Congressional Research Service, before the Subcommittee on Energy Research and Water Resources of the Committee on Energy and Natural Resources of the United States, Idaho Falls. Idaho, February 21, 1977, RG 115, Office of the Chief Engineer, Teton Dam Records Related to Dam Failure, 1976-1981, ACC# 8NS-115-97-434, Box 12, Teton Dam Failure; Subcommittee on Energy Research and Water Resources, *Teton Dam*, 18; United States House of Representatives, Subcommittee of the Committee on Government Operations, *Teton Dam Disaster*, Hearings, 94th Cong., 2nd sess., August 5, 6, and 31, 1976 (Washington, D.C.: United States Government Printing Office, 1976), 3, RG 115, Office of the Chief Engineer, Teton Dam Records Related to Dam Failure, 1976-1981, ACC# 8NS-115-97-434, Box 8, Teton Dam Records.

Teton Dam. The House Committee on Government Operations reported that the "momentum theory" provided a false sense of security; a belief that engineering can overcome every problem; a reluctance to halt construction should problems arise and finally a determination to see a project through even in the face of dangers to the public safety. According to this committee, the collapse of Teton Dam resulted from Reclamation's adherence to this mindset. It concluded, "While warnings clearly pointed to safety hazards that could affect life and property downstream, the Committee believes that these warnings were not properly identified as such by the Bureau, and were largely ignored while construction of the dam proceeded."[378]

In its conclusion, the Subcommittee of the Committee on Government Operations found that there were serious weaknesses in the processes of the Bureau of Reclamation that went far beyond the mere rush to finish projects. Because of Teton, the report claimed that safety problems existed in dams then currently in the design stage or already constructed that had similar geological or seismic conditions to those found at the Teton damsite. The committee also concluded that Reclamation failed to heed or seek professional advice from USGS geologists or "any outside professional geologists or panels." In short, the committee's report blamed much of the failure of Teton Dam on Reclamation's bureaucratic hubris, asserting, "The overconfidence of the Bureau is regarded by some critics as an attitude bordering on arrogance. To this day, the official position of the Bureau is that what happened at Teton was 'impossible'." This scathing rebuke of Reclamation programs and policies damaged the prestige and reputation of the agency. It helped to establish in a public consciousness, already growing wary over the environmental effects of dam building, a new belief—Bureau of Reclamation dams were unsafe.[379]

Anger about the failure of Teton Dam appeared in congressional reports in the immediate months following the event. The engineering investigations of the event were still works in progress. As noted earlier, Commissioner Stamm formed two separate panels to investigate the causes of the dam's failure. The commissioner brought together a team of highly respected engineering professionals known as the Independent Panel to Review Cause of Teton Dam Failure, or the Independent Panel. He also formed a Department of the Interior in-house team of engineers called the Teton Dam Failure Group. Both panels presented their findings long after Congress completed its

[378] Committee on Government Operations, *Teton Dam Disaster*, 16.

[379] *Ibid.*, 32-3, 16.

investigations and in a less emotional style, but their conclusions were no less damaging to the reputation of the Bureau of Reclamation.

Released in December 1976, The Independent Panel's report examined all aspects of the dam's design, construction, and other factors that may have contributed to failure of Teton Dam. It found that Bureau of Reclamation records regarding the site selection and geological reports were in order and that in its opinion Reclamation took no short cuts. It was with the design, engineering, and construction of the dam that the Panel took issue with Reclamation practices. Concurring with initial Reclamation reports, the Panel acknowledged that the damsite was difficult because of the "highly permeable and moderately to intensely jointed" rock. This factor, according to the report, allowed water to flow freely through areas of the foundation where grouting failed to provide an effective seal. This, in turn, caused a "piping effect" that permitted water to flow "through some channel in the embankment section." The Independent Panel's report also questioned the use of certain material in the key trench where Reclamation opted to use "highly erodible" clay silts next to "the heavily jointed rock of the abutment." Thus, a number of contributing factors led to the failure of Teton Dam. Two important reasons appeared: (1) inability to get an all-encompassing seal with the grout curtain; (2) the use of material that could easily wash away. The Panel concluded that

> under difficult conditions that called for the best judgment and experience of the engineering profession, an unfortunate choice of design measures together with less than conventional precautions was taken to ensure adequate functioning of the Teton Dam, and these circumstances ultimately led to its failure.[380]

Not surprisingly, the Bureau of Reclamation objected to the conclusions of the Independent Panel. Reclamation's director of the office of design and construction, H. G. Arthur, saw some validity in portions of the Panel's report but had concerns about the report's conclusion. In a memorandum to the deputy assistant secretary in the Department of the Interior, Arthur conceded that there were obvious defects "in design, or construction, or the site

---

[380] Independent Panel to Review Cause of Teton Dam Failure, "Report to the U.S. Department of the Interior and the State of Idaho on the Failure of Teton Dam," Idaho Falls, Idaho, December 1976, RG 115, Office of the Chief Engineer, Teton Dam Records Related to Dam Failure, 1976-1981, ACC# 8NS-115-97-434, Box 8, Teton Dam Records; for a summary of the Independent Panel's report see Philip M. Boffey, "Teton Dam Verdict: A Foul-Up by the Engineers," *Science*, 195:4275 (January 21, 1977): 270-2.

which led to the failure of the dam," but he did defend Teton Dam designers maintaining that they "did a much better job than the Independent Panel gives them credit for." Arthur believed the Independent Panel's report was hasty and premature. The Bureau of Reclamation's formal response outlined its concerns with the Independent Panel's findings, arguing that Reclamation designers and engineers utilized "suitable material" in the construction of Teton Dam. Reclamation also maintained, "Thoughtful consideration *was* given to differing and unusually difficult geological conditions at the Teton damsite. The design was tailor made for that specific site." The Bureau admitted that there "*may* have been significant shortcoming in the design and construction of Teton Dam" caused by the failure to adequately protect "the downstream side of the key trench against piping," but hedged a bit by claiming that "since positive identification of the cause of the failure has not been made, the Bureau is not satisfied with the extent and depth of the investigations and analysis performed to date."[381]

If the Bureau of Reclamation was looking for a more thorough investigation to clear its good name, it faced disappointment. The Department of the Interior's Teton Dam Failure Group presented an "Interim Report" in July 1976. These early findings echoed those of the Independent Panel noting that Teton Dam failed as a result of interior erosion. Though the Group refused to pinpoint the exact cause of the failure, it did offer a number of possible explanations that included: the pervious and jointed rock, the clay silt used for the dam's core, the narrow width of the key joint, and the difficulty in attaining an impervious grout curtain. In April 1977 the Group issued "A Report of Findings" holding to its initial findings, but adding critiques of Reclamation's engineering decisions. It claimed, "Defensive measures were within the state-of-the-art of dam design at the time Teton Dam was designed, and should have been used." In its *Final Report*, released in January 1980, the Teton Failure Review Group admitted that the "physical mode" of the collapse of Teton Dam will never be known. As to the "<u>cause</u>" of the disaster, the report was unequivocal. It argued that "the dam failure remains 'inadequate protection of the zone 1 impervious core from internal erosion,'" and concluded:

---

[381] Memorandum, H. G. Arthur, Director of Design and Construction to Deputy Assistant Secretary, Land and Water Resources, Department of the Interior, Subject: Office of Design and Construction Comments on the Independent Panel Report on the Failure of Teton Dam; italics in the original, see also "Comments on the Report of the Independent Panel to Review Cause of the Teton Dam Failure," prepared by Office of Design and Construction Engineering and Research Center Bureau of Reclamation, April 15, 1977, RG 115, Office of the Chief Engineer, Teton Dam Records Related to Dam Failure, 1976-1981, ACC# 8NS-115-97-434, Box 8, Teton Dam Records.

> This cause is related to the basic design of the embankment and its foundation. Had the available defensive design concepts of filtration, drainage, and rock foundation surface treatment been employed to provide protection of zone 1, a safe dam could have been constructed at the site.

In short, the Department of the Interior's in-house review concluded that Bureau of Reclamation engineers and designers fumbled the opportunity to construct a safe dam at the Teton site. They failed to utilize the best and most up-to-date engineering practices available at the time.[382]

Still the question persisted as to why the Bureau of Reclamation had made such grievous mistakes. At the time of Teton Dam's failure, Reclamation had constructed well over three hundred dams, of which 250 were earthfill structures. Teton was the first and only failure, and according to all the reports, the Bureau of Reclamation made flawed decisions. Some suggested that Reclamation found itself under pressure to build dams on less than optimal sites. According to critic Marc Reisner, "the Bureau was being forced to build on sites it had rejected forty, fifty, or sixty years earlier. It was building on them because while the ideal damsites had rapidly disappeared, the demand for projects had not." Reisner's observation raised questions about the Bureau of Reclamation's technical expertise and Reclamation policy. No doubt, sixty years previously, engineers would not have viewed the site favorably. The technical know-how to construct a dam at such a questionable site did not exist. On the political and policy side, Daniel McCool's "iron triangles" theory points to powerful forces that joined together to push for authorization and construction funding of Teton Dam.[383]

McCool suggests that after the Second World War, the Bureau of Reclamation joined forces with a cadre of water resource development interests that pushed for the expansion of Reclamation programs. Throughout the 1950s and 1960s, the aggressive lobbying of this consortium helped expand

---

382 United States Department of the Interior, Teton Dam Failure Review Group, "Interim Report on the Teton Dam Failure," July 14, 1976, 1-1, RG 115, Office of the Chief Engineer, Teton Dam Records Related to Dam Failure, 1976-1981, ACC# 8NS-115-97-434, Box 9, Teton Dam Reports; emphasis in the original; United States Department of the Interior, Teton Dam Failure Review Group, "Failure of Teton Dam: A Report of Findings," April 1977, iii; RG 115, Office of the Chief Engineer, Teton Dam Records Related to Dam Failure, 1976-1981, ACC# 8NS-115-97-434, Box 9, Teton Dam Reports; United States Department of the Interior, Teton Dam Failure Review Group, *Failure of Teton Dam: Final Report*, January 1980, 7-1.

383 Reisner, *Cadillac Desert*, 382; see also Pisani, "Federal Reclamation and the American West in the Twentieth Century," 407; McCool, *Command of the Waters*, 5.

4 Colo. Water Projects Out; Evans to Fight

By BILL MYERS

Denver Post Staff Writer

The Carter administration announced Friday night it won't include in its fiscal 1978 budget requests any money for four long-planned Colorado water projects, drawing a promise of a fight from U.S. Rep. Frank Evans, D-Colo.

Evans issued an angry statement Saturday after learning that the administration's budget requests won't include money for the Dolores and Fruitland Mesa Projects in his 3rd District and for Savery-Pot Hook on the northwest Colorado-southern Wyoming border and the Narrows Project northeast of Denver.

12.13. This February 20, 1977, clipping headline, soon after Jimmy Carter became president, indicates the concern Coloradoans felt about threats to the water projects in Colorado.

Reclamation activities. Although at times parties were at odds within the consortium, as in the battle over the Central Arizona Project, there was never any real doubt about the reciprocity in the process of authorizing and financing Reclamation projects. The scarce water resources of the West demanded that individual states, water districts, or communities lobby hard for their project lest someone else claim their water. It was an ideology of circumstance that demanded continued development, and the Bureau of Reclamation happily joined the effort. Reclamation gained not just a supportive and powerful constituency but increasing appropriations that allowed it to become one of the more formidable civilian bureaus of the federal government. In a sense, it was the "momentum to build" that California Congressman Leo Ryan criticized so vehemently in the investigations of the Teton Dam failure. That "momentum" came not only from Reclamation's dam builders, but also from the politicians and water users who sought and demanded protection for that most valuable of western resources: water. Historian Donald Pisani's 2003 interpretation represents a widely held belief among scholars when he says, "The Teton Dam was not just built in a bad place. Its costs far outweighed its benefits, and like many projects built during the 1960s and 1970s, it served relatively few water users."[384] The collapse of Teton Dam was a scar on Reclamation's reputation as one of the world's leading engineering and construction organizations. After Teton, the safety of all Reclamation dams came into question in some minds. As historian Donald Worster points out, "The best designs of the best engineers … could fail, not only all at once, with the thunderous impact as in Idaho, but slowly too, wearing out, falling to disrepair, becoming impossible to salvage."[385]

---

[384] Pisani, "Federal Reclamation and the American West in the Twentieth Century," 409.

[385] Worster, *Rivers of Empire*, 309.

Conclusion: Carter's Hit List

From its beginning as the U.S. Reclamation Service in 1902, the Bureau of Reclamation faced critics. The early detractors sounded the mantra that irrigation programs in the West benefitted only a few and were nothing more than "pork barrel." They decried what they regarded as the subsidies for western water resource development that shifted the burden of paying for Reclamation projects from water users to the American taxpayer. In the 1950s, governmental and political commentator Raymond Moley, author of *What Price Federal Reclamation?* (1955) amplified his anti–New Deal critique to ridicule Reclamation. He was baffled that the charade of the Reclamation program still continued given its dubious accomplishments. Moley lambasted all the proposed benefits that irrigation policy purported to deliver; including Reclamation's benefit/cost analysis that opened "the way to almost unbelievable abuses in fabricating a case for the feasibility of a project." By the late 1960s and into the 1970s, similar criticisms intensified and now included wide-ranging and damning environmental disapproval. Whereas Moley's attacks, focused primarily on the free-spending economic aspects of Reclamation policy, had limited appeal, a more definitive political ideology arising from the environmental movement captured the imagination of many Americans.[386]

The Teton Dam disaster exposed the Bureau of Reclamation to criticism of its untarnished reputation as one of the world's foremost dam builders. Criticism could be expected about the economics of Reclamation policy and from the standpoint of environmental concerns, but rarely had critics spoken of Reclamation's engineering and technical expertise. Teton changed all that. When Jimmy Carter became president in 1977, one of the items on his legislative agenda was reform of water resource development policy. Carter came into office facing a faltering national economy and a federal government still reeling from the scandal of Watergate, the trauma of Vietnam, as well as trying to pay for it. His administration was environmentally conscious and economy-minded. In many ways he held views similar to Raymond Moley regarding the economic folly of the nation's water policy. Carter determined to cut waste in the federal budget and saw water resource development programs as a prime example of that waste.[387]

---

386 Raymond Moley, *What Price Federal Reclamation?* (Washington, D.C.: American Enterprise Association, Inc., 1955).

387 Reisner, *Cadillac Desert*, 308; Jeanne Nienaber Clark and Daniel McCool, *Staking Out the Terrain: Power Differentials Among Natural Resource Management Agencies* (Albany: State University of New York, 1985), 99.

Shortly after taking the oath of office, President Carter announced his decision to make significant cuts in the water resource development programs of the Army Corps of Engineers, the Bureau of Reclamation, and the Tennessee Valley Authority. In outlining this cost-cutting measure, Carter maintained, "Activities which are wasteful, unsafe, or uneconomical or environmentally unsound simply cannot be pursued." The president sought to eliminate or modify fourteen Corps of Engineers projects and seven Bureau of Reclamation projects, with a total savings to the American taxpayer of nearly $4 billion. Carter based these cuts on dubious benefit/cost analysis, safety concerns, and "environmental values ... to ensure that irreplaceable natural resources are protected from needless degradation or destruction." The task of designating which projects the administration intended to cut fell to the respective executive agencies, which for Bureau of Reclamation projects meant the Department of the Interior. In the press and to his political opponents, however, the projects selected became popularly known as Carter's hit list.[388]

Carter's proposal was an attack on water resource development. The list put together by Department of the Interior personnel contained neatly reasoned and succinctly written arguments calling for the elimination or modification of specific projects. The authors questioned the methods utilized by the Bureau of Reclamation to plan projects and provided detailed analyses defending their findings. This was an attack on the long-held precepts of western water development, and the responses were immediate and thunderous. According to popular journalist and historian Philip Fradkin, the list was "the first serious challenge by a president to the West's primal shibboleth—its essential aridity, and the need for more dams and ditches to assure a dependable living and some measure of prosperity in a dry land."[389]

The authors of Department of the Interior's analysis of Bureau of Reclamation projects noted the narrow interest groups they benefitted. As President Carter maintained, "The beneficiaries of federal water projects do not bear a fair share of the enormous capital and operating costs." Noting this discrepancy on the Garrison Diversion Unit in North Dakota, the report observed that project farmers were only obligated to pay $77 per acre on a federal investment of $1,992 per acre. The findings were similar on the Oahe

---

388 The White House, Office of the White House Secretary, "Statement on Water Projects," April 18, 1977, 1-2, "Carter's Hit List," Binder of Primary Source Documents, Bureau of Reclamation History Department, Denver, Colorado; hereafter cited as "Carter Hit List;" see also Reisner, *Cadillac Desert*, 317; Fradkin, *A River No More*, 3-9.

389 Fradkin, *A River No More*, 3.

Project in South Dakota where only $176 per acre was due from farmers compared to a $2,247 commitment by the federal government. To fix these inconsistencies, Secretary of the Interior Cecil Andrus proposed increasing the cost-sharing amongst federal and non-federal irrigation districts and other water users. Andrus argued, "There can be no doubt that the availability of Federal money ... had led to the construction of projects that would not have been built if inclusion of significant amounts of non-Federal money were a prerequisite to federal financing."[390]

Many Reclamation projects were economically viable only because of the revenues produced by hydroelectric facilities incorporated in Reclamation dams. This practice, or subsidy, was a longstanding hallmark of Reclamation policy. The Department of the Interior's report questioned the practice. Once again the report singled out the Garrison Diversion Unit: "Of the reimbursable investment of $497,989,000 allocated for irrigation, only $19,504,000 will be repaid by the irrigators. The remaining $497,485,000 will be repaid by surplus power revenues from the Pick-Sloan Missouri Basin Program." This meant that power users throughout the Missouri River basin heavily subsidized irrigation in the Garrison Diversion Unit. According to the president and his advisors, here was another blatant example of how irrigation projects benefitted a special interest group. This situation demanded reform in the shape of greater cost-sharing between the federal government and water users.[391]

Carter's hit list also took aim at much larger projects that claimed to spread benefits to a wider constituency. Two such projects on the hit list were the Central Arizona Project and the Central Utah Project. Of course the high costs of these projects made them obvious targets. The president's report noted that the price tag for the Central Utah Project was already at $862 million, and that it had the "distinct possibility that operational plans and commitments cannot be met within the authorized limits." While not recommending complete elimination of CUP, the report suggested a major scaling back of project works which would lower the total cost to $27.8 million. Carter made similar recommendations regarding the Central Arizona Project, estimated to cost almost 1.3 billion dollars to finish, and proposed elimination of Orme, Hooker,

---

390 United States Department of the Interior, "Water Project Review, Garrison Diversion Unit, North Dakota," April 1977, 9; United States Department of the Interior, "Water Project Review, Oahe Project, South Dakota, April 1977, 7; Cecil D. Andrus, Secretary of the Interior to President of the United States, April 13, 1977 in "Carter Hit List."

391 Department of the Interior, "Water Project Review, Garrison Diversion Unit, 9; see also Department of the Interior, "Water Project Review, Oahe Project, 4 in "Carter Hit List."

and Charleston dams. These modifications alone, it was believed, offered savings of well over $300 million from CAP's budget.[392]

Critiquing the economics of Bureau of Reclamation projects was not new. Carter's hit list narrative went much further and sought far-reaching reforms. The Carter administration introduced new criteria that demanded Reclamation and other water resource development promoters take into consideration environmental aspects when reviewing projects, placing greater emphasis on wildlife protection, fish enhancement, and water quality controls. These reforms called for a new way of thinking about irrigation projects that put less reliance on structural controls, such as dams and canals, and more emphasis upon developing ecologically sound measures to accomplish desired goals. Carter's report also showed greater concern for and recognition of Indian water rights, arguing that settlement of this issue was mandatory for some projects to move forward. Finally, the hit list questioned one of the founding tenets of Reclamation policy by attacking the doctrine of prior appropriation, which some argued was the reason so many uneconomical projects got authorized. Secretary Andrus claimed, "The 'use or lose' syndrome created by State water law and administrative doctrines has led to Federal projects which may be constructed prematurely primarily to permit a State to capture water supplies before another State is able to acquire them." The secretary also suggested that some project water users used water inefficiently solely "to maintain an embedded right." Many of the president's arguments for elimination or modification of Reclamation projects went beyond mere cost-cutting and attacked the culture of federal water resource development as an archaic spoils system that must come to an end.[393]

Many in Congress from the West vehemently condemned the president's actions, arguing that he was attacking "congressional prerogatives." Others were more direct. Arizona Congressman Morris Udall labeled Carter's list as "a Washington Day ambush," while the press called it Carter's "War on the West." Contemporary pundits observed that Carter's hit list was "a Quixotic tilting at windmills that will gain him nothing except the ill will of powerful senators and representatives who can make trouble for the rest of his

---

[392] The White House, "Statement on Water Projects," see also United States Department of the Interior, "Water Projects Review, Central Arizona Project Arizona," April 1977, 1-32; United States Department of the Interior, "Water Projects Review, Bonneville, CUP, Utah," April 1977, 1-16 in "Carter Hit List."

[393] United States Department of the Interior, "Water Policy Reform Proposals," April 1977, 5 in "Carter's Hit List;" see also Fradkin, *A River No More*, 3; McCool, *Command of the Waters*, 197.

legislative program." The prediction rang true. Congress ignored the report's recommendations and voted in favor of authorizing all but one of the projects on the list. In the end, Carter could not overcome the unity of western congressmen, and both sides reached a compromise wherein nine of the projects were removed from the 1977 Public Works Bill.[394]

Carter's hit list, coming as it did on the heels of the Teton Dam disaster, marked a turning point in the history of the Bureau of Reclamation. Although Congress reinserted the nine projects into the 1979 Appropriations Bill, a presidential veto eliminated them for good. For the remainder of the twentieth century, Congress continued to authorize Reclamation projects, while the executive branch derailed them because of budgetary considerations. The great era of Bureau of Reclamation dam building closed. Skyrocketing costs, tight budgets, lack of good damsites, and opposition from environmental groups all contributed to the end. But the end ushered in a new transition for the Bureau of Reclamation with inducements to move away from construction into the management of water in the West. This meant water flowing through Reclamation dams, canals, and powerplants must be shared amongst irrigators, growing urban metropolises, and recreational needs, all according to demanding environmental standards.

[394] Donald W. Carson and James W. Johnson, *Mo: The Life and Times of Morris K. Udall* (Tucson: University of Arizona Press, 2001), 129-30; Luther J. Carter, "Water Projects Dispute: Carter and Congress Near a Showdown," *Science*, 196:4296 (June 17, 1977): 1303-1305; Reisner, *Cadillac Desert*, 316; for an overview of the hit list controversy see Fradkin, *A River No More*, 3-13; McCool, *Command of the Waters*, 197-206.

CHAPTER 13:

A NEW ERA FOR WATER IN THE WEST: BUREAU OF RECLAMATION, 1980-2000

Introduction

By the late twentieth century water from Reclamation facilities served almost 5 million acres of irrigated farmland, while delivering supplemental water to another 6 million acres. According to one Senate report, this represented "about 3 percent of Nation's Farmland and 25 percent of its irrigated farmland," and the crops grown on these farms grossed an estimated $4.4 billion. In addition, hydroelectric power from Bureau of Reclamation hydroelectric facilities underwrote western industrialization and urbanization. Now a leader in water and power resource management at the end of the twentieth century, Reclamation's charge far surpassed the tasks undertaken by the U.S. Reclamation Service at the beginning of the century.[395]

Despite successes, the Bureau of Reclamation faced daunting challenges and severe criticism. Backlash from the failure of Teton Dam in 1976 scarred Reclamation's reputation, raised uncomfortable questions about its engineering expertise, and provided ammunition to opponents of its programs. Furthermore, the so-called Carter hit list criticized benefit/cost analyses used to justify projects and ultimately the subsidies to Reclamation projects. Critics and supporters sharpened their focus on program benefits, and Congress under fiscal constraints became reluctant to approve construction monies. Moreover, as environmental concerns grew, dam building became widely unpopular. In 1979 further evidence of pressure for a transition in Reclamation's mission occurred when Commissioner R. Keith Higginson changed the Bureau's name to the Water and Power Resources Service. Although not fully articulated at the time, the heyday of the Bureau of Reclamation's construction era and grandiose plans for western regional development was drawing to an end.

As in many previous instances during its history, circumstances compelled the Bureau of Reclamation to transform itself in response to changes occurring throughout the West and American society and culture. Demo-

[395] United States Senate, Committee on Energy and Natural Resources, *Reclamation Reform Act of 1979*, S. Rpt. No. 96-235, 96th Cong., 1st sess., June 21, 1979, 7.

graphic changes and a shift in political power in the "New West" prompted re-examination of acreage limitation in light of economies of scale in late-twentieth century agriculture. Urban water needs called for thoughtful reallocation of the region's limited water resources. During this period, a concerted effort to finally resolve long-running disputes over Native American water rights led to intense negotiations among tribal communities, the federal government, and water users.

Perhaps the single most significant challenge to the Bureau of Reclamation's new role as a water resources management bureau came in confronting the power of the prior appropriation doctrine. This doctrine of "first in time, first in right" was a cornerstone of water resource development in the West. With most western rivers already over-appropriated, this dogma of western water law vexed Reclamation and reformers' efforts at every turn. Water transfers, Indian water settlements, and water for environmental mitigations openly challenged this guiding principle of western water law. Questions arose as to whether or not the federal government was superseding state law to implement programs to meet the water needs of the modern American West. Leading this effort for the Bureau of Reclamation during the early 1990s was Commissioner Daniel P. Beard, who sought to change the water resource policy of the United States to meet the West's demographic and cultural "aspirations."[396]

After Teton

The failure of Teton Dam was a defining moment for the Bureau of Reclamation. For many Reclamation personnel, morale fell in the aftermath of the tragedy. J. Neil Stessman, one of Reclamation's Teton claims managers, recalled that "people around the Bureau sort of felt their own guilt about it or their association with it." Stessman noted too that the Teton failure caused many within the ranks to have doubts about the future and question the mission of the Bureau of Reclamation. The Teton disaster also markedly affected the public's attitude toward Reclamation and its various endeavors. A reconsideration of the benefits occurred as well as a reshuffling of priorities concerning water resource development in the West.[397]

---

[396] Daniel P. Beard, "New Directions for the Bureau of Reclamation," in *Water Laws: Trends, Policies, and Practices*, Kathleen Marion Carr and James D. Crammond, editors (Chicago: ABA Section of Natural Resources, Energy, and Environmental Law, 1995), 357.
[397] Stessman, *Oral History Interview*, 104-5.

A storm of criticism came at Reclamation from all sides. Commissioner Gilbert Stamm referred to the Bureau of Reclamation's "traditional" opposition, primarily environmental groups who not only fought construction of Teton Dam in 1971, but who were also achieving greater success in slowing down initiation and construction of Reclamation projects. Engineering reports from the Teton Dam Failure Review Group and harsh criticism from former congressional allies raised ominous questions about the future of Reclamation activities. Congressional reports sharply criticized the Bureau of Reclamation, accusing it of being driven by "bureaucratic momentum" to take on questionable projects, though few in Congress willingly admitted their role in creating that momentum. Moreover, the engineering reports called into question the dependability of all Reclamation structures. One historian of Reclamation expressed a commonly held public attitude:

> The implications of the Teton Dam collapse were enormous. Many western streams had been plugged so many times that the collapse of one dam could take out a series of structures, producing massive floods, extensive damage, and thousands of deaths.[398]

The Teton Dam failure hurried the Bureau of Reclamation into a new era. Former Bureau of Reclamation Commissioner John W. Keys III recalled the trepidation and uncertainty that hung over Reclamation in the days and months following Teton. He claimed, "Teton was the low point … It told people that we were human, and that was pretty hard to handle for Reclamation. It changed the way we did business forever."[399]

Jimmy Carter's so-called hit list, coming so soon after the Teton Dam failure, was another blow to the Bureau of Reclamation's morale. The Carter presidency's avowed commitments were to environmental ideals and a determination to cut "pork" out of the federal budget. Cancelling scheduled dam construction in April 1977 was consistent with both policies. Selected to guide Reclamation through this transitory period was a new secretary of the interior, Cecil D. Andrus, and a new commissioner of the Bureau of Reclama-

---

398 Pisani, "Federal Reclamation and the American West in the Twentieth Century," 408.

399 Committee on Government Operations, *Teton Dam Disaster*, 32-3; Reisner, *Cadillac Desert*, 386;; John W. Keys III, *Oral History Interview*, Transcript of tape-recorded Bureau of Reclamation Oral History Interview conducted by Brit Allan Storey, senior historian, Bureau of Reclamation, from 1994 to 2006, in Denver, Colorado; Boise, Idaho; Washington, D.C. and Moab, Utah, 89.

tion, R. Keith Higginson. Both men through their political and professional careers had developed close working relationships with Reclamation: Andrus as Governor of Idaho, and Higginson as the Director of Idaho's Department of Water Resources. Each entered office with an entirely different view of the Teton disaster than Reclamation and they were responsible for implementing the administration's water policy directives. Higginson later recalled that after becoming commissioner he confronted efforts from both inside and outside the federal government to either reform the Bureau of Reclamation or eliminate it outright. He responded to these pressures by creating an internal review panel made up of Reclamation personnel to examine the Bureau's "entire design, construction, and maintenance process" and make recommendations for improvement. One of the review's significant recommendations asked that private design experts check Reclamation's designs before approval. Though Higginson later maintained that neither Andrus nor he came into office with the intent to alter Reclamation culture, others disagreed. John W. Keys III suggests that because of Teton Higginson entered Reclamation with a "chip on his shoulder," and "Keith Higginson and Cec Andrus never forgave us for that."[400]

Uncertainty prevailed in the Bureau of Reclamation as it acquainted itself with the new administration. The hit list announced that change was in the air with new policies demanding greater attention to environmental concerns and economic accountability. Reclamation's response remained uncertain. Significantly, 1977 also marked the 75th anniversary of the 1902 Reclamation Act. In its long-standing publication, *Reclamation Era*, the Bureau of Reclamation took time to look back on its history and ponder how the past could best serve an uncertain future. Similar to other anniversary issues, *Reclamation Era* focused on past achievements, celebrating Reclamation's heritage and historical legacy. Not surprisingly social groups or organizations routinely construct interpretations of the past that offer continuity from past to present and into the future. At the release of the anniversary issue, the future of the Bureau of Reclamation was still unclear, and by recalling past

---

400 For background information on Higginson prior to becoming Commissioner see R. Keith Higginson, *Oral History Interview*, Transcript of tape-recorded Bureau of Reclamation Oral History Interview conducted by Brit Allan Storey, senior historian, Bureau of Reclamation, March 22, 1995 and April 19, 1995 in Boise, Idaho, 3, 49; *Public Papers of the Presidents of the United States: Jimmy Carter, 1977*, Book I—January 20 to June 24, 1977 (Washington, D.C.: United States Government Printing Office, 1997), 565; for information on internal review panel see Higginson, *Oral History Interview*, 50-1; see also Keys, *Oral History Interviews*, 87, 89.

13.1. The 75th Anniversary Cover of *Reclamation Era.*

exploits the issue served as a rallying point to demonstrate Reclamation still had a purpose in serving the West's water needs.[401]

Two articles in the anniversary issue discussed Reclamation's historical legacy through documentation of cultural resources. Reclamation's participation in cultural resource management activities resulted largely from new laws and regulations: the National Historic Preservation Act of 1966, Executive Order 11593 "Protection and Enhancement of the Cultural Environment," the Historical and Archaeological Preservation Act of 1974, and the National Environmental Policy Act of 1969. Yet earlier during the 1950s and 1960s, Reclamation played a significant role in cultural resource investigations at the Glen Canyon damsite and on the Pick-Sloan Missouri Basin Program. In 1974 the Bureau of Reclamation developed a formal set of procedures and programs for cultural resources studies. On the Central Arizona Project (CAP) alone, Reclamation's Historic Preservation Officer, Ward Weakly, noted that the Bureau had spent almost $4 million on cultural resources, "with a potential expenditure in the range of $28 to 30 million."[402]

While CAP investigations centered on Native American and early "historic sites," the anniversary articles in *Reclamation Era* focused on Bureau of Reclamation structures and facilities. One article noted, "Many Reclamation projects are now considered part of our Nation's cultural resource legacy. They are recognized as significant structures in the field of industrial archaeology and listed on the *National Register of Historic Places*." The recognition predominantly included Reclamation Service structures such as Theodore Roosevelt Dam, Derby Dam, Arrowrock Dam, and Buffalo Bill Dam and project facilities on the Carlsbad Project and Boise Project. One author compared Reclamation structures—Theodore Roosevelt Dam, Buffalo Bill Dam, and Pathfinder Dam—as having historical relevance or importance similar to other national icons such as the Liberty Bell and Independence Hall. In light of the collapse of Teton Dam and the brewing controversy over the hit list, the 75th Anniversary Issue of *Reclamation Era* offered a morale boost to the Bureau

401 David Lowenthal, *Possessed by the Past: The Heritage Crusade and the Spoils of History* (New York: The Free Press, 1996), 128.

402 Ward F. Weakly, Historic Preservation Officer, Bureau of Reclamation, Denver, Colorado, "Dolores Project and Central Arizona Project," 4-5, RG 115, Binder: Special Lists, Vol. II, an Index of Cultural Resources Reports Located at the National Archives, ACC# 8NS-115-95-122, Box 2; "National Environmental Policy Act of 1969," Sec 101 (b) (4) in United States Department of the Interior, Bureau of Reclamation *Federal Reclamation and Related Laws*, Volume IV of Four Volumes, Louis D. Mauro and Richard K. Pelz, editors (Denver: United States Government Printing Office, 1989), 2493.

13.2. Theodore Roosevelt Dam after completion in 1911.

13.3. Derby Diversion Dam during the dedication ceremony on June 17, 1905.

13.4. Buffalo Bill Dam after completion in 1918.

of Reclamation. As one author put it, "Now after 75 years, public awareness and understanding of Reclamation's mission is still an important concept, and understanding always seems to be sharpened when placed in an historical perspective."[403]

13.5. R. Keith Higginson served as Commissioner of the Bureau of Reclamation from 1977 until 1981.

With Reclamation's historic legacy as a foundation, other articles spoke about new challenges facing the Bureau of Reclamation. Commissioner R. Keith Higginson outlined the water resources policy objectives of the Carter administration which emphasized "safety, environmental protection, economic efficiency, fair distribution of project benefits, and water conservation." Within these policy goals, however, the commissioner also hinted that Reclamation was embarking on a new era in its history and warned that those "who are not ready for change" could threaten its ability to succeed in these changing times. Higginson's article set the cornerstone for Reclamation's future as a water resources management agency. Other authors reaffirmed new directions by outlining Reclamation's efforts toward fish and wildlife enhancement and activities to expand its constituent base through public meetings. All represented the "new" Bureau of Reclamation that actively pursued environmental protection and listened to and responded to concerns of the public. Higginson summed up the changing role for Reclamation by declaring that "we may no longer be in a position of trying to reclaim public lands in the West. But we are still in a position of needing to develop our natural resources, including water."[404]

Congress pushed back against the hit list preventing the budget cuts Carter requested and reauthorized many of the projects on the list. Still some

403 Ward Weakly, "Reclamation's Roots Discovered through Archaeological Studies," *Reclamation Era*, 63:1 and 2 (1977): 84-5; Diane Gelburd, "Reclamation's Cultural Resource Legacy," *Reclamation Era*, 63:1 and 2 (1977): 78.

404 R. Keith Higginson, "New Direction for Reclamation," *Reclamation Era*, 63:1 and 2 (1977): 48-52; Carol Prochaska, "Reclamation's Concern for Fish and Wildlife," *Reclamation Era*, 63:1 and 2 (1977): 44-7; Darrell Adams, "Wanted Public Opinion," *Reclamation Era*, 63:1 and 2 (1977): 56-60.

projects were scaled back and some eliminated in accord with the president's policies. The Central Arizona Project faced redirection; Hooker and Charleston dams were removed, along with Orme Dam on the Fort McDowell Indian Reservation. Significantly, Orme Dam's elimination occurred because it was targeted by the administration and because of Indian opposition due to the dam flooding a large portion of the reservation. This was a great achievement for the Fort McDowell Indians, and the cancellation of Orme Dam is celebrated annually in November with "Orme Dam Victory Days" on the reservation. In California, construction of the controversial Auburn Dam was put on hold because of safety concerns. While the Central Utah Project eventually met President Carter's economic, environmental, and safety criteria, delays occurred. Cost concerns over its estimated billion dollar-plus price tag slowed congressional authorizations and appropriations, while water right disputes with the Ute Tribe hampered construction progress.[405]

The Carter administration's attempt to curb expensive water resources development programs was sporadic. As with other presidents, Carter was unable to rein in the spiraling costs of water projects, despite Congress's general inclination to reduce appropriation funding. In contrast to past administrations, Carter sought to remove some of the subsidies granted to water users. In a 1978 message to Congress, the president issued directives "to improve the implementation of irrigation repayment and water service contracting procedures" of the Bureau of Reclamation. He called for provisions "for recalculation and renegotiation of water rates every five years" in new water contracts, replacing the previous practice of 40-year contracts. Possibly spurred by the specter of faulty long-term contracts on the Central Valley Project, he also sought assurances that new contracts covered operation and maintenance costs, and asked Congress to "more precisely recalculate and implement the 'ability to pay' provision in existing law which governs recovery of a portion of project capital cost." Finally, the president asked Congress to establish a cost sharing plan to compel states to contribute more resources toward water project funding. In short, President Carter's proposed reforms in Reclamation policy

---

[405] For information on the dams removed from planning for the Central Arizona Project see "Plan of Study proposed by the Arizona Project Office, Water and Power Resources Service Assisted by Los Angeles District U.S. Army Corps of Engineers, Central Arizona Water control Study," January 1980, RG 115, Office of the Chief Engineer, Bureau of Reclamation Project Reports, Acc# 8NS-115-95-083, Box 69, Project Record 1910-1995; Earl Zarbin, *Let the Record Show ...: Gila River Indian Reservation Water Rights and the Central Arizona Project* (Tempe, Arizona: Earl Zarbin, 2004), 66; McCool, *Native Waters*, 111; for information on construction delays on the Central Utah Project see USDOI, BR, Upper Colorado Region, *Central Utah Project, Bonneville Unit: Municipal and Industrial System*, 2, 5, 28-9.

were attempts to lessen the burden of water projects on the general taxpayer by passing more economic responsibility on to water users. Some of these compromises and controversies foreshadowed issues and challenges the Bureau of Reclamation encountered as it moved ever closer to becoming a water resource management bureau as opposed to a construction oriented bureau.[406]

By the end of the Carter administration in 1981, there was little doubt that the Bureau of Reclamation was in the process of once again transforming itself to meet changing times and attitudes. Teton Dam's collapse and the hit list played crucial roles in bringing about that transition. In 1979 Reclamation initiated a revised and strengthened dam safety program designed to ensure the structural integrity of its dams. This program included stronger review measures during the design process "for evaluating the risks and probability of failure of proposed dams" and "stronger emergency preparedness plans." Reclamation engineers worked more closely with outside engineering firms throughout the entire construction process, continually reviewing geological conditions and engineering practices. Reclamation's changing mission appeared in a new name in 1979: the Water and Power Resources Service. Commissioner Higginson defended the new name as reflecting Reclamation's response to changing times, which included "national pressures … concerns about energy, and a growing awareness of what in the near future will be a water crisis, and changing government roles in resource management."[407]

The "Water and Power Resources Service" title was short-lived and led to no long-term organizational or program changes. Shortly after Ronald Reagan became president, new Secretary of the Interior James Watt reversed the decision of his predecessor and resurrected the name Bureau of Reclamation. In his May 20, 1981, press release announcing the title change, Secretary Watt asserted,

> The name Bureau of Reclamation is one of historical significance as well as a symbol of excellence. Changing the name to Water and Power Resources Service was a mistake. The public we serve did not like it, nor did the employees who loyally worked for it.

In retrospect the name change was of little significance, but it represented the turmoil and uncertainty the Bureau of Reclamation experienced after the

---

406 Federal Water Policy, Message to the Congress, June 6, 1979, in *Public Papers of the Presidents of the United States: Jimmy Carter, 1978*, Book 1—January 1 to June 30, 1978, 1045-50.
407 "New Name Change for Bureau of Reclamation," *Reclamation Era*, 65:2 (1979): 1; "Interior Agency Changes Name," *New York Times*, November 8, 1979.

Teton Dam failure and during the hit list debate. While ongoing construction projects continued, new authorizations were difficult to obtain. Uncertainty prevailed as the Bureau of Reclamation entered the 1980s at a time when lingering questions from the past came to the fore. The 160 acre limitation and the viability of the family farm became issues for Reclamation policy makers as well as Congress.[408]

160 acre Limitation and the Family Farm

Ronald Reagan's election in 1980 revived hopes in western water circles for renewed support for water resource development. Many westerners perceived Reagan as one of their own. Despite increasing costs, environmental restrictions, and the questionable economic benefits that Carter's hit list revealed, western boosters stood by the standard doctrine that water projects advanced western prosperity. Indeed, Secretary of the Interior James Watt's order to reinstate the name of the Bureau of Reclamation prompted expectations that the new Reagan administration meant a return to business as usual. Furthermore, many western congressmen remained politically bound to water projects in their states. Representatives from Arizona and Utah fought hard for continued funding for the Central Arizona Project and the Central Utah Project. Similar struggles continued throughout the West as congressmen vigorously supported smaller projects that remained popular with their constituents. Yet because of Teton and the hit list, there was little support for water projects outside a congressman's or senator's particular district. After Reagan's election in 1980, environmental concerns along with rising costs further stymied support for new development.

If supporters of western water projects believed the new administration stood ready to reverse Carter administration policies, they were sorely mistaken. Ronald Reagan entered the presidency on the promise of reducing the size and influence of the federal government. Renewed interest in water resource development was simply not part of the administration's plans. As some observers pointed out, "Ronald Reagan proved as uninterested in more water development as Carter was opposed to it, and Reagan in fact achieved only what Carter had sought." The budget-conscious administration never actually pursued a proactive water policy. Rather it sought the basic reforms

---

408 Bureau of Reclamation, Information Public Affairs Center Press Release, "Secretary of the Interior, James Watt, announced the name of the Water and Power Resources Service back to Bureau of Reclamation," May 20, 1981, in "Carter Hit List;" see also "The Bureau is the Bureau ... Again," *Reclamation Era*, 66:2 and 3 (Spring/Summer 1981): 1.

initiated by the previous president, most significantly, in the area of state/federal cost sharing.[409]

By 1980 the West itself had changed. Urbanization radically altered the political and cultural climate. Urbanites, with decidedly different views on what constituted beneficial use of western water, now exercised power. In addition, science and technology played greater roles in farming, fostering the growth of corporate agribusiness and rearranging the economics of agriculture. The transformations challenged the ideal of the family farm and raised questions about who benefits from western water projects. Still the Bureau of Reclamation continued to champion the family farm ideal in its rhetoric. Over the years, Reclamation's defense of the family farm ideal, at least in its upper echelons, succeeded in keeping alive the 160 acre limitation on the Central Valley Project in California. Yet, under Commissioner Michael Straus, the Bureau of Reclamation gave ground—yielding to the "technical compliance" standard on the 160 acre rule. By the 1980s, however, economies of scale and industrial agriculture further undermined the ideal of the family farm, especially in California's Central Valley.[410]

Throughout much of its history, the Bureau of Reclamation offered various exceptions to the 160 acre limitation rule. For example, a husband and wife could each own 160 acres and add another 160 acres for every child; farmers or corporations could own 160 acres in each Reclamation project; a "joint venture" could own 160 acres per individual member; or a farmer could add to acreage through leasing. These illustrations of "technical compliance" seriously damaged the family farm ideal, lessening its importance in Reclamation policy and eroding the utilitarian foundation (greatest good for the greatest number) of Reclamation law. Furthermore, departmental and congressional removal of the acreage limitation requirement on specified projects, i.e., Colorado-Big Thompson Project, the Imperial Valley, and the Truckee-Carson Project, created exceptions within Reclamation law that led to more confusion. These inconsistencies in policy caused some to call for reforms in Reclamation policy to at least remove the impression that Reclamation projects favored large landowners.[411]

---

409 Reisner, and Bates, *Overtapped Oasis*, 197.

410 Pisani, *Water and American Government*, 2; Rowley, *The Bureau of Reclamation*, 47-8; Hogan, *The Acreage Limitation in the Federal Reclamation Program*, 5.

411 National Water Commission, *Water Policies for the Future: Final Report to the President and the Congress of the United States by the National Water Commission* (Washington, D.C.: United States Government Printing Office, June 1973), 144; United States Senate, Select Committee on Small Business, *Will the Family Farm Survive in America?: Federal Reclamation*

Both critics and supporters of Reclamation policy recognized that loopholes in Reclamation law attacked the basic premise of small family farms espoused in the 1902 Reclamation Act. For some people acreage limitation provisions in the 1902 law fell short of intended purposes: to dot the landscape with 160 acre family sized farms and prevent land monopoly and speculation. Others argued that a 160 acre farm was impractical and unsustainable in an era of mechanized agriculture. Both suggested the need existed for a total revamping of the law. One advocate for reform was the 1972 National Water Commission publication *The Acreage Limitation in the Federal Reclamation Program* by Harry J. Hogan. Hogan began his examination arguing that the fundamental values that drove the Reclamation program were centered in the "agrarian myth." According to Hogan, Reclamation policy was part of the continuing process of nation-building, which moved toward "full application of technology in the development of natural and human resources." Hogan declared the idea never considered market influences and that Reclamation policy acted from an ethic of social planning now out of step with modern trends. Historian Donald Pisani states that in 1902 "no one understood how far the federal government would go beyond building dams and canals; many expected it to engage in large-scale social planning."[412]

Hogan perceived this conflict between the "agrarian myth" ideal and market considerations as a prelude to trouble for Reclamation policy. He observed that the "social argument for irrigation investment" had lost much of its relevancy over time. He also noted the enthusiasm for subsidized irrigation to support western settlement and small farms had waned. For Hogan,

---

Policy, S. Rpt. No. 95-702, 95th Cong., 2nd sess. (Washington, D.C.: United States Government Printing Office, 1978), 9; for information on Secretary of the Interior Ray Lyman Wilbur's letter exempting the land limitation requirement in the Imperial Valley in 1933 see Norris Hundley, Jr., *The Great Thirst: Californians and Water, A History,* Revised Edition (Berkeley: University of California Press, 2001), 225-6; Rowley, *The Bureau of Reclamation,* 349; Worster, *Rivers of Empire,* 212; for information on lifting the restriction on the Nevada projects in 1940 see Gerald R. Ogden, compiler, *The Excess Lands Provisions of Federal Reclamation Law: A Bibliography and Chronology,* Agricultural History Center University of California, Davis, July 1980, transcript, 94; Alfred R. Golzé, *Reclamation in the United States* (New York: McGraw-Hill Book Co., Inc, 1952), 67; Rowley, *The Bureau of Reclamation,* 346 for information on lifting the restriction on the Colorado-Big Thompson Project in 1938 see Dan Tyler, *The Last Water Hole in the West: The Colorado-Big Thompson Project and the Northern Colorado Water Conservancy District* (Boulder: University Press of Colorado, 1992), 94; Rowley, The Bureau of Reclamation, 29; Golzé, *Reclamation in the United States,* 67; Ogden, *The Excess Lands Provisions of Federal Reclamation Law,* 93.

412 Hogan, *The Acreage Limitation in the Federal Reclamation Program*, 5; Pisani, *Water and American Government*, 1.

the transformation toward a market economy demanded an end to subsidies on water for western irrigators. He concluded:

> To the extent that those subsidies have been justified by the Agrarian Myth, they should be discontinued both because, if it did exist, there is no evidence that it produces the desired civic virtue. To the extent that those subsidies have been justified by the desire to obtain full development of natural resources, they should be discontinued. The Nation has now reached a stage of maturity in which the West is no longer in need of such a development program and additional agricultural production is of less concern than other uses for the same water resources.

In short, social planning, if it worked at all, was no longer desirable, and the water requirements for the settled West called for a reconfiguration of Reclamation policy.

The underlying argument called for an end to the federal government's investment in irrigation facilities in the West. The 160 acre limitation for the family farm required subsidies to succeed, and minus such should be forced to compete in an open market, freeing water resources to a wider assortment of uses. Hogan claimed,

> The essential aspect of acreage limitation is that in project planning it elevated irrigation into a primary position because the family farm as a national purpose was primary. It came ahead of hydroelectric development, municipal and industrial uses, recreational uses and all other uses.

In essence, Hogan surmised that competition without the benefit of federal subsidies would force some farmers out of business, thereby releasing water resources to uses that truly reflected the needs of an urbanized American West. Nevertheless, he realized that eliminating the acreage limitation faced two competing interests which clung to the family farm ideal. Large landholders found the family farm concept a useful rhetorical tool because "it justifies Federal investment in programs to provide cheap water." On the other hand, "The family farm ideal distracts and confuses conservationist sentiment—basically liberal, humanistic, and anti-corporation—from analysis of environmental problems."[413]

---

[413] Hogan, *The Acreage Limitation in the Federal Reclamation Program*, 21-2, 293-6.

13.6. Harry Hogan prepared *The Acreage Limitation in the Federal Reclamation Program* for the National Water Commission in 1972.

Although abuses and lack of enforcement of the 160 acre rule were widespread, the most intense debate occurred about California's Central Valley Project. California assumed battleground status on acreage limitation in the early 1950s when Commissioner Michael Straus invoked the notion of "technical compliance." But enforcement of the 160 acre rule was even more difficult because of earlier addendums made to Reclamation law. For instance, the 1926 Omnibus Adjustment Act permitted delivery of water to landowners of more than 160 acres, "if the owner executed a recordable contract with the Department of the Interior in which he promised to sell the excess lands in ten years." This provision allowed farmers with excess lands to benefit from Reclamation irrigation works without penalty. In addition, there was no procedure that checked to determine the relationship between the buyer and seller of excess lands. Furthermore, Reclamation law was ambiguous on the subject of leasing of project lands, or on the distribution of project benefits to leased lands. Past practice of the Bureau of Reclamation permitted individuals, joint ventures, and corporations to lease 160 acre tracts and still receive the "benefit from the subsidy." In both cases, it was entirely possible for a single entity to operate multiple farms of 160 acres and receive the same rewards as an individual owning 160 acres or less.[414]

During the 1970s, criticism of the Bureau of Reclamation's lax enforcement and generous application of the 160 acre rule in California increased. Critics charged that the method Reclamation used to approve excess land sales contributed to expansion of agribusiness on the Central Valley Project at the expense of those wishing to purchase family farms. In 1976 a California organization called National Land for People sued the Bureau for failing "to govern its operations by published regulations rather than … solicitor's opinion." In particular, National Land for People objected to the practice that permitted excess land sales to go to relatives or business associates of sellers. Consequently Reclamation permitted landholdings much greater than 160 acres to receive subsidized water. The objective of the suit was to make purchasing a family farm easier for those who desired one by publicizing the process by which Reclamation approved excess sales. The suit also revealed the distribution of land ownership in the Central Valley, of which Reclamation had scant records. On August 9, 1976, the U.S District Court ruled in favor of National Land for People stating that the Bureau of Reclamation failed to comply with the Administrative Procedures Act, and enjoined

---

414 Ray G. Huffaker and B. Delworth Gardener, "The Distributions of Economic Rents Arising from Subsidized Water When Land is Leased," *American Journal of Agricultural Economics* 68:2 (May 1986): 306; Charlotte Robertson, "*Peterson v. Department of the Interior*: Are Contract Rights Ever Property Rights Under the Reclamation Reform Act?" *Golden Gate University Law Review* 21:1 (Spring 1991): 183-4; Alexandra M. Shafer, "The Reclamation Reform Act of 1982: Reform or Replacement," *University of Pittsburgh Law Review* 45:3 (Spring 1984): 653-4.

Reclamation from permitting any further land sales until it published the required rules and regulations.[415]

Reclamation stood at fault for not adhering to the spirit of the 1902 Reclamation Act. Critics raised the larger issue of whether or not corporate agriculture should benefit from the taxpayers' largess in the form of cheap water rates. One provision of Reclamation law, which critics utilized to support their argument about Reclamation catering to large landowners, was the residency requirement of the 1902 Act. The Act mandated that eligible water recipients must be "bona fide residents upon the land or within the neighborhood." In 1909 the Department of the Interior defined the term "in the neighborhood" to mean living "within 50 miles" of the irrigated farm. Yet, as with many aspects of Reclamation law, critics charged that the Bureau of Reclamation "virtually ignored" this requirement and thus, "hastened the trend toward huge farms such as those in central California."[416]

Inconsistencies and contradictions in Reclamation policy persisted. Reclamation had done little to address problems. When Jimmy Carter assumed the presidency in 1977, proponents of changes in the nation's water resource development program had high hopes for the new administration. Carter's record and rhetoric on environmental issues along with the quick release of the hit list were positive signs. During the Carter administration, the Bureau of Reclamation responded by tackling the acreage limitation issue especially in California's Central Valley Project. According to one Senate report investigating survival of the family farm, "Secretary Andrus has indicated publicly that he favored a return to the family farm practice in Reclamation projects, and would actively take steps to implement this policy." Nevertheless, past practices blocked Andrus's promises, and, much like the hit list, it made a lot of noise but accomplished little in the way of reform.[417]

By the 1970s a consensus emerged: it was time to address to the discrepancies in Reclamation law regarding acreage limitation. Both the need

---

415 U.S. Senate, Select Committee on Small Business, *Will the Family Farm Survive in America?*, 19; Robertson, "*Peterson v. Department of the Interior*", 185; Shafer, "The Reclamation Reform Act of 1982." 653-4.

416 Shafer, "The Reclamation Reform Act of 1982," 659-60; U.S. Senate, Select Committee on Small Business, *Will the Family Farm Survive in America?*, 12; "Reclamation Act of 1902," in USDOI, BR, *Federal Reclamation and Related Laws*, Volume I of Three through 1942, 67.

417 U.S. Senate, Select Committee on Small Business, *Will the Family Farm Survive in America?*, 19; Gladwin Hill, "U.S. Proposes Protecting Family-Sized Farms by Enforcing Acreage Limits in 1902 Act," *New York Times*, February 2, 1977; Seth S. King, "U.S. to Enforce Irrigation Land Rule," *New York Times*, November 11, 1977.

to change the law to conform to modern agricultural practices and to remove some of the loopholes in subsidy benefits pushed reform efforts. With the federal government hemorrhaging money on some Reclamation projects, especially the Central Valley Project, reconsideration of water subsidies became a *cause célèbre*. A 1978 article from the *New York Times* stated, "An extensive audit of the Central Valley Project ... by the Department of the Interior disclosed ... that unless rates the project now charges for Federally supplied electricity and irrigation water are raised, it will incur a deficit of nearly $8 billion in the next twenty years." This issue stemmed from the facts that the forty-year contracts were not only interest free but also reflected prices set in the late 1940s and early 1950s and contained no provisions to adjust for inflation in operations costs.[418]

In 1979 Congress began the painstaking process of reforming Reclamation law. The first attempt, the proposed Reclamation Reform Act of 1979, sought "to provide a modern statement of congressional policy on several aspects of Reclamation law, to resolve ambiguities, and to conform the law to the current practical considerations of farm practices and economics." This legislation recommended expansion of acreage limitation to 1,280 acres, recognized the right of joint ventures of 25 individuals or more to receive project water, and eliminated the residency requirement. For all intents and purposes, the 1979 legislation took direct aim at the family farm ideal, recognizing it as irrelevant in modern agriculture. The Department of the Interior frowned on these changes, especially the move to lift residency requirements. "The Department considers a strong residency requirement to be the best means of assuring the owner-operated farms that are envisioned by the reclamation program." To a certain extent, this opinion echoed Bureau of Reclamation values associated with the family farm ideology.[419]

By 1982 Congress came closer to an agreement on Reclamation law reforms, particularly on eliminating the residency requirement and allowing delivery of subsidized water to joint ventures. Major controversies still remained about the acreage amount eligible for subsidized project water and the issue of leasing. Earlier proposals suggested an acreage limitation of 1,280 acres "for a qualified recipient and 640 acres to a limited recipient"—

---

418 Seth S. King, "Central Valley Data Point to a Deficit; Interior Department says Utility Faces an $8 Billion Loss in Next 20 Years Unless Rates Increase," *New York Times*, February 2, 1978; Richard W. Wahl, *Markets for Federal Water: Subsidies, Property Rights, and the Bureau of Reclamation* (Washington, D.C.: Resources for the Future, 1989), 47-53.

419 United States Senate, Committee on Energy and Natural Resources, *Reclamation Reform Act of 1979*, S. Rpt. No. 96-235, 96th Cong., 1st sess., June 21, 1979, 1-2, 20-1.

any legal entity benefitting more than 25 individuals. A qualified recipient could also lease up to 800 acres and still receive water at the subsidized rate, bringing the total acreage allowed to 2,080 acres. However, the legislation stipulated that anyone leasing 801 acres or more would pay "full cost" for irrigation water. This particular proposal was an admission that Reclamation projects gave inordinate benefits to large landowners with the new legislation offering a provision to close that loophole. Nonetheless, some members of Congress objected to the generous acreage limitation in the legislation. For some the 2,080 acres defeated the purpose of the Reclamation program—to provide benefits to the greatest number of recipients possible. Senator Mark Hatfield from Oregon argued, "A less liberal acreage would assure the preservation of the 'small farm' concept and prevent potential abuse and exploitation of a lucrative federal subsidy." In a similar manner, Washington Senator Henry "Scoop" Jackson also felt that 2,080 acres was far too generous, and "that it is important that we do not abandon the policy of broad distribution."[420]

On October 12, 1982, after some three years of negotiations in Congress, President Reagan signed into law the Reclamation Reform Act of 1982. Reagan proclaimed, "While preserving the basic objectives of the original program, this legislation provides a new direction for the federal role in Reclamation—one that will, I believe, prove to be a significant step forward on our road to economic recovery in the 1980's [*sic*]." In essence, the Reclamation Reform Act adjusted Reclamation law to the realities of modern agricultural economies. It raised the individual acreage limitation for receiving federal irrigation water at the non-full cost rate to 960 acres for individuals and legal entities benefiting twenty-five or fewer persons, while establishing a smaller "entitlement" for legal entities benefiting more than twenty-five persons. It also removed the residency requirement and allowed owners of excess acreage, who had already placed their own land in a recordable contract, a 10-year grace period to sell those lands, but in the meantime, still receive project water. The Act addressed discrepancies in water delivery contracts to ensure that eventually the price for irrigation water "shall be at least sufficient to recover all operation and maintenance charges which the district is obligated to pay ... " Provisions in the law also allowed those who wished to remain under the old law the ability to do so, with the provision that after April 12, 1987, leased lands in excess of 160 acres must pay the full cost rate. Finally, the leasing restriction with regard to the pric-

[420] United States Senate, Committee on Energy and Natural Resources, *Reclamation Reform Act of 1982*, S. Rpt. No. 97-373, 97th Cong., 2nd sess., April 29, 1982, 7, 14, 32, 36.

ing of water also applied to those lessees who became subject to the RRA at set acreage levels, thereby closing the leasing loophole.[421]

As one *Reclamation Era* article exclaimed, the Reclamation Reform Act of 1982 was "a quantum leap forward," opening "a new chapter in the Reclamation story in the West." It reflected the changing *milieu* of the American West, as homesteading and family farms gave way to great metropolises, suburban sprawl, and large-scale agribusiness. While the Act "closed the leasing loophole," some believed that the Reclamation Reform Act eliminated a vital component of the family farm ideal by removing the residency requirement. For some critics the loss of the family farm ideal was the most grievous product of the law. In an article for the *University of Pittsburgh Law Review*, Alexandra M. Shafer argued, "In that abandonment the foundation of the family farm is discarded and with it is discarded the original purpose of the 1902 Act–anti-monopolization." When Congress passed the Reclamation Reform Act, few considered whether or not raising the acreage limitation would free up water for other purposes, as proposed by Harry Hogan in 1972. But one thing was certain: during the 1980s and 1990s, new demands for water were emerging and threatening irrigation's hegemony. These new considerations and constituents speeded the Bureau of Reclamation's transition from a construction agency to a water management agency.[422]

The Bureau of Reclamation faced the daunting task of implementing the reforms stipulated in the Reclamation Reform Act of 1982. Much of this effort went into formulating the rules and regulations of the new law. Water users also required time to bring their operations into compliance with the Act, primarily the divesture of excess lands. During the same period, the Reagan administration attempted to execute other policy changes to help defer some of the federal government's financial burden. Similar to his predecessor, Jimmy Carter, Reagan's Department of the Interior sought to "impose local and state cost sharing on any new project." The efforts encountered overwhelming challenges. Water users holding long-term contracts were not inclined to alter their already beneficial situation, and still powerful western water resource development alliances resisted. A 1985 *Los Angeles Times* article neatly summed up

---

421 Mitchell Snow, "A Quantum Leap Forward: The Reclamation Reform Act of 1982," *Reclamation Era*, 67:4 (1982): 21; USDOI, BR, *Federal Reclamation and Related Laws Annotated*, Volume IV 1967-1982, 3332-448; United States Department of the Interior, Bureau of Reclamation, *Federal Reclamation and Related Laws Annotated: Reclamation Reform Act Compilation 1982-1988*, Paul B. Smith, editor (Denver, Colorado: United States Government Printing Office, 1988), 5-6; Wahl, *Markets for Federal Water*, 80-6.

422 Shafer, "The Reclamation Reform Act of 1982," 665-6.

the multiple issues Reclamation encountered in its efforts to modify irrigation culture on the Central Valley Project.

> The federal government provides water to wealthy San Joaquin Valley farmers to grow subsidized products that contribute to crop surpluses. Government buys high-cost private power to pump the subsidized water to the subsidized farms to grow subsidized crops. Contaminated irrigation runoff threatens environmental destruction and requires a costly clean-up program.

Of all of these issues, the Reclamation Reform Act addressed only subsidized water, but it was a start in the process of rethinking water usage in the West.[423]

Named to lead the Bureau of Reclamation through this period was C. Dale Duvall. Duvall, an accountant by trade and from the state of Washington, was a western political advisor for the Reagan-Bush campaign in 1980. As a political appointee, Duvall had no obvious connection with the Bureau of Reclamation or experience with water resources development in the West. He was also the first commissioner to face Senate confirmation, under the new requirements of the Reclamation Reform Act of 1982. Duvall acknowledged that the "era of multiple billion dollar projects … probably is past." He foresaw that urbanization of the West must result in the movement of water toward municipal and industrial uses and away from agriculture, and that environmental concerns would affect future water usage. Duvall also stressed the need to maintain Reclamation facilities and argued against the tendency to transfer funds away from O&M to new construction projects. These beliefs expressed by the new commissioner led some to suppose that Duvall might "endear" himself to water project critics and environmental organizations. Yet there remained a wait-and-see attitude concerning the new commissioner. One observer wrote that "many of those critics write off the statements to naiveté, contending that the commissioner is a novice in the byzantine world of water politics." Others were not so generous claiming "He has no political acumen, and he knows nothing about water policy. The bureau is rudderless."[424]

---

423 "Water in the Pork Barrel," *Los Angeles Times*, August 30, 1985. In 1983 federal agencies began working to correct water contamination problems with drainage from the San Luis Unit of the Central Valley Project flowing into Kesterson National Wildlife Refuge in the San Joaquin Valley. Toxic levels of selenium were entrained in the water flowing into the refuge.

424 "Which Way for Reclamation," *Los Angeles Times*, November 25, 1985; Cass Peterson, "C. Dale Duvall; In the Welter of Water Politics at Bureau of Reclamation," *Washington Post*, February 2, 1987; C. Dale Duvall, *Oral History Interview*, Transcript of tape-recorded Bureau of Reclamation Oral History Interview conducted by Brit Allan Storey, senior historian, Bureau of Reclamation, January 23, 1993, in Washington, D.C., 28.

Duvall took up the opportunity to show the direction for the Bureau of Reclamation under his leadership when Reclamation published the revised rules and regulations concerning acreage limitation stipulated in the Reclamation Reform Act of 1982. It soon became apparent that the commissioner was unwilling to move toward what many Reclamation Reform Act (RRA) proponents had hoped in the breaking-up of large landholdings. Duval stated, "I don't think it is incumbent on the Bureau of Reclamation to demonstrate more courage in limiting operations than Congress did. We intend to carry out the intent of Congress, but we are not taking it on ourselves to crusade." In other words, the commissioner was reluctant to upset the status quo. When the Bureau of Reclamation released the revised RRA regulations in April 1987, there was an instantaneous uproar decrying the policy as a "double cross." According to a *New York Times* article, the issue was the "farm management arrangement," whereby "farmers could operate an unlimited number of small farms receiving subsidized water as long as the farms were legally owned by others." Earlier regulations written by Reclamation in November 1986 did not approve what the Department of the Interior termed "imaginative management arrangements." Duvall defended his decision to change the regulation as reflecting Reclamation's "commitment to uphold the law and provide a reasonable and realistic regulatory framework to govern the reclamation program throughout the West."[425]

For critics of the new regulations, the controversy represented the Bureau of Reclamation's historic commitment to its more powerful constituents and was a prime example of the adage that "water flows to money." California Congressman George Miller charged, "The Secretary of the Interior has taken a dive for no other reason than to pander to a couple of hundred growers who think they have a God-given right to Federal money." To be sure, the larger landowners saw the controversy in a completely different light, arguing that the rules "were an intrusion into their right to make their own decisions on how to run their own farms." The issue of course was not the family farm ideal, but rather that multiple land holdings of up to 960 acres, though owned individually, being operated as a single unit and still receiving subsidized water. Yet the concept of the family farm entered into the discussion by way of reflecting the economic realities of farming in the late twentieth century where large landholdings were a requirement for success. Duvall accepted these "farm management arrangements" because he did not view a family holding—

425 Philip Shabecoff, "U.S. Issues Rules on Water for the West," *New York Times*, April 10, 1987; Cass Peterson, "Interior Dept. Reverses Water Rules; Farmers Could Avoid Acreage Limits on Subsidized Water," *Washington Post*, April 10, 1987.

consisting of an extended family of uncles, aunts, grandparents, etc.—of more than 960 acres as contrary to the family farm ideal. Reclamation policy prior to passage of the RRA gave quiet acquiescence to the formation and continuation of large landholdings, establishing the precedent of not interfering with the rights of property owners. Duvall's desire not to get the Bureau of Reclamation involved in a "crusade" to force the dissolving of huge farm operations mirrored this conviction. In 1993, the former commissioner later explained his thinking on the RRA and its impact on the family farm:

> And then the Government comes along and passes a law that says we don't care about this funny ownership and equity arrangement that you had in order to farm 4,500 acres as you've done in the past. Under this new law you've got to change everything around, you've got to break up your units.... Those things coming down on traditional law-abiding family units from a bureaucratic Federal agency are dramatic. And I don't blame the agricultural community for being up in arms.

This quote helps to explain why Duvall reversed Reclamation's earlier rules prohibiting "imaginative management arrangements," believing that many large farming units were indeed family farms.[426] Duvall was perhaps not alone in his contention that a 4,500 acre farm was not particularly at odds with the family farm ideal. Yet the debate about how to regulate the new law was just one of many discussions occurring simultaneously on how best to utilize the West's limited water supply.

Water Transfers: From Farms to Cities

In 1987 the Bureau of Reclamation delivered 29.9 million acre-feet of water: "25.5 million for irrigation, 3.2 million for municipal and industrial use, and 1.1 million for other non-agricultural use." Throughout the West, anywhere from 85 to 90 percent of all water went to agriculture. Demographic, economic, and social trends dictated that new avenues for reallocating water in the West demanded investigation. With its vast plumbing apparatus in the West, the Bureau of Reclamation possessed the expertise and means to implement these changes.[427]

---

426 Shabecoff, "U.S. Issues Rules on Water for the West," *New York Times*, April 10, 1987; Duvall, *Oral History Interview*, 16-7.

427 Reisner and Bates, *Overtapped Oasis*, 27; Committee on the Future of Irrigation in the Face of Competing Demands, Water Science Technology Board, Commission on Geoscience,

That same year the Bureau of Reclamation announced that because of budget deficits and shrinking resources "the era of constructing large federally financed water projects is drawing to a close." Reclamation made this announcement in a report entitled *Assessment '87 ... A New Direction for the Bureau of Reclamation* that outlined the Bureau's future goals. A team was put together to focus on ways the Bureau of Reclamation could best serve the public under these changing conditions. The "Assessment Team" looked at "three primary areas of opportunity: improve water operations, improve power operations, and partnership potentials for future structural and nonstructural projects." The report marked a dramatic departure for Reclamation by recognizing that the construction era had ended, and that Reclamation needed to discover new and innovative methods of conserving and better utilizing water in the West. The *Assessment* concluded, "the Bureau's mission must change from one based on federally supported construction to one based on effective and environmentally sensitive resource management."[428]

According to the report, the Bureau of Reclamation would accomplish its transformation by using its already established engineering and technical resources in a new manner to better operate existing facilities. In addition, proposed new projects must be drastically smaller in scale in comparison to past projects and rely more upon funding from "non-Federal finances." Reclamation planned to work more closely with other federal, state, and local agencies on myriad water resource problems, especially those concerning environmental protection and water conservation. One criticism leveled against Reclamation involved its narrow commitment to construction. Once it completed construction, the Bureau moved on to the next project. The critique asserted that Reclamation neglected its responsibilities to water users after project completion. The report proposed that the Bureau of Reclamation become more managerial and involved in the day-to-day operations of water projects, enforcing federal regulations, and asserting a visible presence. The end result pointed to an expansion of Bureau of Reclamation responsibilities in western water resource affairs.

Environment and Resources, National Research Council, *A New Era for Irrigation* (Washington, D.C.: National Academy Press, 1996), 48.

428 United States Department of the Interior, Bureau of Reclamation, *Assessment '87 ... A New Direction for the Bureau of Reclamation*, 1987, i-ii; see also Reisner and Bates, *Overtapped Oasis*, 26; Committee on the Future of Irrigation in the Face of Competing Demands, et al., *A New Era for Irrigation*, 111; Lawrence J. MacDonald, *Facilitating Voluntary Transfers of Bureau of Reclamation-Supplied Water*, Volume I (Boulder, Colorado: Natural Resources Law Center, July 1991), 99.

Reclamation's *Assessment '87* suggested that the new role for the Bureau of Reclamation meant reaching out to new constituencies and embracing development of other water uses including urban and environmental water needs. New roles did not mean abandoning the traditional agricultural customers. In assuming the managerial role, the Bureau of Reclamation had to reassure "agricultural interests" that it still considered them important partners in water resources development programs. Despite these assurances, the report's tone indicated that the pursuit of agricultural development through irrigation works no longer held a high priority for Reclamation. The report suggested an overhaul of Reclamation law and policies to correspond to changing water needs of the West. The report stressed the need for congressional and policy reviews on basic "principles" such as the "ability to pay for water, restrictions on ownership and alienation of projects, and less than market interest rates on repayment."

The 1987 report portended a bold departure from past practices as environmental protection and water conservation moved up in importance in Reclamation's mission. It reflected the Bureau's recognition of changing national attitudes concerning these matters. Nonetheless, the document steered Reclamation toward new challenges, especially western water transfers. Rising demands for water transfers revealed that the new western *milieu* required innovative rethinking about water allocation. The report asked whether the policy to supply subsidized water for agriculture was the most "beneficial use" of the resource, or should market considerations apply in water allocation, warning once again that the reforms necessary to implement such a program called for amending Reclamation law. It stated, "From a legal perspective, two of the most obvious constraints are project service areas and authorized project purposes. From a policy perspective, there is little or no policy on permitting transfers from project water users to other users."[429]

By 1989 the Bureau of Reclamation made tremendous strides in achieving some of the goals laid out in *Assessment '87*. Marc Reisner and Sarah Bates noted that Reclamation's 1989 budget request "proposed eleven new planning studies to increase the efficiency of existing projects, protect groundwater quality, and look into nonstructural water development and management alternatives." They also observed that Reclamation sought closer collaboration with other federal, state, and local agencies on drought management, while "experimenting" with methods to line earthen canals without draining them. It had also taken an active role in environmental issues such

---

[429] USDOI, BR, *Assessment '87*, 2, 4-5.

as restoring fisheries and addressing salinity and drainage problems. Yet on the issue of water transfers, the Bureau of Reclamation made little progress. During the western droughts of the late 1980s, Reclamation managed short-term water transfers, but these efforts afforded little opportunity to develop permanent policies.[430]

Water transfers, while an obvious need existed, were a relatively novel concept. Water use transfers presenting challenges to both buyers and sellers raised questions about what constituted equitable and fair compensation, especially when considering subsidized project water. During the western droughts of the 1980s the Bureau of Reclamation executed water transfers through authority it received from the Emergency Drought Relief Measures enacted in April 1977. While the Emergency Act provided a model, it also revealed some of the difficulties involved in implementing a water transfer program. A fundamental question concerned the price buyers paid for the water and the profits eligible for the seller. On this item, the 1977 law stated, "Payments for water acquired from willing sellers will be at a negotiated price, but will not confer any undue benefit or profit to any person or persons compared to what would have been realized if the water had been used in the normal irrigation of crops." Congress apparently meant to insure that sellers did not exploit their advantage to the detriment of the buyer. Yet this provision introduced a number of difficult questions including the rights of a seller to make a profit in a market economy; how much profit should a water user, willing to transfer water to other uses, earn when that water is already subsidized. According to Lawrence J. MacDonald from the Natural Resources Law Center,

> The U.S. built reclamation facilities using general tax revenues. The direct beneficiaries of these facilities, especially irrigators, have returned only a fraction of the real cost of these facilities to the U.S. Treasury. There is understandable concern about allowing those who have enjoyed substantial benefits from these facilities to further benefit from transfer of water the facilities provide.

This issue presented a significant impediment to a well-rounded water transfer policy. The procedure meant restrictions on an open market for water by limit-

---

[430] Reisner and Bates, *Overtapped Oasis*, 59-60; Wahl, *Markets for Federal Water*, 137; USDOI, BR, *Federal Reclamation and Related Laws Annotated*, Volume IV, 3032-7.

ing benefits to sellers and necessitating government involvement in the market to ensure the protection of government facilities.[431]

In December 1988 the Department of the Interior attempted to address this conundrum. It published a set of "principles" on the transfer of water from government-owned facilities. These guidelines expressed the department's desire to help facilitate water transfers, but recognized circumstances when government involvement must safeguard the rights of all water users. These situations included protecting the water rights of Native Americans, or when transactions might adversely affect third parties or "subdivisions," or when "it is proposed to use Federally-owned storage or conveyance capacity to facilitate the transaction." The department's "governing principles" also included ensuring the mitigation of any environmental effects from proposed transfers. In short, the "principles" described a prominent role for the federal government in any water transfer proposal, and, in some cases, made government the final arbiter in deciding the propriety of water transactions. Most pointedly, the document was silent on the topic of profits or benefits sellers might accrue from water transfers, making no rules or proposals regarding the relationship between buyer and seller in the transaction. On financial matters, the Department of the Interior's "objectives will be to ensure that the Federal government is in an acceptable financial, operational, and contractual position following accomplishment of a transaction under this policy."[432]

Questions concerning benefits from water transfers previewed an even larger discussion that went to the very heart of western water development by inquiring into what constituted a beneficial use. For many observers, a reconsideration of irrigation's priority over other water uses in the West was long overdue. Similar to Harry Hogan, geographer Fred Quinn argued in 1968 that market forces were the best means to determine beneficial use. He claimed, "All evidence points to the ability of municipal and industrial users to outbid agriculturalists for water rights under unrestricted market conditions." Quinn noted studies conducted in Arizona and New Mexico definitively showed "that nonagricultural uses yield many times more income per acre-foot than agricultural does." This easily recognizable observation did little to address the overriding problem of how to equitably transfer agricultural water to urban uses. On the other hand, Harry Hogan, in 1977, saw the transfer of water on a

---

431 USDOI, BR, *Federal Reclamation and Related Laws Annotated*, Volume IV, 3032-7; MacDonald, *Facilitating Voluntary Transfers of Bureau of Reclamation-Supplied Water*, 86.

432 U.S. Department of the Interior, "Principles Governing Voluntary Water Transfers that Involve or Affect Facilities Owned or Operated by the Department of the Interior," Washington, D.C., December 16,1988 in the Appendix of Wahl, *Markets for Federal Water*, 298-9.

strictly market basis as a great economic gain for farmers. But he also noted that market forces alone gave municipalities and industry a distinct advantage over isolated "water transactions from water uses that are less able to command money support." In other words, water on an open market might free up resources for industrial and municipal uses, but it could leave environmental and other uses in the lurch because they lacked the funds to purchase water in an open market. The governing principles outlined by the Department of the Interior in 1988 sought to protect those types of interests by ensuring that water transfers would not interfere with environmental policy, but the principles also implied a large federal presence in water reallocation in the West. As water-hungry metropolises looked with envy on their water-rich agricultural neighbors, western communities faced the troubling question of determining a modern concept of beneficial use.[433]

Moving water from agricultural uses to urban uses was a relatively straightforward activity in terms of conveyance and distribution systems. Yet market forces alone could not accomplish the goals of water transfers. Observers such as Quinn and Hogan, who viewed water reallocation simply in economic terms, failed to recognize the legal, historical, and traditional considerations of western water rights. In the West, the doctrine of prior appropriation strengthened agriculture's water rights. The prior appropriation water doctrine gave the highest claim on water to those who first put the water to use, and in most cases, the first use of water in the American West went to agriculture. But this idea of "first-in-time-first-in-right" had a caveat to insure that the water was put to "reasonable use." In order for an individual to keep the water right, it must be put to beneficial use. By the late twentieth century, the idea of reasonable use came under intense scrutiny. Advocates of water transfers suggested doing away with the doctrine or redefining beneficial use to change water allocation patterns that better met the needs of an urbanized and suburbanized West. Nonetheless, this doctrine befuddled water transfer advocates, because the right does not lie in the ownership of water itself, but in the right to use the water. In *Overtapped Oasis*, Marc Reisner and Sarah Bates explain, "Despite the implication of the term, 'water marketing,' water is not simply a commodity to be bought, sold, and traded.... Only the legal interest in the water (appropriative water rights) may be 'owned' and transferred to others."[434]

---

433 Frank Quinn, "Water Transfers: Must the American West Be Won Again," *Geographical Review* 58:1 (January 1968): 123; Hogan, *The Acreage Limitation in the Federal Reclamation Program*, 266.

434 Rowley, *The Bureau of Reclamation*, 114; Worster, *Rivers of Empire*, 90-1; see also Reisner and Bates, *Overtapped Oasis*, 82.

There was little doubt among proponents that water transfers offered the best opportunity to reallocate water and provide the most efficient use of the resource. Advocates of water redistribution preferred not to focus on the water right itself, but instead emphasized the idea of beneficial use. Ironically, this argument tended to invoke the Progressive concept of wise-use of natural resources to prevent waste and inefficiency. Just as dams were once idealized structures that prevented water from wasting to the sea, water transfer advocates suggested that some agricultural water applications were wasteful. The National Research Council among others argued,

> The doctrine of beneficial use is understood to preclude the *waste* of water. These concepts on the duty of water and waste reflect the concerns of irrigators with regard to the importance to the larger community that might place limits on private action. In practice, these principles have rarely been invoked to question established water users.

According to this opinion, waste abounded in irrigation farming as priority water rights went to crops that already received "price supports," or to "lower value" crops such as alfalfa that consumed great quantities of water. These practices also had debilitating effects on the environment when agricultural runoff polluted streams and rivers. The Council concluded that "western water law—with its emphasis on 'use-it-or-lose-it'—remains in need of revision to provide more efficient water use."[435]

By the end of the 1980s, discussions about methods to make water usage more efficient took on a greater sense of urgency and even became more confrontational. Droughts during the 1980s underlined the precariousness of the water situation in the West's urban centers, while environmental concerns drove debates about beneficial use and water allocation. Despite its proactive position regarding water transfers in *Assessment '87*, some critics claimed that the Bureau of Reclamation impeded reallocation of water from rural to urban uses. Much criticism stemmed from Reclamation rules and regulations governing the 1982 Reclamation Reform Act and the loopholes they contained that allowed large landholdings to stay in operation and receive subsidized water. Primarily centered in California's Central Valley, the issue

---

[435] Committee on the Future of Irrigation in the Face of Competing Demands, et al., *A New Era for Irrigation*, 30, 68, 4; see also Kenneth D. Frederick, foreword, in Wahl, *Markets for Federal Water*, xiv; Todd G. Glass, "The Omnibus Water Act: Three Rubrics of Reclamation Reform," *Ecology Law Quarterly* 22:1 (February 1995): 182.

focused on the economic benefits to agribusinesses afforded by federally supplied water.

The underlying theme was the beneficial use of water. For instance, between 1987 and 1988 an issue arose over a Bureau of Reclamation study proposing to market 1.5 million acre feet of CVP yield, an action opponents claimed Reclamation undertook. An outcry ensued from not only environmental organizations, but also from the Environmental Protection Agency, concerned with maintaining water quality in the Sacramento/San Joaquin Delta, and the U.S. Fish and Wildlife Service, which claimed that "it is clear that the Bureau of Reclamation's goal is to sell the remaining Central Valley Project Water without first giving attention to fish and wildlife." Much of the protest concerning Bureau of Reclamation's marketing of CVP water came about because it appeared to counter the Coordinating Operating Agreement, executed in 1986, wherein Reclamation joined efforts with the state of California to improve water quality in the Delta. According to the agreement, the Bureau was "to reserve 25 percent of the uncommitted yield until completion of studies on the water needs of wildlife refuges and wetlands." Furthermore, and to the consternation of many, Reclamation began renewing water delivery contracts under terms of post-World War II contracts without determining "whether alternative provisions would provide better water management."[436]

Many perceived the Coordinating Operating Agreement as a breakthrough in state/federal relations in managing water resources to meet the diverse needs of urban users, agriculture, and the environment. The revelation of Reclamation's intentions came at the same time that California's Water Resource Board was in the process of conducting a review of water exports from the delta of the Sacramento and San Joaquin rivers. Critics asked that Reclamation hold off on signing any long-term contracts until the state completed its study in 1990. Though environmental interests raised the greatest alarm, water marketing was the underlying issue. Both environmentalists and CVP water users decried what they alleged was Reclamation selling water

---

[436] The U.S. Fish and Wildlife Service comment was taken from a draft of an unsent correspondence dated March 23, 1987 see United States House of Representatives, Subcommittee on Water and Power Resources of the Committee on Interior and Insular Affairs, *Implementation of the Coordinating Operating Agreement (COA): Implications for Water Quality in Sacramento-San Joaquin Delta and San Francisco Bay*, Hearings, 100th Cong., 1st sess., April 3, 1987, Concord, CA (Washington, D.C.: United States Government Printing Office, 1987) 62-4; see also Dana Sebren Cooper and D. Michael Harvey, "An Upstream Swim: The Crafting and Passage of the Central Valley Project Improvement Act," in *Water Law: Trends, Politics, and Practices*, Kathleen Marion Carr and James D. Crammond, editors (Chicago: ABA Section of Natural Resources, Energy, and Environmental Law, 1995), 255.

under strict market terms. Stuart Somach, a representative of the Federal Central Valley Project Contractors, argued that "the United States should not and should never place itself in a position of selling water simply to the highest bidder, ignoring the fundamental purpose for which the Central Valley Project was authorized." On the other hand, Laura King of the National Resources Defense Council claimed that "the Bureau appears unwilling to face up to the environmental consequences of selling that water for it is proposing to segment the environmental and planning review of its market activities in a way that will prevent a meaningful assessment of the resulting impact."[437]

Both statements reflected the difficulty in instituting a fair and equitable water marketing program. Somach brought to the fore the historic relationship between the Bureau of Reclamation and irrigation agriculture. At the same time, environmentalists decried what they saw as Reclamation's attempt to market CVP water without first examining the environmental impact or needs. Both inferred threats to their particular views of what constituted beneficial use, and the problem of allocating water on strictly market terms. By the end of the decade, these issues remained unresolved, and Congress stepped forward to ease conflict over the diverse water needs in the American West.

On May 1, 1990, California Congressman George Miller introduced the Central Valley Project Fish and Wildlife Restoration Act to force changes in water allocation in the Central Valley. Miller, chairman of the Subcommittee on Water and Power of the House Interior Committee, consistently supported environmental causes and mirrored his Contra Costa County constituents' concern about water availability in an arid region—which at times caused tension with Central Valley agricultural interests. His legislation sought to limit "existing CVP contract renewals to one year and direct the Secretary of the Interior to implement a program to restore fish and wildlife species to pre-CVP levels." New Jersey Senator Bill Bradley later joined Miller's crusade to reform the Central Valley Project by introducing the Central Valley Project Improvement Act on February 26, 1991. Bradley's bill mirrored Miller's legislation in many ways by putting restrictions on contract renewals and diverting water for fish and wildlife enhancements. Both bills tapped into an emerging belief that the time had arrived to diversify the West's limited water supply for other uses.[438]

---

437 U.S. House, Subcommittee on Water and Power Resources of the Committee on Interior and Insular Affairs, *Implementation of the Coordinating Operating Agreement (COA)*, April 3, 1987, 81, 39; see also "What's the Rush," *Los Angeles Times*, May 14, 1988; "Winking at the Law," *Los Angeles Times*, December 12, 1988.

438 Cooper and Harvey, "An Upstream Swim," 255-6; Glenn Bunting, "Seymour Stalls Senate

On October 30, 1992, after four years of intense discussion and compromise, President George H. W. Bush signed the Reclamation Projects Authorization and Adjustment Act of 1992 (Public Law 102-375, 106 Stat. 4600). The Act authorized over $2.4 billion for multiple Reclamation projects throughout the 17 western states, of which the Central Valley Project Improvement Act was one title. The omnibus Act contained items that had long-term effects on the manner in which the Bureau of Reclamation conducted business. These included new points of emphasis for the Bureau of Reclamation's transformation into a water resource management agency, such as Indian water settlements, funding for recreation facilities improvement, amendments to the National Historic Preservation Act, and fish and wildlife enhancement. The Act offered something for everyone, providing enough incentives to forestall major opposition to the bill.[439]

Sections of the act covered more traditional Reclamation activities such as upgrading the powerhouse at Buffalo Bill Dam in Wyoming. One of the more significant of these traditional pursuits was the Central Utah Project Completion Act. Its primary purpose was to provide funds for completion of the Central Utah Project, along with provisions mandating environmental mitigation and enhancements, and increasing water supplies for municipal and industrial uses. This $924 million dollar authorization, however, did not reflect well on the Bureau of Reclamation's management of the project. One 1990 House Committee on Interior and Insular Affairs report accused Reclamation of misleading Congress on cost overruns, "borrowing" funds from other Colorado River Storage Project participating projects, and failing to act on "mitigating fish and wildlife impacts." As a result of these charges, Congress turned over the management of all project construction activities to the Central Utah Water Conservancy District. Congress justified its decision claiming, "The Bureau's history of mismanagement … has caused delays in the project's completion and added to the costs borne by the federal government and those project beneficiaries who have the responsibility to repay these costs."[440]

Action on Water Bill," *Los Angeles Times*, October 1, 1992; Glenn Bunting, "Water Bill Cleared for Vote in Senate Legislation: Landmark Measure that Would Reform the Central Valley Project May Face Presidential Veto, Compromise by Seymour puts Alternative Plan into Play," *Los Angeles Times*, October 7, 1992; Glass, "The Omnibus Water Act: Three Rubrics of Reclamation Reform," 148.

439 Glass, "The Omnibus Water Act," 146.

440 United States House of Representatives, Committee on Interior and Insular Affairs, *Central Utah Project Completion Act and Reclamation Projects Authorization and Adjustment Act of 1990*, H. Rpt. No. 101-764, 101st Cong., 2nd sess., September 27, 1990, 44-60.

Similar concerns about how the Bureau of Reclamation conducted business resulted in the addition of the Grand Canyon Protection Act of 1992 to the omnibus legislation. This Act instructed the Secretary of the Interior to "operate Glen Canyon Dam ... in such a manner as to protect, mitigate adverse impacts to, and improve the values for which Grand Canyon National Park and Glen Canyon National Recreation Area were established." Congress noted that during the 1970s environmental groups filed three separate lawsuits "challenging the Bureau of Reclamation's failure to evaluate Glen Canyon Dam operations under the National Environmental Policy Act." The courts dismissed these suits on the promise that Reclamation would "review" the environmental effects stemming from dam operations. Congressional investigations found that Reclamation failed to conduct these investigations, continuing to allow further degradation to the Grand Canyon. The Bureau of Reclamation defended its operations at the dam claiming that power production and storage regulation and water delivery were higher priorities than environmental mitigation. Members of the Committee on Interior and Insular Affairs disagreed, noting that sections of both the Colorado River Storage Project Act and Colorado River Basin Project Act "clearly state that power production is incidental to other purposes including fish and wildlife and recreation."[441]

These two events exemplify the slow and apparently painful transition that the Bureau of Reclamation experienced moving toward becoming a resource management agency. Although Reclamation made some progress instituting some environmental programs, as discussed in the 75th Anniversary issue of *Reclamation Era*, for many it had not done enough. Long-time Reclamation critic Marc Reisner observed in 1990:

> The Bureau of Reclamation ... is still very much like the Bureau of Reclamation: its 'new mission' largely undefined, its priorities still unclear or greatly hampered by policies it has not moved to change. And western states ... have not responded well ... to the sweeping change of the past two decades: the shriveling importance of the agricultural economy, the explosive growth of water-short cities, the desperate deterioration of water-dependent ecosystems, and the environmental concerns that the vast majority of their own citizens now share.

---

[441] "Grand Canyon Protection Act of 1992," in *Federal Reclamation and Related Laws Annotated*, Volume V 1983-1998, 3891; United States House of Representatives, Committee on Interior and Insular Affairs, *Grand Canyon Protection Act of 1990*, H. Rpt. No. 101-641, 101st Cong., 2nd sess., July 30, 1990, 4-10.

Reisner's analysis was close to the mark. Reclamation took no active role in laying out either environmental or water transfer policies. In *Assessment '87*, the Bureau of Reclamation related that drastic reforms in Reclamation law were essential for change to occur. And despite the rhetoric of Reclamation commissioners from the mid 1970s to the early 1990s, no one actively lobbied for the necessary amendments to the law.[442]

13.7. Senator Bill Bradley, a Democrat from New Jersey, served in the Senate from 1979 to 1997. Courtesy of the U.S. Senate History Office.

Reisner's comment on the slow response of state governments "to sweeping changes" in part explained Reclamation's hesitancy, but more importantly it represented the lingering influence and power agricultural interests at all levels had on lawmakers. Political pressure prevented Reclamation officials from earnestly pursuing reform measures. For instance, in comments on an early version of the Central Valley Project Improvement Act, Reclamation Commissioner Dennis B. Underwood argued that "the legislation could result in substantial frustration of the original purpose of the Central Valley Project, without accomplishing its intended purpose." Underwood's greatest concern was the legislation's emphasis on fish and wildlife restoration and warned Congress that this aspect threatened "state primacy" regarding water law. The commissioner later recounted that he sought a middle ground among the multiple water uses and "did not want to sacrifice one economic use for another." Nevertheless, Underwood's observation recognized the presence of the prior appropriation doctrine and the concern about larger federal involvement in the state's water resource management. The commissioner's position also underscored the paramount issue of beneficial use and who would decide its meaning.[443]

---

[442] Reisner and Bates, *Overtapped Oasis*, 147.

[443] United States House of Representatives, Subcommittee on Water, Power, and Offshore Energy Resources of the Committee on Interior and Insular Affairs, *Management and Operation of the Central Valley Project*, Hearings 102nd Cong., 2nd sess., May 14, 1992 (Washington, D.C.:

Underwood's stance on water redistribution did not necessarily refute the emphasis on environmental protection and restoration that some members of Congress sought. Rather the commissioner sought a slower approach to Central Valley Project reform to avoid upsetting existing conditions on the project. While testifying before Congress, he explained some measures implemented by the Bureau of Reclamation, which he believed accomplished many of the goals the legislation sought. Underwood explained that Reclamation was conducting discussions with the National Marine Fisheries Service over "long-term Central Valley Project operations to protect winter-run Chinook salmon and with the U.S. Fish and Wildlife Service concerning the Bald Eagle." In addition, the commissioner stated that the Bureau of Reclamation had opened a water conservation office in Sacramento to work with water users, along with entering into an agreement with the California Department of Water Resources to update conservation plans. Finally, Underwood discussed how Reclamation was rewriting water delivery contracts that included water transfer language. Overall, the proposals outlined by Underwood appeared to mirror the vision of CVP reformers, but at a slower pace. Congress, however, demanded immediate change. As Louisiana Senator J. Bennett Johnson stated, "We, as a country, have invested billions and billions of dollars in this project, and the people of this country demand that those billions of dollars not be used to spoil the environment of this great State."[444]

Congress's reluctance to follow the path laid out by the Bureau of Reclamation reflected its own disenchantment with Reclamation's promise to deliver on this task. As Senator Bill Bradley asserted, "the Bureau of Reclamation's devotion to its agribusiness constituency has caused the agency to work against California's broader interests." Bradley's statement was an indictment of Reclamation's overall record on fish and wildlife mitigation and environmental protection. Indeed, Congress passed both the Central Utah Project Completion Act and the Grand Canyon Protection Act partly because it believed that Reclamation failed to follow congressional intent to view environmental concerns as equal to its other activities. Reclamation was slow to embrace the environmental thinking that emerged during the 1960s and 1970s

---

United States Government Printing Office, 1992), 356-61; Dennis B. Underwood, *Oral History Interview*, Transcript of Tape-recorded Bureau of Reclamation Oral History Interview conducted by Brit Allan Storey, senior historian, Bureau of Reclamation, from 1995 to 1998, in Los Angeles and Ontario, California, 18.

444 U.S. House of Representatives, Subcommittee on Water, Power, *Management and Operation of the Central Valley Project*, 361; United States Senate, Committee on Energy and Natural Resources, *Central Valley Improvement Act*, Hearings, 102nd Cong., 1st sess., San Francisco, September 4, 1991 (Washington, D.C.: United States Government Printing Office, 1991), 2.

13.8. Dennis B. Underwood served as Commissioner of the Bureau of Reclamation from 1989 until 1993 during the presidency of George H. W. Bush.

and had become a national priority by the 1980s and 1990s.[445]

The Central Valley Project Improvement Act was a landmark piece of legislation because it reorganized operation of the Central Valley Project by mandating that environmental concerns receive a greater share of project water. It forced the Bureau of Reclamation to accept responsibilities it had long discussed but failed to implement. According to Dana Sebren Cooper and D. Michael Harvey, the Act compelled the Department of the Interior and Reclamation to "move beyond its traditional emphasis on serving the interests of irrigated agriculture." Instead the Act sought to assure water management on the CVP to serve "the broadest possible range of public purposes, including environmental mitigation, protection, and restoration as well as serving the needs of California's urban areas." Among its multiple provisions affecting water redistribution in California was allocation of 800,000 acre-feet of water "dedicated" to fish and wildlife enhancement. Another was the removal of barriers that blocked implementation of water transfers.[446]

---

445 Bill Bradley, "California Commentary U.S. Is Part of the Water Problem Use of federal water should be held to the same environmental and economic standards as use of state water," *Los Angeles Times* February 6, 1991; Committee on Interior and Insular Affairs, *Central Utah Project Completion Act and Reclamation Projects Authorization and Adjustment Act of 1990*, 51; Committee on Interior and Insular Affairs, *Grand Canyon Protection Act of 1990*, 7-9.

446 Glen Bunting, "Bush Ok's Water Policy Overhaul Legislation: President, in a Blow to Wilson and Seymour, signs bill loosening California Farmers' Grip on Federal Water," *Los Angeles Times*, October 31, 1992; Robert Reinhold, "New Age for Western Water Policy: Less for the Farm, More for the City," *New York Times*, October 11, 1992; Cooper and Harvey, "An Upstream Swim," 258; MacDonald, *Facilitating Voluntary Transfers of Bureau of Reclamation-Supplied Water*, 86-7; Committee on the Future of Irrigation in the Face of Competing Demands, et al., *A New Era for Irrigation*, 74, 113-4; "Reclamation Projects and Authorization Act," in United States Department of the Interior, Bureau of Reclamation, *Federal Reclamation and Related Laws Annotated*, Volume V 1983-1998, Donald L. Walker, editor (Denver, Colorado: United States Government Printing Office, 2001), 3932-61.

The Central Valley Project Improvement Act was significant for its broader effects on water concerns throughout the West, besides being an impetus that propelled the Bureau of Reclamation toward a water resource management function. All 17 western states that Reclamation served watched implementation carefully. With the Central Valley Project being arguably the largest and most successful Reclamation project, the impact of this legislation had a ripple effect throughout the West. Cooper and Harvey contended that "proper management of the CVP is vital to protection of other western states' interest in their own water resources." If the reallocation of water in California imposed in the Act failed to meet the state's needs, there remained the longstanding fear that California might look elsewhere for water. More importantly, the unstated but certainly implied redefinition of beneficial use did not bode well for other western states attempting to fully develop their water resources. By 1992 many states in the arid West experienced similar contestations over reallocation of water resources such as had occurred in California and sought settlements that provided equitable redistribution. In nearly all cases, irrigation agriculture relinquished some of its hold on western water through agreements or compromises. For example in Nevada, the Fallon Paiute-Shoshone Indian Tribes Water Rights Settlement Act of 1990 and Truckee-Carson/Pyramid Lake Water Rights Settlement Act allowed the federal government to purchase Truckee River water from Newlands Project farmers in order to restore Native American fisheries and rehabilitate wetlands.[447]

Fish versus Dams

By the 1990s no single issue better represented the failure of federal bureaus to grasp the importance American society began to place on environmental protection than the problem of sustaining salmon runs in the Pacific Northwest. On environmentalist agendas, the issue came down to a simplistic argument of fish versus dams. The elevation of salmon to an iconic symbol of Pacific Northwest health and prosperity threatened power interests and industrial enterprises that depended upon the Columbia River and Snake River dams. The controversy questioned the direction of regional development long based upon the consumption of cheap electricity that had been available since construction of Grand Coulee and Bonneville dams during the 1930s. More importantly to the history of the Bureau of Reclamation, the salmon contro-

---

[447] Cooper and Harvey, "An Upstream Swim," 253; Sabrina Isé and David L. Sunding, "Reallocating Water from Agriculture to the Environment under a Voluntary Purchase Plan," *Review of Agricultural Economics* 20:1 (Spring-Summer, 1998): 214-26.

versy aligned with the larger discussion occurring throughout the West about the redistribution of water and beneficial use.

By the time the Bureau of Reclamation completed Grand Coulee Dam in 1941 salmon runs in the Pacific Northwest were already in trouble. Since the beginning of Euro-American settlement of the Pacific Northwest in the mid-nineteenth century, observers noticed decreasing runs of salmon, presumably as a result of human activity. In *Making Salmon*, Joseph Taylor asserts, "Trapping, farming, mining, irrigation, logging, urbanization, and industrialization reshaped habitat, while industrial fisheries altered harvesting in ways that increased pressure on some runs but reduced it on others." He notes further that some early activities had devastating effects: mining activities eliminating salmon runs on the Boise River by 1865, and irrigation development on the Umatilla, Deschutes, and Yakima rivers drastically reducing runs by the late nineteenth century. Along with degradation of salmon habitat and spawning grounds, the rise of an aggressive commercial fishing industry played an important role in further diminishing salmon runs in the Pacific Northwest.[448]

Reclamation's giant, and only, dam on the Columbia River, Grand Coulee, closed off the river's upper regions to spawning salmon. While few actively opposed major dams on the Columbia River, many realized that construction meant reduced salmon runs. With passage of the 1934 Fish and Wildlife Coordination Act, Congress compelled the Bureau of Reclamation and the U.S. Army Corps of Engineers to mitigate the damage caused by their dams on fish migration. In 1939 Congress provided legislation "for the conservation of fishery resources on the Columbia River" and appropriated "$500,000 to carry out investigations and construct devices for the protection and improvement of feeding and spawning conditions." Both agencies moved to install fish ladders that assisted migrating salmon over dams; yet the sheer size of Grand Coulee Dam, according to then-current wisdom, made fish ladders impractical. Instead the Bureau of Reclamation operated a program to truck trapped salmon to streams closer to the Pacific Ocean.[449]

---

448 Joseph E. Taylor III, *Making Salmon: An Environmental History of the Northwest Fisheries Crisis* (Seattle: University of Washington Press, 1999), 51-3, 7; David P. Billington, Donald C. Jackson, Martin V. Melosi, *The History of Large Federal Dams: Planning, Design, and Construction in the Era of Big Dams* (Denver, Colorado: U.S. Department of the Interior, Bureau of Reclamation, 2005), 393.

449 Billington, et al., *The History of Large Federal Dams*, 393-4; Tim Palmer, *Endangered Rivers and the Conservation Movement* (Berkeley: University of California Press, 1986), 60; Rowley, *The Bureau of Reclamation*, 338-9; Taylor, *Making Salmon*, 228-9.

While there was a conscious effort on the part of dam builders to lessen impacts on salmon runs, the results were modest. Federal and other entities established rarely questioned priorities that favored dams over fish. The costs associated with attempts to save salmon grew exponentially as dam building continued unabated on the Columbia River and its tributaries. Both the Corps and Reclamation expressed confidence that science and technology held solutions for bringing dams and healthy salmon runs into harmony. This meant funding for hatcheries, along with other mitigating activities such as habitat restoration and improving fish passage. Despite these efforts, all evidence pointed to the end of salmon runs in the Pacific Northwest. Taylor reports that at one time the Bureau of Reclamation timidly proposed reserving some streams within the Columbia River watershed as "fish refuges," but "it had neither the power nor the will to enforce the suggestion."[450]

From the late 1940s through the 1970s, the federal government expended enormous sums on the restoration of salmon runs, with little to show for the effort. Observers noted in 1960 that an estimated $127 million "has been invested in facilities to protect Columbia River salmon since Rock Creek Dam was constructed in 1933." By 1973 there was no improvement. The federal commissioner of fisheries reported that despite all efforts government programs to maintain and propagate salmon fisheries had little effect. According to the commissioner, the primary reason for the failure was the still limited understanding of salmon behavior and biology. Reliance on hatcheries as a panacea for diminishing numbers of salmon was unsuccessful, and there was growing awareness that dams were the major cause of declining salmon runs. To mitigate that particular issue, the Corps of Engineers instituted a program of transporting juvenile salmon around dams on barges to ease their migration to the ocean. The tactic had mixed results and drew criticism as conflicting studies argued both over the program's success rate and its effects on the fish populations.[451]

On the mainstem of the Columbia River, eleven major dams transformed the river into a series of slack-water pools. Lacking the current of the unimpeded river, the pools severely slowed downstream salmon migra-

---

[450] Taylor, *Making Salmon*, 227-9.

[451] M. E. Marts and W. R. D. Sewell, "The Conflict between Fish and Power in the Pacific Northwest," *Annals of the Association of American Geographers* 50:1 (March 1960): 47; Federal Commissioner of Fisheries, Directed by Congress to Address the Effects of Bonneville Dam on Salmon Runs, "U.S. Commissioner of Fisheries Report," 1973 in *The Northwest Salmon Crisis: A Documentary History*, Joseph Cone and Sandy Ridlington, editors (Corvallis, Oregon: Oregon State University Press, 1996), 101; Taylor, *Making Salmon*, 245.

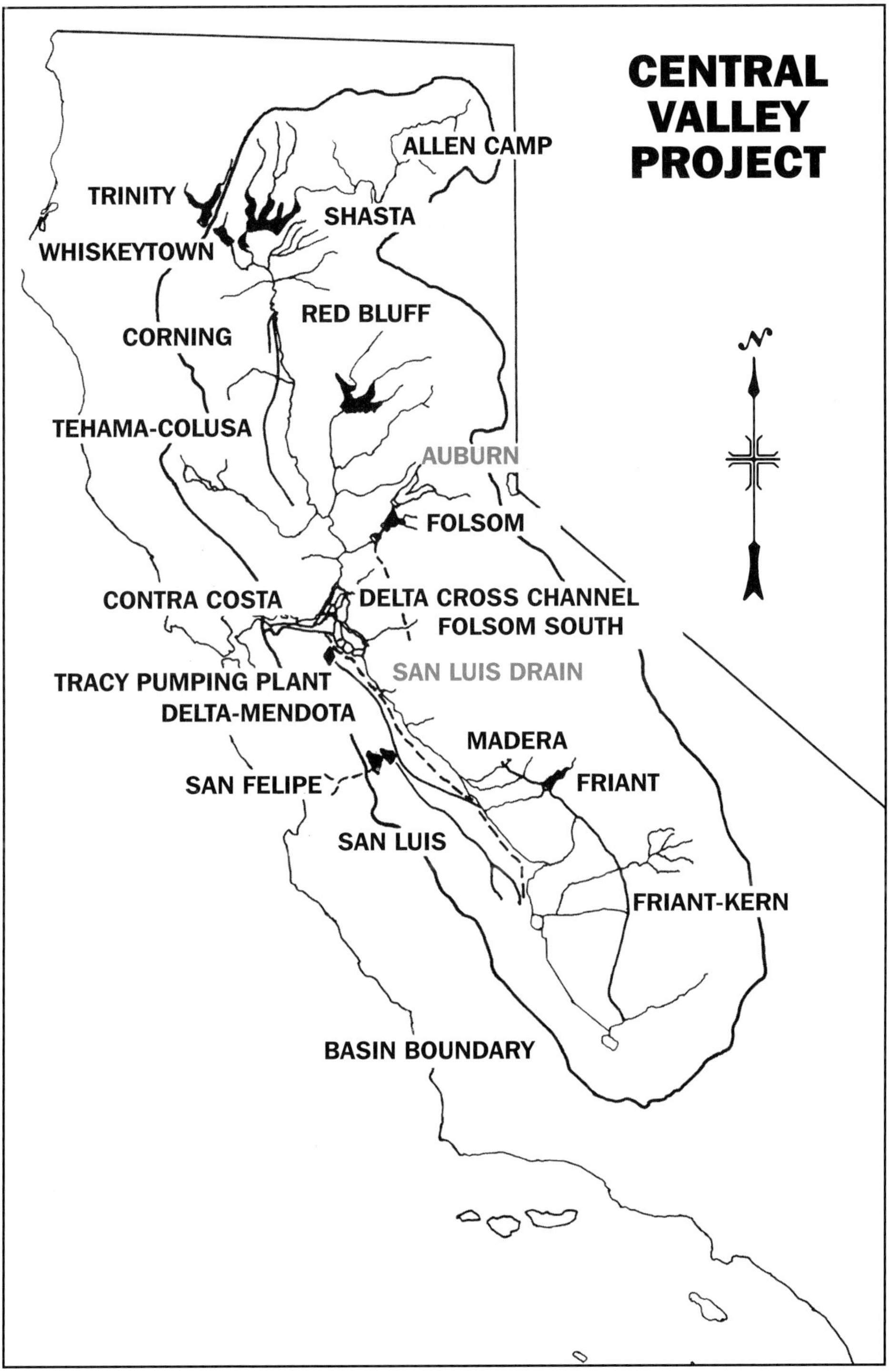

13.9. The major components of the Central Valley Project in California as envisioned in 1977. Note that the San Luis Drain has never been completed and Auburn Dam has not been built.

13.10. Coleman National Fish Hatchery in the northern Central Valley Project, 2009.

tion. Juveniles became easy prey for predators, and some fish adapted to salt water before arriving at the ocean, resulting in major fish kills. In addition, the highly nitrogenated water released from dam turbines caused "gas bubble disease," responsible for the deaths of thousands of downstream migrating salmon. By 1980 the outcry over the salmon caused Congress to pass the Northwest Power Act, to elevate "the importance of salmon in regional power planning, and make fish and wildlife 'a coequal partner with other uses' of the Columbia River." The Act also created the Northwest Power Planning Council (NPPC), made up of a diverse group—Native Americans, environmentalists, power producers, and others with interests in the Columbia River watershed from Washington, Oregon, Idaho, and Montana—to, according to Joseph Taylor, "ensure that dam operators give equal consideration to salmon [while] managing rivers."[452]

The legislation recognized the failure of past efforts to protect salmon runs and compelled dam operators to be more responsive to the salmon crisis. Through the intervening years, however, efforts were made to direct dam builders and operators to give greater consideration to the impact of their operations on salmon. In 1938 Congress passed the Mitchell Act, which authorized the secretary of the interior to conduct studies and experiments to "facilitate conservation of the fishery resources of the Columbia River and its tributaries." This early legislation directed the secretary to improve habitat, to take necessary precautions to protect fish from irrigation projects, and to ease "migration of fish over obstructions." In 1958 Congress amended the Wildlife

[452] "Northwest Power Act," in Cone and Ridlington, *The Northwest Salmon Crisis*, 255-8; Taylor, *Making Salmon*, 246; Lichatowich, *Salmon Without Rivers*, 202-3.

13.11. One of the raceways in the Coleman National Fish Hatchery, northern Central Valley Project, 2009.

Coordination Act to require in all water resource development projects that fish and wildlife be given "equal consideration" in project planning. According to Keith Peterson, in *River of Life, Channel of Death*, this legislation was a "major milestone in the American conservation movement, it provided that fish and wildlife not only be considered but actually enhanced at federal water projects." In 1974 the courts entered the fray with the release of what became known as the Boldt Decision. Although this ruling pertained to Native American fishing rights on the Columbia River and its tributaries, it culminated in the forging of an alliance among Native Americans and state fish and wildlife agencies. This coalition began to apply pressure on hydroelectric operators to improve salmon habitat protection, which eventually led to the Northwest Power Act. According to one source "it is no exaggeration to suggest that the era of modern salmon management began with this historic decree."[453]

[453] "The Mitchell Act," in Cone and Ridlington, *The Northwest Salmon Crisis*, 103; "Columbia River Fishery Development," in USDOI, BR, *Federal Reclamation and Related Laws Annotated*, Volume I, 602-3; for information on the 1958 Wildlife Coordination Act see Peterson, *River of Life*, 134; "Name and Amend Fish and Wildlife Coordination Act, " in USDOI, BR, *Federal Reclamation and Related Laws Annotated*, Volume II, 1434-39; MacDonald, *Facilitating Voluntary Transfers of Bureau of Reclamation-Supplied Water*, 21; for information on the Boldt Decision see Cone and Ridlington, *The Northwest Salmon Crisis*, 185, 193; Taylor, *Making Salmon*,

Scholars argue the rhetoric espoused by federal authorities and legislation about overseeing water development projects to protect fish and wildlife never matched results. Michael C. Blumm and F. Lorraine Bodi argue, "These standards were never taken seriously by project managers over the years. They became empty promises, forgotten in the increasing emphasis on generating every possible kilowatt from Columbia Basin dams." There is little doubt power production and irrigation development received higher priority in the Pacific Northwest because these were the traditional and primary authorized missions of federal projects, and because the huge federal investment in infrastructure militated against any proposed solution contemplating either dam removal or operational changes that would result in reductions of water and power deliveries, flood control, or navigation. This was as much a result of institutional priorities, i.e. the Bureau of Reclamation and the U.S. Army Corps of Engineers, as it was the contested debate over economic development of the region. In 1978 Idaho conservationist Ed Chaney wrote that "there is nothing even approaching general unanimity on what constitutes a balanced use of the region's water resources … Particularly if *balanced use* means altering long-term plans for hydropower production … in order to protect salmon and steelhead runs of the upper basin." One could easily replace "balanced use" with beneficial use in the debates over water reallocation taking place across the American West during this period. In the Northwest Power Act and the creation of the Northwest Power Planning Council, Congress intended to settle both of these issues by providing the framework which allowed the people of the Pacific Northwest to come to an equitable resolution to the fish versus dams debate.[454]

The Northwest Power Planning Council was hesitant to institute any major reforms to the operation of dams on the Columbia River, and it turned to the historic reliance on hatcheries as the primary means of supplementing salmon runs. Initially there was little discussion about examining the impact of dams on fish migration or finding the resource balance that Ed Chaney advocated. On the failure to act decisively Blumm and Bodi assert, "The promise of power interests that they were anxious to accommodate salmon mitigation requirements seemed to wane rapidly after enactment of the statute. Neither Congress nor the Council expressed much interest in seeing to it

---

243-4; Anthony Netboy, *The Columbia River and Steelhead Trout: Their Fight for Survival* (Seattle: University of Washington Press, 1980), 115-6.

454 Michael C. Blumm and F. Loraine Bodi, "Commentary," in Cone and Ridlington, *The Northwest Salmon Crisis*, 121; Ed Chaney, "A Question of Balance," (1978) in Cone and Ridlington, *The Northwest Salmon Crisis*, italics in the original, 251.

that these promises were carried out." In response, environmental organizations and Indian tribes generated vocal opposition to the status quo, challenging established interests in the courts. For instance in 1984 the Ninth Circuit Court of Appeals ruled in *Yakima Indian Nation v. Federal Energy Regulatory Commission* that the commission "failed its Federal Power Act obligation to consider fishery issues prior to licensing, it concomitantly failed to meet its obligation to give fish … 'equitable treatment' under the [Northwest Power Act]." Because of this ruling the courts began to play an increasingly influential role in ensuring that federal agencies enforced "equitable treatment."[455]

For the most part, the Bureau of Reclamation stayed under the radar as the salmon versus dams debate focused on Corps and private dams on the Columbia and lower Snake rivers. But it was not left out of the conversation since critics routinely recalled that Grand Coulee Dam eliminated "70 percent of the original spawning area of the Columbia River." Over time, it was discovered that Grand Coulee also contributed to fish kills by raising gas levels in the river when water spilled over the dam. Reclamation also encountered criticism when it came to commentary about its handling of fish and wildlife matters in other places in the West. On the Central Utah Project, Adam Eastman notes that there were concerns over Reclamation plans to divert streams on the project that "would dry up 245 miles of streams" that wildlife experts estimated would result in enormous fish losses. Similar issues arose on the Animas-La Plata Project, along the Colorado/New Mexico border, that required the Bureau of Reclamation to alter and improve its original plans for "fish and wildlife features." In addition, the listing of the Pyramid Lake cui-ui as endangered and the Lahontan cutthroat as threatened in 1989 forced Reclamation to commit Stampede Reservoir water in California away from the Newlands Project in Nevada to raise water levels in Pyramid Lake.[456]

While never reaching the level of the public controversy in the Pacific Northwest, Bureau of Reclamation involvement with the dams versus salmon debate also occurred in California. Salmon runs in California suffered from development that began with Euro-American settlement. Mining, agriculture, and overfishing played havoc with salmon runs, creating severe strains on fish

---

455 Blumm and Bodi, "Commentary," in Cone and Ridlington, *The Northwest Salmon Crisis*, 263-4; Ninth Circuit Court of Appeals, "Yakima Indian Tribe v. Federal Energy Regulatory Committee," 1984, in Cone and Ridlington, *The Northwest Salmon Crisis*, 271-2.

456 Eastman, "From Cadillac to Chevy, 85; McCool, *Native Waters*, 91-2; National Research Council, *Water and Choice in the Colorado Basin: An Example of Alternatives in Water Management*, 1968, 127.

populations. Reclamation dams on the Sacramento, San Joaquin, and Trinity rivers along with the Central Valley Project further diminished runs. And similar to what occurred in the Pacific Northwest, little was done to mitigate the effects of water resource development projects on fish. In 1975 a California Department of Fish and Game study decried the lack of federal support for solving the fish problems in California streams. The report targeted Reclamation's Trinity River Project as "a prime example of a water development which has caused tremendous damage to a fishery, yet at the federal level no serious attempt has been made to identify and correct the problem."[457]

Federal and state fish agencies, similar to their counterparts in the Pacific Northwest, tended to rely on hatcheries to stem the tide of declining salmon runs. Reclamation's primary forum, *Reclamation Era*, discussed measures implemented by the California Department of Fish and Game and the U.S. Fish and Wildlife Service to assist salmon fisheries during a 1976-1977 California drought. According to one article, drought conditions resulted in lower stream flows and declining reservoirs that increased water temperature, which led to egg mortalities. State and federal wildlife agencies sought to mitigate drought effects by blocking upstream migration at Red Bluff Diversion Dam to either "trap and transport" salmon to "cooler adjacent tributaries," or "force" salmon to spawn in the cooler waters below the dam. While the article admitted that the results of these efforts remained unknown, they provided an example of a concerted effort to be proactive on the fishery issue. Nevertheless, by the late 1980s, little had been accomplished in stemming the decline in California salmon runs. In 1989 the National Marine Fisheries Service (NMFS) declared the Sacramento River's winter Chinook salmon threatened under the Endangered Species Act and in 1994 upgraded the designation to endangered.[458]

In 1991 the Pacific Northwest experienced similar problems. The Northwest Power Planning Council had accomplished little in improving fish runs in the Pacific Northwest. The federal government reacted by listing the Snake River sockeye and Chinook runs as endangered under the Endangered Species Act. Despite the listing, there was little uproar in the Pacific North-

---

[457] California Department of Fish and Game, Citizens Advisory Committee on Salmon and Steelhead Trout, *Report on California Salmon and Steelhead Trout: The Time is Now!*, 1975, 30.
[458] Steve Rademacher, "California Drought Effects on Fish and Wild Life," *Reclamation Era*, 63:3 and 4 (1977): 1-2; see also Prochaska, "Reclamation's Concern for Fish and Wildlife," 44; Lichatowich, *Salmon Without Rivers*, 202; Cone, "Commentary," in Cone and Ridlington, *The Northwest Salmon Crisis*, 297.

west because earlier reports such as the Endangered Species Committee of the American Fisheries Society's "Pacific Salmon at the Crossroads" had already sounded the alarm. Joseph Cone writes that "the findings of the study alerted the entire Northwest Public … to the very serious danger that salmon populations faced." The article listed the usual suspects associated with salmon population depletion—habitat loss, dam obstructions, agriculture, and logging—and emphatically called for new and innovative methods to save the salmon. The committee frankly argued that supplementing salmon populations through hatcheries was not working and charged that hatcheries probably contributed to salmon destruction by weakening the genetic makeup of the species. What was needed, the committee argued, was a program to rejuvenate wild salmon runs.[459]

Understood in this suggestion was the realization that restoring salmon runs required huge sacrifices impacting everyone throughout the region. The dams that lined the main stems of the Columbia and lower Snake rivers created a trade thoroughfare from Lewiston, Idaho, to the Pacific Rim. Reclamation's Columbia Basin Project served one of the largest and most successful irrigation projects in the West. Hydroelectricity produced by the region's dams resulted in the lowest utility rates in the nation and helped to spur the region's urban and industrial development. Saving salmon meant these economic interests would bear the brunt of the effort. In 1994 the Northwest Power Planning Council proposed a major readjustment in dam operations along the Columbia and Snake rivers that affected every one of these enterprises. The plan called for turning the Snake River back into a free-flowing river by drawing down four federal dams during particular times of the year "to speed passage of juvenile salmon to the ocean." This proposal, according to Michael C. Blumm and F. Lorraine Bodi, was innovative "because it started from the premise of accomplishing a biological objective … and then investigate how to accomplish the objective in an economic and technically feasible manner."[460]

Not surprisingly, the NPPC's proposed drawdowns generated a storm of protest from irrigators, barge operators, power companies, and aluminum manufacturers. They argued that the plan placed an undue burden on their

---

[459] Cone, "Commentary," in Cone and Ridlington, *The Northwest Salmon Crisis*, 296-7; Lichatowich, *Salmon Without Rivers*, 204; Taylor, *Making Salmon*, 247; Tom Kenworthy, "U.S. Designates Sockeye Salmon an Endangered Species," *Washington Post*, November 15, 1991.
[460] Tom Kenworthy, "Plan to Save Salmon Roils Northwest; Change Seen Causing Ripples in Economy," *Washington Post*, December 15, 1994; Blumm and Bodi, "Commentary," in Cone and Ridlington, *The Northwest Salmon Crisis*, 318.

13.12. The Red Bluff Diversion Dam (1964) on the Sacramento River at Red Bluff, California. The diversion dam now has been bypassed by the Red Bluff Fish Passage Improvement Project, a series of pumps and fish screens (August 2012), which delivers water to the Corning Canal and Tehama-Colusa Canal. This improvement allows dam gates to be kept permanently open for improved fish passage in the river. May 8, 2010. Photographer: Loredana Potter.

enterprises to the detriment of the Pacific Northwest economy. On the other hand, conservationists applauded the proposal and "hailed the plan as an important milestone in the long process of replenishing salmon runs." These diametrically opposed positions in the debate provided another example of the ongoing question of what constituted beneficial use of water in the American West as the twenty-first century dawned. Old habits and old alliances die hard. In 1995 the National Marine Fisheries Services (NMFS) offered its own plan to restore fish populations. It called for further studies on the drawdown proposal and relied "more heavily on physical improvements to dams to ease salmon passage, more effective transportation of fish around dams and less drastic changes in dam operations to provide additional water for migrating fish when they need it." In short, the NMFS plan was another attempt to utilize technological fixes without incurring any inordinate hardship to traditional water users.[461]

The issue of salmon versus dams assumes importance in the context of the history of the Bureau of Reclamation because it speaks to the role

[461] Kenworthy, "Plan to Save Salmon Roils Northwest; Change Seen Causing Ripples in Economy," see also Tom Kenworthy, "Northwest Salmon Plan Is Outlined," *Washington Post*, June 26, 1995.

of the federal government in water resource management. Joseph Taylor's *Making Salmon* reveals the government's long-term interest in salmon restoration in the Northwest that relied on science and technology primarily through improvements in hatcheries. Early government attempts to reverse the destruction of salmon runs, as outlined by Taylor, also encouraged greater efforts in habitat restoration and easing fish migration around obstacles. These early attempts to save salmon runs, however, avoided impeding the economic growth of the region and were at times counterproductive to preserving fish runs. In other words, while endeavoring to salvage fish habitat and ease migration, dam building, irrigation farming, logging, and mining continued unabated, destroying more habitat. The Bureau of Reclamation inherited this legacy during development of the Columbia Basin Project. Reclamation projects always espoused benefits that would lead to greater prosperity throughout the Pacific Northwest. Even when Congress intervened directing both Reclamation and the Corps of Engineers to give fish and wildlife "equal consideration" compared to other project goals, dam-building agencies hesitated. They, in effect, followed a tradition that embraced technical answers such as fish ladders, passageways, and hatcheries without sacrificing economic development.

Despite incredible sums spent to mitigate the problem, these efforts never achieved successes, and salmon runs continued to decline. Taylor notes that since 1981 "the region has invested three billion dollars to save these fish, and the only thing everyone can agree upon is that the effort has largely failed." Nevertheless, the over one hundred-year effort to save the salmon is astonishing. It reflects a cultural value that people of the Pacific Northwest invested in the fish, beginning in the nineteenth century. It came to full flower with the rise of the environmental movement of the 1960s and 1970s and eventually morphed into an aesthetic within the region as the salmon became a cultural symbol of the region in the "far corner" of the United States. Ironically long before the fish versus dams debate reached its peak in the 1990s, dams filled a similar place in the imagination of the Pacific Northwest. Regional boosters during the 1940s, such as Leland Olds, proclaimed to the people of the Northwest that there was nothing preventing dams and fish from existing in harmony, and that the economic growth provided by dams would not affect that other valuable resource: the "silver horde." Of course, Olds was wrong. He never considered the arrival of the anti-dam sentiments that emerged with the environmental movement that convinced many Americans to reconsider their perceptions of dams.[462]

---

[462] Leland Olds Address, RG 48, Entry 779, Box 15; for information on the cultural significance of the salmon to the Pacific Northwest see Lichatowich, *Salmon Without Rivers*, 222-30;

13.13. Daniel P. Beard served as Commissioner of the Bureau of Reclamation from 1993 until 1995.

The question of whether dams or fish have more economic, social, or cultural value speaks to the question of water allocation that reverberated throughout the American West in the late twentieth century. Ed Chaney, director of the Northwest Information Center, argued that "existing water supplies are not sufficient to serve all simultaneous demands during most years, and the shortfall is growing rapidly." It was the overriding issue faced by the Bureau of Reclamation in its new mission as a water resource management agency. Traditional water users who had benefitted most from government policies favored their vision of the West. Opposed to them was a new viewpoint that favored values that could not be measured in kilowatts or agricultural production.[463] During the early 1990s, the Bureau of Reclamation faced these challenges head-on when a new commissioner came into office determined to see Reclamation complete the transition to a water management bureau.

Daniel Beard and the "New" Bureau of Reclamation

In 1993 Daniel Beard became the Bill Clinton administration's commissioner of the Bureau of Reclamation. Beard, similar to many Reclamation commissioners between 1980 and 2000, did not come from the ranks of Reclamation. Indeed, Beard's record in water resource politics marked him as an ardent adversary of many traditional Reclamation activities and constituencies. While not directly involved, Beard was part of Jimmy Carter's transition team in 1976-1977 that put together the so-called hit list. He later became deputy assistant secretary of the interior for Land and Water Resources from 1977 to 1980. In addition, Beard was the staff director for the Subcommittee on Water and Power of the House Interior Committee from 1985 to 1990, and later was staff director of the House Interior Committee when George Miller

---

Taylor, *Making Salmon*, 246-7.

[463] Chaney, "A Question of Balance," in Cone and Ridlington, *The Northwest Salmon Crisis*, 251.

replaced Morris Udall as chairman in 1991. He was a key player, along with Congressman Miller in developing and passing the legislation that eventually became the Central Valley Project Improvement Act of 1992. In short, Beard came to the Bureau of Reclamation with a reputation as a reformer who sought diversity in Reclamation policy that met agricultural, environmental, and urban water needs.[464]

Similar to many twentieth-century presidents, Bill Clinton entered the oval office with no set water policy, especially one that focused on water needs in the West, and a slight concern for environmental issues. Clinton, however, came into office with a determination to streamline the federal government to reduce waste and spending. Under the slogan "reinventing government," the administration called on all federal agencies and bureaus to look for ways to trim the federal budget. The new Department of the Interior Secretary Bruce Babbitt and Commissioner Beard embraced the administration's objectives, and for both men the Bureau of Reclamation was prime ground for reform. As Arizona governor, Babbitt was a vocal critic of Reclamation claiming, "Its practices have been the most environmentally destructive of all the public-land agencies." Both Beard and Babbitt sought to turn the Bureau of Reclamation around, making it more sensitive to environmental issues, and to move the Bureau away from its construction-orientated past. Beard summed up his vision of Reclamation's new role as moving "toward new environmental priorities" and clarifying "its role in water management."[465]

There is little doubt that the Bureau of Reclamation was well on its way toward becoming a water resource management agency before Dan Beard's arrival. Beard, however, was the first commissioner to pursue institutionalization of this transition. In 1995 he declared:

> Water resource policy in the United States, and particularly in the West, is undergoing fundamental changes. These changes are driven not so much by political issues as they are by

---

[464] Daniel P. Beard, *Oral History Interview*, Transcript of tape-recorded Bureau of Reclamation Oral History Interviews conducted by Brit Allan Storey, senior historian, Bureau of Reclamation, from 1993 to 1995, in Washington, D.C., 31-6; Daniel P. Beard, *Passage of the Central Valley Project Improvement Act, 1991-1992: The Role of George Miller*, Oral History Interview conducted by Malca Chall in 1995 (Berkeley: Regional Oral History Office, Bancroft Library, University of California, 1996), iv.

[465] "Behind Standoff on Water Rights: A Century of Policy and Habit," *Christian Science Monitor*, November 9, 1993; Tom Kenworthy, "A Leaner and Greener Bureau of Reclamation; Water Management Being Redefined," *Washington Post*, November 2, 1993.

desires and aspirations of an urban population with an abiding interest in economic growth and environmental sensitivity.

Statements such as this one perhaps did not bode well for Reclamation's traditional constituents, but it did recognize the American West's changing demography and cultural values.[466]

Beard laid out changes for the Bureau of Reclamation in his "Blueprint for Reform, The Commissioner's Plan for Reinventing Reclamation," released in November 1993. The Blueprint proclaimed that Reclamation's new mission was, "To manage, develop, and protect water and related resources in an environmentally and economically sound manner in the interest of the American public." Beard's goals for Reclamation sought to ease the reallocation of water resources within the parameters of state water law, to improve existing facilities and their management, and to encourage water conservation. In addition, Beard believed Reclamation must work more closely with other federal, state, and local interests and "conduct ourselves in a fiscally responsible manner and ensure the use of sound business practices." To a great extent, Beard's "Plan" was a reassertion of the principles described in Reclamation's *Assessment '87*. Beard, to his credit, however, provided the leadership to ensure that reforms occurred. Former Bureau of Reclamation Commissioner John W. Keys III stated that Dan Beard "accomplished something that we have been trying to get done in Reclamation for about ten years, and that is to shift our direction away from big old projects and the high dam thing."[467]

Beard believed water resource development over the past twenty years did not reflect the changing economic, societal, and cultural *milieu* of the West, and past Reclamation policy was partially to blame. He argued that "we have not placed enough emphasis on water conservation and greater efficiency of water use, nor have those who benefit from our program." Beard realized that the reforms he envisioned could not totally abandon irrigation agriculture. Nonetheless, Beard was determined to broaden the Bureau of Reclamation's responsibilities, which he also saw as serving "the needs of Western urban communities and Native Americans." Implied in Beard's proclamation of reform was the understanding that the current political and economic climate

---

466 Daniel P. Beard, "New Directions for the Bureau of Reclamation," 57.

467 Dan P. Beard, Commissioner of Reclamation, "Blueprint for Reform: The Commissioner's Plan for Reinventing Reclamation," November 1, 1993, 1; Keys, *Oral History Interviews*, 85

of the early 1990s necessitated a cultural change within Reclamation and the American West.[468]

For employees of the Bureau of Reclamation, the reforms of the commissioner were swift and perhaps painful. First of all, Beard announced that Reclamation would no longer develop, plan, or seek large-scale construction projects, but instead would look for "non-structural solutions" to water resource problems. While this too echoed earlier Reclamation proclamations, the commissioner actively implemented departmental policies to assure its realization. Beard slashed budgets, reorganized the regional offices granting them greater autonomy, consolidated many project and construction offices, and reduced the engineering and research staff. "Reinventing" Reclamation also cut top-level management positions, reworked administrative practices, which abolished one hundred positions, and eliminated, according to Beard, "one column of Bureaucracy in Denver." In 1995 Beard proudly reported to Congress,

> Reclamation today is a leaner, more efficient organization than it was two years ago. We have reduced our work force from 8,100 to 6,600 in less than two years and have signed buyout agreements with more than 700 new workers. We have adopted new organizational structures designed to empower frontline employees.[469]

Beard's reforms appeared draconian to some, but they reflected a larger goal; one that succeeded in establishing a permanent break with the past. Reclamation not only had to conform to a new mission, but also a new identity.

Part of Beard's plan called for the Bureau of Reclamation to place greater emphasis on the water needs of Native Americans—a departure from historical practices. Reclamation's track record regarding Indian projects—Navajo Indian Irrigation Project, Animas-La Plata Project, and disputes with the Ute Indian Tribe on the Central Utah Project—was less than stellar. Critics describe Reclamation's relationship with Native Americans two ways: the first condemns the Bureau for ignoring Indian water needs and water

---

[468] Beard, "Blueprint for Reform," 2-3.

[469] "Opening Statement of the Honorable Daniel P. Beard Commissioner of Reclamation before the Subcommittee on Energy and Water Development of the Senate Committee on Appropriations," May 2, 1995, Federal Document Clearing House Congressional Testimony, LexisNexis Congressional, http://0-web.lexis-nexis.com.

rights in favor of non-Indian water users; the second identifies Reclamation's advocacy of Indian projects as the "Indian blanket" that married Indian projects to non-Indian projects usually at the expense of tribal communities. Beard's announcement meant greater efforts by the federal government to quantify Indian water rights through settlement negotiations rather than litigation. Indeed, the push to settle Indian water claims began in earnest during the Reagan administration, which hoped to avoid the costly process of lawsuits. Between 1982 and 1993, Congress approved fourteen Indian water right settlements, and, in 1993 alone, the federal government was involved in fifteen others.[470]

The movement toward settlement stemmed from the desire of the federal government, Native American communities, and western states to quantify Indian water rights in order to resolve long-standing disputes. The U.S. Supreme Court asserted Indian water rights in its 1908 ruling in *Winters v. United States* and later strengthened those claims in *Arizona v. California* in 1963. In the latter case, the Supreme Court held that "a tribe is entitled to enough water to irrigate all 'practically irrigable acreage' on its reservation." Although the ruling reaffirmed the priority of Indian water rights, it did little to provide tribes with "wet water." Nonetheless, "practical irrigable acreage" became the rubric utilized by negotiators in determining the water allocation for a given reservation. Indian water settlements joined other emerging discussions—environmental protections/enhancement and rural/urban water transfers—over reallocation of the West's limited water supply.[471]

By the time Beard became Reclamation's commissioner, the character of western agriculture had change dramatically from what it was when *Winters* was handed down in 1908 or the *Arizona* ruling came out in 1963. While "practical irrigable acreage" may have provided the measuring stick from which to allot water to Native American tribes, urbanization, environmental concerns, and corporate agriculture had altered water usage and needs. Indian water settlements reflected this change, and negotiators established basic elements upon which to conduct negotiations. Settlements included a federal

---

[470] David H. Getches, "Indian Water Rights Conflicts in Perspective," in *Indian Water in the New West*, Thomas R. McGuire, William B. Lord, and Mary G. Wallace, editors (Tucson: University of Arizona Press, 1993), 20-1; McCool, *Native Waters*, 46.

[471] Jacobsen, "A Promise Made," 93; Elizabeth Checchio and Bonnie G. Colby, *Indian Water Rights: Negotiating the Future* (Tucson: Water Resources Research Center, University of Arizona College of Agriculture, June 1993), 13; David H. Dejong, "'The Sword of Damocles?' The Gila River Indian Community Water Settlement Act of 2004 in Historical Perspective," *Wicazo Sa Review* 22:2 (Fall 2007): 69.

investment in water resource facilities, such as additions to the Central Arizona Project to supply water to Indian reservations. Often settlements established a cost-sharing fund where federal, state, and local governments contributed to allow tribes to develop "wet water" for use on reservations. Water settlements also included water-marketing language to permit Native American communities to gain some economic benefits from their water resources. One authority argued that water marketing revenues "could be used ... to foster the intent of *Winters*: to create economically viable tribal homelands."[472]

As with many other issues that involved the reallocation of water resources in the West, discussions involved the prior appropriation doctrine and the idea of beneficial use. Of course, legal conversations about prior appropriation and the Winters Doctrine regarding Native American "reserved rights" had been going on since the Supreme Court made the *Winters* decision in 1908. In *Native Waters*, Daniel McCool writes that

> the Winters Doctrine created a concept of water acquisition and ownership wholly at odds with the prevailing western doctrine of prior appropriation.... prior appropriation rights are based on diversion and use, while reserved rights do not require use; prior appropriation rights are based on state law, while reserved rights are created by the courts to meet the purpose of federal reservations.

For many non-Indian water users, recognition of a "reserved right" threatened the very fabric of western water policy. On the other hand, settlement proponents contended that Indian water rights were superior to those of non-Indians because Congress or executive orders confirmed those rights when establishing reservations. This was the basis of the *Winters* decision, but there were also instances when Congress recognized Indian ownership of land, such as those held by the Pueblo people, that bestowed a water right to those lands.[473]

---

472 Getches, "Indian Water rights Conflicts in Perspective," 21; Dejong, "'The Sword of Damocles?'" 77-8; Daniel McCool, "Negotiating Water Settlements: Ten Common Themes," in McGuire, et al., *Indian Water in the New West*, 94; McCool *Native Waters*, 118-9; Bonnie G. Colby, John E. Thorson, Sarah Britton, *Negotiating Tribal Water Rights: Fulfilling Promises in the West* (Tucson: University of Arizona Press, 2005), 31; Thomas R. McGuire, "Introduction: Notes on Context and Finality," in McGuire, et al., *Indian Water in the New West*, 3.

473 McCool *Native Waters*, 19; Charles T. DuMars, Marilyn O'Leary, Albert E. Utton, *Pueblo Indian Water Rights: Struggles for a Precious Resource* (Tucson: University of Arizona Press, 1984), 21; David Getches, foreword, in Colby, et al., *Negotiating Tribal Water Rights*, xiii; Checchio and Colby, *Indian Water Rights*, 10.

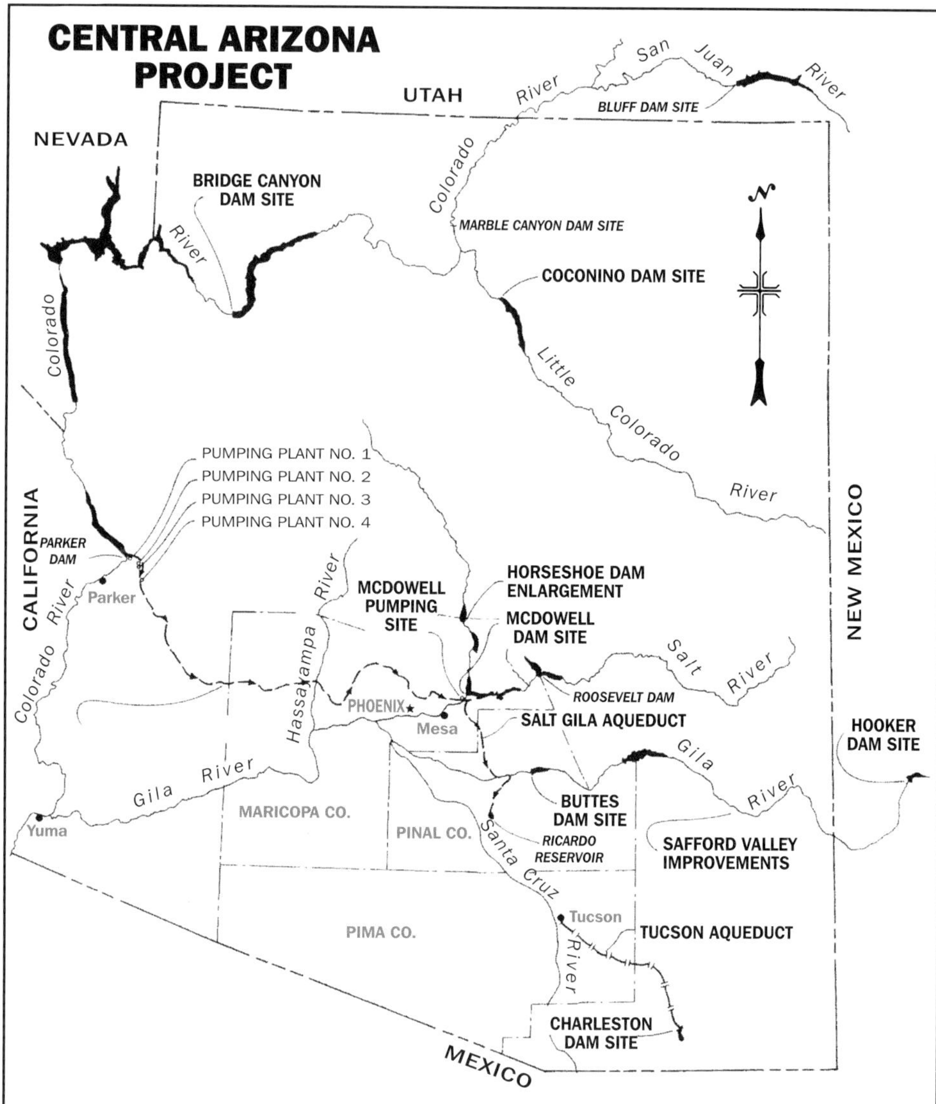

13.14. The Central Arizona Project as conceived in 1945. Note that, as then planned, Tucson water comes from the Charleston Dam on the San Pedro River and there is no direct connection to Colorado River water by aqueduct.

Uncertainty about the priority of water rights was an overriding concern among non-Indian water users who feared it could result in Native Americans controlling the bulk of western water resources. There is a degree of truth in this assertion, but only so far as Indian water rights affected junior appropriated rights. However, in many drainage areas virtually all water users were junior to the Indian reserved right. Opponents of Indian settlements also argued that there were no restrictions on how tribal communities utilized the water. Instead of using water for agricultural purposes, they maintained, Native Americans were free to use water for environmental enhancement or fish and wildlife benefits. From a more cynical perspective, supporters of Indian water settlements maintained that if Native Americans gained control over a significant allotment of water, non-Indian water users would have to give up, or "pay for water that they have used freely for generations." Some proponents of Indian water settlements claimed that this aspect was the primary reason traditional water users opposed the marketing of Indian water.[474]

Despite Reclamation's changing culture, other events rooted in earlier commitments to large water development projects brought embarrassment and controversy. In 1994 the Central Arizona Project was essentially complete, with Reclamation finishing construction on the 336-mile aqueduct from the Colorado River to Tucson. After over fifty years of dreaming and fighting for it, there was little celebration in Arizona. As had been predicted by some, the estimated $4.7 billion for the project cost made the water for irrigators too expensive. This unexpected turn of events left many wondering who was going to pay for the project. A *Los Angeles Times* article reported, "Often propelled by fears of water scarcity, the CAP's major problem is 'underutilization.'" According to the article, project farmers refused to purchase the water "and whole irrigation districts were defaulting on their repayment obligations." For the Bureau of Reclamation the problem revealed the changing face of water reallocation in the American West. Perhaps for the first time in Reclamation's history, it had constructed a project where no one appeared to want the water. Dan Beard later commented on this irony stating, "Who would have thought that we would finish the project and nobody needed water for agriculture—too expensive."[475]

With the near-completion of CAP in the early 1990s, the issue of who was going to pay for the project revealed agriculture's weakened claims to the

---

474 McGuire, "Introduction: Notes on Context and Finality," in McGuire, et al., *Indian Water in the New West*, 3; Joseph R. Membrino, "A Federal Perspective," in McGuire, et al., *Indian Water in the New West*, 66; John B. Weldon, Jr., "Non-Indian Water Users' Goals: More is Better, All is Best," in McGuire, et al., *Indian Water in the New West*, 83.

475 Tom Kenworthy, "Arizona's Federal Water Project a Bitter, Costly Disappointment the West: Cheap, Plentiful Supplies were Supposed to Wean Farmers Off Underground Water, but with a High Price Tag, It Isn't Working as Planned. California Could Benefit," *Los Angeles Times*, January 30, 1994; Louis Sahagun, "Civil War Over Water: Although a Federal Project Has Provided a Wealth of Colorado Flow, Growers Can't Afford It. And Cites Are Asked to Pay the Bill," *Los Angeles Times*, December 26, 1992; Beard, *Oral History Interview*, 203.

Assessment '87

. . . A New Direction for the Bureau of Reclamation

UNITED STATES
DEPARTMENT OF THE INTERIOR
BUREAU OF RECLAMATION

13.15. Cover of the document instrumental in Reclamation's reorganization in 1987-1988.

West's water. According to one 1999 source, "The project was originally proposed at a time when agriculture, ranching and mining were dominant industries. Today, although agriculture still uses 74% of Arizona's available water, it contributes only 5% to 15% to the state gross product." A similar situation occurred in southern California in 1999 when the Metropolitan Water District asked the secretary of the interior "to reconsider the 1931 agreement that gives farmers the lion's share of the Colorado River." This overt attack on agricultural water went to the heart of Reclamation history by asking for basic reviews of "the law of the river" and Reclamation law in general. If undertaken, it would necessitate re-examining the 1922 Colorado River Compact, the Boulder Canyon Project Act, and all other legislation governing allocation of water on the Colorado River. More importantly, it would force western states to re-evaluate state water laws and their interpretations of the prior appropriation doctrine.[476]

It is difficult to identify the accomplishments and failures of Dan Beard's brief two-year tenure as commissioner of the Bureau of Reclamation. What is evident, however, is that Beard finalized the process of transforming Reclamation into a water resource management agency. In conjunction

[476] Sahagun, "Civil War Over Water;" Tony Perry, "California and the West; Battle Lines Drawn Over Water Rights; Growth: Metropolitan District's Decision to Seek Reconsideration of Colorado River Split is Met with Saber-Rattling by Farm Interests," *Los Angeles Times*, January 13, 1999.

with the commissioner's assertions that Reclamation was getting out of the construction business, Beard announced the Bureau of Reclamation's withdrawal from the Three Gorges Dam project in China. While the commissioner cited environmental concerns for his decision, Reclamation's departure ended a fifty-year intermittent relationship with Three Gorges that began in the mid-1940s. Moreover, this new streamlined Bureau of Reclamation was sensitive to environmental concerns and groups that objected to American participation in the destruction of the Yangtze River. In California Reclamation instituted guidelines for water transfers as directed by the Central Valley Project Improvement Act "to provide a more efficient and effective use of the water supply developed by the Central Valley Project." Reclamation also showed greater environmental awareness in 1996 when it opened the gates of Glen Canyon Dam to recreate the natural periodic flooding of the Grand Canyon for the purpose of "rebuilding beaches and restoring slack backwaters that are the biological heart of the canyon." In addition, Reclamation provided Central Valley Project water to improve wetlands and installed a "temperature-control device at Shasta Dam to aid Sacramento River salmon."[477]

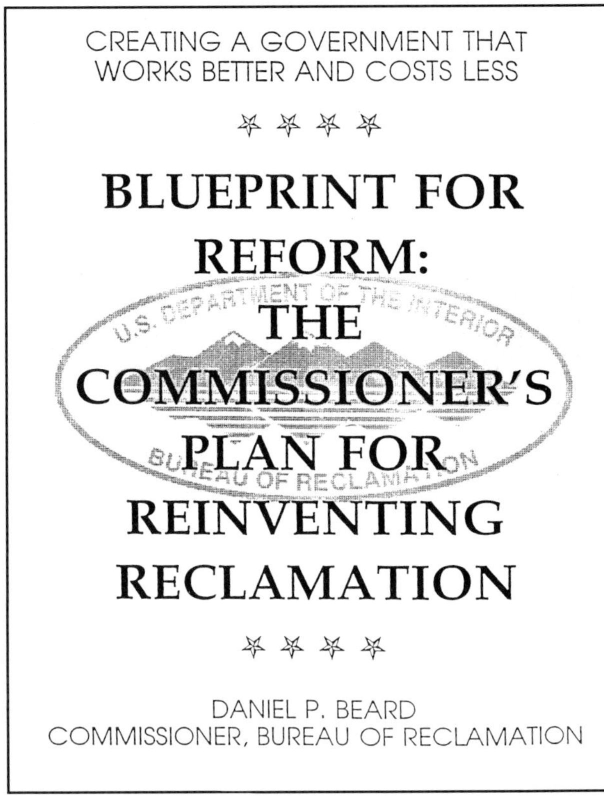
CREATING A GOVERNMENT THAT WORKS BETTER AND COSTS LESS

BLUEPRINT FOR REFORM: THE COMMISSIONER'S PLAN FOR REINVENTING RECLAMATION

DANIEL P. BEARD
COMMISSIONER, BUREAU OF RECLAMATION

13.16. The Blueprint for Reform embodied reorganization efforts while Dan Beard served as Commissioner of the Bureau of Reclamation.

---

477 Memorandum, Regional Director, Mid-Pacific Regional Office to All Interested Persons, Organizations, and Agencies, Subject: Bureau of Reclamation's Interim Guidelines for Implementation of Water Transfers Under Title XXXIV of Public Law 102-575 (Water Transfer); Brad Knickerbocker, "Artificial Deluge Used for Natural Ends By Unleashing Billions of Gallons of Water From Dam, Scientist Hope to Restore the Grand Canyon's Beaches and Wildlife

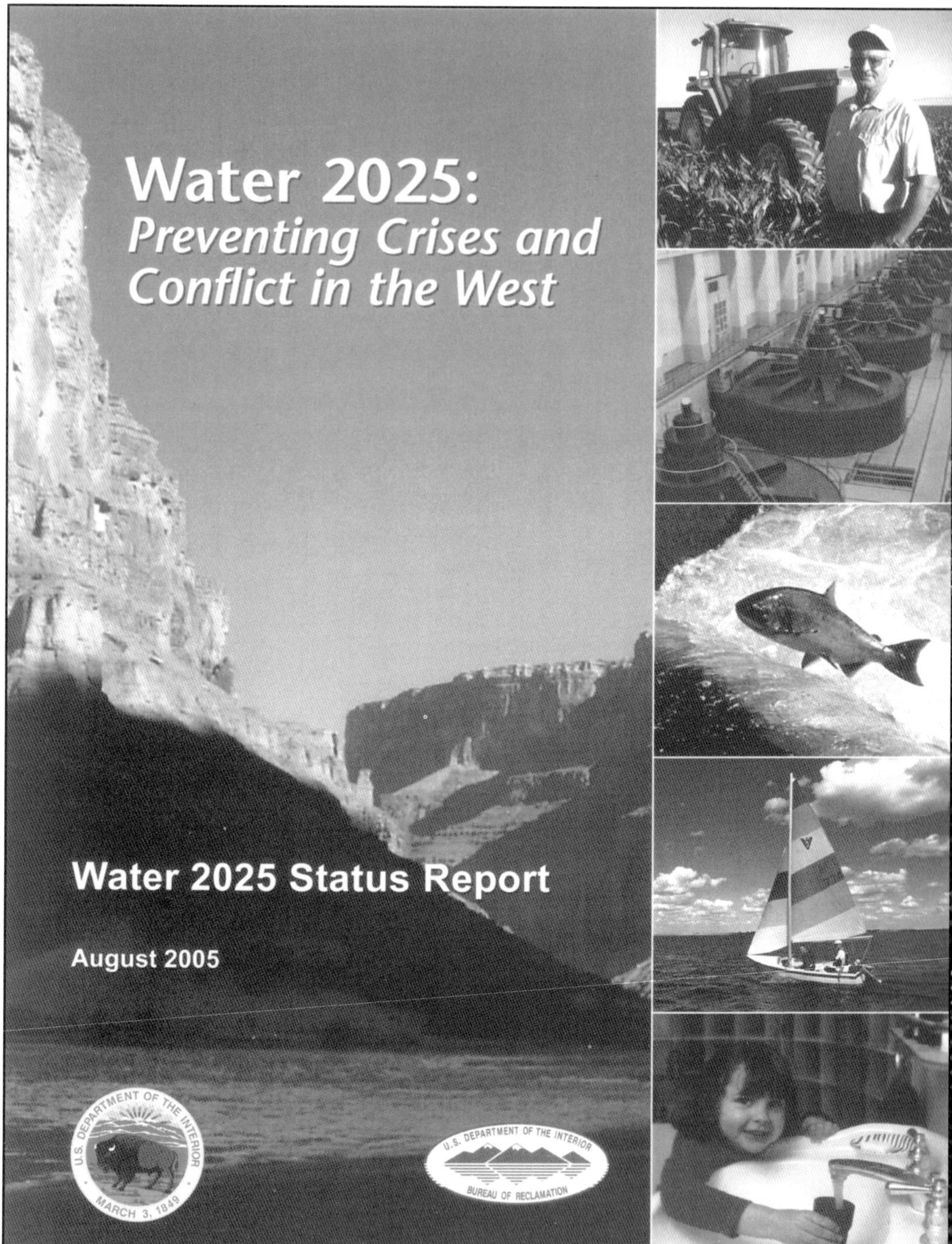

13.17. In August 2003 a booklet for the Water 2025 initiative under Commissioner John W. Keys III appeared.

In May 2003 the Bureau of Reclamation published "Water 2025: Preventing Crisis and Conflict in the West." The report outlined the water problems facing the American West as it entered the twenty-first century and offered possible solutions to address the West's water needs. Included was a long list of water controversies among competing interests vying for the region's limited water resources. Growing urban communities, agriculture, Native American tribal communities, and the environment all asserted claims to this limited resource; *Water 2025* proposed policies and programs designed to mitigate conflict among disparate parties. The document acknowledged that the federal government's major role in water resources development in the West had changed since the 1980s. Reclamation recognized "that state and local governments should have a leading role in meeting these challenges," and that Reclamation must focus "on areas where scarce federal dollars can provide the greatest benefits to the West and the rest of the nation."

Water 2025 restated Reclamation's objectives previously expressed in *Assessment '87* and Dan Beard's "Blueprint for Reform," which emphasized the Bureau of Reclamation's transformation from a construction agency to a water management agency. As in those previous documents, Reclamation reasserted the conviction that altering water usage in the West began with state law. Past goals contained similar basic "principles," such as protecting water rights long-established under the plethora of water laws, compacts, and Supreme Court "decrees." The guiding standards included modernizing existing infrastructure, enhancing water conservation, easing procedural restrictions for water transfers, and "eliminating institutional barriers for storage and delivery of water to other uses." The difference between previous Reclamation statements and *Water 2025* was the emphasis upon the urgency of changing water usage in the West. *Assessment '87* and the "Blueprint for Reform" announced new policy and program directions for the Bureau of Reclamation, while *Water 2025* called for a greater effort throughout the West to recognize the need to change views and attitudes regarding water redistribution.[478]

According to *Water 2025*, the Bureau of Reclamation had made tremendous strides in redistributing the water resources of the West to meet

---

Habitats," *Christian Science Monitor*, March 28, 1996; Tom Kenworthy, "High Tide in Grand Canyon; Opening Dam, Interior Dept. Creates Spring Flood to Restore Habitat," *Washington Post*, March 27, 1996; Mitchell Benson, "After 5 Years, Water Law Has Few Fans," *Wall Street Journal*, October 1, 1997.

[478] United States Department of the Interior, Bureau of Reclamation, "Water 2025: Preventing Crisis and Conflict in the West," May 5, 2003, 2-3; see also USDOI, BR, *Assessment '87*; Beard, "Blueprint for Reform."

13.18. The Nevada generating floor during Reclamation's centennial birthday party at Hoover Dam. June 17, 2002. Courtesy of Linda C. Arko.

13.19. (Above) Secretary of the Interior Gale Norton with the Native American veterans color guard at Reclamation's centennial birthday party. June 17, 2002. Courtesy of Linda C. Arko.

13.20. (Right) Three guests patriotically dressed for Reclamation's centennial celebration at Hoover Dam on June 17, 2002. Courtesy of Linda C. Arko.

13.21. John W. Keys III, formerly regional director of the Pacific Northwest Region in Boise, 1986-1998, served as Commissioner of the Bureau of Reclamation from 2001 until 2006.

multiple demands for a precious resource. Reclamation noted achievements in protection of endangered species on the upper Colorado River and in California's Central Valley through the cooperative efforts of water users, state agencies, and Reclamation. In addition, equitable development of water markets and water banks occurred in California, Colorado, and Idaho to the benefit of both farmers and municipalities. These examples showed that many different water users understood the dire water conditions faced by westerners and evidenced their willingness to seek solutions. Despite these apparent successes, the prior appropriation doctrine remained a powerful impediment to water reallocation in the West. While "Water 2025" recognized that the authority to change western water law rested with the states, the report suggested possible avenues for change, through collaboration, to end the often bitter disputes. It maintained, "Collaborative processes that are based on recognition of rights and interests of stakeholders allow the problem solving that maximizes the opportunity for innovation and creativity."[479]

Conclusion

In June 2002 the Bureau of Reclamation celebrated the 100th anniversary of the signing of the Reclamation Act in 1902. The Water for the West Foundation hosted a banquet attended by high-ranking officials of the Department of the Interior and the Bureau of Reclamation, along with thousands of guests that included many former and current Reclamation employees. Celebration planners selected the powerhouse at Hoover Dam as the location for the festivities. Probably more than any other Reclamation structure,

479 USDOI, BR, "Water 2025," 14-20.

13.22. President George H. W. Bush signed the Reclamation Projects Authorization and Adjustment Act of 1992. White House portrait by Herbert Abrams.

Hoover Dam represented the height of Reclamation accomplishments. Hoover Dam's symbolic and iconic status lent itself to the commemorative spirit of the evening. For many, the dam still stands as a testament to American ingenuity and technical prowess. According to a 2006 *Wall Street Journal* article, Hoover Dam "is at once an engineering marvel, a national landmark, a major tourist attraction, and a magnificent piece of public art—a perfect example of form following function." Hoover Dam's grandeur notwithstanding, the 100th anniversary highlighted Reclamation's transformation over the last twenty years. Reclamation Commissioner John W. Keys III stated during his remarks, "We are employing such strategies as water banking, voluntary water transfers, improving water treatment technologies, and contingency planning for drought." Keys went on to emphasize Reclamation's efforts in developing "new, environmentally sound water supplies in the process." With Hoover Dam as his backdrop, Keys's statement reflected the momentous changes the Bureau of Reclamation experienced during its one hundred year history. The dam embraced the traditions of the old Bureau of Reclamation and served as a monument to the men and women who contributed to building the modern West.

13.23. Robert (Bob) W. Johnson, formerly regional director of the Lower Colorado Region in Boulder City from 1995 to 2006, served as Commissioner of the Bureau of Reclamation from 2006 until 2009.

13.24. This classic early Reclamation photograph shows settler success on a project. "Shoshone project. Cherry Ranch. Old home of 1909 and new 1914 home. Came to project with $80. Has 40-acre ranch all in cultivation." September 1914. Photographer: Henry T. Cowling.

On the other hand, Keys's statement provided the framework for the future of Reclamation and its mission to mediate remedies for the diverse water users in the American West.[480]

As part of its centennial activities, the Bureau of Reclamation sponsored a history symposium, which brought together scholars from across the country. They presented their latest research covering the gamut of Reclamation activities such as construction and engineering history, project histories examining Reclamation's sometimes tempestuous relationship with water users, and biographical sketches of important personalities in Reclamation history. In a sense, the symposium was similar to the 75th anniversary issue of *Reclamation Era*. Though perhaps more critical than those found in *Reclamation Era* articles, the symposium essays nonetheless allowed Reclamation opportunities to reflect on the past and contemplate its present mission. A major theme in the scholarship emphasized the impacts and contributions,

[480] Joseph Stevens, "Masterpiece: Harnessing the Power and Awe; The Hoover Dam Embodies the American Dream of Limitless Possibility," *Wall Street Journal*, August 19, 2006; Brad Knickerbocker, "An Agency that Built the West—And Controversy; On Its Centennial, Bureau of Reclamation Stirs Reverence and Wrath for Water Policies," *Christian Science Monitor*, June 20, 2002.

both good and bad, by the Bureau of Reclamation on western American settlement and development. In his introduction to the published version of the symposium, released in 2008, Reclamation Commissioner Robert W. Johnson highlighted Reclamation's still highly visible and influential presence in the West. Johnson wrote,

> Today, Reclamation provides one out of every five farmers with water for 10 million irrigated acres. These farmlands produce sixty percent of the nation's vegetables and twenty-five percent of its fruits and nuts. We are the largest electric utility in the seventeen western states (operating 58 hydropower plants) and the nation's largest wholesale water supplier, administering 348 reservoirs with a total storage capacity of 245 million acre-feet. Nearly 30 million people all over the West depend on Reclamation projects for their municipal, industrial, and domestic water supply.[481]

Commissioner Johnson's statement pronounced, not only Reclamation's legacy, but also confirmed the still vital role the Bureau of Reclamation plays in water resources issues.

In 1893 John Wesley Powell disappointed attendees at the Irrigation Congress in Los Angeles by emphasizing the limitations of the West's water.[482] By 2000, nothing had changed in terms of scarcity, and, indeed, the situation had grown worse. Population growth, industrialization, and increased agricultural production stretched limited water resources. Added to these variables were new demands on water for environmental enhancement. While heightened demand places new premiums on water, the dams, canals, and reservoirs constructed by the Bureau of Reclamation form a foundation for water security in the West. Reclamation's new mission today seeks to achieve some measure of equity to satisfy competing demands. It actively pursues programs to promote water transfers, to improve water conservation methods, and to utilize its facilities for environmental restoration and protection. All reflect Reclamation's ability to transform its culture to meet the demands of changing times and values. A reconfigured Bureau of Reclamation still serves the water needs of the American West, as was its purpose throughout the twentieth century.

---

481 *The Bureau of Reclamation: History Essays from the Centennial Symposium*, Volumes I and II (Denver, Colorado: Bureau of Reclamation, U.S. Department of the Interior, 2008), xxvii.

482 Rowley, *The Bureau of Reclamation*, 64; Worster, *Rivers of Empire*, 132; Billington, et al., *The History of Large Federal Dams*, 24.

CHAPTER 14:

SELLING RECLAMATION: THE BUREAU OF RECLAMATION IN PHOTOGRAPHS, ART, AND FILM

Introduction

As a public service bureau, the Bureau of Reclamation recognized the value of recording and portraying its activities to mold positive perceptions of its mission. Artistic representations occurred across a variety of media. The works of photographers, filmmakers, sculptors, and painters contained an unmistakable message of progress through water development—dams, reservoirs, hydroelectricity. Bureau of Reclamation achievements presented in these various formats reflected the can-do spirit of Reclamation employees, and through them, all Americans.

Generally, photography was and remains the medium through which Reclamation advertised its activities. Many of these photographs appeared in the Bureau of Reclamation's publication *Reclamation Era*. Images also appeared in the popular periodicals of the day such as *National Geographic*, *Life*, and *Fortune* magazines. Photographers often went into the field with engineers during the early years of the Reclamation Service to collect images for use in lantern slide shows and lectures to attract settlers and inspire congressional and public support. These images documented progress occurring on irrigation projects and assisted engineers in their study and observation of construction techniques. During the large dam era (1930s-1960s), renowned photographers sought out Bureau of Reclamation works as subjects. They saw in dams and hydroelectric powerplants representations of what was termed "machine aesthetic." The profusion of images depicting Bureau of Reclamation structures, Hoover, Grand Coulee, and Glen Canyon dams, promoted the fame of these structures as some of the most prominent technological icons of the twentieth century. The images reinforced Reclamation's reputation as one of the world's leading engineering organizations.

While still pictures served the early public relations goals of Reclamation, motion pictures offered even more vivid portrayals of its good works. As early as the 1910s, the Reclamation Service produced and distributed films documenting the accomplishments of both project engineers and farmers. These films reflected the hostile environments surrounding many projects and celebrated Reclamation's ability to alter nature, build homes and communities, and create a livable West for Americans. Reclamation filmmaking eventually highlighted the major undertakings of Hoover, Grand Coulee, and Shasta dams— their construction, completion, and operation. The enormity of these projects appeared in film, especially construction phases with huge explosions tearing away rocky cliffs, the ballet-like movements of high-scalers drilling and chipping away loose rocks, and the tremendous number of men and machinery mobilized to build these massive structures. As Reclamation moved beyond building big dams in the late twentieth century, promotional films produced by the Bureau of Reclamation stressed its new emphasis on water management and environmental enhancement.

Bureau of Reclamation dam designers and engineers valued both functional and aesthetically pleasing structures. Designers adorned dams and powerhouses with paintings, sculptures, and floor terrazzo to accentuate the structural qualities of both large and small projects. The sculptures at Hoover Dam sought to capture the purposes of the dam—power, water, irrigation, navigation, flood control. While the images were in the styles of the 1930s social realism, they were symbolic of the historic ideals of nineteenth-century reclamation advocates, reflecting dominance over the chaotic forces of nature. In the late 1960s and early 1970s the Bureau of Reclamation Art Program sought to immortalize Reclamation projects by commissioning American artists to produce an amazing collection of images. In what some called a return to the landscape art of the nineteenth century, scenes of Bureau of Reclamation projects and structures appeared to blend with the surrounding scenery.

In the media of photography, motion pictures, and art, the history and contributions of the Bureau of Reclamation unfold in a vast representative panorama. The effort supplemented the tradition of western boosters who had long promoted western growth and development. The representations of dams, canals, reservoirs, and bountiful farmlands adorned depictions of accomplishments by the Bureau of Reclamation. Project communities or irrigation districts also added to the multiple narratives of Reclamation history through the production of documentary films that highlighted their own growth and prosperity. While many of these looked at the Bureau's efforts to reclaim desert lands from a positive point of view, some criticized the federal government for not following through on its commitments spelled out in the 1902 Reclamation Act. In a larger scope, however, the images reveal the transformation of the

American West from a nineteenth-century arid wasteland to a region of great urban centers and desert transformed into a garden.

Photography and the Photographers

Since the mid-nineteenth century, photography has played an important role in promotion and settlement of the American West. For instance, John C. Frémont took a daguerreotype camera on three of his five expeditions into America's western frontier. For the most part, Frémont had little success in capturing images of the West. It was not until his fifth and final expedition in 1853 and his collaboration with Solomon Carvalho that Frémont brought back daguerreotypes of the prairies and mountains of Kansas, Colorado, and Utah. Also in 1853, painter and daguerreotypist John Stanley Mix recorded images of Isaac Ingalls Stevens's Pacific railroad survey from Lake Superior to Puget Sound.[483] These initial attempts at capturing images of western landscapes continued throughout the remainder of the nineteenth century.

With the advent of new photographic technology and the mass production of paper prints, photographs of the West reached much larger audiences. After the Civil War, the federal government funded a number of scientific surveys of the American West to create topographical maps and provide general information on its economic potential. Survey leaders Ferdinand V. Hayden, John Wesley Powell, Clarence King, and George M. Wheeler also understood the value of photographs as a promotional tool for not only the region but their own personal achievements and ambitions. Historian Martha Sandweiss notes, however, that photographs alone were unable to convey the whole story of western benefits, but combined with a narrative, the potential for depicting the continuation of national progress in western lands appeared more promising. The photographs inserted in government survey reports, along with descriptive narratives, spoke to the future of the United States and the challenges of settling the West. Images of western places also had the broader power to implant a mental picture of the West into the minds of many Americans. In *Print the Legend: Photography and the American West*, Sandweiss states,

> Photographs of western landscapes and the native peoples of the West would move into American parlors and schools, work

[483] Martha A. Sandweiss, *Print the Legend: Photography and the American West* (New Haven: Yale University Press, 2002), 108; Weston J. Naef and James N. Wood, *Era of Exploration: Rise of Landscape Photography in the American West, 1860-1885* (Buffalo: Albright-Knox Art Gallery; Boston: distributed by the New York Graphic Society, 1975), 29, 77.

places and exhibition halls; and the words affixed to them would help to ensure their importance as a popular form of art that would simultaneously reaffirm and create broadly held ideas about the place of the West in national life.[484]

The popular press published excerpts of the government surveys of Powell, King, Hayden, and Wheeler and brought to the American public vivid descriptions of a place imbued with marvelous colors, incredible landscapes, and economic potential.

Westerners too saw the benefits that photographs played in promoting their communities and attracting new citizens and investors. Civic boosters portrayed their cities as models of modernity, emphasizing new roads and buildings with all the modern conveniences. Photographs of western mining areas revealed the vast investment opportunities for eastern and foreign capitalists, while picturesque landscape images lured both developers and tourists. Martha Sandweiss points out the omission of certain landscapes and environments that did not fit an optimistic regional vision. Few photographs revealed deserted towns fallen into disrepair or left to ruin, nor did photographs portray the "foreboding deserts of the American West."[485] It was a glorious and profitable future that these photographs evoked rather than an untamed wilderness full of dangers and hardships. In 1902 the Reclamation Act brought to many western communities a promise of growth and prosperity. The Reclamation Service quickly took up the practice of using photographs accompanied by romantic narratives to encourage settlement on its irrigation projects and, perhaps most importantly, promote Reclamation activities to a national audience.

The photographic department within the Reclamation Service fell under the direction of Clarence John Blanchard, commonly referred to as C. J. Blanchard, lead statistician and the person in charge of the Service's Settlement Section. Blanchard not only brought to the Reclamation Service engineering experience through his work as the deputy state oil examiner of Iowa and as a member of various irrigation surveys, but also an understanding of public relations gained through his tenure as a newspaper reporter in Central America in 1892 and 1893.[486] Blanchard enthusiastically embraced the task of publicizing

---

484 Sandweiss, *Print the Legend*, 126; see also, Naef and Wood, *Era of Exploration*, 32.

485 Sandweiss, *Print the Legend*, 328.

486 *Sixteenth Annual Report of the Reclamation Service: 1916-1917* (Washington, D.C.: Government Printing Office, 1917), 14-5; for biographical information on Blanchard see, *Who was*

14.1. A hand-colored lantern slide used to give talks on Reclamation and its projects. This 1910 slide taken at Conconully Dam, on the Okanogan Project, shows Reclamation using a now discredited modified hydraulic earthfill method of dam construction.

Reclamation activities. He eagerly joined forces with other western boosters and government agencies to promote western development. Relatively unknown in Reclamation history, Blanchard's essays, with photographic illustrations, in the Reclamation Service's publication *Reclamation Record* and popular journals of the time reveal an ardent advocate of the Reclamation Service's mission.

One of the many methods Blanchard used to promote activities of the Reclamation Service was lectures accompanied with stereopticon slides. The mixture of images and narratives brought to life the progress and accomplishments occurring on irrigation projects. His narratives told of Reclamation engineers' heroic accomplishments, while slides revealed the massive tasks of digging irrigation canals and building dams. Blanchard described Reclamation's tremendous undertaking as a benefit for the whole nation and attempted to calm criticism that it was only a benefit to the West. In 1906 Blanchard maintained

> Uncle Sam is today the largest owner of the Great American Desert, no doubt because it is not worth stealing. For many years the sentiment has been growing that the government

---

Who in America Vol. IV, 1961-1968 (Chicago: Marquis—Who's Who Inc., 1968), 93.

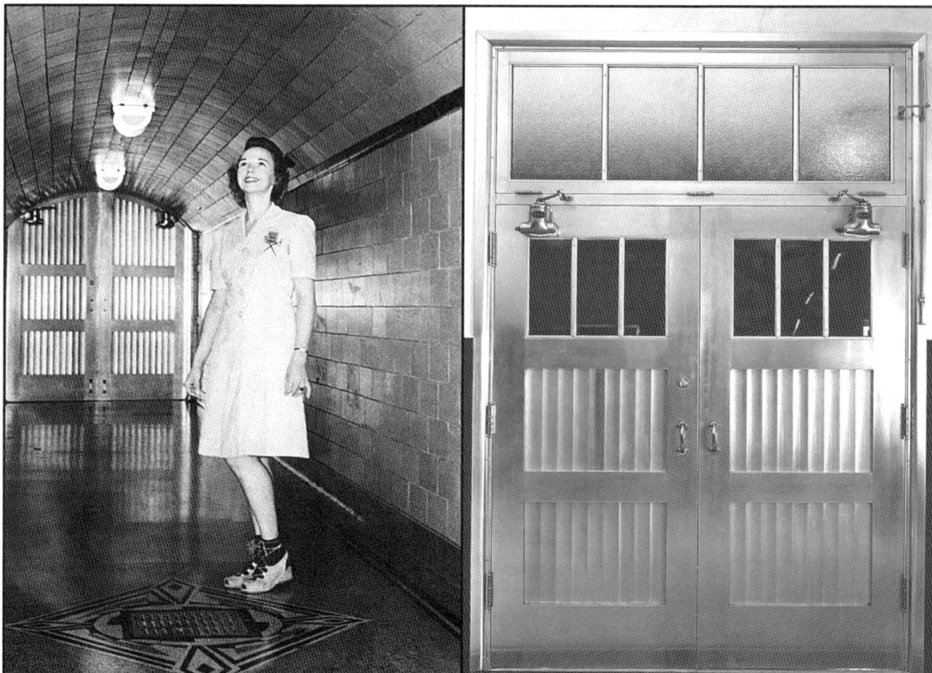

14.2. High visitation at Hoover Dam, from the very beginning of construction work, caused Reclamation to develop impressive public facilities using materials like terrazzo floors, tile lined visitor galleries, marble elevator lobbies with bronze doors and indirect lighting, and use of aluminum detailing in doors, light fixtures, and office furniture. This was government construction in an era when a large part of the American public wanted to see the government strong and stable.

> should make habitable this vast empire which is so great potentially.

Like nineteenth-century boosters before him, Blanchard looked to growth in the American West as a national question demanding the attention of the federal government.

Blanchard's lectures usually related a summary of the progress of the Reclamation Service—making it a point to emphasize the difficult engineering obstacles encountered on some of the projects and the resourcefulness of Reclamation engineers. Digging the Gunnison Tunnel on the Uncompahgre Project in 1906 was especially worthy of note because it was "*where courage was even more necessary than engineering skill*." Photos showed workers boring through the mountain to connect the waters of the Gunnison River with those of the Uncompahgre River in Colorado. With unabashed admiration, Blanchard noted that in western Colorado "we find the engineers of the Reclamation Service constructing one of the most spectacular works ever attempted in the West, a great tunnel nearly six miles in length and passing under a mountain 2,000 feet high."

Images of fantastic engineering feats accompanied photos showing fields of abundant crops as proof of the economic benefits occurring on Reclamation projects. In order to emphasize irrigation's economic potential, Blanchard's lectures paid particularly close attention to the rising land values associated with projects. His slides cleverly showed the amazing transition that was taking place in the West as sagebrush gave way to fields of crops, homes, and communities—which became a mainstay of Reclamation photography. And in 1906, this was only the beginning. Blanchard proclaimed that "out of the desert, now the haunt of the skulking coyote and jack-rabbit, fair cities shall rise, and in the midst of islands of emerald smoke thousands of industries shall rise."[487]

The early work of the Reclamation Service competed for national attention with construction of the Panama Canal as well as massive subway systems in America's urban centers. In this era of the engineer, Blanchard equated land reclamation with other large-scale engineering projects. Much of Reclamation's progress, according to Blanchard, derived from innovative engineering plans that utilized the natural resources in proximity to projects and that overcame obstacles to construction. Government engineers not only built dams and canals but constructed roads to remote areas, cement plants to supply material, and machine shops and other necessary appurtenances. Much of the power came from hydroelectric plants. At Roosevelt damsite on the Salt River Project in Arizona, Blanchard noted that "the government cement mill gives noisy evidence that Uncle Sam is a manufacturer. Night and day his

[487] C. J. Blanchard, "Winning the West: An Account of the Marvelous Progress of Our Reclamation Service in Reclaiming the Desert," *National Geographic Magazine*, 17 (February 1906): 83, 89-90, 98; italics in the original.

plant is grinding the best cement ever made, and ere the mill has finished its work 240,000 barrels will have been used in the dam."[488] Blanchard's photographs and others provided chronicles of the construction used and shared throughout the Service as learning tools. Whether used in-house or for publicity purposes, the images were available for all to see Reclamation's efforts to build a more prosperous American West.

Blanchard's lectures and slides included discussions on the current condition of project farmers and their future prospects. For Blanchard and indeed all Reclamation supporters, the greatest value of the federal reclamation policy derived from the creation of homes for American families. Blanchard shared the concern of many Americans about the problems arising from a rapidly urbanizing and industrializing America. Many saw western irrigation projects as a possible safety valve to relieve urban congestion and possibly social turmoil. Blanchard, as did others, argued that crowded urban living was not conducive to establishing American ideals, and that there was "no national stability in a citizen born and reared in tenements." He also maintained that it was the government's mission to ensure its citizens had the opportunity to acquire homes and "that it is a national duty to render the acquirements of homes as early as possible."[489] The arguments for western reclamation increasingly returned to the safety valve thesis that saw western arid land agriculture as part of a solution to Americanize the new immigrant population in the cities. His belief that federal reclamation helped Americans acquire homes updated the nineteenth-century homestead ideal embodied in the 1862 Homestead Act. The idea played well with many eastern Progressives and rallied support for Reclamation in those circles.

Blanchard's slide shows revealed not only how irrigation projects relieved urban crowding but also showed that Reclamation Service projects nourished a prosperous and energetic citizenry. One of the many, and perhaps the most powerful, methods Blanchard utilized was "before" and "after" pictures. Where once stood a barren sagebrush desert, photos revealed growing vibrant communities that reflected both the resourcefulness of settlers and the accomplishments of the Reclamation Service. The images depicting this rapid transformation of the arid West impressed audiences. Blanchard assured

---

[488] C. J. Blanchard, "Millions for Moisture: An Account of the Work of the U. S. Reclamation Service," *National Geographic Magazine*, 18 (April 1907): 225.

[489] C. J. Blanchard, "Home-Making by the Government: An Account of Eleven Immense Irrigation Projects to be Opened in 1908," *National Geographic Magazine*, 19 (April 1908): 250; see also Blanchard, "Millions for Moisture," 230.

his listeners that these achievements meant economic benefits in the form of rising land values and the growth of western enterprise. To dispel the notion that farm life was a lonely venture, Blanchard emphasized the modern features found in communities springing up in and around Reclamation projects. His images revealed towns with all the modern conveniences of newspapers, telephone lines, schools, libraries, "and trolley lines to the towns serving to bring the desert farmer within the stimulating current of the world's thoughts."[490]

Through words and pictures, Blanchard depicted settlers on irrigation farms as similar in circumstances to colonial era farmers imbued with courage and patriotism which, in Jeffersonian fashion, created democratic communities of independent landowners. His idealism almost echoed the hyperbole of irrigation advocate William E. Smythe, who in *Constructive Democracy* (1905), declared that land ownership "offered both prosperity and the highest form of democratic government to the common man." Indeed, according to Blanchard, irrigation farming brought communities together to work for common goals as opposed to the "provincialism" found in eastern agricultural areas. In pointing out these differences, he attempted to impart to his audiences the unifying aspects of agriculture under the ditch, held together by the "irrigation canal … which binds the community together."[491] These were no ordinary farms, and neither were these ordinary farm communities. Blanchard's slides showed settlers entering a new frontier fraught with many perils who, with the help of the Reclamation Service, were creating bountiful farms and revealing to the world the strength and fortitude of the American people.

There was more going on here than homemaking. According to Blanchard, these settlers were continuing the process of nation building, "which enabled our ancestors to wrest a commonwealth from the New England wilderness." Propelled by this ideology, settlers were "imbued with high ideals and noble purposes, and by their achievements are establishing us more firmly in our place among the greatest nations of the earth."[492] Blanchard's photographs and rhetoric, however, belied the conditions many settlers encountered upon entering Reclamation projects. The photographs of a barren, empty

490 C. J. Blanchard, "The Call of the West: Homes are being made for Millions of People in the Arid West," *National Geographic Magazine*, 20 (May 1909): 434.

491 As quoted in Rowley, *The Bureau of Reclamation*, 79; C. J. Blanchard, "The Spirit of the West: The Wonderful Agricultural Development Since the Dawn of Irrigation," *National Geographic Magazine*, 21 (April 1910): 335; see also William E. Smythe, *Construction of Democracy: The Economics of a Square Deal* (New York: The Macmillan Company, 1905); William E. Smythe, *The Conquest of Arid America* (New York: The Macmillan Company, 1905).

492 Blanchard, "The Call of the West", 407.

14.3. A "1 horsepower" mine car carries excavation material from the Gunnison Diversion Tunnel on the Uncompahgre Project in Colorado.

14.4. Hole-through in the Gunnison Diversion Tunnel.

14.5. C. J. Blanchard, on the right, with J. H. Quinton, supervising engineer of the Uncompahgre Project, at the west portal of the Gunnison Diversion Tunnel while it was still under construction.

wasteland alone may have made many potential settlers wonder about their chances for success. Blanchard himself even understood the difficulties involved. Yet, Blanchard's images along with an optimistic promising narrative portrayed a bright future courtesy of the intrepid and enterprising spirit of the Reclamation Service.

It was essential for Blanchard and other irrigation supporters to maintain public support for Reclamation by downplaying the fact that irrigation projects primarily benefited the West. Blanchard emphasized the economic potential of western reclamation to the rest of the nation to prove that irrigation projects were more than just a regional subsidy—they were a national endeavor. He argued, "Western development means additional markets for eastern manufactured products.... With enormous increase in demand for such products, the manufacturers will be compelled to enlarge their plants and add to the number of employees."[493] Blanchard's images presented emerging towns complete with merchants ready to sell the latest commodities; with the added benefit of hydroelectric power, farm families were ready and eager to own all the modern household conveniences. Photographs showed farmers using modern agricultural equipment that meant emerging markets for eastern goods.

Another source of national pride proclaimed in Blanchard's lectures and slides was the transformation of the environment brought about through the work of the Reclamation Service. This, of course, is more apparent in Blanchard's romantic praise of the engineer's ability to control the forces of nature. "The engineer," Blanchard claimed,

[493] Blanchard, "Millions for Moisture," 218-21.

14.6. On August 5, 1910, Walter J. Lubken captured Reclamation's early ideal of agriculture while visiting the Uncompahgre Project near Montrose, Colorado—small acreage scientifically and intensively farmed with a variety of crops like orchards, truck gardens, and hay fields.

> finds no field more attractive than this for his energies. He curbs the stream with masonry dams and lifts the water into huge canals. Water and land long divorced are wedded, and wavering fields of grains and orchards prolific beyond comparison replace the wastes of sand and sagebrush.

Material improvements also portended richer spiritual lives. Both Smythe and Blanchard stressed that desert living stood to bring out the best qualities of men, claiming that the desert offered opportunities to break away from "the dead level of mediocrity which prevails where people are overcrowded and underfed." Instead, Blanchard maintained, "the perpetual sunshine and the individual home" would lead "us back from the material to the spiritual into ways of gentleness and simple living."[494]

Blanchard's lectures and images painted a glowing future for the American West and, with it, the entire nation. The photographs that he used reinforced the notion of humanity's ability to dominate nature and profit from it. On the Klamath Project in Oregon for example, Blanchard related how

[494] Blanchard, "The Call of the West," 407; Blanchard, "Millions for Moisture," 239-43.

14.7. Alfalfa field on the A. W. Emberg Ranch, Belle Fourche Project, South Dakota. July 30, 1910. Photographer: Walter J. Lubken.

14.8. This field of tomatoes, near Dixon, California, on the Solano Project, was laid out to allow use of machinery for both cultivation and harvest. May 21, 1984. Photographer: D. M. Westphal.

14.9. Walter J. Lubken images taken on February 21, 1905, and March 29, 1905, document the progress of excavation on a rock cut on the Apache Trail between the railroad east of Phoenix and Theodore Roosevelt damsite on the Salt River Project.

Reclamation Service engineers planned to drain two lakes and reclaim their exposed beds for agriculture. The Reclamation Service even challenged the mighty Colorado River by sinking the 600,000 ton Laguna Diversion Dam into the silt laden river bed to irrigate fields on the Yuma Project in Arizona. Also in Arizona, the Salt River Project offered the Reclamation Service's crowning achievement in the first decade of the Reclamation Act. It represented all the positive facets of land reclamation in regards to increased agricultural production and hydroelectric power output. In addition, it showed off the Reclamation Service's ability to control nature to suit the needs of the human community. Construction photographs revealed the progress being made on Roosevelt Dam as it slowly rose in the midst of the Arizona desert to "create the largest artificial lake in the world and furnish 200,000 acres of land with water."[495] Blanchard's, and through him the Reclamation Service's, slides and lectures presented powerful evidence of American technological prowess.

Along with lectures and slide presentations, the Reclamation Service relied on an array of venues to promote its activities and encourage settlement. In many instances, C. J. Blanchard was at the center of these efforts. The idea was to get the message out to as many people as possible, and the many expositions that occurred during the first two decades of the twentieth century offered marvelous opportunities. With the large numbers of spectators these expositions attracted, the Reclamation Service was able to educate the public on western reclamation and, perhaps interest prospective project settlers. In 1906 J. C. Boykin of the United States Government Board for the Jamestown Tercentennial Exposition (1906-1907) expressed similar sentiments to Reclamation

[495] Blanchard, "Millions for Moisture," 226.

14.10. On April 16, 1910, Walter J. Lubken captured this view of the operating cement mill at Roosevelt dam. This plant provided cement for construction at both Granite Reef Diversion Dam east of Phoenix and Theodore Roosevelt Dam.

Service Chief Engineer F. H. Newell. Boykin thought the Exposition presented the Reclamation Service an excellent venue "to bring its work to the attention of those who were not previously familiar with it." He also reminded Newell of the self-serving importance of such advertising because "we are dependent upon people to take up our lands, to repay our expenditures and to make our continued existence possible."[496] Along with the Jamestown Tercentennial Exposition, the Reclamation Service created exhibits for the 1905 Lewis and Clark Exposition in Portland, the 1908 Alaska-Yukon-Pacific Exposition in Seattle, and the 1915 Panama-Pacific Exposition in San Francisco.

These exhibits consisted of a wide range of visual material depicting the work of the Reclamation Service including models of Roosevelt Dam, Laguna Dam, and the Yuma Project. Photographic images—consisting of transparencies and bromides—played an important role in presenting not only the Service's accomplishments but also views of western landscapes. The photographs showed exposition visitors the rate of progress made on Reclama-

[496] J. C. Boykin to F. H. Newell, December 10, 1906, RG 115, Entry 3, Box 244, Folder 823 "General Exhibits by the Bureau of Reclamation at Expositions."

tion projects with images of "desert wheat fields" juxtaposed against "mesquite growth." Photographs of Arizona's Salt River in flood and its destructive nature reinforced in viewer's minds the need for the work of the Reclamation Service. Views of reservoirs and damsites implied that these threats to western farmers were coming under control. Pictures of powerplants, steam dredges, and diamond drills revealed the technical prowess of the Reclamation Service, which served to project the enormity of its monumental task in the arid West, and its ability to succeed.[497]

In 1914 the Director of the Reclamation Service, Frederick H. Newell, authorized Blanchard to travel west to secure information from state land agents, railroad immigration agents, and water users' associations on the conditions and opportunities for home seekers on Reclamation projects. Part of Newell's objectives for this trip were not only to gather settlement data, but for Blanchard, along with Reclamation Service photographer Henry T. Cowling, to obtain "photographic material as may be needed in illustrating the present condition of reclamation projects particularly for use in the Panama-Pacific Exposition at San Francisco in 1915." The Reclamation Service's exhibit in San Francisco used a wide assortment of images including illustrated lectures with films and slides along with a large diorama. Blanchard praised the exhibit for its ability to attract attention to the Service's settlement work and the opportunity it presented to meet face to face with prospective settlers. The exhibit won numerous awards including a gold medal for "Photographs and Enlargements" and an honorable mention for "Motion Picture Drama."[498] The Panama-Pacific Exposition offered a platform—a high profile event—for the Reclamation Service to show off its accomplishments and to advertise the benefits of land reclamation. Blanchard also noted that Reclamation now had a multi-media format through which to promote land reclamation in the nation.

Newell's instructions to Blanchard underlined the Reclamation Service's endeavors to attract and cultivate good relations with water users' associations and important western settlement boosters, railroads, and local chambers of commerce—all in an effort to attract settlers to the projects. For

---

[497] A. P. Davis to Edward M. Dawson, September 1, 1905, RG 115, Entry 3, Box 244, Folder 823 "General Exhibits by the Bureau of Reclamation at Exposition."

[498] F. H. Newell to C. J. Blanchard, June 5, 1914, RG 115, Entry 3, Box 102; C. J. Blanchard, "Current Comments," *Reclamation Record*, 7 (February 1916): 51-2; C. J. Boykin to the Secretary of the Interior, July 17, 1915, RG 115, Entry 3, Box 244, Folder 823 "General Exhibits by the Bureau of Reclamation at Expositions;" see also C. J. Blanchard, "Current Comments," *Reclamation Record*, 6 (June 1915): 248-9; "Panama-Pacific Prizes," *Reclamation Record*, 6 (September 1915): 394.

Blanchard, these relationships were critical to the success of his promotional activities. Because of constant shortages of funds for publicity campaigns, Blanchard often solicited donations from all three to help pay for the production of promotional material. He used flattery and promises of future economic benefits to appeal for funds. In Orland, California, Blanchard cajoled the local water users' association by noting that "in view of the very great development which is taking place on the Orland project ... I do not think you should have any difficulty in securing the cooperation of the people of Orland to the full amount of $1,000 for this purpose."[499]

Blanchard's enthusiasm for promotional documentaries led him beyond land reclamation and irrigation projects. In a letter to the Los Angeles Chamber of Commerce, Blanchard proposed a venture to produce a "government release" titled "Pathway of the Padres.... The pictures I have in mind would illustrate a tour of California from San Diego to San Francisco, and would take in the principal missions and the best scenic spots along the coast." As with chambers of commerce, railroads "interest" became a major source of funding for promotional purposes. For a proposed image gathering foray on the North Platte Project, Blanchard reminded Union Pacific executives of their economic stake in the project. He asked them to "not forget that all future work on the Nebraska side of the North Platte Project is to be exclusively on the U.P. side of the river, and that many thousands of acres from Gering to Northport are to be reclaimed." Though Blanchard's primary task was to promote settlement on Reclamation projects, he viewed his responsibilities broadly as part of the larger picture of advancing western development. Eventually Blanchard's extensive work came to the attention of Secretary of the Interior Franklin K. Lane who by 1915 was critical of Reclamation and in particular Director Newell. Lane saw Blanchard's work with outside entities beyond his responsibilities as statistician for the Reclamation Service and curtailed his work.[500]

By 1915, however, collaboration among multiple interests who sought to promote the West was fairly standard practice. In *Promised Lands: Promotion, Memory, and the Creation of the American West*, David Wrobel emphasizes this very point. Wrobel argues various promoters with an eco-

---

[499] C. J. Blanchard to P. D. Dodd, President of the Orland Water Users' Association, February 18, 1920, RG 115, Entry 7, Box 17.

[500] C. J. Blanchard to Frank Wiggins, Secretary of the Los Angeles Chamber of Commerce, June 18, 1920, RG 115, Entry 7, Box 78; C. J. Blanchard to A. L. Craig, Union Pacific Railroad, September 26, 1921, RG 115, Entry 7, Box 78; Rowley, *The Bureau of Reclamation*, 198.

14.11. A tractor disk plow clearing raw sagebrush land on the Kittitas Division of the Yakima Project. July 9, 1929. Photographer: J. E. Stimson.

nomic interest in developing the West sold "places like they would any other commodity." As "Pathway of the Padres" suggests, Blanchard proposed to do more than just lure prospective farmers onto Reclamation projects. He was selling an image of the West. In this case it turned out to be the romance of Spanish California. As Secretary Lane indicated, such undertakings were well beyond the bounds of Blanchard's responsibilities with the Reclamation Service. Yet, as Blanchard's correspondence with the Orland Water Users' Association and the Union Pacific railroad suggest, the notion of an agency of the federal government working in unison with local organizations was not outside the realm of possibility. The intersection of "interior and exterior" boosterism was commonplace in the West. Wrobel notes the two worked together "to promote the flow of settlement and capital to western places."[501]

Many of the photographs taken during the early years of the Reclamation Service found their way onto the pages of the Service's own publication *Reclamation Record*. Others depicted the endeavors of western irrigation in popular magazines such as *National Geographic*, *The Mentor*, and *Scientific American*. In Blanchard's articles and lectures, the actual photographers who took the images received little or no public credit. Despite the presence of cameras on most Reclamation projects, few Bureau employees received much public recognition for the work they did in documenting the progress and achievements accomplished on Reclamation projects. The primary explanation of the mystery surrounding Reclamation Service/Bureau of Reclamation photographers is

[501] David M. Wrobel, *Promised Lands: Promotion, Memory, and the Creation of the American West* (Lawrence: University Press of Kansas, 2002), 7.

the result of its policy of controlling the publication and distribution of photographic material. For example, in her study on photographic chronicling of the construction of Hoover Dam, Barbara Vilander maintains that the Bureau's complete control over photographic images ensured that the published photographs only portrayed positive aspects of the dam's construction.[502]

Reclamation Service records and correspondence, however, place some names on the numerous photographs taken during the first twenty years. Reclamation Service photographer Henry T. Cowling accompanied Blanchard on his trip to gather images for the exhibit at the Panama-Pacific Exposition. Blanchard praised the work of Ray B. Dame, who assisted him in amassing photographic material on a trip to the Minidoka Project in 1920. Dame continued to work as a photographer for the Department of the Interior, and in 1936, achieved the position of associate chief of the Division of Motion Pictures.[503] Interoffice correspondence and memos offer a brief glimpse of a few of Rec-

14.12. May 8, 1910, Walter J. Lubken photographed Laguna Dam's regulating and sluice gates. Yuma Project, California and Arizona.

[502] Barbara Vilander, *Hoover Dam: The Photographs of Ben Glaha* (Tucson: University of Arizona Press, 1999), 19.

[503] F. H. Newell to C. J. Blanchard, June 5, 1914, RG 115, Entry 3, Box 102; see also William Harper Dean, "The Master Movie Maker," *Technical World Magazine*, 22 (January 1915): 730-2, 788; see also C. J. Blanchard to The Director, U. S. Reclamation Service, September 6, 1920, RG 115, Entry 7, Box 17; "Reclamation Organization Activities and Project Visitors," *Reclamation Era*, 26 (June 1936): 155.

14.13. Frederick H. Newell was chief engineer of Reclamation from 1902 to 1907 and director of the U.S. Reclamation Service from 1907 to 1915.

lamation's employees who helped create the Bureau of Reclamation's vast photographic archive.

A few Bureau of Reclamation photographers, however, have achieved a measure of outside recognition. Their talent as photographers reflected in the artistic quality of their work awarded them recognition beyond their careers with the Bureau of Reclamation. Walter J. Lubken, Reclamation Service photographer from 1903 to 1917, accompanied Blanchard in documenting the construction and growth of the Reclamation Service's projects throughout the West. In his fourteen years with the Reclamation Service, Lubken photographed the emergence of towns and homes on all twenty-five projects in seventeen western states. Lubken's efforts are reminiscent of the work performed by nineteenth-century photographers William Henry Jackson and Timothy O'Sullivan. In Lubken's images, the arid West remained a wild mysterious frontier. Yet, juxtaposed against this beautiful but hostile environment were the people and accomplishments of the Reclamation Service who brought civilization and progress. If the images of William Henry Jackson helped bring the mysteries and wonders of the West to many Americans, Lubken's photographs revealed the taming of the West through American ingenuity and determination.

The artistic quality of Lubken's photographs has been recognized for their balance of light, shadows, and texture. More importantly, his images gained acclaim for capturing the spirit and energy of the time. According to the *Salt River Project Pulse*, the images' greatest asset is in their content and how events "that shaped American history were faithfully recorded by Lubken." Bruce I. Bustrad of the National Archives echoed this praise and argued that Lubken's images not only capture the engineering feats of the Reclamation Service but also scenes of everyday life in the early twentieth century. In *Picturing the Century: One Hundred Years of Photography from the National Archives*, Bustard states that Lubken's optimistic images impressed viewers with technical and social advances made by westerners. They make the point that progress and community had come to barren places and that abundant opportunities awaited those willing to move west and work hard on reclaimed land.[504]

Lubken left the Reclamation Service in 1917 and seems for a while to have disappeared from the photography stage until 1930 when he became the official photographer for the Six Companies, documenting its part in the construction of Hoover Dam. In this endeavor, Lubken and managers of the Six Companies departed from the tradition of portraying triumphal, heroic dam construction. Lubken's images depicted dangerous working conditions and blistering heat at the damsite. Photographic historian Barbara Vilander maintains that Six Companies officials "did not hesitate to publish images that graphically portrayed the extremely hot temperatures and physical dangers the workers confronted."[505] Nevertheless, Lubken's photography record linked the history of the Bureau of Reclamation and its capacity to engineer gigantic dams to a heroic tradition of monumental dam building.

Others outside the Bureau of Reclamation also contributed to the photographic history of the Reclamation Service. They revealed a keen interest on the part of local communities in the development of irrigation projects. Like the Bureau of Reclamation, project communities welcomed the promise of a sustained economy and the benefits derived from completion of irrigation works. Unlike official Reclamation advocates, these communities did not necessarily see irrigation projects as purely a national accomplishment. While eagerly accepting contributions of the federal government in constructing dams and canals, these communities perceived Reclamation works as their own. Local photographers played up the role of local communities on various Bureau of Reclamation projects.

George Edward Anderson was one of many local photographers who created images that documented the growth of the West. In *The Utah Photographs of George Edward Anderson*, Rell G. Francis relates how Anderson's photographs recorded the growth and development of Mormon towns, mining

504 "Extraordinary Photographer Captures History in the Making," *Salt River Project Pulse* (January 28, 1982): 4; Bruce I. Bustard, *Picturing the Century: One Hundred Years of Photography from the National Archives* (Washington, D.C.: National Archives and Records Administration in Association with the University of Washington Press, 1999), 47.

505 Vilander, *Hoover Dam*, 84; see also Bustard, *Picturing the Century*, 47.

Photography of Henry T. Cowling

14.14. 1923. Stacking alfalfa near Sunnyside on the Yakima Project.

14.15. June 16, 1916. Carlsbad Project, New Mexico.

14.16. September 14, 1914. Huntley Project, Montana.

14.17. June of 1922. Laguna Dam, gates, spillway, and head of the Main Canal. Yuma Project, Arizona and California.

14.18. June 1922. Yuma Project. Interior of Sanguinetti home. Yuma Project.

14.20. August 23, 1913. Grand Valley Project, Colorado.

14.19. July 1, 1914. Salt River Project, Arizona.

14.21. August 12, 1924. Thrashing alfalfa seed on the A. M. Shields ranch. Yuma Project.

14.22. 1920. The Merced River with Half Dome on the right Yosemite National Park.

14.23. June 1923. Mt. Rainier National Park showing Reflection Lake and Mount Rainier.

14.24. June 1923. Mt. Rainier National Park and Glacier Point Highway.

Photography of Walter J. Lubken

14.25. (Top) June 24, 1910. Government mules posing for Walter J. Lubken during a break from working on Willow Creek Dam on the Sun River Project, Montana.

14.26. (Left) September 29, 1909. Drying fruit at the Ashenfelter ranch, near Montrose, Colorado, on the Uncompahgre Project.

14.27. (Below) August 27, 1908. Growing first crops under the Umatilla Project, Oregon.

14.28. September 29, 1901. Packing apples on the Ashenfelter ranch, near Montrose, Colorado, on the Uncompahgre Project.

14.29. September 1910. Irrigating young alfalfa on the Umatilla Project.

14.30. August 28, 1908. Digging the first crop of potatoes on the L. H. Furnas ranch near Hermiston, Oregon, Umatilla Project.

14.31. October 1, 1904. A drill testing to determine foundation preparation requirements at the Shoshone (later Buffalo Bill) damsite on the Shoshone River near Cody, Wyoming. Shoshone Project.

14.32. March 27, 1908. Irrigating an orange grove on the Salt River Project near Phoenix.

14.33. (Top) May 24, 1909. Shoshone project. Government demonstration farm at Powell, Wyoming, on the Shoshone Project.

14.34. (Center) February 14, 1907. Typical country farm house on the J. P. Ivy Ranch, near Phoenix, Arizona. Salt River Project.

14.35. (Left) July 2, 1905. The Reclamation office building at the east end of the Gunnison Diversion Tunnel on the Uncompahgre Project, Colorado.

interests, and farms. Anderson's photographs display a vast array of images including construction of the Reclamation Service's Strawberry Valley Project. Unfortunately, the names of many local photographers who helped document the development of Reclamation projects during this period are unknown. Still their images give visual evidence of the transformation of the West and, as Francis maintains, "recorded the shifting frontier as it gave way to urbanization."[506]

By 1925 C. J. Blanchard had left the Bureau of Reclamation, a departure caused by a reorganization prompted by the 1925 Fact Finders' Commission Report. Secretary of the Interior Hubert Work called upon the Fact Finders' Commission, created in 1923, to investigate the numerous complaints and problems on Reclamation projects and to make recommendations for corrective legislation. Many of the "successes" of Reclamation portrayed by Blanchard in films and photographs became matters for concern by the Fact Finders' Commission. One of Blanchard's favorite images and topics was the engineering achievements of the Reclamation Service. The Commission's report critically noted that the engineering side of the projects received more emphasis than agriculture to the detriment of project farmers. Through pictures and rhetoric, Blanchard tried to praise the independent nature of project farmers, while many settlers to the contrary complained of the Bureau's paternalism and saw themselves as "wards of the Government."[507] Ironically, the very aspects that Blanchard proclaimed as successes for the Reclamation Service became sources of criticism about the way in which the Reclamation Service conducted business.

Blanchard's career after he left the Bureau of Reclamation is sketchy. Some evidence suggests that for a short time he continued as an advocate of land reclamation. In December 1927 the Department of the Interior hosted a "Southern Reclamation Conference" to promote "a better rural condition and a more advanced type of agriculture in the South." Though there is no mention of a prominent role for him at the conference, C. J. Blanchard represented Hardee County, Florida.[508]

---

506 Rell G. Francis, *The Utah Photographs of George Edward Anderson* (Lincoln: University of Nebraska Press, 1979), 4; for more information on George Edward Anderson, see Rell G. Francis, "Views from Mormon Country: The Life and Photographs of George Edward Anderson," *The American West* 15 (November/December 1978): 14-29.

507 Brian Q. Cannon, "'We Are Now Entering a New Era:' Federal Reclamation and the Fact Finding Commission, 1923-1924," *Pacific Historical Review* 66 (May 1997): 185-207; see also Pisani, *Water and the American Government*, 139-42; see also "Impending Reclamation Disaster May Be Averted," *New Reclamation Era*, 15 (May 1925): 68.

508 "Southern Reclamation Conference Makes Plans for the Future," *New Reclamation Era*, 19 (January 1928): 4.

Though seldom mentioned in historical accounts of the Reclamation Service, Blanchard left a long-lasting legacy. For over twenty years, Blanchard helped bring the work and accomplishments of the Reclamation Service, through both words and images to the American public. The photographs and films he helped produce reflected some of the idealism of a progressive age and its commitment to federal reclamation. Blanchard, in a sense, set the public relations tone that the Bureau of Reclamation adhered to throughout the twentieth century. He saw federal reclamation through the eyes of a romantic who believed in the mission of the Reclamation Service and saw in its achievements the fulfillment of the nation's manifest destiny. He wrote:

> The history of national reclamation is as interesting and romantic as a tale from the Arabian Nights. Romance colored the vision of the builders that saw in the sparkling streamlets, unchecked floods, the wide, free plains and the vacant mountain valleys a promise of independence, happy homes and laughing children.... As they toil in the fastness of the mountains, in abysmal canyons or far out in the voiceless desert, through the blazing summer heat of the Southwest or fierce blizzards of the northern plains, this thought was uppermost, "By this work we shall make the desert bloom."[509]

Finally, Blanchard's promotional efforts on behalf of the Reclamation Service had the added benefit of introducing many Americans to new and exciting images of the American West. Nineteenth-century government survey photographs mirrored that century's understanding of and appreciation for images of the sublime or picturesque. With their narratives, these images also spoke of a bright and prosperous future for the American West. At the same time they offered a backward glance at a nineteenth-century vision of the idyllic life on the land with a promising future. There is also this similarity in the photographs Blanchard produced. Blanchard's images showed the excitement of a nation on the move, where new frontiers awaited. His photographs portrayed the West in transition: a place where families could find homes to fulfill the promise of American progress.

509 C. J. Blanchard, "Reclaiming the Desert," *The Mentor*, 6 (October 15, 1918): 5.

Large Dams and the "Machine Aesthetic"

Construction of Hoover Dam in the 1930s occurred at a time when photographers sought to capture the aesthetic qualities of machines and other manmade objects. Photographers of the early twentieth century embraced modernism's technological achievements as well as its concepts and theories. In his essay, "The Machine Between Cult Object and Merchandise: Photography and the Industrial Aesthetic in the United States During the Interwar Years," Olivier Lugon states:

> The very stamp of modern works—with their displayed objectivity, extreme precision in rendition, glossy surfaces—was there to remind us and to glorify its mechanical origin. This affirmation of basic equivalence with the industrial world culminated in the close-up of the machine, where the mechanical device was celebrated both as an object and as a medium; revealing the aesthetic richness of the former also validated the latter.[510]

Hoover Dam nicely fell into this photographic genre. The New Deal and its political leaders quickly understood how huge structures—Hoover Dam and the Tennessee Valley Authority's dams—symbolized the energy and achievements of the federal government under the guidance of New Deal recovery and reform policies.

For professional photographers, however, the combination of great dams and their powerplants into industrial landscapes provided attractive objects for artistic expression. Lugon maintains that "the engineers' projects now encompassed geographical space and acquired an almost geological quality through their combinations with rocks, river, and sky. Industry and nature seemed to be reconciled here." This fascination with science and technology as an artistic object became known as the "machine aesthetic" or the "cult of the machine." Artist/photographer Charles Sheeler and others sought out the beauty of an urbanized and industrialized America, while simultaneously fos-

---

[510] Olivier Lugon, "The Machine Between Cult Object and Merchandise: Photography and the Industrial Aesthetic in the United States During the Interwar Years," in *American Art 1807-1947 from Winslow Homer to Jackson Pollock*, Eric de Chassey, Editor (Reunion des Musees Nationaux; distributed by Harry N. Abrams, 2002), 106; see also David P. Peeler, *The Illuminating Mind in American Photography: Stieglitz, Strand, Weston, Adams* (Rochester, New York: University of Rochester Press, 2001), 41.

Photography of Ansel Adams

14.36. Ansel Adams photographed Boulder Dam, later Hoover Dam, scenes between October 1941 and October 1942 while working for the National Park Service on photographs to be used in development of murals for the new Department of the Interior Building in Washington, D.C.

14.37. Harold Arthur, the Director of the Office of Design and Construction during construction of the Third Powerhouse at Grand Coulee Dam, recalled that the architectural design for the powerhouse was influenced by Ladybird Johnson's campaign for beautification of America.

tering an idealized view of the nation's technical prowess. In *Charles Sheeler and the Cult of the Machine*, Karen Lucic writes Sheeler "embraced the iconography of the machine and produced images … that also provocatively invoked America's well-established technological and industrial prestige."[511] Hence, dams, transmission lines, factories, and towering skyscrapers represented a positive example of American progress and, as such, were proper subjects for artistic interpretation.

Photographer Paul Strand also perceived the machine age as a major turning point in human history. Strand embellished technology with god-like qualities and even posited that machines had replaced God. Machines, he argued, now did "the work of a thousand men that altered the natural world to conform to human desires." With a reliance on technology, "men con-

[511] Lugon, "The Machine Between Cult Object and Merchandise," 113-4; for information on Bureau of Reclamation officers' understanding of "machine imagery" see Vilander, *Hoover Dam*, 65; see also Karen Lucic, *Charles Sheeler and the Cult of the Machine* (Cambridge: Harvard University Press, 1991), 13.

summated a new creative act, a new Trinity: God the machine, Materialistic Empiricism the Son, and Science the Holy Ghost." While fascinated with technology, Strand expressed concern about machines controlling human beings. He argued, however, that photography and the photographer could help man regain command over machines through "the creative control of one form of the machine, the camera."[512]

Such reflections were not limited to artists and art critics but also occurred in intellectual circles and amongst industrial leaders. In *Human-Built World*, Thomas P. Hughes maintains that the noted economic historian Charles Beard held similar views praising the machine, the engineer who created it, and the technicians who kept them running. Beard "likened them to the all-knowing and powerful Creator giving shape to inanimate, chaotic matter." According to Hughes, Beard saw engineers able to help society create even greater accomplishments—"if free to fulfill their own set of values."[513] Strand's and Beard's observations hark back to the fascination and concern expressed by Henry Adams in "The Dynamo and the Virgin," with his portrayal of technology as a new creative and aesthetic force in American society and culture. Casting aside Adam's ambivalent concerns, many Americans mostly praised America's technological achievement "for its supposed potential to create a perfectly functioning, efficient, and just society."[514]

The Boulder Canyon Project lent itself to high technological praise and even hubris. Hoover Dam quickly became an icon of the Machine Age for both artists and government bureaucracies. The Bureau of Reclamation saw in the achievement the opportunity to distance itself from many of the criticisms of its past expressed in the Fact Finders Report. The dam opened a new era in which the Bureau of Reclamation's influence in the development of the West became ever-present. Unlike land resources management bureaus—BLM or Forest Service—Reclamation's large-scale construction projects brought a disproportionate share of public attention to its activities. This fortunate circumstance helped Reclamation capture the imagination of photographers and artists, and, in turn, the nation.

---

[512] Paul Strand, "Photography and the New God," in *Photographers on Photography: A Critical Anthology*, Nathan Lyons, editor (Englewood Cliffs, New Jersey in Collaboration with the George Eastman House, Rochester, New York, 1996), 138; Peeler, *The Illuminating Mind in American Photography*, 112-3.

[513] Thomas P. Hughes, *Human-Built World: How to Think About Technology and Culture* (Chicago: The University of Chicago Press, 2004), 73.

[514] Lucic, *Charles Sheeler and the Cult of the Machine*, 16; Henry Adams, *The Education of Henry Adams* (New York: The Modern Library, 1931).

14.38. This massive Unit 22 turbine runner shows the huge size of the units in the new Third Powerhouse at Grand Coulee Dam. May 15, 1976. Photographer: H. S. Holmes.

When construction began on Hoover Dam, the Bureau of Reclamation was fortunate to have an employee who appreciated and understood the "machine aesthetic" and had the talent to produce striking images. Ben Glaha initially began working on the Boulder Canyon project in 1931, in the Designs and Drafting department. Although having only limited experience as a professional photographer, Glaha's background as an engineering draftsman influenced the artistic specifics of his images of Hoover Dam. Glaha's photographic accomplishments have been well documented in Barbara Vilander's *Hoover Dam: The Photographs of Ben Glaha*. In this study, Vilander notes that Glaha was artistically inclined and admired the artists who popularized the machine aesthetic and "was a devotee of the works of Charles Sheeler and was acquainted with several renowned art photographers of the day, among them Ansel Adams and Margaret Bourke-White." Glaha's photographs trace the various phases of dam construction with an artistic quality unsurpassed in other Bureau of Reclamation photographs.

Glaha stands apart from other Bureau of Reclamation photographers in the notoriety he achieved while a Bureau employee; due in part to Reclamation's "proactive stance" in promoting the public display of his images. According to Vilander, "The Bureau's reasoning in allowing Glaha to display his images in artistic venues was that the exposure was positive publicity for the Bureau and for the projects." Indeed, artistic venues also provided the Bureau of Reclamation a unique opportunity to showcase its activities in an entirely new light and to diverse audiences. In January and February of 1935 some 250,000 visitors viewed Glaha's exhibition at the De Young Museum of Art in San Francisco. During this period, his work was also the subject of a lecture given by noted photographer Ansel Adams. In an extended tour, the exhibit traveled to the Fine Arts Gallery in San Diego and then to the Haggin Museum in Stockton, California. Finally, in August 1935, Glaha's images went on display at the National Museum in Washington, D.C. Vilander notes that the exhibitions received excellent reviews, and earned Glaha acclaim as an artist of immense talent while gaining favorable press for the Bureau of Reclamation.[515]

Vilander argues that the artistic quality of Glaha's photographs was ideal for the Bureau's propaganda activities. She states,

> The aesthetic components of his images was integral to their value to the Bureau because of their persuasive power; photographs that were well crafted, visually dynamic, and aesthetically pleasing were, as the Bureau propagandists were well aware, more likely to capture viewers' attention and, ultimately, affect their perception.

Glaha and the Bureau of Reclamation were by no means alone in "affecting perceptions." Hoover Dam was the subject of images made by other renowned photographers, Edward Weston and Ansel Adams. Edward Weston's 1941 photograph of Boulder Dam was part of larger series of landscapes in which Weston "concentrated on landscapes that had been transformed by the human hand." Also in 1941 Adams's image of Boulder Dam was included in a mural project for the new Department of the Interior building in Washington, D.C. Along with photographs of national parks and Indian reservations, Adams wished to "display the benefits of conservation, good administration, and care-

---

[515] Vilander, *Hoover Dam*, 10-4, 80-2; Vilander's study is a comprehensive examination of the photography of Ben Glaha, and how his photographs helped to accomplish the Bureau of Reclamation's promotional goals regarding the construction of Hoover Dam; see also "Glaha Complimented," *Reclamation Era*, 25 (July 1935): 152.

Photography of Ben Glaha

14.39. Ben Glaha's official Reclamation photographs from Hoover Dam are famous as documentation of construction work there and have been analyzed by Barbara Vilander in her book: *Hoover Dam: The Photographs of Ben Glaha.*

ful long-term planning."[516] Their photographs not only illuminated the dam's aesthetic beauty, but also served to reinforce and embellish what the artists believed to be the ideals of the Bureau of Reclamation.

An important aspect of those ideals was the Bureau of Reclamation's effort to sustain a belief in the positive benefits of irrigation. Like other Reclamation photographers, Glaha's images served as evidence of the Bureau of Reclamation's efforts to transform the landscape. Vilander maintains that Glaha's photographs fall under William Slott's definition of social documentation, which "tends to advocate social improvement." Glaha's work, she argues, was "an attempt, on behalf of the Bureau, to show the public that the dam's construction would improve the quality of life for the people living downstream by controlling flooding and providing a reliable source of irrigation water and electricity." In this regard, Glaha produced panoramic images of the areas directly affected by the rising waters of Lake Mead by making a photographic record of the deserted town of St. Thomas, Nevada. As the reservoir filled, Glaha traveled by boat to photograph places and landmarks that previously had been almost inaccessible "through which he meant to proclaim a newfound scenic heritage and to emphasize the harmony of nature and engineering."

A more immediate transformation of the western landscape as a direct result of Hoover Dam was the construction of Boulder City. Glaha chronicled the growth of the town which illustrated the Bureau of Reclamation's ability to "make the desert bloom." According to Vilander, Glaha's images are evidence of not only his devotion to the machine aesthetic, but also his belief in the dam's purpose to transform the arid regions of the American West. Vilander argues:

> Better than most of his time, his photographs conveyed, and in doing so spread, the belief that the land required mankind's intervention before its natural resources could be utilized. In Glaha's images, arid lands could, and therefore must, be made green. It was not acceptable, given the availability of modern technology, to coexist with arid terrain. Instead, the land found its fulfillment only when subjugated to mankind's will.

Once again, Vilander's assertions reiterate views by both Bureau of Reclamation veterans and western boosters. The ideas expressed by Vilander regarding

---

516 Vilander, *Hoover Dam*, 97; Peeler, *The Illuminating Mind in American Photography*, 259; Anne Hammond, *Ansel Adams: Divine Performance* (New Haven: Yale University Press, 2002), 110; Jonathan Spaulding, *Ansel Adams and the American Landscape: A Biography* (Berkeley: University of California Press, 1995), 186.

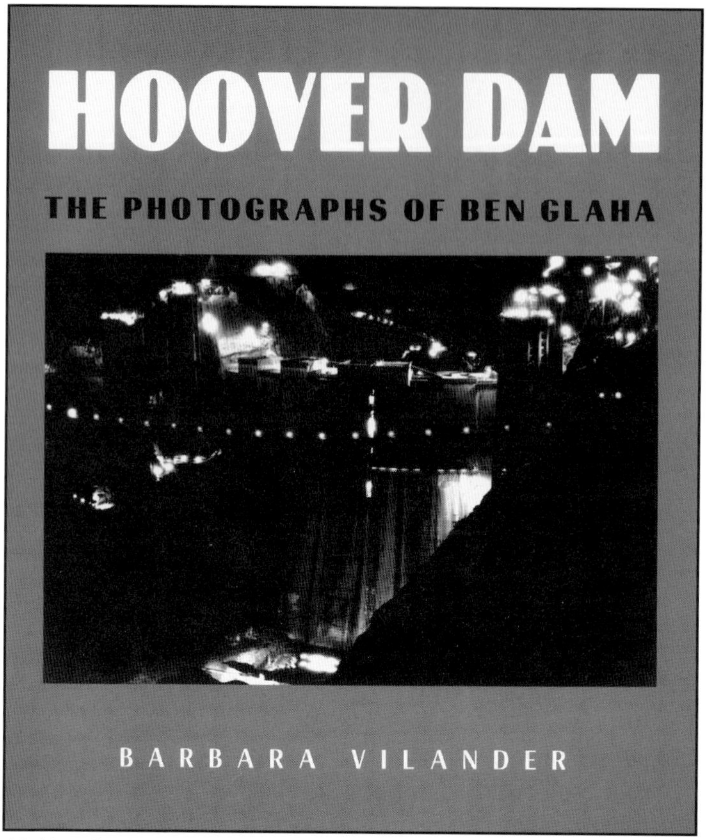

14.40. The dustjacket of Barbara Vilander's study of the photography of Ben Glaha at Hoover Dam.

the social documentary of Glaha's photographs mirror the sentiments extolled by C. J. Blanchard and others during the early years of the Reclamation Service. Glaha's astounding images of the construction of Hoover Dam transcended the standard promotional activities of Blanchard. As Vilander points out, Glaha's photographs stand out "not only because of their subject matter but also because the technical/propagandistic and artistic dichotomy of Glaha's images is unprecedented in the work of a government photographer."[517]

During construction of Hoover Dam, Glaha's images were an important promotional tool for the Bureau of Reclamation's Boulder Canyon Project. Construction photographs of the dam made their way into the nation's leading newspapers and trade journals such as *Electrical West*. A rather romantic portrayal of the men and effort involved with construction of Boulder Dam appeared in an article in the September 1933 issue of *Fortune* magazine. The photographs of both Glaha and Lubken appeared. Boulder Dam images also showed up in unlikely places such as the backdrop for advertisements to sell Camel cigarettes and International Trucks, an activity that continues to the present. As Vilander points out, the Bureau of Reclamation decided to take

[517] Vilander, *Hoover Dam*, 17, 41, 34,134, xvi.

Even before leaving Boulder Dam, Ben Glaha worked at other locations on Reclamation projects.

14.41. Owyhee Dam, September 1935.

14.42. Owyhee Project, 1944.

14.43. Veterans homestead drawing winners Mr. and Mrs. Robert Metz are congratulated on December 18, 1946, by Frederick Lehman. Klamath Project.

advantage of the popular reviews of Glaha's photographic exhibitions and produce "exhibition" prints. Further evidence of the popular and artistic appeal of Glaha's images comes from their appearance in "art-oriented publications" such as *Camera Craft* and *U. S. Camera*.[518] The breadth of interest in these images reveals not only the broad measures the Bureau of Reclamation took to promote the Boulder Canyon Project, but, more importantly, the immense talent of Ben Glaha. Glaha's photographs of Hoover Dam, and the vast array of formats in which they appeared, helped shape the dam's iconic image as a symbol of American technological prowess.

The 1930s and early 1940s encompassed a dramatic period of dam construction for the Bureau of Reclamation. Even before completion of Hoover Dam, construction began on Grand Coulee Dam in Washington; soon followed by Shasta Dam in northern California. Similar to all Bureau of Reclamation projects, photography played a major role in documenting all phases of construction and in promoting these two enormous undertakings. Neither the photographs nor the photographers on these projects attained the same acclaim as Ben Glaha's work at Hoover Dam. By 1936 Glaha transferred from the Boulder Canyon Project to the Central Valley Project, but his subsequent work and Reglamation's use of it failed to give notoriety to the Central Valley Project and its key fixture Shasta Dam.[519] On these projects Reclamation officials followed a policy of anonymity regarding the authorship of photographs, perhaps assuming that the size and scope of both Grand Coulee Dam and the Central Valley Project promoted themselves.

Howard Colby was a photographer who achieved a small measure of recognition documenting the building of Shasta Dam. In 1938 Colby became the official photographer for Pacific Constructors, with which the Bureau of Reclamation contracted to build Shasta Dam. According to Peter E. Palmquist in *Once Upon a Dam Site: Howard Colby's Shasta Dam Photographs*, Project Superintendent Frank Crowe hired Colby because his "photographs showed action and men working." Palmquist also maintained that initially Crowe "didn't want *any* photographers working on the project *because* of an unnamed 'prima donna' photographer who had been associated with his previous post on Boulder Dam"—perhaps a reference to Ben Glaha

[518] "The Dam," *Fortune*, 8:3 (September 1933): 74-88; the advertisements featuring Boulder Dam appeared in *Fortune*, 11:4 (April, 1935): 5 and *Fortune*, 14:4 (October 1936): 5; Vilander, *Hoover Dam*, 87.

[519] "Reclamation Organization Activities and Project Visitors," *Reclamation Era*, 26 (November 1936): 271.

or Walter Lubken. Crowe nevertheless gave the self-taught Colby the opportunity to document the dam's construction. Colby's images served many of the same purposes of previous Bureau of Reclamation photographs in recording the construction process and engineering techniques. In addition, Pacific Constructors used Colby's photographs to promote the company's achievements and enhance its reputation. In 1945 to commemorate the dam's completion, it published *Shasta Dam and Its Builders*. In this rather self-congratulatory work, Colby's photographs provided impressive images of the dam's construction from start to finish. These images qualify as what Barbara Vilander termed social documentation, by extolling the social benefits—flood control, hydroelectric power, and irrigation. Peter Palmquist claimed that "Colby's photographs stand primarily as dramatic symbols of our industrial society at work. Shasta Dam clearly represented a big task on all fronts, and one with enormous social implications. It was built during a world war, and it was built with American know how."[520]

Photographs remained an important aspect of the Bureau of Reclamation's promotional activities throughout the second half of the twentieth century. But a distinct change occurred in how the Bureau of Reclamation employed images. The Boulder Canyon Project Act of 1928 and subsequent Works Progress Administration appropriations during the Great Depression ended Reclamation's dependency on the "revolving fund" stipulated in the original 1902 Reclamation Act. By the end of World War II, Reclamation enjoyed a supportive constituency in the western states. Yearly appropriations in the federal budget now funded its projects and promotional activities. Other modes of media opened for promotional work in films and radio.[521] It was no longer the small struggling government agency that had been the U.S. Reclamation Service. Reclamation had grown into one of the federal government's most powerful bureaucracies with an international reputation as a leader in the field of water development and engineering.

---

[520] Peter E. Palmquist, editor, *Once Upon a Dam Site: Howard Colby's Shasta Dam Photographs* (Redding, California: Redding Museum and Art Center, 1987), 11; Palmquist's study is a somewhat romantic account of Colby's photographic career in the Redding area. The book, however, does contain a number of images demonstrating Colby's photographic talents; Pacific Constructors, Inc., *Shasta Dam and Its Builders* (1945); Palmquist, *Once Upon a Dam Site*, 8.

[521] Department of the Interior, Bureau of Reclamation, *Man is a Giant: The Story of Boulder Dam* (sound recording, n.d.) held by the American Heritage Center, University of Wyoming, Laramie; it is uncertain whether the Bureau of Reclamation produced more radio programs that highlighted its activities. The tone in *Man is a Giant* kept with the Bureau of Reclamation's overall theme of placing the construction of Boulder Dam as a heroic and national effort.

14.44. Mormon pioneers founded St. Thomas, Nevada in 1855. The last residents abandoned town in June 1938 because of the rising waters of Lake Mead. April 11, 1945. Photographer: W. L. Russell

Photographic images from this period and beyond reflect the power and influence of the Bureau of Reclamation as evidenced in its official publication *Reclamation Era*. The immediate postwar years offered new challenges and opportunities. War-delayed projects were renewed with vigor and excitement. Work resumed on the multiple projects designated for the Missouri River basin, and the irrigation portion of the Columbia Basin Project began in earnest. For these projects, articles and photographs in *Reclamation Era* emphasized Reclamation's original mission of creating homes for a new generation of independent farmers.[522] In addition, because of the contributions made by Bureau of Reclamation hydroelectric plants during the war, power production assumed great importance alongside water delivery facilities.

The contribution of Reclamation's dams and powerhouses to the war effort enhanced its reputation and underlined the West's potential for future hydroelectric projects. The New York Museum of Modern Art's show *The Road To Victory* (1942) exhibited images of Shasta Dam and Hoover Dam in murals that highlighted their role in the American war effort. Throughout the second half of the twentieth century, Hoover Dam remained one of the most

[522] J. A. Krug, "My New Job," *Reclamation Era*, 32 (May 1946): 93-4; Sylten, "Pioneering in the Missouri Basin," 98-9; "Columbia Basin Program Speeded," *Reclamation Era*, 32 (July 1946): 148; "Water Planning Pays Off," *Reclamation Era*, 32 (July 1946): 152-3.

14.45. Shasta Dam during construction.

14.46. The generator floor at Glen Canyon Dam in July of 1966.

powerful icons for photographers as a representation of the massive transformation of the American West. Aaron Betsky, in "Emptiness on the Range: Western Spaces," asserted that Hoover Dam is "the most famous … monument to man's desire to completely transform the American West … into homes for millions by harnessing its water."[523] While celebrating the manifest benefits derived from power production, the Bureau of Reclamation still kept alive its original mission to promote farms in the West.

Artists and Representations of Reclamation

Prior to the advent of photography, western boosters and government agents utilized paintings and other forms of art work to visualize and enhance the natural wonders and economic potential of the American West. In the mid-nineteenth century, Emanuel Gottlieb Leutze's painting *Westward the Course of Empire Takes Its Way* (1861) and John Gast's *American Progress* (1872) urged Americans to look to the West to see the nation's future. For many nineteenth-century Americans, periodical illustrations offered the first image of the West and prompted ideas about western opportunities. According to Martha Sandweiss's essay "The Public Life of Western Art," paintings and lithographs were "the visual images that introduced Americans to the physical appearances of the Far West and provided them with a common visual vocabulary for understanding the importance of the West in the national life." Paintings and lithographs, along with photographs, also filled the pages of many western survey reports produced by the federal government. These images not only helped readers visualize the locations and sites discussed in the reports but combined with the text to romanticize western landscapes.[524]

By the beginning of the twentieth century, the use of artwork as a promotional tool for agencies and bureaus of the federal government faded. The Reclamation Service and later the Bureau of Reclamation relied primar-

---

523 Spaulding, *Ansel Adams and the American Landscape*, 196; see also Aaron Betsky, "Emptiness on the Range: Western Spaces," in Sandra Phillips, Richard Rodriguez, Aaron Betsky, and Eldridge M. Moore, *Crossing the Frontier: Photographs of the Developing West, 1849 to the Present* (San Francisco: San Francisco Museum of Modern Art, Chronicle Books, 1996), 55.

524 Martha A. Sandweiss, "The Public Life of Western Art," in Jules David Prown, et al., *Discovered Lands, Invented Pasts: Transforming Visions of the American West* (New Haven: Yale University Art Gallery, 1992), 113; Joni Louise Kinsey, *Thomas Moran and the Surveying of the American West* (Washington, D.C.: Smithsonian Institution Press, 1992), 4-5; Patricia Janis Broder, *The American West: The Modern Vision* (Boston: A New York Graphic Society Book, Little, Brown and Company, 1984), 2.

14.47. Oskar Hansen created the paired Winged Figures of the Republic for a plaza at the west end of Hoover Dam.

ily on photographic images. Painting, much like photography in the early twentieth century, was going through a transition. Influenced by European modernism, "western" artists, Paul Henri, John Sloan, George Bellows, Marsden Hartley, Georgia O'Keeffe, and John Marin, moved away from the nineteenth-century idealized image of the West and "focused instead on the present." They painted a West in transition, exploring "the relationship between individuals and their environment." According to Howard Lamar in "Looking Backward, Looking Forward: Selected Themes in Western Art since 1900," landscapes were "not only not virginal or empty but often filled with Indians, Hispanics, Anglos, houses, factories, and grain elevators."[525] Irrigation or reclamation themes were part of this transitionary view of the West and were especially important subject matter for the artists who participated in the New Deal's various art programs.

The art program of the Works Progress Administration (WPA) also addressed aspects of Reclamation and further celebrated the national benefits derived from dam construction and irrigation works. New Deal promotion of the arts during the Great Depression was first and foremost designed to put unemployed artists to work, but it was also the first significant government sponsorship of the arts on a broad scale in the nation's history. Officials of multiple federal art projects saw opportunities to promote the policies and reform measures of the New Deal. In *Federal Art and National Culture*, Jonathan Harris argues, "In broad terms the guiding proposition ... was that the federal state ... was capable of resolving the economic, political and ideological conflicts of a capitalist soci-

[525] Broder, *The American West*, 1, 9; Howard Lamar, "Looking Backward, Looking Forward: Selected Themes in Western Art Since 1900," in Prown, et al., *Discovered Lands, Invented Pasts*, 183.

ety in crisis."[526] Bureau of Reclamation projects, both large and small, provided concrete (literally) proof of the federal government's efforts to stimulate employment and make life better for all Americans.

Images such as Adrian D. Clem's *Boulder Dam* (1934) and William Gropper's *Construction of the Dam* (1939) emphasized men and technology at work harnessing the multiple powers of western water resources. Many western images found their way to displays throughout the nation, as Reclamation themes became part of the larger lexicon of New Deal activities. Edgar Britton's *Modern Man* (1936-37) fresco, for the Lane Technical High School in Chicago, depicted "a worker and his family looking at the swift flow of waters from Boulder Dam." According to Belisario Contreras in *Tradition and Innovation in New Deal Art*, this image "suggests ... that the New Deal has brought a sense of individual freedom through collective effort."[527]

In a similar vein, the Bureau of Reclamation and the Department of the Interior saw the benefits of using pieces of art to decorate grand and renowned structures. In 1937 the Department of the Interior's newly completed office building in Washington, D.C., was lavishly decorated with murals and sculpture depicting the activities of its many bureaus. William Gropper's mural, *Construction of a Dam,* and Nicolai Cikovsky's mural, *Irrigation* (1938), represented the work of the Bureau of Reclamation. By 1938 sculptures and colorful Native American designs decorated both the exterior and interior of Hoover Dam.[528]

In 1935 the Bureau of Reclamation commissioned Oskar J. W. Hansen to create sculptures and other works to enhance the majesty of Boulder Dam.

---

[526] During the Great Depression New Deal programs sponsored multiple art projects including: Public Works of Art Project (December 1933-June 1934), the Section of Painting and Sculpture (October 1934-June 1943), Treasury Relief Art Project (July, 1935-June 1938), and WPA Federal Art Project (September, 1935-May 1943); see Jonathan Harris, *Federal Art and National Culture: The Policies of Identity in New Deal America* (New York: Cambridge University Press, 1995), 7-8; Belisario R. Contreras, *Tradition and Innovation in New Deal Art* (Lewisburg: Bucknell University Press, London and Toronto: Associated University Presses, 1983), 10-20.

[527] Contreras, *Tradition and Innovation in New Deal Art*, 119; Adrian Clem produced a number of images of Boulder Dam during this period including; *Beginning of Excavation, Boulder Dam, Nev.* (1934), *Blasting Rock in Black Canyon, Boulder Dam, Nevada* (1934), *Loading Trucks, Boulder Dam, Nevada* (1934), *Looking Upstream through Black Canyon, Boulder Damsite, Nevada* (1934), *Nevada Spillway, Boulder Dam, Nev.* (1934).

[528] David W. Look, AIA and Carole L. Perrault, *The Interior Building Its Architecture and Its Art* (Washington, D.C.: U.S. Department of the Interior National Park Service Preservation Assistance Division, 1986), 110-1; see also "Labor's Memorial to Its Dead at Boulder Dam," *Reclamation Era*, 25 (July 1935): 143.

Over the next three years, Hansen produced thirty-six pieces, including two thirty-foot tall bronze statues, a terrazzo star map commemorating the date of the dam's dedication, and a bronze plaque memorializing the eighty-nine workers who died during construction. The plaque carried the inscription: "They Labored That Millions Might See A Brighter Day." Hansen's most imposing works created for the dam were the two thirty-foot tall bronze statues titled *Winged Figures of the Republic*. He made these seated figures using sand molds weighing 492 tons into which the foundry poured over four tons of molten statuary bronze. The two figures bookended the 142-foot high flagpole and conformed to not only the rugged landscape but also the vertical rise of the dam. Hansen explained,

> The distant view of Fortification Mountain and the closer mesas made it desirable to break this vertical composition with the single angular bend of the seated posture. The shape and surface area of the wings were not only calculated to convey potency for flight but to repeat the shadowed wedge areas created by the serried buttresses along the face of the dam, but in reverse order.[529]

The bas-reliefs on the elevator towers bespeak two contrasting and often conflicting aspects of the history of the American West. On the Nevada side, the panels represent the various "purposes served by Reclamation projects"—flood control, navigation, irrigation, water storage, and power. The Arizona side depicts aspects of Native American life and culture prior to the coming of the white man and the ultimate joining of these two cultures in peaceful co-existence. For Hansen, this panel represented the Indians' continuous struggle to maintain their freedom, which he viewed as synonymous with principles shared by all Americans. Hansen claimed, "From the appeal of freedom which existed in the breast of the Red Man as he reaches his hand toward his *Great Spirit* above to building a joint effort for the building of a common destiny in the acts of peace."[530] In a common artistic trope, Hansen idealized both the concept of the "noble savage" in his simple but dignified culture and his eventual and predetermined submission to a vastly superior civilization. In Hansen's bas-

---

529 Julian Rhinehart, "The Grand Dam," http://www.usbr.gov/lc/hooverdam/History/articles/rhinehart1,html; see also "Plaque Erected at Boulder Dam Memorializes Workers," *Reclamation Era*, 28 (January 1938): 6-7, 10; Peggy and Harold Samuels, *Illustrated Biographical Encyclopedia of Artists of the American West* (Garden City, New Jersey: Doubleday, 1976), 209; see also Oskar J. W. Hansen, "From Bones of Water Pipe and Wood: The Sculptures at Boulder Dam, Part III," *Reclamation Era*, 32 (April 1942): 78-9.

530 Hansen, "From Bones of Water Pipe and Wood," 80; italics in the original.

reliefs, the Indian and the Euro-American join forces for the common good of both races supported by shared ideals of freedom and liberty.

Hansen wanted his sculptures to mirror the magnitude and grand scale of Hoover Dam, yet connect with the common man. The sculptor believed the dam represented "the building genius of America in the same sense as the Pyramids represent that of ancient Egypt, the Acropolis of classical Greece, the Colosseum of Imperial Rome, and Chartres Cathedral that of the brooding religious fervor which was Gothic Europe." Just as the ancients adorned magnificent structures with sculptures to create an enduring legacy to the vision and ingenuity of those who built them, Hansen concluded that the "historic mission of sculpture is therefore to evoke a pungent realization of man and to make this realization nearly imperishable against the oblivion of time." More important than Hoover Dam's relationship to the wonders of ancient civilization was its value as a testimony to the creativity of the American people. For Hansen, the *Winged Figures of the Republic* represented the sturdy republican citizenry who planted civilization from coast to coast on the North American continent. He argued,

> There grew up with the settling of this continent a virile type of man, inured through constant adjustment into quickness of wit and beaten by privations and the strong winds of the mountains and plains into a facial physiognomy with the look of eagles.

The *Winged Figures of the Republic*, and by association Hoover Dam, embodied all that made the United States a great and powerful nation. Hansen arranged the placement of the statues and the flagpole to represent the "Winged Figures" as strong and pure protectors of the American Republic. The sculptor explained,

> I wanted to emphasize the common origin of our humanity which under our institutions is expressed in a Bill of Rights that is a law alike for rulers and for people. In jealous guardianship of the sacred entity of individuals lie the potent powers of those who govern. In their common good lies the security of the flag.

Works of art placed throughout the dam exude these ideals. Memorial plaques paid homage to the sacrifices and achievements of American labor, while the

14.48. Southwestern Indian art inspired Allen Tupper True's terrazzo floor designs for Hoover Dam.

bas-reliefs on the Nevada side honored the skill and foresightedness of the engineers and benefits derived from federal reclamation. Even the bas-reliefs depicting Native Americans upheld a romantic notion of a free and natural but eventually conquered frontier. Hansen wrote these statements in early 1942, shortly after the United States entered World War II. The patriotic fervor sweeping the country dominates his rhetoric. He wrote, "The Winged Figures of the Republic give evidence to the thought which preceded the reality of Boulder Dam and to that vigilance which is the price of liberty."[531]

On a purely practical level, Bureau of Reclamation artistic enhancement of Hoover Dam translated into public awe and applause for its work. By 1936 an estimated 300,000 tourists a year were visiting the dam. The public's apparent fascination with Hoover Dam and the popularity of Ben Glaha's photographs encouraged Reclamation to decorate the interior and exterior of the dam's two powerhouses in the style of the machine aesthetic. Western muralist, painter, illustrator, and color consultant Allen Tupper True became the government consultant on decoration and color scheme for the Hoover Dam powerplants. True advocated highlighting both the form and function of Hoover Dam by emphasizing the dam's visual and mechanical qualities. He maintained that "as the machine assumes every day an increasingly prominent place in our lives, it is proper that its housing be considered from the viewpoint of orderly beautification."[532]

---

[531] Oskar J. W. Hansen, "With the Look of Eagles: The Sculptures at Boulder Dam—Part I," *Reclamation Era*, 32 (February 1942): 30, 32.

[532] Boulder Tourist Travel," *Reclamation Era*, 25 (August 1935): 160: "Boulder Dam Tourist Accommodations," *Reclamation Era*, 27 (July, 1937): 149; Arno B. Cammerer, Director, National Parks Service, "Building a Playground at Boulder Dam," *Reclamation Era*, 28

True envisioned the dam as something more than just a gigantic cement plug in the Colorado River. Instead he saw a structural form exhibiting its own beauty. He selected colors that intensified the dam's artistic character. Similar to Hansen, True equated the construction of Hoover Dam with the great engineering feats of the past. Like the polychrome placed on ancient Greek temples, True argued that "when color is applied to Boulder Dam it is with the object of making forms readable, coherent, and as beautiful as the dynamic new beauty of machine forms can be." Also, like Hansen's vision to conform his sculptures to the dam's topography, True selected colors that melded with desert landscape. He, too, chose Native American motifs to decorate Hoover Dam trying to distinguish it as a truly American creation.

A desire to create distinctiveness in American art appeared early as nineteenth-century American artists saw in the American landscape an avenue to separate American art from that of Europe. Similar to those artists who flocked to Santa Fe and Taos, New Mexico, during the early twentieth century, True saw in Native American art an untouched source of inspiration for artistic expression. He argued that "native" artwork found on "pottery, basketry, and sand paintings of the Colorado River watershed" was "superior to the classic Greek, Roman, and Egyptian motifs." In these Native American motifs, Hansen also saw images that corresponded to the mechanical aspects of the Hoover Dam powerplants. From the patterns of Pima Indian basketry, True distinguished similarities in "an engineer's basic diagram of a generator or turbine, with valves, gates, and a suggestion of centrifugal motion." In an effort to accentuate the dam's primary purpose, True also used Pueblo Indian designs representing lightening, clouds and rain.[533] By utilizing Native American symbols and colors associated with the Southwest, True presented an image of the dam that conformed to its desert environment, through the exhibition of designs easily recognizable as being centered in the cultures of the American West.

(November 1938): 222-4; according to *Reclamation Era*, even during construction Boulder Dam had been a tourist destination and noted that between 1933 and 1935 well over 300,000 tourists visited the damsite; Allen Tupper True, "Color and Decoration at the Boulder Dam Power Plant," *Reclamation Era*, 26 (January, 1936): 12; for biographical information on Allen Tupper True see, Samuels, *Illustrated Biographical Encyclopedia of Artists of the American West*, 490-1; True also went on to become the color consultant for both Grand Coulee and Shasta dams.

[533] True, "Color and Decoration at the Boulder Dam Power Plant," 12, 13, 25; polychrome was a style of vase painting developed in Athens in the latter part of the Sixth Century B.C., using various colors to paint decorative figures and other motifs; see also Broder, *The American West*, 9; for perspectives on nineteenth-century American artists see, Nancy K. Anderson, "'Curious Historical Artistic Data:' *Art History and Western American Art*," in Prown, et al., *Discovered Lands, Invented Pasts*.

The "machine aesthetic" is visible in the multiple ways in which True attempted to emphasize form and function at Hoover Dam, through his color and design choices. Part of this has to do with artists' fascination with technology during the first half of the twentieth century, but more importantly, much of the interest centered on the dam itself. Indeed, the Bureau of Reclamation appears to have opened all aspects of the dam to artists of various mediums. In 1934 painter Stanley Wood produced a number of water colors for an article in *Fortune* magazine depicting construction scenes. *Fortune* editors noted that Wood's "work is an exciting and quietly moving record of America busy at work."[534] The works of both Hansen and True, in conjunction with the photographs of Ben Glaha, made Hoover Dam widely recognizable as a true American technological wonder. Charles Sheeler furthered the wonderment in his painting *Conservation—Sky and Earth* (1940) that "portrays a radio transmission tower silhouetted against a brilliant blue sky with the newly built, monumental Hoover Dam in the background." Part of a series of images Sheeler produced for *Fortune* magazine, *Conservation—Sky and Earth* was meant "to portray the concept of power." It was not just a power derived from generators, engines, or machines, but a power that grew from the soul of American ingenuity and resourcefulness.[535]

In 1969 the Bureau of Reclamation began an Art Program to promote its activities and invited a number of prominent American artists to paint various aspects of Reclamation projects. This program included artists Norman Rockwell, Dean and Lynn Fausett, John McCoy, Peter Hurd, Mitchell Jamieson, Lloyd Goff, Edward Laning, Herman Meril, and Ralston Crawford. As opposed to the strict confines of the Bureau of Reclamation's photographic record, each artist enjoyed the freedom to feature any aspect of a Reclamation project they desired and in any style they chose. John DeWitt, program director and the force behind Reclamation's Art Program, maintained that the paintings "will provide valuable insights to a great many Americans who are largely unaware of the productive results of Reclamation's water development projects." Despite the immense photographic record held by the Bureau of Reclamation, DeWitt believed that only through artistic representation could the emotional aspects of the projects be realized. He claimed, "Only an artist, viewing a scene through his own eyes, and capable of synthesizing his intellectual response and his imagination on paper or canvas, can provide the intrinsic graphic statement."[536]

---

534 Stanley Wood, "Boulder Dam: A Portfolio of Water Colors," *Fortune*, 9:5 (May 1934): 92-100.

535 Lucic, *Charles Sheeler and the Cult of the Machine*, 109; Martin Friedman, *Charles Sheeler* (New York: Watson-Guptill Publications, 1975), 129.

536 John DeWitt, "Reclamation Launches Art Program," *Reclamation Era*, 56 (February 1970):

As opposed to prior promotional activities, the Bureau of Reclamation never intended its Art Program to promote settlement on Reclamation projects or new construction activities. DeWitt claimed that the program's purpose was merely to bring the Bureau's accomplishments in transforming the American West to a national audience. Illustrating Reclamation's purpose and achievements was, of course, nothing new. It continued the activities begun under C. J. Blanchard, only now through a different visual medium. The new Bureau of Reclamation Art Program, however, lacked a narrative to accompany the images. Viewers could make their own interpretations.

In "Portraits of Reclamation," Leslie Stinger and Bobbie Ferguson suggest that the Art Program was in response to cultural changes taking place in America during the 1960s. It was a reaction to the emerging environmental movement of the late 1960s, which was "quite vocal in protesting Reclamation projects … viewed as detrimental to the natural environment." The authors made a valid point arguing that DeWitt held to the view that Reclamation projects actually enhanced the natural environment. He argued that "the water the Bureau's development provided has changed the face of the West and brought about new ways of life for its inhabitants. In doing so, the Bureau has created a new environment." As such, the Art Program differed from anything the Bureau of Reclamation had done in the past to promote its activities. In 1973 Secretary of the Interior Rogers C. B. Morton observed that "the paintings succeeded in conveying the spirit and mission in a fashion that had never been done before."[537]

In the impressionistic "spirit" of oil and water paints on canvas, Bureau of Reclamation officials hoped viewers would find a pleasing aesthetic experience to blunt a growing chorus of environmental criticism. As Stinger and Ferguson point out, "There is … the multiplicity of readings inherent in a work of art. Although one message may dominate, there are many other ways to interpret and understand a painting, which allows dialog to occur." Despite various motives, the Bureau of Reclamation found in this medium an effective means to

---

9; Appreciation of Reclamation's art collection can be enhanced by seeing the originals in full color, but, failing that opportunity, many of the works can be seen on the Reclamation website at http://www.usbr.gov/museumproperty/art/.

[537] Leslie Stinger and Bobbie Ferguson, "Portraits of Reclamation," CRM No. 4—1999, www.usbr.gov/museumproperty/art/; John DeWitt, "Reclamation and the Creative Artist," *Reclamation Era*, 59 (Spring 1973): 12; one critic with a sensationalistic bent has applied the term "eco-porn" to Reclamation's art program—see Paul Lundboldt, "From Sublimity to Ecopornography: Assessing the Bureau of Reclamation's Art Collection," *Journal of Ecocriticism* 1:1 (2009): 1-25—however, this article is representative of a small body of interpretation that maintains Reclamation's art program was merely an attempt to show that its programs were not as environmentally damaging as some critics claimed.

deliver its message to new generations of Americans. By 1973 forty-one artists had contributed to the Art Program producing over three hundred paintings, which the Bureau of Reclamation exhibited throughout the country. Promoters of the Reclamation's Art Program likened this activity to the days of nineteenth-century exploration where the federal government enlisted artists to depict the natural wonders of the American West. Douglas MacAgy of the Smithsonian's Hirshhorn Museum stated that the Art Program "revives the earlier historical relationship between artists and the natural setting." Not all the artists, however, selected landscapes as their subjects but chose instead construction scenes or machinery, which were reminiscent of images produced by the WPA's Great Depression Art Program. Construction of the third powerhouse at Grand Coulee Dam and the Bonneville Unit of the Central Utah Project provided some of the artists the opportunity to create paintings and watercolors of men and machinery, perhaps even resurrecting the styles of the "machine aesthetic."[538]

Perhaps to Reclamation officials, the most gratifying representations were those that merged dams and structures with the natural environment. John McCoy's *Shasta Dam* exemplifies this ideal. McCoy created a broad panorama to include the dam, the lake, and mountain all sharing the same name. Likewise, Eugene Kingman's *Roosevelt Dam* allows the viewer to look down on the dam and perceive the perfect fit between the stone masonry on the dam's face and the surrounding rocky cliffs. William Palmer's *Flaming Gorge*, Xavier Gonzalez's *Olympus Dam*, and Ethel Magafan's *Gibson Dam—Sun River Project* permit viewers to absorb an apparently pristine landscape with the direction of the gaze eventually leading to dams that seem neither out of place nor incongruent with the scenery. Here in these images was the grandeur and color of the American West tempered and even improved by the handiwork of civilization. If Stinger and Ferguson are correct in suggesting that the Art Program was a response to an environmentalist critique, Reclamation demonstrated a creative response to its detractors.

Lloyd Goodrich of the Whitney Museum of American Art pointed out the connection between the Bureau of Reclamation's Art Program and the contributions made by nineteenth-century artists to the federal government's survey expeditions. Goodrich also noted that since that time artists had been

---

[538] Stinger and Ferguson, "Portraits of Reclamation," 48; Douglas MacAgy, introduction, in Bureau of Reclamation, *The American Artist and Water Reclamation* (Washington, D. C.: Government Printing Office, 1973); see also "Reclamation Art Exhibit," *Reclamation Era*, 58 (August 1972): 24; DeWitt noted that many of the artists commissioned for the Bureau of Reclamation art program received their first opportunities to create art for public consumption through the WPA's art program; see DeWitt, "Reclamation Launches Art Program," 12-5.

Reclamation Art Program

14.49. (Above) John McCoy, *Shasta Dam.*

14.50. (Left) Eugene Kingman, *Theodore Roosevelt Dam.*

14.51. (Right) William Palmer, *Flaming Gorge Dam.*

14.52. (Top) Ethel Magafan, *Gibson Dam.*
14.53. (Middle) Dean Fausett, *Campsite at Dawn* (on Lake Powell).
14.54. (Bottom) Peter Hurd, *The Elephant Butte and Lake.*

14.55. Norman Rockwell, *Glen Canyon Dam.*

14.56. Fritz Scholder, *Indian Ruin.*

14.57. Billy Morrow Jackson, *Pyramid Lake.*

14.58. Michael Frary, *Irrigation.*

fascinated with "this land of mountains and deserts, of great rivers and spectacular canyons."[539] Indeed many of the images are extremely reminiscent of the nineteenth-century landscape genre, but, because of the subject matter, fall into the tighter niche of western art. Some artists opted to create paintings in which Reclamation structures or the impact of Bureau activities is missing. Dean Fausett's *Campsite at Dawn—Lake Powell* and Peter Hurd's *Elephant Butte Lake* depict an almost virgin landscape in which humans are merely visitors revealed through the boats sitting upon the lakes. While Hurd's painting is suggestive of the nineteenth-century ideal of the picturesque, Fausett's picture gives one a hint of the sublime as storm clouds begin to move into the area.

One popular nineteenth-century subject in the Bureau of Reclamation's Art Program was the ideal of the "pastoral landscape" or the "taming of the wild." Historian William Cronon notes that for nineteenth-century painters of the pastoral "the portrayal of a transformed landscape implied the passing of wilderness, the planting of a garden, the growth of a new civilization."[540] Michael Frary's *Irrigation* and Eugene Kingman's *Along the Gila Valley* best represent this style of painting by featuring bountiful fields that presented concrete evidence of Reclamation's economic value. Images such as these probably elated Bureau of Reclamation officials in the portrayal of the multipurpose aspects of all Reclamation projects.

Another theme that connects the Bureau of Reclamation's Art Program to nineteenth-century American art traditions appeared in the depiction of Native Americans and ancient Indian ruins. One of the most popular artists commissioned for the Art Program was Norman Rockwell, who chose Glen Canyon Dam as his subject. In his creation, Rockwell placed a family of Native Americans with their backs to the viewer looking across a canyon toward the dam from a cliff. Rockwell's juxtaposition of the Native Americans and the dam revives a popular-nineteenth century commemorative allegory. Many nineteenth-century artists saw the Indian as both victim and witness to the rapid advance of American society. Historian Brian Dippie states,

> The most common composition in commemorative allegories places a single Indian or a group of Indians in the foreground

---

[539] Lloyd Goodrich, foreword, in Bureau of Reclamation, *The American Artist and Water Reclamation*; Goodrich was also involved in selecting many of the artists who contributed to the Bureau of Reclamation's Art Program, see DeWitt, "Reclamation and the Creative Artist," 14.

[540] William Cronon, "Telling Tales on Canvas: *Landscapes of the Frontier Change*," in Prown, et al., *Discovered Lands, Invented Pasts*, 67.

high above whatever scene they survey. . . . They brandish no weapons; instead they seem resigned, even fatalistic. . . . The future held only the grave for them, but their unconquerable spirit would provide white Americans a lasting legacy. Thus the second allegorical tradition, a commemorative tradition lamenting the Indian's passing as the passing of a purer, original America.[541]

Rockwell may have contemplated the allegory in his contribution to Reclamation art. According to W. L. "Bud" Rusho, public affairs officer at Glen Canyon Dam, Rockwell's inclusion of the Navajo family was an attempt to add a "human element" to what Rockwell referred to as "a mechanical drawing." Nevertheless, Native American images and places seemed to have been a popular theme in the Bureau of Reclamation's Art Program. Some of the commissioned artists went out of their way to insure the inclusion of Native Americans as part of their representations of Reclamation. Fritz Scholder presented an abstract painting of ancient Indian ruins in *Indian Ruin—Lake Powell*: an interesting choice considering the number of Native American ruins and artifacts that lay beneath the lake's waters. In addition, Fletcher Martin and Billy Morrow Jackson included Pyramid Lake (on the Paiute Indian Reservation in Nevada) in their portrayal of the Newlands Project. The project's adverse impact on Pyramid Lake made disputes on water use a constant focus of legal suits over water allocations between the Pyramid Lake Paiutes and Newlands Project water users.[542]

To a limited extent, the Bureau of Reclamation's Art Program accomplished much of what DeWitt intended. In a sense, it revealed a "New West," transformed by Reclamation's control and management of the waters of the West. Large blue reservoirs set against a seemingly desolate environment indicated landscape improvement in the Great American Desert. Despite the lack of narrative, the art and its message provided the Bureau of Reclamation

---

541 Brian W. Dippie, "Western Art and the Dynamics of Change," in Prown, et al., *Discovered Lands, Invented Pasts*, 99-100.

542 For a concise description of Norman Rockwell's visit to Glen Canyon Dam, see W. L. (Bud) Rusho, *Oral History Interview*, Transcript of tape-recorded Bureau of Reclamation Oral History Interviews, conducted by Brit Allan Storey, senior historian, Bureau of Reclamation, during 1995 in Salt Lake City, Utah, Edited by Brit Allan Storey, 123; for some examples of the Bureau of Reclamation's fascination with Indian cultures see, Albert H. Wathen, "Indian Irrigation," *Reclamation Era*, 31 (December 1941): 322-3; Ben D. Glaha, "The People Who Have Gone Away," *Reclamation Era*, 32 (April 1942): 51-3; Stan Rasmussen, "Adventures in the Glen Canyon of the Colorado," *Reclamation Era*, 44 (May 1958): 41-5.

a powerful justification for its past accomplishments. On the other hand, the inclusion of Native American sites, and issues, highlighted the intrusion of Euro-American culture on the lives of Native peoples. Ambiguities remain about the purpose of the Art Program: was it a reaction to environmentalism, or simply a further explication of the purposes of the Bureau of Reclamation to the American public?

Films

Similar to its photographic efforts, the Bureau of Reclamation employed the medium of moving pictures to promote and explain its activities. Many of its early film productions have not been preserved. Still, some segments of earlier films made during the first two decades of the Reclamation Service have survived. Though it is difficult to determine the provenance of many of these clips, it may be safe to conclude that the Reclamation Service, and later the Bureau, played a role in their production. For example, on the American Memory section of the Library of Congress's web site there is film clip recorded at the 1911 dedication ceremony of Roosevelt Dam featuring former President Theodore Roosevelt. In films documenting construction of Hoover Dam, there are images of the early flooding of the Imperial Valley in California in 1905. These images presented the dangers of an uncontrolled river and argued for a dam somewhere on the Colorado River and its importance to the livelihood of farmers who made a living downstream.[543] During the 1980s the Bureau of Reclamation collaborated with many individual projects to produce films documenting their histories. Each contained some construction footage prior to 1920. These bits and pieces that can be easily viewed appear to be just about all that remain of a time when few Reclamation officials and western boosters recognized the value of this relatively new form of communication.

In discussing the Bureau of Reclamation's film history, C. J. Blanchard and his work as head of the Settlement Division within the Reclamation

---

[543] Carolyn Gentry, editor, *The Roosevelt Dam*, produced by the Roosevelt Memorial Association Film Library, located at the Library of Congress, American Memory, http://memory.loc.gov/ammem/collections/roosevelt_film; see also, www.srpnet.com/about/history/video.aspx accessed 9/12/2012; the SRP web site commemorates the 100th anniversary of the project. Theodore Roosevelt Dam was dedicated on March 18, 1911; see also, "Progress of Work," *Reclamation Record*, 2 (April 1911): 154-5; for information on Reclamation Service film activities prior to 1910 see Richard Dyer MacCann, *The People's Films: A Political History of U. S. Government Motion Pictures* (New York: Hasting House, Publishers, 1973), 44; United States Department of the Interior, *The Story of Hoover Dam*, videocassette, no date; United States Bureau of Reclamation, *Black and White Construction Film—Hoover Dam*, videocassette, no date.

Service plays an important role. Blanchard's almost yearly expeditions out West included making motion pictures to accompany photographs of project development. During a 1915 foray through the Shoshone Project in Wyoming, Blanchard sought to dramatize the everyday life of project farmers. He ended up producing a "photoplay" which offered a sense of the human aspect of life on an irrigation project, while still emphasizing the technical side of farm development under irrigation. Blanchard's story line revolved around a single woman who decides to leave her job as a school teacher in Illinois and start a new life out West. She purchases a farm and becomes romantically involved with her bachelor neighbor whom she hired "to put it into crops." Through a series of successes and tragedies—including the burning down of her house—the two eventually marry and begin a life as hardworking and successful project farmers. Blanchard boasted, "The film, in progressive steps, shows the desert, plowing, leveling, irrigating, seeding, harvesting, and threshing, and in the final chapter, two years later, in the new home, there is a baby."[544]

This early docudrama is testimony to Blanchard's and the Reclamation Service's innovative attempts to communicate with the public. The written description by Blanchard in *Scientific American Supplemental* is the only evidence of the making of this film. Viewers could make the connection between the characters in Blanchard's "photoplay" with tales of pioneering the American frontier during the nineteenth century. The film also portrays the West as a place where opportunities awaited; not to mention what it might be saying about designated gender roles, and what happens to those who deviate from them. Nevertheless, the film reveals Blanchard's willingness to utilize a relatively new technology in popular culture to advertise federal reclamation.

Most Reclamation Service film documentaries tried "to persuade or promote." This is also part of what Richard Dyer MacCann describes as "the process by which social values are constructed and changed." In *The People's Films*, MacCann argues, "The modern public relations campaign, in government or outside, is a formal attempt to build a new addition on the structure of our attitudes toward the outside world." Indeed, Blanchard saw films as a means to disseminate the activities of the Reclamation Service throughout the country. For example, he supplied films to Rollen D.

---

544 C. J. Blanchard, "Current Comments: Gathered from the Project Press and People," *Reclamation Record*, 9 (September 1918): 410-3; see also William Dean Harper, "The Master Movie Maker," *Technical World Magazine*, 22 (January 1915): 730-2, 788; C. J. Blanchard, "The Government Uses Motion Pictures," *Scientific American Supplemental*, 80 (August 1915): 120-1; see also Pisani, *Water and the American Government*, 102.

Salisbury, University of Chicago president, for the purpose of presenting the mission of the Reclamation Service to the nation's school children. In this particular case, Blanchard claimed that film provided the best means to inform the American public about the role of the Reclamation Service. In *Reclamation Record*, he enthused about the opportunity to show films to school children, who, in turn, discussed what they learned about Reclamation activities with their families.[545]

Reclamation Service films from this period reached a nationwide audience through a variety of different methods. In New England, for example, covered trucks carrying projectors and generators screened films in parks and other open spaces to reportedly large audiences. According to Blanchard, even the White House lawn became an impromptu movie theater "using trucks and amplifier apparatus so that talking will be made easier." Blanchard was continually short of funds to finance his film-making operations, but like every good advertising agent, he created methods to secure funding from outside sources. Blanchard solicited donations by promising to prominently display machinery used during project construction in films destined for overseas distribution, by "showing same in actual use by the U.S. Government and with suitable subtitles which of course would indicate by name the manufacturer." This was perhaps one of the first instances of product placement. Blanchard's filmmaking excursions also included producing films for Department of the Interior bureaus and services and the Department of Agriculture.[546]

Blanchard's filmmaking legacy has not been thoroughly appreciated by historians analyzing the history of documentary films. Prior to 1920 both the Reclamation Service and the Department of Agriculture were at the forefront of government produced informational films. Richard Dyer MacCann, in *The People's Films*, acknowledged the early activities of the Reclamation Service in producing nonfiction films to promote irrigation farming, noting that by the 1910s the Service had "made some pictures of large-scale farming

[545] MacCann, *The People's Films*, 4; for information on Renov's definitions of the functions of nonfiction films see Carl R. Plantinga, *Rhetoric and Representation in Nonfiction Film* (Cambridge: Cambridge University Press, 1977), 1; see also C. J. Blanchard, "Current Comments: Gathered from the Project Press and People," *Reclamation Record*, 11 (August 1920): 358; see also C. J. Blanchard, "Current Comments," *Reclamation Record*, 7 (July 1916): 290-1.

[546] C. J. Blanchard, "Current Comments," *Reclamation Record*, 12 (June 1921): 253-4; Memorandum for Mr. Weymouth, Statistician, May 1, 1923, RG 115, Entry 7, Box 17; C. J. Blanchard to Austin Machinery Corporation, March 18, 1922, RG 115, Entry 3, Box 17; C. J. Blanchard, "Current Comments: Gathered from the Project Press and People," *Reclamation Record*, 13 (September 1922): 214.

practices." Despite this brief recognition, MacCann further states that soon after this period the Bureau of Reclamation had "lapsed into inactivity again shortly afterwards, as far as films were concerned." Although Reclamation films became fewer in number, Bureau of Reclamation filmmaking continued throughout the twentieth century, staying fairly consistent to the subject, format, and distribution processes initiated by Blanchard. For example, in 1926, Bureau of Reclamation Commissioner Elwood Mead went on a film expedition in the West with photographer Maurice G. Wicker, who made "a photographic survey" of the Northwest. *New Reclamation Era* reported, "Motion-picture reels will be made up for distribution to educational organizations, chambers of commerce, and others interested in the relation of reclamation development to the economic life of the nation."[547]

By the late 1920s and into the 1930s, Bureau of Reclamation films primarily focused on construction of large dams or educational films to aid farmers with irrigation techniques. Construction of Hoover and Grand Coulee dams became favorite topics of Reclamation films promoting interest and support of New Deal public works enterprises. Many films officially appeared under the auspices of the Department of the Interior rather than the Bureau of Reclamation. This may explain the reduction in the number of films dealing with land reclamation. According to MacCann, the Department of the Interior films covered almost every aspect of the Department's responsibilities such as Indian reservations, cattle ranches, the General Land Office, and the Office of Education. MacCann observed, "The output amounted to about 25 reels per year during the 1930s. In the 1940s, the (film) division had some 69 films available in 16 mm and 35 mm, some silent, one or two reels in length."[548]

During this period, the Bureau of Reclamation produced films concentrated on the needs of project farmers. *Fundamentals of Irrigation* (1940) presented "modern irrigation methods that conserve soil, plant food, and water and pay in the production of higher quality of crops and larger yields." In addition, the film showed "laboratory demonstrations," animated strips illustrating water moving through different soil types, and the problems caused by overwatering. The film *Weeds* (1940) warned farmers of the dangers to

---

[547] MacCann, *The People's Films*, 44; "Motion Pictures Show Project Development," *New Reclamation Era*, 17 (July 1926): 125.

[548] MacCann, *The People's Films*, 47; for information on the consolidation of Department of the Interior film activities see "Reclamation Organization Activities and Project Visitors," *Reclamation Era*, 26 (June 1936): 155; "Photographic Activities Consolidated," *Reclamation Era*, 29 (August 1939): 217.

livestock and crops if weeds accumulate in irrigation canals. *Weeds* revealed to farmers "the latest methods of eradicating weeds on ditch banks and permanent control measures to keep weeds from regaining a foothold, once they are driven out."[549] Although films produced by the Bureau of Reclamation still promoted its activities, their messages now differed. No longer did the films place an emphasis on homemaking or building communities but rather offered instructions to established project farmers. Also, Reclamation's mission now included the generation of hydroelectric power, flood control, and maintaining and enhancing a reliable water supply for the American West.

Many New Deal art projects and motion pictures dramatized the necessity for public works projects. Two of the most influential films were Pare Lorentz's *The Plow that Broke the Plain* (1935) and *The River* (1937). These films provided powerful visual evidence to support the role of government programs in preventing dust storms on the Great Plains and efforts at flood control in the Mississippi River system. According to William M. Drennen Jr., commissioner of culture and history at the Cultural Center in Charleston, West Virginia, Lorentz's films went well beyond the bounds of simply recording events. Through the combination of images, music, and words, Lorentz transformed the medium of documentary films into a "tool for social change."[550]

The films' topics of soil erosion and flood control easily corresponded to the greater mission of the Bureau of Reclamation. One of the many recommendations to alleviate the distress of dustbowl farmers was to have them move onto irrigation projects in the West.[551] Lorentz's film, *The River*, addressed flood control and water management. It celebrated the achievements of the Tennessee Valley Authority and its efforts to control flood waters in that river valley and to produce cheap electricity for its rural residents. For the Bureau of Reclamation and its proponents, the larger message was a call for large-scale river basin management that only the federal government could accomplish. Such projects offered the promise of putting Americans back

549 "Fundamentals of Irrigation," *Reclamation Era*, 30 (March, 1940): 83; "Weeds," *Reclamation Era*, (May, 1940): 143.

550 William M. Drennen Jr. introduction, in Pare Lorentz, *FDR's Moviemaker: Memoirs & Scripts* (Reno: University of Nevada Press, 1992): 1, 43; Pare Lorentz, writer and director, *The Plow that Broke the Plain* (Culver City, California: Zenger Video, 1936); Farm Security Administration, USDA, Pare Lorentz, writer and director, *The River* (Washington, D.C.: U.S. Department of Agriculture; Capital Heights, Maryland: distributed by National Audiovisual Center, 1937).

551 Pisani, *Water and American Government*, 151; Rowley, *The Bureau of Reclamation*, 316-8.

to work. For Lorentz, the larger message advocated for change in American culture and a new relationship between humans and the environment.

One of Lorentz's projects on behalf of the New Deal directly dealt with the work of the Bureau of Reclamation. In 1938 he produced a radio program about the plight of eastern industrial workers titled *Ecce Homo* [Here is the Man]. In this radio drama, the main character, worker 7790, laid off from his job in an automobile plant, began to head west where he heard that opportunities abounded. Lorentz maintained, "The simple philosophy of 'Ecce Homo' was that with the gigantic industrial equipment and the magnificent amount of arable land in our country, it was stupid to have eleven million to fifteen million unemployed men and women." The hero headed west toward the construction of Grand Coulee Dam where the environment was being transformed and American potential was limitless. In a soliloquy that would have made C. J. Blanchard blush with pride, Lorentz wrote:

> They can make plenty more!
> They can make the desert green!
> Maybe they'll build a green city.
> Maybe they'll start East and build her all over again.
> Maybe there'll be farms for the little man!
> They can move mountains and shove rivers around!
> There's men and machines and there's sun and land and room
>   for a man to turn around in.
> And there's a man-sized job to be done!

Ecce Homo aired only once in the United States: Ford Motor Company, a major sponsor at CBS, voiced displeasure with the progeram's representation of the industry's employment practices.[552] Nevertheless, *Ecce Homo* exemplifies the importance of the Bureau of Reclamation's work to the Roosevelt administration and how that work assumed a vital role in the New Deal scenarios.

Lorentz's funding for both *The Plow that Broke the Plain* and *The River* came from various government agencies, but the success of those two movies led President Roosevelt to create the United States Film Service in 1938, with Lorentz named executive director. The function of the Film Service under Roosevelt's executive order was to coordinate all film productions within the federal government, maintain a film library, and "to distribute and exhibit

[552] Lorentz, *FDR's Moviemaker*, 80; for a complete transcript of *Ecce Homo*, see *Ibid.*, 84-104.

such motion pictures." One of Lorentz's first endeavors as executive director was to produce *Ecce Homo* as a film documentary. In this particular case, Lorentz sought to push the administration's agenda for continued development of the Columbia River basin, which directly concerned the work of the Bureau of Reclamation in the Pacific Northwest. According to Robert Snyder, in *Pare Lorentz and the Documentary Film*, the last page of the script "was based on the regional development program the President was preparing for Congress. This program called for the area development of resources patterned on the TVA." Other important films produced in the brief tenure of the Film Service were Robert Flaherty's *The Land* (1941) and Joris Invens's *Power and the Land* (1940). Although not tied directly to the activities of the Bureau of Reclamation, both films addressed the Reclamation topics of soil erosion and rural electrification.[553] The United States Film Service closed shop in 1940, a victim of both Congress's growing displeasure with an array of New Deal programs that focused on the arts and the Roosevelt administration's increased attention to the war in Europe.

After World War II, the Bureau of Reclamation renewed its motion picture production efforts, with an emphasis on informational movies for project farmers. In 1946 the film *Fighting Weeds* impressed upon irrigators the necessity for controlling weed growth in irrigation canals and ditches. By the early 1950s Bureau of Reclamation films reached beyond the problems of everyday farm life to broader topics. In 1952 Reclamation's film *Water and the West* appeared in the documentary films category at the Edinburgh Film Festival. The Festival recognized *Water and the West* to be "one of the outstanding documentary films of the year," and it won additional praise for its director Ben Glaha. Many of the Bureau of Reclamation's postwar films followed the formula first initiated by C. J. Blanchard and emphasized federal reclamation's economic benefits.[554]

---

553 Robert L. Snyder, *Pare Lorentz and the Documentary Film* (Reno: University of Nevada Press, 1994), 83-4; see also Lorentz, *FDR's Moviemaker*, 77-8; for information on *The Land* see, Snyder, *Pare Lorentz and the Documentary Film*, 131-40; Lorentz, *FDR's Moviemaker*, 141, 152; Richard Barsam, *The Vision of Robert Flaherty: The Artist as Myth and Filmmaker* (Bloomington: Indiana University Press, 1988), 84-98; for information on *Power and the Land* see, Snyder, *Pare Lorentz and the Documentary Film*, 121-30; Lorentz, *FDR's Moviemaker*, 141, 152; Hans Schoots, *Living Dangerously: A Biography of Joris Ivens* (Amsterdam: Amsterdam University Press, 2000), 152-8.

554 "Technicolor Films Available," *Reclamation Era*, 32 (June 1946): 119; "'Water in the West' Goes Global," *Reclamation Era*, 38 (June 1953): 196; "'Water in the West' Honored," *Reclamation Era*, 38 (May 1953): 107.

By the early 1960s Bureau of Reclamation films documented new dam construction and relived past successes with new films commemorating the construction of Hoover and Grand Coulee dams. *Key to the Future* told the story of the Colorado River Storage Project, presenting construction scenes at dams like Glen Canyon, Flaming Gorge, and Navajo, as well as the Curecanti unit. The film reminded viewers of national benefits these efforts provided and "the potential wealth of the upper basin in the form of high crop production and minerals such as oil, natural gas, coal, and uranium whose development awaits the impetus of adequate water and power." Other films of this era followed similar story lines, whether discussing the Colorado-Big Thompson Project or revisiting past glories about the taming of the lower Colorado River, which included commemorating Parker and Davis dams.[555]

By the late 1970s and early 1980s, Bureau of Reclamation films mirrored its movement away from creating greater reservoirs and supplies of western water to a manager and conservator of water. Spurred by the failure of Teton Dam in 1976, lack of new authorizations from Congress, and intense pressure from the environmental movement, the era of new, major dam construction was drawing to an end. The Bureau of Reclamation found itself with a new mission in water management and serving as arbiter amongst competing interest groups over allocation of the West's precious water supply. Western waters under the management of the Bureau of Reclamation became a source of contention among farmers, urban centers, and environmental organizations. Each believed it had a legitimate claim to scarce waters as Reclamation officials attempted to referee and placate these competing groups. In addition, the Bureau of Reclamation had promoted the multipurpose functions of its dams and reservoirs, almost from its inception. During the first half of the twentieth century, multipurpose meant water for irrigation, production of hydroelectric power, and recreation. By the 1980s, multipurpose also included the protection of endangered species, maintaining and increasing water supplies for urban areas, and preservation of the West's diminishing wetlands. Bureau of Reclamation films of this era embraced new responsibilities while still trying to maintain Reclamation's commitment to project farmers and communities.

---

555 "Camera-Eye View of Water Development," *Reclamation Era*, 47 (February 1961): 23; for films on development of the lower Colorado River, see United States Department of the Interior, *The Story of Hoover Dam*, videocassette, c. early 1960s; for other films on Hoover Dam construction see, United States Bureau of Reclamation, *Black and White Construction Film —Hoover Dam*, videocassette, no date; for films on the Colorado-Big Thompson project, see United States Bureau of Reclamation, *The Barrier Between*, videocassette, 1960s; Thomas J. Fields, producer, *Green Fields*, videocassette, 1963.

The Green Echo of Snow (1970s), produced by the Bureau of Reclamation and the Northern Colorado Water Conservancy District, underscored the diverse responsibilities of the Bureau of Reclamation. The film opens with actor Hal Smith, who plays a number of characters and directs the narrative, portraying as Captain Stephen H. Long giving his pronouncement about the Great Plains as the "Great American Desert." The film then presents the history of Euro-Americans in Colorado, beginning with the 1859 Gold Rush, followed by early attempts at irrigation and finally the struggle to get the necessary legislation passed for the Colorado-Big Thompson Project. Through his various characters, Smith takes the viewer through the project's construction history, stressing the engineering feat of bringing water from the Western Slope of the Rocky Mountains to the eastern plains of Colorado. The film emphasizes the standard tropes of Bureau of Reclamation promotion: irrigation, hydroelectric power for home and industry, and now environmental benefits. In contrast to earlier Reclamation films, *The Green Echo of Snow* does not include rhetoric regarding human ability to control nature. Rather, the film takes an enlightened and symbiotic view of the relation of nature and human culture. Still, the economic benefits derived from a reliable water supply and the production of cheap electricity compose the main theme. Footage of melting mountain snow running into picturesque mountain lakes or catchment basins provides project water that benefits both nature and human enterprise.[556]

The Bureau of Reclamation film, *The Snake: Jewel of the Gem State* (1987), offered a similar narrative. The Minidoka Project in Idaho, one of Reclamation's earliest projects, provides the focus of the film. The movie highlights the growth of communities that "literally grew out of the desert." Economic and recreational benefits are, of course, prominently displayed, while the narrative emphasizes the 1.1 million acres of farmland that produced $300 million worth of crops in 1986. Fishermen and campers line the shoreline of the project's six reservoirs, giving the impression of a veritable vacation paradise. Here the compatibility of human technology and nature abounds. Reservoir waters recharge rivers and underground aquifers, while the project's various man-made structures represent "man's ability to adapt to what nature offers."[557]

---

[556] United States Bureau of Reclamation and the Northern Colorado Conservancy District, *The Green Echo of Snow*, videocassette, 1970s; many of the images in this film were originally used in earlier Reclamation films such as *The Barrier Between* and *Green Fields*.

[557] United States Bureau of Reclamation, *The Snake: Jewel of the Gem State*, videocassette, 1987; for other Bureau of Reclamation films discussing similar themes see, United States

Perhaps the best example of how the Bureau of Reclamation promoted its traditional mission of "making the desert bloom," while also implementing its new environmental responsibilities, came in the film *The Great Web of Water: The Central Valley Project.* Produced in the early 1980s, it demonstrated all the practical benefits derived from federal reclamation. *The Great Web of Water* traced the history of the Central Valley Project beginning with construction of Shasta Dam in the late 1930s. Early black and white construction footage showed a dam rising over the waters of the Sacramento River, while the narrative told how this New Deal project gave Depression-era workers on the dam "renewed pride and personal meaning." Through a montage of footage and graphics, viewers take in the enormity of the project, and how water delivered through Reclamation facilities made possible the bountiful and diverse farming in California's great Central Valley. Again reminiscent of C. J. Blanchard's lectures, *The Great Web of Water* stressed the project's technological achievements and emphasized how western irrigation economically benefited the entire nation. The film also explains how the Central Valley Project assisted in the growth of communities through flood control, recreation, and the production of hydroelectricity.

Nevertheless, this CINE (Council on International Non-Theatrical Events) award winning film argued intensely that the Bureau of Reclamation's primary responsibility was to ensure the most efficient and beneficial use of every drop of water. The theme demonstrated the new outlook in the Bureau of Reclamation—it was no longer only a government construction company. *The Great Web of Water* not only demonstrated Reclamation's critical role in the economic prosperity of California, but also showed its ability to adapt to new realities. Familiar Reclamation scenes of bountiful fields and orchards, combined with images of rejuvenated wetlands and pristine mountain streams, gave the impression that water development projects worked for the benefit of both humans and nature.[558]

Bureau of Reclamation, *The Yakima Project—From the Cascades to the Columbia*, videocassette, 1984; United States Bureau of Reclamation, *The Boise Project: Gift of Life*, videocassette, 1983.

[558] United States Department of the Interior, Bureau of Reclamation, *The Great Web of Water: The Central Valley Project*, videocassette, early 1980s; for information on CINE see, "CINE—What it is. What it does," www.cine.org/history; The Council on International Non-Theatrical Events was founded in 1957 "to depict American life and thought realistically for a global audience." The award won by *The Great Web of Water* was a Golden Eagle Award to "recognize and foster" the production of high-quality documentary films.

Alternative Narratives

By the 1980s the Bureau of Reclamation came under severe criticism from several sources. The discord gradually emerged in environmental groups and some Reclamation project communities that offered vastly different narratives about the story of irrigation in the West. And similar to the Bureau of Reclamation, they relied on images in various forms to validate and spread their views. In the late decades of the twentieth century, the Bureau of Reclamation's progressive ideal faded under a wave of criticism.

As early as the mid-1950s, protests occurred against the federal government's seeming desire to dam every river in the West. Even Reclamation's definition of progress was questioned. The proposed Colorado River Storage Project in 1950 sparked opposition. An emerging environmental movement became critical of the Bureau of Reclamation's plan to build dams at Echo Park and Split Mountain near the confluence of the Green and Yampa rivers in Utah. Reclamation officials acknowledged the inundation of parts of Dinosaur National Monument from the construction of these dams.[559] Incensed wilderness advocates argued the dams threatened the sanctity of the nation's national parks and park system. In their efforts to stop construction of the Echo Park and Split Mountain dams, early environmentalists used some of the same forms of image making that previously had been so beneficial to the Bureau of Reclamation in promoting its activities.

In 1955 the Sierra Club, under the leadership of David Brower, led a coalition of conservation groups and organizations in opposition to parts of the Colorado River Storage Project that threatened Dinosaur National Monument. Its activities included testifying before congressional committees, producing articles in some of the nation's leading magazines, and placing ads in major newspapers. An important aspect of the Sierra Club's public relations campaign was publication of *This is Dinosaur: Echo Park Country and Its Magic Rivers* in 1955. *This is Dinosaur* attempted to inform the general public of the natural and cultural resources found at an infrequently visited national monument. According to its editor, Wallace Stegner, the book was not "a fighting document," but rather a broad overview of the archeological, paleontological,

559 "Echo Park—Split Mountain Dams Approved," *Reclamation Era*, 36 (August 1950): 148; for information on the Bureau of Reclamation's Colorado River Storage Project see, Bureau of Reclamation, *The Colorado River*.

and recreational values of Dinosaur National Monument. Despite this disclaimer, Stegner expressed the conviction, "The dams will kill most of this."[560]

In both words and images, *This is Dinosaur* took the reader through the natural wonders and beauty of a little-known area of the American West. The text called into question claims made by the Bureau of Reclamation about of the benefits hydroelectric power and water storage at the cost of sacrificing a beautiful natural environment. The book became an essential tool in challenging the rhetoric supporting the Bureau of Reclamation's planned resource development projects. It also set the stage for future arguments over dam building by establishing an emotional and aesthetic framework of values that stood in stark contrast to the economic and utilitarian motivations behind Reclamation projects. In *A Story that Stands Like a Dam*, Russell Martin asserts that *This is Dinosaur* and the other Sierra Club publications "were bringing a kind of emotion to bear that the conservation movement never had been assisted by before."[561]

The coalition of conservation forces largely succeeded in stopping construction of dams in Dinosaur National Monument because of the arguments and images used in *This is Dinosaur*. Their success was limited. In order to win at Dinosaur National Monument they "sacrificed" another relatively unknown area of the West—Glen Canyon. For the Bureau of Reclamation, Glen Canyon presented an opportunity to build a major component of the CRSP, and it was another large dam on a scale and similar to earlier endeavors, like Hoover, Shasta, and Grand Coulee dams. Images in the pages of *Reclamation Era* revealed the spectacular beauty of Glen Canyon, while narratives discussed the epic task at hand for the Bureau of Reclamation. In particular, these articles discussed Reclamation's efforts in conjunction with academics and scientists to record the natural and cultural history of the canyon before they were lost to the rising waters of the new reservoir.[562]

According to Russell Martin, the Bureau of Reclamation, and especially Commissioner Floyd Dominy, believed that a "publicity barrage was absolutely essential" to maintain public support for Glen Canyon Dam. Utiliz-

---

560 Stegner, *This is Dinosaur: Echo Park Country and Its Magic Rivers*, vi.

561 Martin, *A Story that Stands Like a Dam*, 244.

562 David Brower, *For Earth's Sake: The Life and Times of David Brower* (Salt Lake City: Peregrine Smith Books, 1990), 341; see also Stan Rasmussen, "Adventure in the Glen Canyon of the Colorado," *Reclamation Era*, 44 (May 1958): 41-5; C. Gregory Crampton, "History in Glen Canyon," *Reclamation Era*, 45 (May 1959): 41-4, 52.

ing promotional tactics reminiscent of those of C. J. Blanchard, Martin maintains that Dominy

> was certain that if he got the truth out, if school kids and Kiwanis clubs, Rotarians and Lions and chambers of commerce across the country got the straight stuff about what a boon, what a *blessing* this dam would be, the carping would quiet quickly and Reclamation could get back to the business of building things.

Photographs and motion pictures formed a large part of promotional activities for the dam, including the film *Canyon Conquest* in which construction of the dam was seen through the approving "eyes of Navajo headman Alvin Tsi'najinii."[563]

Dominy had every reason to be concerned. Successful at halting construction of dams at Echo Park and Split Mountain, the infant environmental movement took aim at Bureau of Reclamation's extensive plans for the Colorado River. In 1963 the Sierra Club published Eliot Porter's *The Place No One Knew: Glen Canyon of the Colorado*, a photographic documentary of the beauty and natural wonder that the Reclamation dam stood poised to destroy at Glen Canyon. For David Brower and other conservationists, the book and its marvelous photographs were not only a visual reminder of the consequences of humanity's impact on the environment but also a remorseful recognition of the conservation movement's tacit acquiescence in the dam's construction. He also took the opportunity to call the Bureau of Reclamation and its supporters to task for their shortsightedness, and questioned their definition of progress.[564] Here the Bureau of Reclamation faced a new type of criticism that went beyond protecting the sanctity of America's national parks. It questioned the Bureau and its advocates' very idea of conservation and whether or not it was beneficial to actually make the desert bloom.

For Brower, however, the loss of Glen Canyon meant the disappearance of one of the world's striking natural wonders and an occasion for mourning. In respect of that ritual, Porter's *The Place No One Knew* offers impressive photographs of Glen Canyon that literally explode into a profusion of colors in which one can only marvel at the beauty, while sharing Brower's sadness. Porter's brief narrative echoed the philosophy of wilderness advo-

---

563 Martin, *A Story that Stands Like a Dam*, 185-6.

564 Brower, foreword, in Porter, *The Place No One Knew*, 18.

cates who argued for the necessity of natural places to escape from the ever-pressing demands of the modern world. Porter wrote,

> The world's demands fade the faster, but nonetheless surely your own will shrink to acceptable proportions and cannot sally forth to attack you. In the wilderness of Glen Canyon you do not assail yourself. You glide on into the day unpursued, living, as all good river travelers should, in the present.

Porter's narrative in conjunction with his photographs did not simply portray Glen Canyon's natural beauty but communicated a sense of loss for something of greater value to humanity than benefits derived from water storage and production of hydroelectric power.

Of course, this was the book's primary purpose: to serve as a counter narrative to the publicity efforts of the Bureau of Reclamation and other Reclamation supporters. Ironically just as C. J. Blanchard's photographs revealed an arid environment in transition, Porter's images resurrected the nineteenth-century ideal of nature by showing it at its most sublime and picturesque. Here, humans enter the scene as occasional visitors, and nature controls both time and space. By 1963 the fate of Glen Canyon was determined, but Brower and Porter sought to inform the American public that there were aspects of the natural world that had value beyond material enrichment. Porter wrote:

> The waters impounded by this plug of artificial stone spread back through Glen Canyon and for one hundred eighty-six miles in all, inundated the sparkling river, swallowing its luminous cliffs and tapestry walls, and extinguishing far into the long, dim, distant future everything that gave it life. As the waters creep into the side canyons, enveloping one by one their mirroring pools, drowning their bright flowers, backing up their clear, sweet springs with stale flood waters, a fine opaque silt settles over all, covering rocks and trees alike with a gray slimy ooze. Darkness pervades the canyon. Death and the thickening, umbrageous gloom takes over where life and shimmering light were the glory of the river.[565]

[565] Porter, *The Place No One Knew*, 18, 15.

1 . Wall and river's edge

Past these towering monuments, past these mounded billows of orange sandstone, past these oak-set glens, past these fern-decked alcoves, past these mural curves, we glide hour after hour, stopping now and then, as our attention is arrested by some new wonder.—JOHN WESLEY POWELL, 1869

by ELIOT PORTER

THE PLACE NO ONE KNEW Glen Canyon on the Colorado

EDITED BY DAVID BROWER

Sierra Club • San Francisco

14.59. Eliot Porter's photo essay, *The Place No One Knew: Glen Canyon on the Colorado,* published in 1963, lamenting the loss of the beauty and wilderness that was Glen Canyon, is an iconic publication in the environmental movement.

While the natural beauty of Glen Canyon was irreversibly destroyed, the Sierra Club and other environmentalist organizations were able to alter the Bureau of Reclamation's plans for further development of the Colorado River. *The Place No One Knew* gave the American public the opportunity to see the negative side of water development in the control of nature, and the consequences of an unquestioning faith in material progress. Brower saw it as a larger effort to influence the way Americans thought about the natural world. Sierra Club publications also included Francois Leydet's *Time and the River Flowing* (1968) that specifically focused on the Bureau of Reclamation's plans to build Bridge and Marble Canyon dams along the Colorado River. Similar to *The Place No One Knew, Time and the River Flowing* combines photographs and a strong narrative to send a twofold message revealing the majestic beauty of the Grand Canyon and the river that formed it, while strenuously arguing that the building of the two dams would mean the imminent demise of the canyon. In *Time and the River Flowing*, Leydet attempts to convince the

14.60. Reclamation participated in several of the films and videos about the Colorado-Big Thompson Project advertised on the cover of this anthology disc.

reader not just of the sublime wonder of the Grand Canyon, but that the canyon and the river have a "soul." He argued that the Bureau of Reclamation's two proposed dams at either end of Grand Canyon National Park would effectively "kill" the living force that created and continues to create the canyon.[566]

The Bureau of Reclamation, especially its commissioner, Floyd Dominy, did not take criticism kindly. Dominy took this assault from the budding environmental community as a personal affront to not only the Bureau of Reclamation, but himself as well. He responded utilizing methods similar to those employed by environmentalists—in both images and narratives to

[566] Leydet, *Time and the River Flowing*, 86, 94.

promote and explain how further development of the Colorado River stood to benefit all Americans. The film *Clear Water on the Colorado* and a small booklet *Lake Powell: Jewel of the Colorado*—in which the Commissioner was the photographer—emphasized the recreational benefits that Reclamation's dams would provide. Dominy argued:

> A blue lake above Bridge Canyon, deep within the inner gorge, would make this spectacular canyon easy of access by boat for millions. Easy of access for millions of Americans who love to boat, fish, and swim and water ski—or just laze in the sun—in God's country. For millions of Americans who would see–for the first time–a new part of their heritage of natural beauty.[567]

Dominy's argument deviated little from past message: those utilitarian efforts of the Bureau of Reclamation would ensure that all Americans had the opportunity to enjoy and appreciate the wonders of the American West.

Still, it was well recognized that recreational benefits were not the primary reason for construction of the proposed Bridge Canyon and Marble Canyon dams. Production of hydroelectric power and management of the waters of the Colorado River were the Bureau of Reclamation's primary aims. Disagreements between the environmental movement and Reclamation supporters went deeper. Both had a powerful message about what constituted the best standard by which to manage the nation's natural resources. Yet both used almost identical methods to promote, advertise, and argue their respective points of view. Americans who had access to the materials produced during this struggle saw similar scenery with images intended to produce a sense of awe and wonder. It is in the narratives, however, that the messages conflicted and the real aims of both sides diverged. For Reclamation supporters, early twentieth-century progressive ideals of utilitarian management of natural resources formed the foundation for their arguments and views. The environmental movement stressed preservation of wilderness and the beauty of natural scenery as necessary for the health of the modern world and human society.

---

567 United States Department of the Interior, Bureau of Reclamation, *Lake Powell: Jewel of the Colorado* (Washington, D.C.: United States Government Printing Office, 1965), 28; see also, Leydet, *Time and the River Flowing*, 133.

By 1997 the Bureau of Reclamation suspended its career of building new dams.[568] Still, the same questions that arose during the late-1950s through the 1960s remained. In 1993 Jon Else produced for public television a documentary series based on Marc Reisner's 1986 inflammatory exposé of water development in the West, *Cadillac Desert*. The television series addressed the tremendous growth of the American West during the twentieth century and the need to develop its meager water resources to insure growth and progress. The series took a critical look at the political intricacies of water management in the West and the overall environmental effects of westerners' water development. *Cadillac Desert* focused on what Reisner conceived of as the corrupt power relationship between the Bureau of Reclamation and those in the West constantly desiring new water projects. It also noted disapprovingly the power of agribusiness to achieve cheap and subsidized water at the expense of the larger society. Finally, the documentary explores an urbanized and industrialized West that, because of unchecked development, is rapidly outgrowing its scant and unreliable water supply.[569]

Through old film clips, photographs, spectacular images of the modern West, the *Cadillac Desert* television series demonstrated the ability of humans to control and manage nature. In one sense the series celebrates American ingenuity and engineering prowess in creating these vast and expensive water works. On the other hand, it questions whether those same abilities and wonders are worth all the environmental costs. Through frank, and at times touching personal recollections, *Cadillac Desert* tells the story of those who effected and were affected by the course of water development in the West. The documentary's depiction of the Bureau of Reclamation applauds its engineering feats; at the same time portraying Reclamation as an uncaring bureaucracy, as a pawn for powerful special interest groups. Most important, *Cadillac Desert* provides an intriguing look at the controversies that have been a part of the history of water development in the West.

Other films and filmmakers offered narratives that veered away from Bureau of Reclamation representations of progress and prosperity and instead focused on those heroic but nameless individuals who successfully settled

---

568 Reclamation still had to build the Ridges Basin Dam in southern Colorado, as part of the construction of the Animas La Plata Project authorized in 1968, and dam safety work on some small dams resulted in the virtual rebuilding of an occasional dam. For practical purposes, however, there were no new large dams in sight for Reclamation.

569 *Cadillac Desert*, produced by KTEH/San Jose, Jon Else, writer, director and producer, Deborah Hoffman, editor, videocassette, 1997; see also Reisner, *Cadillac Desert*.

Reclamation projects. For example, Roger Hansen's *Moving a River: History of the Strawberry Valley Project* looks at the development and growth of Utah's first Reclamation irrigation project from the point of view of Strawberry Valley water users. Some of these films tell a contested story of the relationship between the Bureau of Reclamation and project farmers. Produced by the Churchill County Museum Association, *Turn This Water Into Gold: The Story of the Newlands Project* (1998) is highly critical of the Reclamation Service. The film disparages the Reclamation Service's original optimistic view of the project's potential and reinforces its criticism with dramatic images of farms withering away because of a lack of water and farms ruined by chronic drainage problems. Instead, it idealizes the fortitude and independence of project farmers who, despite all the tremendous difficulties, succeeded in transforming their desert environment.[570]

Dennis Meyers's and Erik Westby's *Water for a Desert Dream: The Newlands Project* (1996) offers a similar perspective examining the trials and tribulations project farmers faced in their attempts to establish a community. This film "traces the development of the first reclamation project in the West ... from its conception in the 1880s to the controversy of recent years." *Water for a Desert Dream* discusses many of the issues found in *Turn This Water Into Gold* and also notes the militancy of Newlands Project farmers. This film is also critical of the federal government, but its criticism focuses on the impact the project had on the Pyramid Lake Paiutes and Paiute Shoshone tribes. Filmmakers describe how tribal members lost their 160 acre allotments and were only given 10-acre plots with the promise of a water delivery system that never materialized. Images depict the poverty on the reservation and the falling water levels of Pyramid Lake due to diversions of Truckee River water to supply the Newlands Project.[571] Ironically, and despite its obvious sympathy for the plight of the Indians and the film's staunch admiration of project farmers, *Water for a Desert Dream* fails to mention the obvious benefits to Newlands Project farmers resulting from the diversion. Both Newlands Project films and Hansen's *Moving a River* provided broader insights into the history of land reclamation in the West from local filmmakers.

---

[570] Roger Hansen, writer, director, and producer, *Moving A River: History of the Strawberry Valley Project*, J. T. V. Productions, videocassette; *Turn This Water Into Gold: The Story of the Newlands Project*, a production of the Churchill County Museum Association, Jane Pieplow, executive producer, Mark Gandolfo and Jane Pieplow, directors, videocassette, 1998; see also John M. Townley, *Turn This Water Into Gold: The Story of the Newlands Project* (Reno: Nevada Historical Society, 1998; originally published in 1973).

[571] *Water for a Desert Dream: The Newlands Project*, Dennis Meyers and Erik Westby, producers, Curt Daniels, director, videocassette, 1996.

14.61. *Lake Powell: Jewel of the Colorado* was Commissioner Floyd Dominy's 1965 response to Eliot Porter's *The Place No One Knew.*

14.62. In 1986 journalist Marc Reisner published *Cadillac Desert: The American West and its Disappearing Water*, a popular exposé regarding issues surrounding the development and use of water in the West.

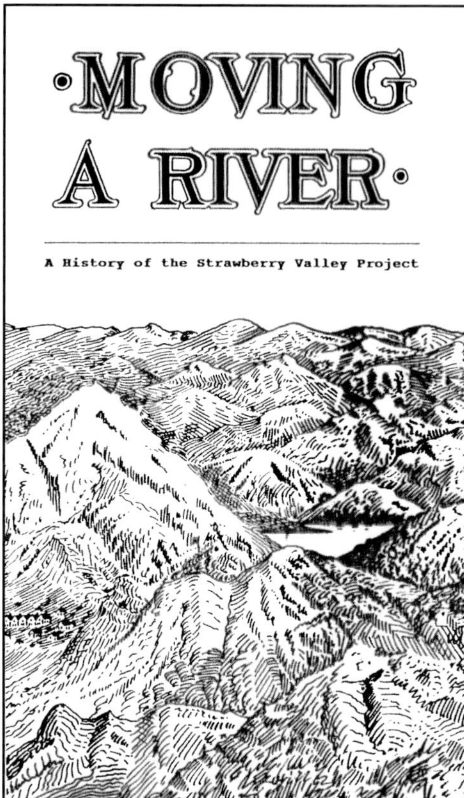

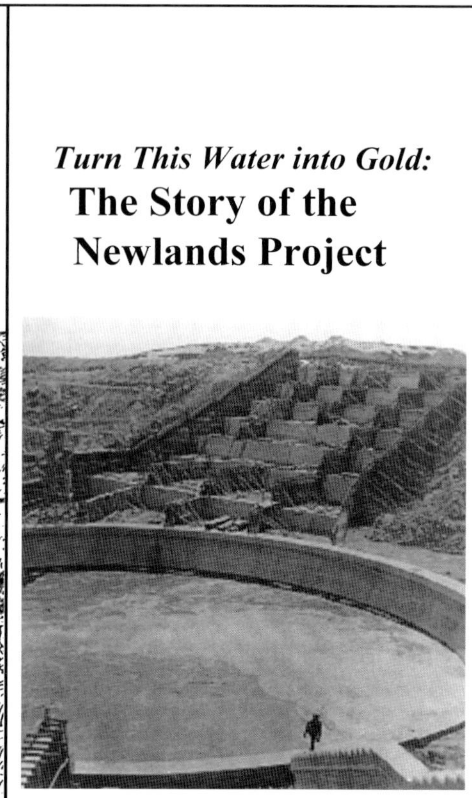

14.63. (Above) Roger Hansen's 28 minute long video history of the Strawberry Valley Project was titled *Moving a River*.

14.64. (Above Right) The 1998 video *Turn This Water into Gold: The Story of the Newlands Project* was a collaboration of the Churchill County Historical Society, the Truckee-Carson Irrigation District, and the Nevada State Historical Society.

14.65. (Right) The Water Education Foundation's *Healing the Water* was another of the videos focused on water issues of the Newlands Project.

In 1997 the Water Education Foundation, in Sacramento, California, produced *Healing the Water*, which encompasses many of the modern challenges that the Bureau of Reclamation faced in its attempts to reconcile all the demands placed on the West's finite water resources. Once again, the focus of this documentary is on the Newlands Project in northwestern Nevada. *Healing the Water* discusses the long-term impacts of Truckee River water diversions on the environmental health of Pyramid Lake and on the lives of the Pyramid Lake Paiute who traditionally relied on the lake's fisheries. This episode in western water controversies is a convoluted story intertwined with ideals and goals of the 1902 Reclamation Act, the federal government's trust responsibilities to Native Americans, the aspirations of Newlands Project farmers, environmental issues, and water law. The documentary traces the history of this seemingly irreconcilable dispute, beginning with the Orr Ditch Decree in 1944 followed by further demands placed on the Truckee River's water because of the 1972 Endangered Species Act. The narrative of *Healing the Water* suggests that the grievances of all parties might be met in negotiations made possible by passage of the Truckee-Carson Water Settlement Act by Congress in 1990. Better than most, this documentary depicts the arguments and concerns of the competing forces—Indian tribes, farmers, growing urban centers, and power companies—who vie for the precious waters of the Truckee River. *Healing the Water* portrays the Bureau of Reclamation as a federal agency attempting to placate the demands of contending interests dependent upon Truckee River water. It speaks of Reclamation's new role as a water arbiter in the West. More importantly, the documentary reveals that the history of the development and management of western water involves much more than the narrow historical representation first brought to the public by the Bureau of Reclamation earlier in the twentieth century. [572]

Conclusion

In 2002 the Bureau of Reclamation produced two short films: *Looking Back on a Century of Water* and *A Century of Water for the West, 1902-2002*. Both films were produced as part of the 100th anniversary of the signing of the Reclamation Act in June 1902. They emphasized the Bureau of Reclamation's continuing role in contributing to the growth and prosperity of the American West and celebrate the contributions of the men and women of the Bureau of Reclamation. *Looking Back on a Century of Water* asserts that Reclamation

---

572 *Healing the Water*, a KNPB public television special, Sue Pearson Atkins, producer/writer, produced by the Water Education Foundation, videocassette, 1997.

14.66. A video developed for Reclamation's centennial in 2002.

14.67. Reclamation developed *A Century of Water for the West* for the centennial in 2002.

"started the West on the road to claiming its heritage." *A Century of Water for the West*, with actor Steve Stark playing the role of Theodore Roosevelt, highlights the achievements of the Bureau of Reclamation, and its changing mission over time. This film maintains that the "Bureau of Reclamation whose philosophy of stewardship fulfilled a promise that the lands and waters of the American West exist for the benefit of all."[573] Through sound bites, photographs, and impressive film footage of Reclamation dams, reservoirs, powerplants, and canals, these brief fifteen minute films tell the story of the Bureau of Reclamation.

Reminiscent of the lectures and slide presentations of C. J. Blanchard, both films illustrate the technological innovations of Bureau of Reclamation engineers, the wealth produced on irrigated farms, and the benefits derived from the production of hydroelectric power. Their stories develop chronologically; both films discuss the struggles of western farmers in eking out a living in an arid environment and emphasize the irrigation reports of John Wesley Powell—not surprisingly, stressing Powell's belief that the federal government should play the leading role in water development in the West. Also noted is the Bureau of Reclamation's new role as a leader in water management as an agent of water conservation, fisheries development, and wetlands restoration. According to *A Century of Water for the West, 1902-2002*, the Bureau supplies water for millions of families throughout the West, while its structures prevent billions of dollars in flood damage. With its fifty-eight powerplants, Reclamation provides clean hydroelectric power that helps run the economic engine of an urbanized West. Federally irrigated farms produce sixty-five percent of the vegetables and twenty-five percent of all fruits and nuts grown in the United States. The most important achievement, according to the film, is the commitment of the men and women of the Bureau of Reclamation "to the mission."[574] In short, both films demonstrate the Bureau of Reclamation's continuing influence, in the transformation of the American West.

According to these films, in its one hundred years of existence, the message of the Bureau of Reclamation remained constant and consistent. Photographs, art, and films, celebrated an institution whose sole objective was to

---

573 *Looking Back on a Century of Water*, United States Department of the Interior, Bureau of Reclamation, videocassette, 2002; *A Century of Water for the West, 1902-2002*, United States Department of the Interior, Bureau of Reclamation, videocassette, 2002.

574 *Looking Back on a Century of Water*; *A Century of Water for the West, 1902-2002*; Bureau of Reclamation Mission Statement: "To manage, develop, and protect water and related resources in an environmentally and economically sound manner in the interest of the American public."

improve life in the American West. Bureau of Reclamation officials and engineers took pride in their accomplishments. And well they should when one considers the technical achievements represented in structures of the caliber of Hoover and Grand Coulee dams. In addition, as *A Century of Water for the West, 1902-2002* illustrates, the social and economic impact of the Bureau on development of the American West is undeniable. Even in cases when the representations of Reclamation activities were not sponsored by the Bureau, such as the works commissioned by the New Deal art projects and films produced by the United States Film Service, they promoted the progressive ideology embraced by the Bureau of Reclamation at its inception in 1902.

The visual images produced by the Bureau of Reclamation fall under what film historian Richard Dyer MacCann identifies as "public communication." These images not only informed the American public about Reclamation activities but helped audiences to see benefits, take pride in American ingenuity, and view the marvelous transformation of a desert to agricultural lands. Martha Sandweiss notes that in the 1870s and 1880s, "The seeming emptiness of the arid land ... provided a blank slate upon which Americans could project and inscribe grand narratives of national life, narratives which, in the beginning, inevitably stressed success and growth, never failure and defeat."[575] These Bureau of Reclamation images during the twentieth century testify to the fact that those aspirations remained a powerful force. Reclamation boosters, especially ones as verbose as C. J. Blanchard and William E. Smythe, shared a vision of the West's grand potential that reached beyond the narrow promotion of Bureau of Reclamation activities.

Bureau of Reclamation images always embraced change; they evolved as Reclamation's mission morphed over time. From the 1930s until the mid-1960s, dam building, hydroelectric power, recreation, and flood control appeared to take precedence over irrigation and land settlement. Bureau images highlighted these aspects of Reclamation's national benefits. As American concerns for the environment developed during the latter half of the twentieth century, Bureau photographs, films and even its Art Program reflected those anxieties and emphasized its new role in water management. Yet throughout these transformations, the Bureau of Reclamation-produced images remained consistent in representing American progress and ingenuity: "never failure and defeat."

---

[575] MacCann, *The People's Films*, 4; see also Sandweiss, *Print the Legend*, 6.

APPENDIX A:

BUREAU OF RECLAMATION TIMELINE FOR VOLUME 2

DATE	ITEM
1944	Mexican Water Treaty and Protocol signed. Primary provisions were 1.5 million acre feet a year to be delivered to Mexico on the Colorado River and the name of the International Boundary Commission was changed to the International Boundary and Water Commission
	Arizona ratified the Colorado River Compact and entered into a contract with the secretary of the interior for 2.8 million acre feet of water from the Colorado River.
	Pick-Sloan Missouri Basin Program authorized in the Flood Control Act of 1944.
1948	Fish and Wildlife Coordination Act of August 14, 1946, amended an act of March 10, 1934, providing for wildlife conservation and rehabilitation work.
1949	Upper Colorado River Basin Compact approved by Congress thus paving the way for water developments in the Upper Basin such as the Colorado River Storage Project (1956) and the Colorado River Basin Project Act (1968). Provided: 50,000 acre feet for Arizona; and a percentage of the balance of the Upper Basin's entitlement to Colorado (51.75%), New Mexico (11.25%), Utah (23%), and Wyoming (14%).
1950-1953	The Korean Conflict caused reductions in Reclamation activities as the Congress channeled budget into that war effort.
1953	U.S. Supreme Court granted the State of Arizona leave to file a bill of complaint against the State of California and seven of its public agencies resulting in the famous case *Arizona v. California* which was finally settled in 1963-1964.
	Interior Department Appropriations Act required no monies be spent on a project until land classification and soil survey showed the lands to be served were suitable for agriculture.
1956	Colorado River Storage Project Act (CRSP) passed the Congress and authorized, among others, the Flaming Gorge, Glen Canyon, Navajo, and Wayne Aspinall Units in the Upper Basin.
	Construction began on both the Flaming Gorge and Glen Canyon units of CRSP.

1958	Fish and Wildlife Coordination Act of 1934, as amended 1946, further amended.
1961-1973	American involvement in Vietnam from 1961 to 1973, as well as new social and other programs competed with Reclamation for budget and increased environmental awareness resulted in political and legal action on environmental issues—all of which combined to reduce Reclamation's construction program.
1961	The U.S. Supreme Court received the report of the Special Master in *Arizona v. California* and ordered the report filed.
	Columbia River Treaty signed, but it was implemented in 1964. The treaty cleared the way for development of the Third Powerhouse at Grand Coulee Dam and provided for development of four upstream dams—three in Canada and one in the United States.
1962	Publication of Rachel Carson's *Silent Spring*.
1963	The first Clean Air Act became law.
	U.S. Supreme Court filed its opinion in *Arizona v. California.*
	Reclamation topped out Glen Canyon Dam.
	Eliot Porter published *The Place No One Knew: Glen Canyon on the Colorado* for the Sierra Club.
	Pacific Southwest Water Plan unveiled and eventually the discussion led to passage of the Colorado River Basin Project Act in 1968 (but this was much changed from the plan originally devised.)
1964	Major flood near Great Falls, Montana, including Birch Creek and the Sun River. Flood waters overtopped Reclamation's Gibson Dam, which did not fail.
	Wilderness Act passed.
	The U.S. Supreme Court entered its decree in the case of *Arizona v. California*. Briefly, the decision allowed Arizona 2.8 million acre feet under the Colorado River Compact and ruled that 1 million acre feet in the tributary Gila River was not part of the 2.8 million acre feet and was also an Arizona entitlement. The decision also ruled that 1 million acre feet were reserved for Indian reservations in California, Arizona, and Nevada.
	Public Law 88-278 authorized Federal purchase of lands in the Third Division of the Riverton Project.
1965	In early September Fontanelle Dam on the Green River showed signs of failure and Barney Bellport, Reclamation's chief engineer, ordered it drained until stable.

1966	National Historic Preservation Act passed.
	Glen Canyon Dam dedicated.
	Third Powerplant at Grand Coulee Dam authorized.
1967-1980	Construction of the Third Powerplant and installation, testing, and bringing on-line of generation units.
1967	Clean Air Act revised.
	Secretary of the Interior Stewart Udall, in order to move the Colorado River Basin Project Act forward, proposed removal of Bridge and Marble Canyon Dams from pending legislation, substituting the Navajo Steam Generating Powerplant for the to provide electricity for pumping water on the Central Arizona Project.
1968	Colorado River Basin Project Act passed by the Congress. It authorized Reclamation construction of the Central Arizona Project and the Navajo Steam Generating Powerplant. California supported the bill in return for a guarantee that it would receive its 4.4 million acre feet before Arizona or Nevada received any Colorado River water. California also agreed the Upper Basin on the Colorado River could further develop its water resources so long as certain salinity requirements were met in the Lower Colorado River Basin.
	Congress passed the Clean Water Act.
	According to Theodore Schad's oral history interviews in 1989 with Martin Reuss, the National Water Commission was created to ensure a comprehensive look at national water issues and to head off Reclamation studies of projects to supplement the Colorado River's water supply.
	Congress passed the Wild and Scenic Rivers Act
1970	President Richard M. Nixon signed the National Environmental Policy Act (NEPA) on January 1. The act passed the Congress in 1969.
1970-1976	The Salt River Project Agricultural Improvement and Power District of Arizona built the Navajo Generating Station in Page, Arizona, holding 24.3 percent of the powerplant, 550 megawatts, in trust for the United States to provide the electricity for pumping water on the Central Arizona Project.
1971	The Arizona legislature created the Central Arizona Water Conservation District (CAWCD) in Maricopa, Pinal, and Pima counties to be the local water district for the Central Arizona Project. CAWCD now has changed its name to the Central Arizona Project.

1973	Reclamation began construction of the Central Arizona Project.
	Mexico and the United States reached an agreement on dealing with salinity issues on the lower Colorado River.
	Endangered Species Act became law.
1975	The Oroville Earthquake in California (August 1) sharpened public concern about Reclamation's construction of Auburn Dam above Sacramento on the American River.
1976	On June 5 Reclamation's recently completed Teton Dam on the Snake River failed.
1977	Department of Energy Organization Act created regional power marketing administrations for Federal power. Reclamation's electricity transmission and marketing functions transferred to the Western Area Power Administration except in the Pacific Northwest where the Bonneville Power Administration already had those responsibilities.
	The "Hit List" of President Jimmy Carter was revealed to the public. The Narrows Project on the South Platte River in Colorado was closed down.
	The Federal Power Commission was renamed the Federal Energy Regulatory Commission and placed in the Department of Energy.
	Clean Air Act amended.
1981-1989	The administration of President Ronald Reagan required local cost sharing on Reclamation projects, thus reducing new authorizations.
1982	The Reclamation Reform Act, addressing the realities of modern American agriculture, changed the acreage limitation from 160 acres to 960 acres for subsidized agricultural water deliveries.
1983	Very high volumes of water on the Colorado River resulted in damage to the spillways at Glen Canyon and Hoover dams.
1984	Reclamation completed repairs on the spillways at Glen Canyon and Hoover dams just in time for a second high water year.
1985	The Lower Missouri Region in Denver and the Upper Missouri Region in Billings were combined due to a decline in planning and construction work in the regions. Reclamation located the regional headquarters in Billings and renamed the combined region the Missouri Basin Region.

1986	The first water was delivered along the Granite Reef Aqueduct, now renamed the Hayden-Rhodes Aqueduct, of the Central Arizona Project
1988	Reclamation combined the Southwest Region in Amarillo with the Missouri Basin Region in Billings and put the regional headquarters in Billings. The region was renamed the Great Plains Region.
	Reclamation conducted a study titled *Assessment '88* and undertook a reorganization which tried to consolidate many planning and environmental functions in the Denver office transferring staff in from the regions and Washington, D.C. This was done due to the reduction in planning and construction needs in the regional offices.
1992	The Reclamation Projects Authorization and Adjustment Act of 1992 (Omnibus Water Act) had forty titles, most affecting Reclamation, including Titles II through VI, the Central Utah Project Completion Act; Title XXX, the Western Water Policy Review; and Title XXXIV, the Central Valley Project Improvement Act.
1994	Reorganization of Reclamation during the term of office of Commissioner Daniel P. Beard. The effort was undertaken to flatten Reclamation's organization, broaden constituencies, and direct its efforts more toward water management activities.

APPENDIX B:

COMMISSIONERS OF THE BUREAU OF RECLAMATION, 1945-2012

Harry W. Bashore August 3, 1943-1945	Commissioner
Michael W. Straus 1945-1953	Commissioner
Goodrich W. Lineweaver 1953	Acting Commissioner
Wilbur A. Dexheimer 1953-1959	Commissioner
Floyd E. Dominy 1959-1969	Commissioner
Ellis L. Armstrong 1969-1973	Commissioner
Gilbert G. Stamm 1973-1977	Commissioner
Donald D. Anderson 1977	Acting Commissioner
R. Keith Higginson 1977-1981	Commissioner
Clifford I. Barrett 1981	Acting Commissioner
Robert N. Broadbent 1981-1984	Commissioner
Robert A. Olson December 1984 – August 1985	Acting Commissioner

Clifford I. Barrett August 1985 – December 17, 1985	Acting Commissioner
C. Dale Duvall December 18, 1985-1989	Commissioner
Joe D. Hall 1989	Acting Commissioner
Dennis B. Underwood 1989-1993	Commissioner
Lawrence F. Hancock April 10, 1993 – May 23, 1993	Acting Commissioner
Daniel P. Beard May 24, 1993 – September 9, 1995	Commissioner
Stephen V. Magnussen September 9, 1995 – December 22, 1995	Acting Commissioner
Eluid L. Martinez December 22, 1995 – January 20, 2001	Commissioner
J. William McDonald January 21, 2001 – July 16, 2001	Acting Commissioner
John W. Keys III July16, 2001 – April 15, 2006	Commissioner
William Rinne April 16, 2006 – October 17, 2006	Acting Commissioner
Robert (Bob) W. Johnson September 30, 2006 – January 3, 2009	Commissioner
J. William McDonald January 4, 2009 – May 31, 2009	Acting Commissioner
Michael L. Connor June 1, 2009 –	Commissioner

BIOGRAPHICAL SKETCHES OF THE COMMISSIONERS OF RECLAMATION, 1945–2012

Harry W. Bashore 1943–1945

A Missouri native born in 1880, Bashore graduated from Missouri State University in 1906 with a Bachelor of Science in Civil Engineering.

He joined Reclamation out of college as a "engineering aide" on the North Platte Project. From 1927 to 1933 Bashore served as construction engineer at Lake Minatare, before assuming the superintendent's job on the North Platte Project. Bashore left North Platte to work as a construction and investigative engineer in Oregon, Washington, and California. Bashore returned to Wyoming as an engineer on the Kendrick Project from 1933 to 1939. In 1942, Bashore became assistant commissioner, before his appointment as commissioner on August 3, 1943. During his term, Reclamation's bureaucracy grew with the birth of the regional office system in 1943-1944. In 1944 Congress passed the working agreement between Reclamation and the Corps of Engineers for development of the Missouri River Basin (Pick-Sloan Missouri Basin Program).

Michael W. Straus 1945–1953

Straus was born in Chicago in 1897. He attended the University of Wisconsin in pursuit of a chemical engineering degree. Instead of engineering, Straus began his career as a newspaper reporter, eventually working his way to the managing editor's job of two of the nation's largest newspapers, and Washington bureau chief of the International News Service.

He entered government as an assistant to Secretary of the Interior Harold Ickes. Ickes appointed Straus first assistant secretary of the Interior in 1943. During his term as commissioner, the states signed the Upper Colorado River Basin Compact, and Grand Coulee and Shasta dams as well as Reclamation's Engineering and Research Center at Denver were dedicated. Straus died August 9, 1970.

Wilbur A. Dexheimer 1953–1959

Born in Denver in 1901, Dexheimer graduated from Colorado State University in 1926 with a degree in "civil and irrigation engineering." The Bureau of Reclamation hired Dexheimer as an engineer in 1928. He worked on the Yakima, Shasta, and Hoover projects between 1929 and 1942.

During World War II Dexheimer held the rank of colonel in the U.S. Army Corps of Engineers where he served in India and China as assistant theater engineer on the staff of General Joseph Stilwell. He was with Jack Savage in China during the time Savage was looking at the Three Gorges project at the invitation of Chairman Chiang Kai-shek of the Nationalist Government of China.

Dexheimer returned to Reclamation in 1947 to serve six years as assistant chief construction engineer in the Denver office. President Dwight D. Eisenhower then appointed him commissioner in which post he served from 1953 until 1959. During his term as commissioner, Reclamation obtained congressional approval for the Trinity River Division of the Central Valley Project. Congress also passed the Colorado River Storage Project and Small Reclamation Projects Act, both in 1956.

Floyd E. Dominy 1959–1969

Dominy is easily the most colorful Commissioner in Reclamation's history. Born in 1909 and raised on a Nebraska farm, Dominy grew up realizing the importance of irrigation in the arid West beyond the hundredth meridian. He studied civil engineering at Georgia Tech but eventually dropped out for financial reasons. Returning to the West, he worked as an agricultural laborer and received his Bachelor of Arts degree in Agricultural Economics from the University of Wyoming in 1932.

After college, in the depths of the Great Depression, Dominy worked as a teacher, agricultural agent and, beginning in 1938, in Washington, D.C., as a field agent for the Agricultural Adjustment Administration. He worked two years for Nelson Rockefeller, the coordinator of inter-American affairs implementing programs in Paraguay, Peru, Brazil, Columbia, Venezuela, and Central American countries to obtain raw materials necessary to the war effort and to counter Nazi influence. He joined the Seabees and served in the Pacific, most notably working to establish agriculture on Guam to provide food supplies to American troops.

Probably influenced by a 1937 visit to recently completed Hoover Dam, after the war, he joined Reclamation in 1946 as a land settlement specialist. He supervised the Allocations and Repayment Branch, Division of Irrigation between 1950 and 1957. Dominy rose to assistant commissioner from 1957 to 1958, and was named the associate commissioner from 1958 to 1959. Dominy became commissioner on May 1, 1959.

Notable events during his term as commissioner included completion of Glen Canyon, Flaming Gorge, and Navajo dams of the Colorado River Storage Project. Dominy also played a role in authorization and initiation of construction on the San Luis Unit and completion of the Trinity River Division, both on the Central Valley Project. Congress authorized the massive Third Powerplant at Grand Coulee and Reclamation's last very large authorization, the Colorado River Basin Project Act which included the Central Arizona Project and expanded the Central Utah Project, during his term in office. During his term as commissioner, Reclamation kept tabs on widespread, visionary, public and private planning efforts aimed at supplementing water supplies of the arid West and actually developed the *Pacific Southwest Water Plan* of January 1964.

Commissioner Dominy served under Presidents Eisenhower, Kennedy, Johnson, and Nixon, and some contemporaries said he wielded more influence on Capitol Hill than any Secretary of the Interior. He was a key subject in two influential books focusing on water in the West, Marc Reisner's *Cadillac Desert* and John McPhee's *Encounters with the Archdruid.*

Ellis L. Armstrong 1969–1973

Born in 1914 in Cedar City, Utah, he studied civil engineering at Utah State University. He supervised Civilian Conservation Corps (CCC) enrollees on the Pine River Project in 1934. Armstrong received his degree in "civil and irrigation engineering" from Utah State University in 1936. He went to work in the Salt Lake City office the day after he received his degree. Reclamation put Armstrong to work doing Colorado River hydrology studies and then he went to Deer Creek Dam where he was a materials engineer. He worked for Reclamation on the Moon Lake project, Deer Creek Dam, Anderson Ranch Dam between 1936 and 1954. Armstrong left Reclamation to become project engineer and deputy project manager of the St. Lawrence Power and Seaway Project from 1954 to 1957. The next stop in his career was as commissioner of the U.S. Bureau of Public Roads from 1958 to 1961.

Armstrong left public life to work as a consultant between 1961 and 1968.

He returned to Reclamation as assistant director of Region 4 in Salt Lake City in 1968 when Floyd Dominy decided to groom him as his successor.

He was appointed to the post of commissioner in 1969. During Commissioner Armstrong's tenure he reorganized the Bureau creating four assistant commissioner positions and changed the title of the chief engineer to Director, Office of Design and Construction. Though the position title changed over the years the occupant of the position was unofficially known within Reclamation as the "chief engineer" until the reorganization of Reclamation in 1993 during the term of Commissioner Daniel P. Beard. It was during Armstrong's term of office that the National Environmental Policy Act became law, and Armstrong began the process of integration of environmental issues into Reclamation planning efforts. Most famously, Armstrong hired Warren Fairchild from the state of Nebraska to begin to try to change Reclamation's approach to project planning and Reclamation held a planning conference in Tucson called "Talking Turkey in Tucson" where the organization attempted to figure out how to efficiently implement NEPA responsibilities within Reclamation.

Gilbert G. Stamm 1973–1977

Gilbert Stamm was born in Denver in 1911. He studied engineering and economics at Colorado State University.

Stamm's first Federal job was with the Department of Agriculture's Land and Water Resources Division in 1936.

In 1946 he left Agriculture to join Reclamation's regional office in Boise. In 1959 Stamm advanced to chief of the Division of Irrigation and Land Use in Washington, D.C. His next appointment was as assistant commissioner of legislation and coordination in 1964. As commissioner, Stamm encouraged long-range water resource planning in the West and warned that a lack of planning could push the West, and the Nation, to the brink of a water crisis. Stamm led Reclamation during the failure of the Teton Dam in June 1976.

R. Keith Higginson 1977–1981

Higginson was born in 1930 in Boise Idaho. In 1948 he started work with the Soil Conservation Service [SCS] as a research assistant and snow surveyor. In 1957 Higginson earned his Bachelor of Science degree in civil engineering from Utah State University.

His career in water management began as a water rights engineer in the Utah State Engineer's Office from 1958 to 1965. He left Utah to direct Idaho's Department of Water Resources from 1965 to 1977.

After Higginson's nomination by Secretary of the Interior Cecil Andrus and confirmation as Commissioner in 1977, the Bureau of Reclamation changed its name to the Water and Power Resources Service (WPRS). Higginson found himself in a difficult position since he was a career water manager in the West and had to terminate Assistant Commissioner Harold Arthur's tenure at Reclamation as a token dismissal due to the failure of Reclamation's Teton Dam in 1975, and he had to defend President Jimmy Carter's "hit list" of Western water projects and initiation of policy changes that eventually resulted in the Reclamation Reform Act. After leaving Reclamation at the end of the Jimmy Carter Administration, Higginson went into consulting and eventually returned to direct Idaho's Department of Water Resources until he retired in 1995.

Robert N. Broadbent 1981–1984

Broadbent held a Bachelor of Science in Pharmacology from Idaho State College. He worked as a pharmacist and politician in Clark County, Nevada, from 1950 to 1975. Broadbent was elected the first Mayor of Boulder City, Nevada, and he also directed the Las Vegas Valley Water District. As commissioner he oversaw changing the name of the Water and Power Resources Service back to the Bureau of Reclamation in 1981, and Congress enacted the Reclamation Reform Act in 1982. He subsequently served as assistant secretary of the interior for water and science from 1984 until 1986. After leaving the Department of the Interior he returned to Clark County, Nevada, where he headed the airport commission in Las Vegas.

C. Dale Duvall 1985–1989

Duvall was raised near Grand Coulee Dam. Duvall attended Eastern Washington College of Education, Gonzaga University, and Kinman Business University. He entered the business world as a C.P.A. and partner in a Spokane accounting firm, 1965-1980. Duvall acted as vice president and treasurer of Overseas Private Investment Corp. (OPIC) from 1981 to 1985. Highlights of his term as commissioner at Reclamation include the rehabilitation of the Salt River Project's Theodore Roosevelt Dam, an increased emphasis on dam operation and maintenance, and the 1987/1988 reorganization of Reclamation associated with the "Assessment '87" studies. On July 6, 1989, Duvall left Reclamation to serve in the Department of Veterans Affairs as the chief financial officer in Acquisition & Materiel Management.

Dennis B. Underwood 1989–1993

Born in rural Vermont in 1944, Underwood graduated from Norwich University in 1966 with a degree in civil engineering.

He joined the California Department of Water Resources where he helped to update the State Water Plan.

Then from 1966 to 1969, he served as a commissioned officer with the U.S. Army Corps of Engineers in Thailand, achieving the rank of Captain.

After serving in the Corps of Engineers he returned to the Department of Water Resources. From 1978 to 1989, Underwood served as the executive director of the Colorado River Board of California. He worked extensively with the seven Colorado River Basin States, the International Boundary and Water Commission, and various Federal agencies on developing and managing Colorado River water resources.

After his swearing-in as commissioner on November 14, 1989, Underwood pursued Reclamation's shift from water project builder to water resources manager. During his term at Reclamation he was detail-oriented, and Reclamation studied the potential of groundwater recharge in the 17 western states, produced a comprehensive water reuse initiative for Southern California, and published in June 1992 *Reclamation's Strategic Plan: A long-term framework for Water Resources Management, Development, and Protection.*

After leaving Reclamation in 1993 at the end of the term of President George H. W. Bush, he did consulting work in California and Washington, D.C., and eventually became the head of the Metropolitan Water District of Southern California.

Daniel P. Beard, 1993–1995

Beard was born in 1943 in Bellingham, Washington, and he earned his B.A. from Western Washington University and then attended the University of Washington where he received his M.A., and, in 1973, Ph.D. in geography.

His early career also included environmental policy study, teaching at Dartmouth College, and work at the Congressional Research Service. Beard worked for Congressman Sidney Yates (D-Illinois) on the House Interior and Related Agencies Appropriations Subcommittee from 1975 to 1976.

He served as assistant director of the Domestic Policy Council in the Executive Office of the President, and deputy assistant secretary of the interior for land and water resources from 1977 to 1980 during the administration of President Jimmy Carter.

Beard was staff director of the House Interior Subcommittee on Water and Power from 1985 to 1992, and staff director for the Committee on Natural Resources in 1993.

During his tenure as commissioner, Beard directed reorganization of the Bureau of Reclamation. His reorganization reduced Reclamation's budget by more than $100 million and staff by over 10 percent. Beard pushed for Reclamation's transition from a construction agency to water resources management leader. He also insisted Reclamation take new approaches toward water conservation, environmental issues, and recognizing a broader constituency than Reclamation's traditional water and power users, including the increasing demands of the environment, recreation, and cities on western water.

After leaving Reclamation, he served as chief operating officer and senior vice president for public policy at the National Audubon Society, and he then worked as a senior advisor for the consulting firm Booz Allen Hamilton, Inc. In early 2007 he returned to the House of Representatives to work for

House Speaker Nancy Pelosi as the third Chief Administrative Officer (CAO) of the House of Representatives, a position he held until 2010.

Eluid L. Martinez 1995–2001

Commissioner Eluid L. Martinez was born in 1944 in Cordova, New Mexico, a village long famous for its wood carving traditions. He was raised mostly in Santa Fe and received his degree in civil engineering from New Mexico State University in Las Cruces in 1968.

He first worked for the Bureau of Public Roads in California and then the New Mexico State Highway Department. In 1971 he joined the New Mexico State Engineer's Office then headed by long-time State Engineer Steve Reynolds. Martinez served in the New Mexico Engineer's Office for 23 years, working in several different positions, and he eventually served as the state engineer from 1990 to 1994 and as the secretary of the New Mexico Interstate Council on Water Policy. He retired from the state of New Mexico in 1994 with 31 years of service.

Secretary of the Interior Bruce Babbitt nominated Martinez as commissioner of the Bureau of Reclamation, and he served from 1995 until the end of the Bill Clinton administration in 2001—the longest serving commissioner since Floyd Dominy left Reclamation in 1969. He spent a good deal of time establishing good relations with water and power users.

John W. Keys III 2001–2006

Commissioner John Keys was born in Sheffield, Alabama, in 1942. He received his degree in civil engineering in 1964 from the Georgia Institute of Technology and a master's degree in civil engineering from Brigham Young University in 1971. He spent his entire professional career working with Reclamation throughout the western United States and in Washington, D.C. From 1964 to 1998 he worked as a civil and hydraulic engineer, manager, and executive in the Great Basin, the Missouri River Basin, the Colorado River Basin, and the Columbia River Basin. In 1995 Keys received Interior's highest honor—the Distinguished Service Award. In 1998 he retired from federal service, then having served as regional director in the Pacific Northwest Region for 12 years.

In 2001 Keys returned to Reclamation as its commissioner. While commissioner he devoted a great deal of energy to opening communications with water users and was very proud to be commissioner during Reclamation's centennial year in 2002. During his tenure Reclamation developed its program Water 2025 to assist the West in anticipating and planning for water needs. He again retired from Reclamation in 2006.

A commercial airplane pilot, Keys flew many hours a year for Angel Flight, Air LifeLine, and County Search and Rescue, based out of Moab, Utah. In addition, he was a college football referee between 1970 and 2001 as well as a high school referee beginning in 1962.

Keys was a registered professional engineer in the states of Colorado, Wyoming, Montana, and North Dakota.

He died when the plane he was piloting crashed in Canyonlands National Park in 2008.

Robert W. Johnson 2006–2009

Born in Lovelock, Nevada, in 1951, Bob Johnson grew up on a farm that received its irrigation water from a Reclamation project. A graduate of the University of Nevada, Reno, Johnson received his bachelor's in 1973 and his master's degree in 1977—both in agriculture and resource economics.

Johnson joined the Bureau of Reclamation in 1975 at Reclamation's Mid-Pacific Region, in Sacramento, California, and completed his master's degree while working there. From Sacramento, he moved to Reclamation's Boulder City, Nevada, office, in 1979, and he then moved on to Washington, D.C., in 1987, where he served in the Commissioner's Office as the chief of the Contracts and Repayment Branch. He returned to Boulder City from Washington, first as the chief of the region's Water, Land, and Power Operations Division and then as the deputy regional director before serving as regional director from 1995 to September 2006. As regional director of Reclamation's Lower Colorado Region, he served as "watermaster" of the lower Colorado River on behalf of the Secretary of the Interior until he was appointed Commissioner of the Bureau of Reclamation. Johnson's tenure as regional director and commissioner was guided by an overarching vision of providing flexibility and creativity in solving water problems through planning, negotiation, and consensus building.

Michael L. Connor 2009–

Michael Connor was born in Granger, Utah, in 1963, and within a few years his family moved to New Mexico. He received his degree in chemical engineering at New Mexico State University in Las Cruces in 1986. He first worked for General Electric in Louisville, Kentucky, and then Denver. In 1990 he began work at the University of Colorado's Law School where he was particularly interested in environmental, natural resources, and water law. While in the law school he was accepted into the Solicitor's Honors Program at the Department of the Interior, and he received his law degree in 1993.

In the fall of 1993 Connor joined the Interior solicitor's staff in Washington, D.C., where he worked on a number of issues, but particularly dealt with Indian and water rights issues. In May of 1997 he transferred to the Southwest Regional Solicitor's office in Albuquerque and continued to work on a variety of topics including Indian and Fish and Wildlife Service issues. In 1998 he returned to Washington, D.C., as deputy director of the Secretary of the Interior's Indian Water Rights Office. In 1999 Connor became director of the Indian Water Rights Office and was involved in two major settlements, the Colorado Ute and the interaction of Central Arizona Project repayment issues with Indian water rights settlements.

Having previously worked closely with various staff members and senators, in June of 2001 Connor transferred to be counsel for the Water and Power Subcommittee in the U.S. Senate and then became Counsel for the Energy and Natural Resources Committee. Senator Ken Salazar, after becoming the secretary of the interior, nominated Connor to be commissioner of the Bureau of Reclamation.

APPENDIX C:

CHIEF ENGINEERS AND SUCCESSOR POSITIONS IN RECLAMATION

Appendix C: Chief Engineers and Successor Positions in the Bureau of Reclamation 1945-1993

Walker (Brig) R. Young January 1945 – June 1948	Chief Engineer
Leslie N. McClellan July 1948 – May 1958	Assistant Commissioner and Chief Engineer
Grant Bloodgood May 1958 – February 1963	Assistant Commissioner and Chief Engineer
Bernard (Barney) P. Bellport February 1963 – September 1970	Director, Office of Design and Construction
Bernard (Barney) P. Bellport September 1970 – April 1972	Director, Office of Design and Construction Chief Engineer
Harold G. Arthur April 1972 – July 1977	Director, Office of Design and Construction[1]
Robert B. Jansen September 1977 – February 1978	Director, Office of Design and Construction[2]
Robert B. Jansen February 1978 – December 1979	Assistant Commissioner for Engineering and Research

1 In spite of the official title change in April 1972, until 1994 this position's incumbent was unofficially widely known in Reclamation as the "chief engineer."

2 After failure of Teton Dam in early June of 1975, construction contracting and construction oversight responsibilities in Reclamation were shifted away from the Director of the Office of Design and Construction and distributed among the regional directors. In addition, Commissioner Higginson chose Robert B. Jansen, the first non-Reclamation employee to ever fill the post, to supervise Denver office technical functions.

Rodney J. Vissia January 1980 – March 1982	Assistant Commissioner for Engineering and Research
Darrell W. Webber	Assistant Commissioner for Engineering and Research
Felix W. Cook Sr. September 1993 – October 1994	Assistant Commissioner for Engineering and Research[1]

1 As part of the 1994 realignment of Reclamation under Commissioner Daniel P. Beard, Reclamation abolished this position.

APPENDIX D:

SECRETARIES OF THE INTERIOR, DEPARTMENT OF THE INTERIOR, 1945-2012

Secretary	President
Harold L. Ickes March 4, 1933 – February 15, 1946[1]	Franklin D. Roosevelt Harry S. Truman
Julius A. Krug March 18, 1946 – December 1, 1949	Harry S. Truman
Oscar L. Chapman December 1, 1949 – January 20, 1953	Harry S. Truman
Douglas McKay January 21, 1953 – April 15, 1956	Dwight D. Eisenhower
Fred A. Seaton June 8, 1956 – January 20, 1961	Dwight D. Eisenhower
Stewart L. Udall January 21, 1961 – January 20, 1969	John F. Kennedy Lyndon B. Johnson
Walter J. Hickel January 24, 1969 – November 25, 1970	Richard M. Nixon
Rogers C. B. Morton January 29, 1971 – April 30. 1975	Richard M. Nixon
Stanley K. Hathaway June 12, 1975 – October 9, 1975	Gerald R. Ford
Thomas S. Kleppe October 17, 1975 – January 20, 1977	Gerald R. Ford
Cecil B. Andrus January 23, 1977 – January 20, 1981	Jimmy Carter

[1] Unless otherwise noted, dates are time in office—not confirmation dates.

James Watt January 23, 1981 – November 8, 1983	Ronald Reagan
William P. Clark November 18, 1983 – February 7, 1985	Ronald Reagan
Donald P. Hodel February 8, 1985 – January 20, 1989	Ronald Reagan
Manuel Lujan Jr. February 3, 1989 – January 20, 1993	George H. W. Bush
Bruce Babbitt January 22, 1993 – January 2, 2001	Bill Clinton
Gale A. Norton January 31, 2001 – March 31, 2006	George H. Bush
Dirk Kempthorne May 29, 2006 – January 19, 2009	George H. Bush
Ken Salazar Confirmed by the U.S. Senate January 20, 2009	Barack Obama

APPENDIX E:

RECLAMATION POWERPLANTS

E.1. Reclamation Owned and Operated Powerplants

Powerplants Owned/ Operated by Reclamation in 2012	Reclamation Project/Nearby Town	Initial Date of Service	Current Generating Capacity kW	Original Generating Capacity kW
Alcova Powerplant	Kendrick Project, Alcova, Wyoming	1955	41,400	36,000
Anderson Ranch Powerplant	Boise Project, Mountain Home, Idaho	1950	40,000	27,000
Big Thompson Powerplant	Colorado-Big Thompson Project, Loveland, Colorado	1959	4,500	4,500
Black Canyon Powerplant	Boise Project, Emmet, Idaho	1925	10,200	8,000
Blue Mesa Powerplant	Colorado River Storage Project, Gunnison, Colorado	1967	86,400	60,000
Boise River Diversion Powerplant	Boise Project, Boise, Idaho	1912	3,450	1,500
Boysen Powerplant	Boysen Unit: P-SMBP, Thermopolis, Wyoming	1952	15,000	15,000
Buffalo Bill Powerplant	Shoshone Project, Cody, Wyoming	1992	18,000	18,000
Canyon Ferry Powerplant	Canyon Ferry Unit: P-SMBP, Helena, Montana	1953	50,000	50,000
Chandler Powerplant	Yakima Project, Benton City, Washington	1956	12,000	12,000
Crystal Powerplant	Colorado River Storage Project, Montrose, Colorado	1978	31,500	28,000
Davis Powerplant	Parker-Davis Project, Bullhead City, Arizona	1951	255,000	225,000
Elephant Butte Powerplant	Rio Grande Project, Truth or Consequences, New Mexico	1940	27,945	24,300
Estes Powerplant	Colorado-Big Thompson Project, Estes Park, Colorado	1950	45,000	45,000
Flaming Gorge Powerplant	Colorado River Storage Project, Dutch John, Utah	1963	151,950	108,000
Flatiron Powerplant	Colorado-Big Thompson Project, Loveland, Colorado	1954	94,500	73,158
Folsom Powerplant	Folsom and Sly Park Units: Central Valley Project, Folsom, California	1955	198,720	162,000

Powerplants Owned/ Operated by Reclamation in 2012	Reclamation Project/Nearby Town	Initial Date of Service	Current Generating Capacity kW	Original Generating Capacity kW
Fontenelle Powerplant	Seedskadee Project, La Barge, Wyoming	1968	10,000	10,000
Fremont Canyon Powerplant	Glendo Unit: P-SMBP, Alcova, Wyoming	1960	66,800	48,000
Glen Canyon Powerplant	Colorado River Storage Project, Page, Arizona	1964	1,320,000	950,000
Glendo Powerplant	Glendo Unit: P-SMBP, Glendo, Wyoming	1958	38,000	24,000
Grand Coulee Powerplant	Columbia Basin Project, Grand Coulee, Washington	1941	6,809,000	2,288,000
Green Mountain Powerplant	Colorado-Big Thompson Project, Kremmling, Colorado	1943	26,000	21,600
Green Springs Powerplant	Rogue River Basin Project, Ashland, Oregon	1960	17,290	16,000
Guernsey Powerplant	North Platte Project, Guernsey, Wyoming	1927	6,400	4,800
Heart Mountain Powerplant	Shoshone Project, Cody, Wyoming	1948	5,000	5,000
Hoover Powerplant	Boulder Canyon Project, Boulder City, Nevada	1936	2,078,800	1,344,800
Hungry Horse Powerplant	Hungry Horse Project, Columbia Falls, Montana	1952	428,000	285,000
Judge Francis Carr Powerplant	Shasta/Trinity River Division: Central Valley Project, French Gulch, California	1963	154,400	141,444
Keswick Powerplant	Shasta/Trinity River Division: Central Valley Project, Redding, California	1949	117,000	75,000
Kortes Powerplant	Kortes Unit: P-SMBP, Sinclair, Wyoming	1950	36,000	36,000
Lewiston Powerplant	Shasta/Trinity River Division: Central Valley Project, Lewiston, California	1964	350	350
Lower Molina Powerplant	Collbran Project, Molina, Colorado	1962	4,860	4,860
Marys Lake Powerplant	Colorado-Big Thompson Project, Estes Park, Colorado	1951	8,100	8,100
Minidoka Powerplant	Minidoka Project, Rupert, Idaho	1909	27,700	13,400
Morrow Point Powerplant	Colorado River Storage Project, Montrose, Colorado	1970	173,334	120,000
Mt. Elbert Pumped-Storage Powerplant	Fryingpan-Arkansas Project, Twin Lakes, Colorado	1981	200,000	200,000

Powerplants Owned/ Operated by Reclamation in 2012	Reclamation Project/Nearby Town	Initial Date of Service	Current Generating Capacity kW	Original Generating Capacity kW
New Melones Powerplant	New Melones Unit Project, Jamestown, California	1979	300,000	300,000
Nimbus Powerplant	Folsom and Sly Park Units: Central Valley Project, Folsom, California	1955	13,500	13,500
Palisades Powerplant	Palisades Project, Palisades, Idaho	1957	176,564	114,000
Parker Powerplant	Parker-Davis Project, Parker Dam, Arizona	1942	120,000	120,000
Pilot Butte Powerplant	Riverton Unit, P-SMBP, Morton, Wyoming	1925	1,600	1,600
Pole Hill Powerplant	Colorado-Big Thompson Project, Loveland, Colorado	1954	38,238	33,250
Roza Powerplant	Yakima Project, Yakima, Washington	1958	12,937	11,250
Seminoe Powerplant	Kendrick Project, Sinclair, Wyoming	1939	51,750	32,400
Shasta Powerplant	Shasta/Trinity River Division: Central Valley Project, Redding, California	1944	714,000	379,000
Shoshone Powerplant	Shoshone Project, Cody, Wyoming	1922	3,000	8,600
Spirit Mountain Powerplant	Shoshone Project, Cody, Wyoming	1994	4,500	4,500
Spring Creek Powerplant	Shasta/Trinity River Division: Central Valley Project, Redding, California	1964	180,000	150,000
Stampede Powerplant	Washoe Project, Truckee, California	1988	3,650	3,650
Trinity Powerplant	Shasta/Trinity River Division: Central Valley Project, Redding, California	1964	140,000	100,000
Upper Molina Powerplant	Collbran Project, Molina, Colorado	1962	8,640	8,640
Yellowtail Powerplant	Yellowtail Unit: P-SMBP, Hardin, Montana	1966	250,000	250,000

E.2. Reclamation Owned Powerplants That Have Been Retired or Transferred

Retired Reclamation Powerplants	Project/Nearby Town	Retirement/ Transfer Date
Gateway (Transferred O&M)	Weber Basin Project, Ogden, Utah	October 1, 1968
Siphon Drop (Transferred O&M)	Yuma Project, Yuma, Arizona	December 31, 1962
Prosser (Retired)	Yakima Project, Washington	May 17,1955
Angostura (Retired)	Rapid City, South Dakota	December 1, 1966
Eklutna (Transferred Title)	Eklutna Project, Anchorage, Alaska	July 1, 1967
Wanship (Transferred O&M)	Weber Basin Project, Wanship, Utah	October 1, 1968
Lingle (Retired)	North Platte Project, Lingle, Wyoming	April 1, 1956
Senator Wash (Transferred O&M)	Colorado River Front Work, Yuma, Arizona	May 1, 1977
Medicine Bow (Wind, Retired)	Wyoming	January 1, 1987

E.3. Reclamation Owned Powerplants Operated by other Entities

Powerplant Name	Project	State	Capacity (kW)	Date	Operating Entity
Arizona Falls Powerplant	Salt River Project	Phoenix, Arizona	750	2003	Salt River Valley Water User's Association
Causey Powerplant	Weber Basin Project	Ogden, Utah	1900	1999	Weber Basin Water Conservancy District
Cross Cut Powerplant	Salt River Project	Tempe, Arizona	3000	1914	Salt River Valley Water User's Association
Deer Creek Powerplant	Provo River Project	Heber, Utah	4,950	1958	Provo River Water Users Association
Gateway Powerplant	Weber Basin Project	Ogden, Utah	4000	1958	Weber Basin Water Conservancy District
Grand Valley Powerplant	Grand Valley Project	Grand Junction, Colorado	3000	1933	Grand Valley Water Users Association
Horse Mesa Powerplant	Salt River Project	Phoenix, Arizona	129580	1927	Salt River Valley Water User's Association
Lahontan Powerplant	Newlands Project	Hazen, Nevada	2400	1911	Truckee-Carson Irrigation District
Lower Spanish Fork Powerplant	Strawberry Valley Project	Spanish Fork, Utah	250	1937	Strawberry Water User's Association
McPhee Powerplant	Dolores Project	Cortez, Colorado	1,283	1992	Dolores Water Conservancy District
Mormon Flat Powerplant	Salt River Project	Phoenix, Arizona	57850	1926	Salt River Valley Water User's Association
New Waddell Pump/Generating Plant	Central Arizona Project	Phoenix, Arizona	36000	1993	Central Arizona Water Conservation District
Navajo Generating Station (coal fired) (partial Reclamation ownership)	Central Arizona Project	Page, Arizona	24.3% of 2,250,000 kW or 546.75 kW	1974	Salt River Project Agricultural Improvement and Power District (SRP)
Olmsted Powerplant		Utah	10300	1904	Purchased from PacifiCorp in 1990
O'Neill Pumping-Generating Plant	San Luis Unit, Central Valley Project	Los Banos, California	25,200	1967	San Luis Delta-Mendota Water Authority

Powerplant Name	Project	State	Capacity (kW)	Date	Operating Entity
Payson Powerplant	Strawberry Valley Project	Spanish Fork, Utah	400	1941	Strawberry Water User's Association
San Luis Pumping-Generating Plant	San Luis Unit, Central Valley Project	Los Banos, California	202,000	1939	California Department of Water Resources
Senator Wash Pump/Generating Plant	Colorado River Front Work and Levee System	California	7200	1966	Imperial Irrigation District
Siphon Drop Powerplant	Boulder Canyon Project	All American Canal System, Arizona	4600	1987	Yuma County Water User's Association
South Consolidated Powerplant	Salt River Project	Mesa, Arizona	1400	1912	Salt River Valley Water User's Association
Stewart Mountain Powerplant	Salt River Project	Phoenix, Arizona	10400	1930	Salt River Valley Water User's Association
Theodore Roosevelt Powerplant	Salt River Project	Phoenix, Arizona	36020	1973	Salt River Valley Water User's Association
Towaco Powerplant	Dolores Project	Cortez, Colorado	11,495	1993	Dolores Water Conservancy District
Upper Spanish Fork	Strawberry Valley Project	Spanish Fork, Utah	3900	1983	Strawberry Water User's Association
Wanship Powerplant	Weber Basin Project	Wanship, Utah	1900	1958	Weber Basin Water Conservancy District

E.4. Powerplants on Reclamation Projects Owned and Operated by other Entities

Operated on Reclamation Facilities and Built by Other Entities	Reclamation Facility	State	Capacity (kW)	FERC License Number	Operating Entity and Comments
East Side	Link River Diversion Dam	Oregon	3200	2082	Scottish Power (Pacificorp)
West Side	"	"	600	"	"
American Falls	American Falls Dam	Idaho	92,400	2736	Idaho Power Co
Monticello	Monticello Dam	California	11,500	2780	Solano I.D.
Potholes East Canal Headworks	O'Sullivan Dam	Washington	6,650	2840	Grand Coulee Project Hydroelectric Authority
Cascade Hydroelectric	Cascade Dam	Idaho	12,420	2848	Idaho Power Co.
Main Canal Headworks	Dry Falls Dam	Washington	26,000	2849	East, Quincy, & South, Columbia Basin I.D.'s
Whiskey Dam Power Project	Whiskeytown Dam	California	3,530	2888	City of Redding
Friant Power	Friant Dam	California	27,360	2892	Friant Power Authority
Russel D Smith	Potholes East Canal	Washington	6,100	2926	East, Quincy, & South, Columbia Basin I.D.'s
Quincy Chute Hydroelectric	Quincy Chute	Washington	9,367	2937	East, Quincy, & South, Columbia Basin I.D.'s
Madera Canal Water Power	Madera Canal	California	3,645	2958	Madera-Chowchilla Water & Power Authority
Island Park Hydroelectric	Island Park Dam	Idaho	4,800	2973	Fall River Rural Electric
Garland Canal	Garland Canal	Wyoming	2,610	3031	Shoshone I.D.
Vallecito	Vallecito Dam	Colorado	5,844	3174	Ptarmigan Resources & Energy, Inc.
Stony Gorge Hydroelectric	Stony Gorge Dam	California	4,900	3,193	Santa Clara, City of
Summer Falls	Summer Falls on Main Canal	Idaho	92,000	3295	East, Quincy, & South, Columbia Basin I.D.'s
Ruedi	Ruedi Dam	Colorado	5,052	3603	City of Aspen

Operated on Reclamation Facilities and Built by Other Entities	Reclamation Facility	State	Capacity (kW)	FERC License Number	Operating Entity and Comments
Tieton	Tieton Dam	Washington	13,600	3701	Yakima-Tieton Irrigation District
Echo	Echo Dam	Utah	4,500	3755	City of Bountiful
Sugarloaf	Sugarloaf Dam	Colorado	3,830	3819	STS Hydropower Ltd.
Eltopia Branch Canal 4.6	Eltopia Branch Canal	Washington	2,200	3842	East, Quincy, & South, Columbia Basin ID's
Potholes East Canal 66.0	Potholes East Canal	Washington	2,400	3843	East, Quincy, & South, Columbia Basin ID's
Owyhee Tunnel No. 1	Owyhee Tunnel No. 1	Oregon	8,120	4359	Gem ID et. al.
Pine View	Pineview Dam	Utah	1,800	4597	Weber-Box Elder Conservancy District
Arrowrock Dam	Arrowrock Dam	Idaho	15,000	4656	Big Bend Irrigation District, et. al.
Navajo	Navajo Dam	New Mexico	30,000	4720	City of Farmington
El Vado	El Vado Dam	New Mexico	8,000	5226	County of Los Alamos
Mitchell Butte Power	Mitchell Butte Canal Drop	Oregon	1,880	5357	Owyhee ID et. al.
Madera Canal	Madera Canal	California	440	5765	Madera-Chowchilla Water & Power Authority
High Line Canal	High Line Canal	California	530	7252	Santa Clara
Cowiche	Yakima-Tieton Pipeline	Washington	1,470	7337	Yakima-Tieton ID
Orchard Avenue	Yakima-Tieton Pipeline, Orchard	Washington	1,441	7338	Yakima-Tieton ID
New Lahontan	Lahontan Dam	Nevada	4,000	7828	Truckee-Carson I.D.
McGee Creek Dam	McGee Creek Dam	Oklahoma	85	8492	McGee Creek Authority
Dietrich Drop	Dietrich Drop	Idaho	4,770	8909	Big Wood Canal Company
C. C. Craigin Dam and Powerplant	C.C. Craigin Dam	Arizona	3,000	2304	Salt River Project
Felt Hydro	Teton River	Idaho	7,450	5089	Fall River Rural Cooperative
Little Wood Hydro	Little Wood Dam	Idaho	1,925	7427	Little Wood River Irrigation District
Mile 28	Milner Gooding Canal	Idaho	1,500	10522	Contractor's Power Group

Operated on Reclamation Facilities and Built by Other Entities	Reclamation Facility	State	Capacity (kW)	FERC License Number	Operating Entity and Comments
Mora Drop Hydro	Mora Canal	Idaho	1,900	3403	Boise Kuna Irrigation District et. al
Owyhee Dam	Owyhee Dam	Oregon	4,340	4354	Gem I.D., Owyhee I.D., & Ridgeview I.D
Tiber Dam and Powerplant	Tiber Dam	Idaho	7,500	3574	Tiber Montana, LLC

APPENDIX F:

BRIEFING PAPER FROM PRESIDENT JIMMY CARTER'S "STATEMENT ON WATER PROJECTS" OF APRIL 18, 1977

EMBARGOED FOR USE AFTER BRIEFING APRIL 18, 1977

Office of the White House Press Secretary

THE WHITE HOUSE

STATEMENT ON WATER PROJECTS

Today I am announcing my decisions on federal water resource programs:

--I am recommending the deletion of funds for 18 projects, at a total savings of over $2.5 billion.

--I am recommending modifications of 5 projects, at a total savings of almost $1.5 billion.

--I am recommending the continuation of 9 projects without modification.

--I am recommending the development of major policy reforms in the following areas:

1. more realistic project evaluation criteria;
2. dam safety;
3. cost sharing for federal projects;
4. water conservation; and
5. redirected public works programs.

In balancing the budget, cutting back on inflation and making the federal government more responsive to the needs of the people, difficult choices have to be made. Activities which are wasteful, unsafe or economically or environmentally unsound simply cannot be pursued. Water resource development programs of the Corps of Engineers, the Bureau of Reclamation and the Tennessee Valley Authority are a case in point.

In my budget recommendations to the Congress last February I initiated a major review of ongoing water resource projects. The review is now complete and I have specific recommendations for the Congress on the 32 projects which were subject to public hearings. They are based on reviews by the Interior Department, the Corps of Engineers and the Tennessee Valley Authority, with assistance from the Office of Management and Budget and the Council on Environmental Quality.

My decision on individual projects was a difficult one. I have tried to be fair and to give the benefit of the doubt on some projects which would certainly not be justified if they were proposed today. However, I have not hesitated to recommend termination or modification of projects which appeared justified when they were originally authorized.

In consultation with the Congress, state and local governments and the public, I intend to develop detailed policy recommendations to insure that our water-related needs are met in the best manner, and to use realistic criteria for water project evaluation. The review process I started during the first days of my Administration is not going to stop here; further work needs to be done and fundamental improvements needs to be made in our water policies and programs.

The drought in the West and recent severe flooding in the East have shown us that despite the massive numbers of federally-funded water projects in existence, we are still as susceptible as ever to the ravages of the weather. Instead of proceeding down the same road of more and bigger structural projects, we need to rethink our policies.

(more)

In particular, I will work with Congress to develop policy reforms in the following areas:

1. Realistic assessment of both economic and environmental costs and benefits

I will work with Congress to establish more realistic criteria and procedures to insure that initial development decisions are wise:

--A more realistic interest rate must be used in calculating the costs and benefits of projects. Many of the projects I reviewed were authorized at such low rates that even though we are building them today, we are pretending that the cost of capital is still the same as it was many years ago. In times of a tight budget, we must be realistic about what it is actually costing the taxpayers of the nation to build these projects.

--We must be more realistic in initial cost estimates for projects, to avoid the enormous cost overruns typical of so many water projects. Some projects are ending up costing many times what they were estimated to cost when the Congress originally authorized them.

--We must scrutinize the beneficiaries of the projects to make sure that the general public is benefiting from projects, not merely narrow or special interests. One project I reviewed would have benefited only two companies; another would have spent over $1 million per landowner benefited with little repayment. Yet such projects are typically described as providing broad public benefits or helping family farmers.

--Demonstrated need for projects must precede authorization and funding. Too often, exaggerated "benefits" and questionable claims of recreation value, fish and wildlife enhancement or area redevelopment have been used to justify otherwise marginal projects All too often, valuable river recreation and fish and wildlife habitat have been destroyed in the name of "enhancement."

--Alternatives, especially non-structural or small-scale solutions to specific problems such as floods, should always be investigated as substitutes for expensive and damaging projects which often do not provide effective solutions anyway. Interagency cooperation and encouragement of local solutions to local problems need to be an integral part of every water project analysis.

Through each aspect of analysis, environmental values must be a primary concern, to insure that irreplaceable natural resources are protected from needless degradation or destruction.

2. Dam safety

I am taking action to upgrade our federal dam safety and inspection program, and I will work with the Congress to develop legislation to insure that every state has an adequate dam safety program. The recent Teton Dam tragedy indicates the importance of this problem and several of the projects examined during the review raised significant safety questions. This is a critical consideration for both existing and proposed dams.

3. Cost sharing for federal water projects

The beneficiaries of federal water projects do not bear a fair share of the enormous capital and operating costs. An example of this problem is that the users of the nation's waterways pay nothing for their construction or maintenance. Today I am recommending continuation of some waterway projects, but I will work with the Congress to develop a system to recoup the costs from the beneficiaries. It is essential as a test of economic demand for existing and future facilities and in assuring a balanced transportation system that the beneficiaries of waterway projects pay their fair share of both construction and operating costs. I will also be recommending comprehensive reforms in other cost-sharing formulas. This action is essential to genuine water program reforms.

(more)

4. Water conservation

In the arid West and across the entire nation, we must begin to recognize that water is not free--it is a precious resource. As with our energy problem, the cornerstones of future water policy should be wise management and conservation. Irrigation efficiency, water pricing, ground-water management and thoughtful land use decisions will help institute lasting protection from drought and lessen the need for expensive new water projects. Some of the 32 projects would bring water to areas where water use is not even metered and where there are no state groundwater management programs. And the General Accounting Office has recently shown that over half of the water delivered through Bureau of Reclamation Irrigation systems is completely wasted. This is unacceptable.

5. Redirected public works program

The current heavy emphasis on expensive water projects is counter to the need for a more balanced public works program providing jobs where they are needed the most, at a cost we can afford, accomplishing necessary work. Water projects provide more expensive jobs than other government spending programs ($25,000 per job), and the current pattern of water project distribution is contributing to the federal dollar drain out of the heavily populated Northeast where economic stimulus is needed. Many of our water projects simply shift economic development for no apparent policy reason. I have proposed reforms in this area as part of my economic stimulus program, and I will also be developing suggested redirections for the Corps of Engineers and the Bureau of Reclamation.

My specific recommendations follow:

(more)

		Original FY78 Request (millions)	Total Savings (millions)
1.	Applegate Lake, Oregon (COE)	7.4	74.0
2.	Atchafalaya River and Bayous Boeuf, Black & Chene, Louisiana (COE)	5.1	10.1
3.	Bayou Bodcau, Louisiana (COE)	2.4	10.0
4.	Cache Basin, Arkansas (COE)	2.0	88.5
5.	Grove Lake, Kansas (COE)	1.0	83.9
6.	Hillsdale Lake, Kansas (COE)	14.0	38.6
7.	LaFarge Lake, Wisconsin (COE)	2.0	36.9
8.	Lukfata Lake, Oklahoma (COE)	0.2	29.6
9.	Meramec Park Lake, Missouri (COE)	10.0	88.7
10.	Richard B. Russell, Georgia South Carolina (COE)	21.0	254.9
11.	Tallahala Creek, Mississippi (COE)	5.0	52.0
12.	Yatesville, Kentucky (COE)	7.2	42.7
13.	Columbia Dam, Tennessee (TVA)	20.0	110.4
14.	*Auburn, California (BR)	39.7	898.6
15.	Fruitland Mesa, Colorado (BR)	7.7	82.5
16.	*Narrows Unit, Colorado (BR)	9.7	139.2
17.	*Oahe, South Dakota (BR)	17.0	414.4
18.	Savery-Pot Hook, Colorado, Wyoming (BR)	6.0	71.3
	TOTAL SAVINGS	$ 177.4	$ 2,526.3

*Further analysis might eventually lead to reinstatement or modification — see specific recommendations.

Recommendation for Modification

		Cost to Complete Original Plan (millions)	Estimated Savings Due to Modification (millions)
1.	Mississippi River, Gulf Outlet Louisiana (COE)	282.8	24.0
2.	Tensas Basin, Arkansas and Louisiana (COE)	186.3	135.0
3.	Bonneville Unit, Central Utah Project, Utah (BR)	687.6	659.8
4.	Central Arizona Project, Arizona (BR)	1,280.3	333.0
5.	Garrison Diversion, North Dakota (BR)	436.4	302.0
		$ 2,873.4	$ 1,453.8

Recommendation for Continued Funding

		FY1978 Request (millions)	Remaining Federal Cost (millions)
1.	Dayton, Kentucky (COE)	2.9	7.2
2.	Fulton, Illinois (COE)	4.2	12.0
3.	Red River Waterway, Louisiana (COE)	26.0	815.9
4.	Tennessee-Tombigbee Waterway, Alabama and Mississippi (COE)	157.0	1,144.4
5.	Tyrone, Pennsylvania (COE)	1.5	27.5
6.	Bear Creek, Alabama and Mississippi (TVA)	18.6	22.4
7.	Dallas Creek, Colorado (BR)	12.2	46.3
8.	Dolores, Colorado (BR)	5.7	183.4
9.	Lyman, Wyoming (BR)	4.1	9.9
		$232.2	$ 2,269.0

I intend to cooperate with the Congress in accomplishing reform in the water resources area, and I hope that the Congress will cooperate with me in eliminating wasteful and destructive spending on water projects.

Individual sheets detailing the recommendations and other information on each of the 32 projects follow.

Auburn
(Bureau of Reclamation, California)

Project description:

700-foot high arch dam, reservoir and canals to provide full irrigation to 29,000 acres, supplemental water to 388,000 acres and 332,000 acre feet of M&I water supply annually. Initially project will also produce 522 million kilowatt hours of power annually.

Major benefits claimed:

Irrigation	74%
Electric Power	11%
Recreation	10%

Benefit/cost ratios:

Remaining Costs and Benefits @ current rate (6 3/8%): 1.0
Total Costs and benefits @ authorized rate (3 1/8%): 2.6

Financial data: (in millions)

Estimated Federal Cost:	$ 1,131
Allocated to Date:	233
Estimated Local Share:	0

Irrigation investment per acre: $1,626

Total project is 20% complete.
Construction is 19% complete.

Factors in decision:

- Economic questions over irrigation benefits remain unresolved: There are no firm contracts for the irrigation water, and many other questions exist concerning the Central Valley Project Account.

- An active earthquake fault may underlie Auburn Dam site. The adequacy of the dam design is under review. Collapse of the dam would imperil the lives of 750,000 persons downstream.

- Reservoir would inundate 10,000 acres of wildlife habitat, inundate 48 miles of free flowing stream and destroy 20 historic and 17 archeological sites.

- Downstream flows would be periodically reduced, adversely affecting fish and wildlife resources in the lower American River.

President's recommendation:

Delete funding and reinstate only if the safety questions are resolved and not until; firm water contracts are negotiated; an audit of the CVP power account is performed; and every effort is made to reach an agreement with the state of California on water uses and streamflows in the American River Basin.

Fruitland Mesa
(Bureau of Reclamation, Colorado)

Project description:

Soap Park Dam and conveyance system to provide irrigation to 11,940 acres of dry farmed land and supplemental irrigation to 6,310 acres.

Major benefits claimed:

Irrigation 93%

Benefit/cost ratios:

Remaining Costs and Benefits @ current rate (6 3/8%): 0.3
Total Costs and Benefits @ authorized rate (3 1/8%): 0.5

Financial data: (in millions)

Estimated Federal Cost:	$ 87.9
Allocated to Date:	5.4
Estimated Local Share:	0

Irrigation investment per acre: $4,615.00

Total project is 6% complete.
Construction is 7% complete.

Factors in decision:

- Benefit/cost ratio at both the current and authorized rate is below unity.
- The irrigation investment per acre is very high at $4,615, of which only $208 will be repaid by irrigators.
- The project area is farmed or ranched by only 69 landowners; this amounts to an irrigation investment per landowner of about $1.2 million.
- Increased salinity from irrigation return flows would decrease water quality in the lower Colorado River Basin.
- The project would convert 12,000 acres of rangeland to irrigated cropland, inundate 584 acres of land and 4.5 miles of good fishing stream.
- The lands benefitted by the project are located at high altitudes -- six to eight thousand feet -- with a short growing season.

President's recommendation:

Delete funding and deauthorize project.

more

Narrows Unit
(Bureau of Reclamation, Colorado)

Project description:

Earthfill dam on South Platte River to provide supplemental irrigation water to 287,000 acres.

Major benefits claimed:

Irrigation	56%
Recreation	35%

Benefit/cost ratios:

Remaining costs and benefits @ current rate (6 3/8%): 0.9
Total @ authorized rate (3 1/4%): 1.4

Financial data: (in millions)

Total estimated Federal cost:	$ 145.7
Allocated to date:	6.5
Estimated local share:	.01
Allocated to date:	.01

Total project is 5% complete.
Construction is 4% complete.
Irrigation investment per acre: $291

Factors in decision:

- Reservoir would dislocate 642 people and their community.
- Water quality problems in reservoir might negate recreation benefits.
- Safety/seepage issues remain unresolved; the right abutment of the dam is located on sand and gravel.
- Over 40 historical and archaeological sites would be destroyed.
- 15,000 acres of agricultural and wildlife lands would be regularly inundated by the reservoir; 15 miles of free-flowing river would be lost.
- Flood control benefits are questionable.

President's recommendation:

Delete funding pending re-evaluation of project, including study of the alternative of a groundwater recharge system, determination of the effects of the project on the Platte River system, resolution of water quality issues, resolution of the safety and seepage questions, and re-evaluation of the claimed flood control benefits.

more

Oahe
(Bureau of Reclamation, South Dakota)

Project description:

The project plan provides for the construction of over one hundred miles of canals and storage reservoirs to divert water from an existing reservoir on the Missouri River to irrigate 190,000 acres of currently dry-farmed land; other project purposes include municipal and industrial water supply and recreation.

Major benefits claimed:

Irrigation	90%
Recreation	4%
M&I water supply	3%
Unemployment	2%

Benefit/cost ratios:

Remaining costs and benefits @ current rate (6 3/8%): 1.0
Total @ authorized rate (3 1/4%): 1.6

Financial data:	(in millions)
Estimated Federal cost:	$ 504.6
Allocated to date:	89.1
Estimated local share:	0
Allocated to date:	0

Total project is 18% complete including lands.
Construction is 18% complete.

Factors in decision:

- Water quality and quantity problems resulting from large irrigation return flows into the James River. 50% increase in total dissolved solids; 200% increase in average annual streamflow.
- Loss of wildlife habitat, particularly prairie wetlands for migratory wildfowl; up to 23,800 acres of prairie wetlands would be inundated.
- Channelization of the James River -- a 120-mile stretch would be reduced to 54 miles of channel.
- Uncertainty of local commitment to the project.
- The large investment per farm unit -- $735,655 per farm unit; $183,913 per person.
- Potential conflict with Indian water rights.
- Marginal economic benefits and high cost of extensive supply works necessary to move water 100 miles east to Lake Plain service area.

President's recommendation:

Delete funding and reinstate only if local assurances are firm and if project is modified to eliminate the East Plain service area and associated supply works.

more

Savery-Pot Hook
(Bureau of Reclamation, Colorado)

Project description:

2 dams and conveyance systems to supply irrigation water on Little Snake River.

Major benefits claimed:

Irrigation 95%

Benefit/cost ratios:

Remaining Costs and Benefits @ current rate (6 3/8%): 0.4
Total Cost and Benefits @ authorized rate (3 1/8%): 0.7

Financial data:	(in millions)
Estimated Federal Cost:	$ 75.8
Allocated to Date:	4.1
Estimated Local Share:	0

Irrigation investment per acre: $2,563.00 ($4,008 on a full-service equivalency basis)

Total project is 5% complete.
Construction is 6% complete.

Factors in decision:

- The benefit/cost ratio at both the current and authorized rate is below unity.
- The project as planned will benefit only 106 existing farms, at an average irrigation investment per farm of almost $700,000.
- High dissolved salt concentrations in project return flows would increase existing water quality problems downstream in the Colorado River.
- Project lands are at 6,500 feet altitude, and thus have a short growing season.
- Approximately 20,100 acres of wildlife habitat would be modified or eliminated, affecting deer, elk, antelope and sage grouse. While these losses are substantially mitigated by development of additional habitat, significant losses in sage grouse and antelope population would occur.
- There is a concern that energy developers will build project for coal development use if federal government does not build it for agriculture.

President's recommendation:

Delete funding and deauthorize project.

more

Bonneville Unit, Central Utah Project
(Bureau of Reclamation, Utah)

Project description:

The project consists of a series of 10 new reservoirs, enlargement of 2 existing reservoirs, 140 miles of aqueducts, 3 power plants, and 9 pumping plants to divert water from the Colorado Basin to the Great Salt Lake Basin. The project would provide a full water supply to irrigate 29,370 acres of new land, a supplemental water supply to partially irrigate 213,170 acres, and 99,000 acre feet of water annually for municipal and industrial use, and 320 million KW hours of electric power annually.

Major benefits claimed:

Irrigation	25%
Power	21%
M&I Water	46%

Benefit/cost ratios:

Remaining Costs and Benefits at Current Rate (6 3/8%): 1.0
Total Costs and Benefits at Authorized Rate (3 1/8%): 1.3

Financial data:	(in millions)
Estimated Federal Cost	$862.7
Allocated to Date:	154.4
Estimated Local Share:	40.3
Allocated to Date:	18.1

Irrigation investment per acre: $1,434 ($4,295 at full service equivalent)

Total project is 19% complete including lands.
Construction is 19% complete.

Factors in decision:

- High cost of the project -- $862 million at current estimates and distinct possibility that operational plans and commitments cannot be met within authorization limits.
- Marginal economic value -- ratio 1.0:1 on remaining costs and benefits at current discount rate -- and high cost of irrigation investment -- $4295 per acre at full service equivalent.
- Losses of high quality stream fisheries in Uinta Basin, habitat losses on Utah Lake, including destruction of 57 miles of high quality trout streams and 25,000 acres of marsh and water habitat.
- Salinity impacts on downstream users in Colorado River Basin.
- Potential impact on Indian water rights and interests, especially the Ute Tribe.

more

Bonneville (cont'd)

- Questionable need for project and possible alternative water supply development in Bonneville Basin.

President's recommendation:

Modify the project to complete only the Currant Creek Reservoir and rehabilitation of Strawberry Tunnel and Provo Reservoir Canal, permitting the development of 30,000-36,000 acre feet of water for irrigation in the Utah Lake area and municipal use in the Salt Lake City area. The best way to fulfill obligations to the Ute Indian Tribe are to be determined by the Interior Department and the Tribe. The total cost of this modification is estimated to be $27.8 million.

more

Central Arizona Project
(Bureau of Reclamation, Arizona)

Project description:

A major dam, a 400-mile system of drains, aqueducts and pumping stations, 3 smaller dams to divert 1.2 million acre feet of water from the Colorado River for irrigation and municipal water supplies in Central Arizona.

Major benefits claimed:

M&I water	48%
Electric Power	23%
Irrigation	21%

Benefit/cost ratios:

Remaining Costs and Benefits @ current rate (6 3/8%): 1.4
Total Costs and Benefits @ authorized rate (3 1/4%): 1.6

Financial data:	(in millions)
Total Estimated Federal Cost:	$ 1,679.1
Allocated to Date:	399.0
Estimated Local Share:	1.0
Allocated to Date:	1.0

Total project is 25% complete.
Construction is 24% complete.

Irrigation investment per acre: $757.00

Factors in decision:

- The relative high cost of project water in relation to groundwater costs ($106 per acre foot vs. $10 to $40 per acre foot).
- The lack of adequate controls on groundwater pumping and consideration that groundwater supplies might be adequate.
- The lack of water conservation in the State (per capita consumption is well above the national average).
- The placement of Orme Dam on the Ft. McDowell Indian Reservation, its effects on the Salt and Verde Rivers and endangered species and archeological sites, and possible safety question.

  The allocation of project water to Indian tribes.
- The relative inefficiency of the auxiliary dams: Orme, Hooker, Charleston and Buttes.
- The possible inadequacy of water supplies in the Colorado River and anticipated salinity increases in Colorado River (involving commitments to Mexico).

President's recommendation:

Modify the project to eliminate the Orme, Hooker and Charleston dams and make further federal funding contingent upon further study of groundwater supplies and institution of groundwater regulation and management by the state of Arizona. The cost of the three dams to be deleted would have been $333 million.

more

Garrison Diversion Project

(Bureau of Reclamation, North Dakota)

Project description:

The project consists of Lonetree Reservoir, two dams, a pumping plant, and the McClusky, New Rockford, Warwick and Velva canals extending several hundred miles from a large existing reservoir on the Missouri River to North Central, Eastern and Southeastern North Dakota; designed to irrigate 250,000 acres now dry farmed; some recreation and municipal and industrial water supply would also be provided.

Major benefits claimed:

Irrigation	83%
M&I Water	4%
Recreation and Fish & Wildlife	12%
Flood control	1%

Benefit/cost ratios:

Remaining Costs and Benefits @ current rate (6 3/8%): 1.36
Total @ authorized rate (3 1/8%): 1.7

Financial data:	**(in millions)**
Estimated Federal Cost:	$ 609.7
Allocated to Date:	172.0
Estimated Local Share:	5.6
Allocated to Date:	3.8

Irrigation investment per acre: $ 1,992

Total project is 29% complete including lands.
Construction is 30% complete.

Factors in decision:

- Water quality problems resulting from irrigation return flows in five rivers in the Northern Great Plains -- Souris, Red, James, Wild River, and Cheyenne. Total dissolved solids (TDS) concentrations will increase in all rivers with large impacts on the Souris River flowing into Canada.
- Loss of wildlife habitat, particularly prairie wetlands for migratory waterfowl. Over 50,000 acres of prairie wetlands and 50,000 acres of natural prairie land would be lost to project construction, excluding losses from channelization.
- Adverse effects on eight National Wildlife Refuges by increased flows and/or rough fish introduction.
- The Canadian Government's objections to the project on the grounds that it violates treaty obligations. Recommendations from the International Joint Commission are expected in June.
- Conversion or destruction of 220,000 acres of productive land to irrigate only 250,000.

more

- Project would consume over 150 million KWH of electricity annually and reduce Missouri River hydropower by more than 300 million KWH annually.
- The investment per farm unit of $389,390 (for 1,250 farms).
- Potential conflict with Indian water rights.
- Relative high cost of extensive supply works necessary to move long distances from the Garrison Reservoir into the three major service areas.
- Litigation is presently pending on U. S. District Court contesting the adequacy of the process used to comply with the National Environmental Policy Act (NEPA).

President's recommendation:

Modify to eliminate the Souris, Karlsruhe, Warwick-McVille and New Rockford irrigation areas and associated supply works, with the exception of reaches 1 and 2 of the New Rockford Canal, reducing the size of Lonetree Reservoir and the width of the New Rockford Canal reaches.

Estimated cost of this option is $134 million. The recommendations of the upcoming International Joint Commission Report will be fully considered and the modification will undergo further analysis. Project modifications will in no way alter our commitment to fulfill all treaty obligations with Canada.

Dallas Creek
(Bureau of Reclamation, Colorado)

Project description:

Ridgway Reservoir on the Uncompahgre River would provide water supply for municipal and industrial uses and supplemental irrigation for 64,660 acres of currently irrigated land.

Major benefits claimed:

M&I Water	61%
Recreation	24%
Irrigation	13%

Benefit/cost ratios:

Remaining Costs and Benefits @ current rate (6 3/8%): 0.9
Total Costs and Benefits @ authorized rate (3 1/4%): 1.4

Financial data:	(in millions)
Estimated Federal Cost:	$ 53.7
Allocated to Date:	7.4
Estimated Local Share:	0

Irrigation investment per acre: $732.00

Total project is 14% complete.
Construction is 7% complete.

Factors in decision:

- Economic development of area.
- Water quality problems, especially toxic chemicals and heavy metals, which might limit water use for domestic purposes, can be mitigated by an exchange which would allow use of Gunnison River water for domestic consumption.
- Question of need for supplemental irrigation water, if improved water management techniques were employed.
- Downstream salinity impact on the Colorado River.

President's recommendation:

Continue funding.

more

Dolores
(Bureau of Reclamation, Colorado)

Project description:

Dam and reservoir on the Dolores River for irrigation and municipal water supply, including irrigation benefits to Ute Indians.

Major benefits claimed:

Irrigation	57%
M&I Water Supply	19%
Recreation	16%

Benefit/cost ratios:

Remaining Costs and Benefits @ current rate (6 3/8%): 0.6
Total Costs and Benefits @ authorized rate (3 1/4%): 1.1

Financial data:	(in millions)
Estimated Federal Cost:	$ 187.3
Allocated to Date:	3.9
Estimated Local Share:	0
Irrigation investment per acre:	$2,397.00

Total Project is 2% complete.
Construction is 2% complete.

Factors in decision:

- Benefits to the Ute Mountain Ute Indian Tribe and no guarantee of implementing other alternatives for Indian economic development.
- High investment per acre to be subsidized from power revenues.
- Environmental effects, including the potential increase in Colorado River salinity, loss of white-water recreation, inundation of archeological sites; project would inundate 11 miles of mountain-alpine and canyon-land river and inundate 12,000 acres of big game habitat.

President's recommendation:

Continue funding.

more

Lyman
(Bureau of Reclamation, Wyoming)

Project description:

Project consists of 2 reservoirs, one of which is completed. The Stateline reservoir is under construction. It will provide supplemental irrigation of 36,000 acres of land and municipal and industrial water supply for Lyman and Mountain View communities.

Major benefits claimed:

Irrigation	67%
Recreation	25%
M&I Water Supply	8%

Benefit/cost ratios:

Remaining Costs and Benefits @ current rate (6 3/8%): 0.5
Total Costs and Benefits @ authorized rate (2 7/8%): 0.9

Financial data:	(in millions)
Estimated Federal Cost:	$ 29.2
Allocated to Date:	19.2
Estimated Local Share:	0

Irrigation investment per acre: $518.00

Total project is 66% complete.
Construction is 67% complete.

Factors in decision:

- The increasing need for domestic water supplies due to mineral development and population growth.
- The percentage of completion of the project (the sunk costs of the project) and fact that irrigation repayment depends on completion.
- The increase in salinity in the Colorado River due to irrigation depletions and salt loading.

President's recommendation:

Continue funding.

# # # # #

APPENDIX G:

THREE SAMPLES OF RECLAMATION DOCUMENTATION OF FOREIGN VISITORS AND TRAINEES OVER TIME

G.1 Trainees from Thailand in the period 1947 to 1959

Name	Title at time of training	Present title	Sponsored by	
Group of 30 RID Engineers (1st Thai participants associated with USBR--Years 1947-1948) (Names and Titles not complete)			Thai Gov't.	
Sathien Bhadrasiri	Mechanical Engineer (L.E)	-1st Grade Eng., Royal Irrigation Department	Thai Gov't.	1949-54
Udom Panyaphol	-	Chief of Operations & Maintenance Div., RID	Thai Gov't.	1949-51
Chalit Vejjajiva	Electrical Engineer	-Superintending Eng., Head of Hydro-Energy Division	Thai Gov't.	1953-54
M. L. Jeong Jan Kambhu	-	Chief Engineer, RID	Thai Gov't.	1953
Visudh Chutima	-	Inspection Engineer, RID	Thai Gov't.	1953
Ananda Indrathongmon	Irrigation Engineer	-1st Grade Technician, State Irrigation Division	Thai Gov't.	1949-51
M. L. Pilandh Malakul	Civil Engineer	-Superintending Eng., Head of Survey Division	Thai Gov't.	1949-51
Chien Bhumisawasdi	Irrigation Engineer	-1st Grade Technician, Bhumiphol Dam Const. Proj.	Thai Gov't.	1949-51
Padung Hansakunatai	Civil Engineer	-1st Grade Eng., Bhumiphol Dam Construction Project	Thai Gov't.	1959
Phol Chakkaew	Irrigation Engineer	-1st Grade Technician, Construction Division	Thai Gov't.	1959
Surajita Rimphati	Irrigation Engineer	-1st Grade Technician, Construction Division	Thai Gov't.	1959
Chalaw Rutanasen	Irrigation Engineer	-1st Grade Technician, Construction Division	Thai Gov't.	1959
Samai Suravallopa	Civil Engineer	-1st Grade Eng., Constr. Division	Thai Gov't.	1959
Sa-ad Makakkagarnjana	Civil Engineer	-1st Grade Eng., Constr. Division	Thai Gov't.	1959
Nathi Menabandhu	Civil Engineer	-Deceased	Thai Gov't.	1959
Prayad Brahmakom	Irrigation Engineer	-2nd Grade Technician, Bhumiphol Dam Constr. Proj.	Thai Gov't.	1959
Chalerm Bunsuwan	Irrigation Engineer	-1st Grade Technician	Thai Gov't.	1959
U[illegible] loysungvarn	Irrigation Engineer	-2nd Grade Technician, Construction Division	Thai Gov't.	1959

Name	Title at time of training	Present title	Sponsored by	
Saard Siddhikol	Electrical Engineer	-	Thai Gov't.	1952-54
Niyom Vachanonda	Electrical Engineer	-2nd Grade Eng., Bhumiphol Dam Const. Project	Thai Gov't.	1952-54
Prayura Chaneleudefa	Electrical Engineer	-	Thai Gov't.	1952-54
Sahus Ratanakul	Civil Engineer	-1st Grade Eng., Bhumiphol Dam Const. Project	Thai Gov't.	1955-57
M. L. Pijit Kambhu	Electrical Engineer	-	Thai Gov't.	1952-54
Leck Jindasnguan	Hydrologist	-2nd Grade Technician, Hydrology Section	Thai Gov't.	1961-62
Sa-ard Srinkapaibulaya	Civil Engineer	-1st Grade Eng., Technical Division	Thai Gov't.	1955
Mrs. Sumrit Rojanasoonthon	Foreign Secretary	-1st Grade Officer, Office of the Secretary	Thai Gov't.	1958
Pratut Jayapani	Architect	-2nd Grade Arch., Technical Division	Thai Gov't.	1956-57
Sandis Virathian	Civil Engineer	-1st Grade Eng., Technical Division	Thai Gov't.	1953
Lek Thanomkulbutra	Civil Engineer	-1st Grade Eng., Bhumiphol Dam Design Sec., Tech. Div.	Thai Gov't.	1955-57
Chareuk Nonthathum	Civil Engineer	-1st Grade Eng., Bhumiphol Dam Construction Proj.	Thai Gov't.	1955-58
Srid Aphaiphuminart	Electrical Engineer	-1st Grade Eng., Hydro-Energy Division	Thai Gov't.	1955-58
Prathai Phisphumvidhi	Civil Engineer	-1st Grade Eng., Hydro-Energy Division	Thai Gov't.	1955-58
Boonyok Vadhanaphuti	Electrical Engineer	-2nd Grade Eng., Graduate Student	Thai Gov't.	1956
Pokai Theppisai	Electrical Engineer	-1st Grade Eng., Hydro-Energy Division	Thai Gov't.	1955-58
Prakaiprock Stutananda	Electrical Engineer	-1st Grade Eng., Hydro-Energy Division	Thai Gov't.	1955-58
Phijit Jalichandra	Electrical Engineer	-	Thai Gov't.	1957-58
Maitri Poolsup	Civil Engineer	-2nd Grade Eng., Tech. Division	Thai Gov't.	1962
Kamthorn Sangkhavasi	Civil Engineer	-2nd Grade Eng., Bhumiphol Dam Const. Project	Thai Gov't.	1959-60

Name	Title at time of training	Present title	Sponsored by	
Suthon Muenraksa	Electrical Engineer	-2nd Grade Eng., Hydro-Energy Division	Thai Gov't.	1962
S'Nguan Jamprawit	Civil Engineer	-2nd Grade Eng.	Thai Gov't.	1962
Manit Komaladat	Design Engineer	-2nd Grade Eng., Bhumiphol Design Sec., Tech. Div.	Thai Gov't.	1961-62
Kaiwan Tephasdin Na Ayduhya	Civil Engineer	-1st Grade Eng., Construction Division	Thai Gov't.	1959-60
Pramote Boonsuvun	Civil Engineer	-2nd Grade Eng., Bhumiphol Design Sec., Tech. Div.	Thai Gov't.	1962
Jeruay Lekavanich	Civil Engineer	-2nd Grade Eng., Tech. Div.	Thai Gov't.	1962
Thavil Chuenchusilpa	Chief Cashier	-1st Grade Officer, Finance Division	Thai Gov't.	1959-60
Prame Chotivanich	Electrical Engineer	-2nd Grade Engineer, Hydro-Energy Division	Thai Gov't.	1961
Prateeb Monyakula	Civil Engineer	-1st Grade Engineer, Hydro-Energy Division	Thai Gov't.	1955-58
Taweechai Mackaman	Civil Engineer	-1st Grade Engineer, Hydro-Energy Division	Thai Gov't.	1955-57
Prabhansak Bhengbon	Civil Engineer	-2nd Grade Engineer, Tech. Division	Thai Gov't.	1959
Damrong Jaraswathana	Civil Engineer	-1st Grade Eng., Hydrology Sec., Survey Division	Thai Gov't.	1953
Chamnan Pradithavanij	Civil Engineer	-1st Grade Engineer, Tech. Division	Thai Gov't.	1955-56
Pravit Ruyabhorn	Civil Engineer	-	Thai Gov't.	1955
Suwitt Sthantraiphop	Chief of Purchasing Section, Stores Division	-1st Grade Officer, Store Division	Thai Gov't.	1955
Phorn Kanpetch	Chief of Section, Personal & Records	-	Thai Gov't.	1955
Krasae Subbasiddhi	Civil Engineer	-1st Grade Eng., Royal Irrigation Department	Thai Gov't.	1953
Swarz Suttrowatt	Mechanical Engineer	-	Thai Gov't.	1952-54
Santad G. Suwan	Electrical Engineer	-2nd Grade Eng., Construction Division	Thai Gov't.	1952-54

Name	Title at time of training	Present title	Sponsored by	
Sotsukh Kanchanagom	Geologist	-1st Grade Eng., Research & Lab. Sec., Tech. Div.	Thai Gov't.	1958-59
Komain Unhanand	Civil Engineer	-1st Grade Eng., Graduate Student	Thai Gov't.	1955
Kasame Chatikavanij	Electrical Engineer (L.E.)	-	Thai Gov't.	1950-51 1956-57
Somchate Tindhukasiri	Civil Engineer	-2nd Grade Eng., Hydro-Energy Division	Thai Gov't.	1961
Taweesak Mahasandana	Civil Engineer	-2nd Grade Engineer, Graduate Student	Thai Gov't.	1961-62
Tip Wanakansobhon	Civil Engineer	-2nd Grade Technician, Technical Division	Thai Gov't.	1961
Chalongbhan Komarakul	Civil Engineer	-3rd Grade Technician, Hydro Energy Division	Thai Gov't.	1962
Chamnien Soomsvasdi	Mechanical Engineer	-2nd Grade Eng., Hydro-Energy Division	Thai Gov't.	1955-58
Burirom Jatikarat	Chief of Administration Section	-2nd Grade Officer, Hydro-Energy Division	Thai Gov't.	1961-62
Anant S. Narupon	Geologist	-2nd Grade Eng., Trainer Eng., U.S.B.R.	Thai Gov't.	*1958 1962-

*Called home, resumed program

Note: Training programs were from 3 months to 2 years with majority at 1 year.

Thai government sponsorship mostly by Royal Irrigation Department. Perhaps 6 sponsored by Thai Civil Service Commission.

Name	Title at time of training	Present title	Sponsored by	
Apichai Lamchiagdhase	Irrigation Engineer RID	-2nd Grade Technician, State Irrigation Division	MSA-Dept. Agriculture 3 months	1953
Parkit Onvimala	Irrigation Engineer	-1st Grade Technician, Construction Division	Same	
Boonyong Sirivatana	Irrigation Engineer	-	Same	1953
Leph Tanthasri	Irrigation Engineer	-1st Grade Technician, Construction Division	Same	1953
Suthep Tingsabhat	Irrigation Engineer	-1st Grade Eng., Hydrology Section	Same	1953
Suttichai Kraikunasai	Civil Engineer, NEA	-	AID 3 months plus	1962
Pira Vathakanon	Civil Engineer, RID	-1st Grade Eng., Steam Powerplant	ICA 1 year	1957-58
Miss Sakon Changsanit	Administrative Asst., NEA	-	ICA 3 months	1960
Abhai Prayurasook	Electrical Engineer, RID-	-2nd Grade Eng., Mechanical Division	ICA 1 year	1958-59
Uthai Rugrachagarn	Civil Fngineer, NEA	-	FOA 1 year	1955
Uthai Charmreonprucksa	Civil Engineer, RID	-1st Grade Technician, Construction Division	MSA 1 year	1954-55
Prapath Prammani	Geologist, NEA	-	FOA 15 months	1955-56
Prateep Nudakarnchanakula	Civil Engineer, RID	-1st Grade Eng., Technical Division	MSA 1 year	1954-55
Prasong Patradhamkul	Mechanical Engineer, RID-	-2nd Grade Eng., Technical Division	MSA 1 year	1954-55
Banpot Siripanich	Civil Engineer, RID	-1st Grade Eng., Bhumiphol Dam Construction Project	MSA 1 year	1954-55
Suree Sornsonboon	Civil Engineer, RID	-2nd Grade Technician, State Irrigation Division	MSA 1 year	1954-55
Jit Pitchakul	Civil Engineer, RID	-1st Grade Eng., Technical Division	MSA 1 year	1954-55
Chuen Ketupama	Chief of Admn. Section, Accounts Division	-Chief Accountant for Bhumiphol Project	FOA 9 months	1956-57

Name	Title at time of training	Present title	Sponsored by	
Prasom Bandurat	Engineer, Cooperative Land Development Dept., RID	-	FOA 6 months	1955
Prasarn Tavivatana	-Chief of Finance Division	-	MSA 5 months	1952
Khien Suwanakamjaya	Chief of Central Section of Construction, RID	-Secretary of R.I.D.	MSA 6 months	1952
Kio Nuanplaud	Chief of Stores Division, RID	-Retired	MSA 8 months	1952
L. Rochanasiri Warindr	Chief of Central Administration Division, RID	-Retired	MSA 8 months	1952

Name	Title at time of training	Present title	Sponsored by	
Chalaw Buranabhan	Irrigation Engineer	-1st Grade Technician, Bhumiphol Dam Const.Proj.	MSA 1 year	1951-53
Banchong Pisalbutra	Geologist	-1st Grade Eng., Bhumiphol Dam Construction Project	MSA 1 year	1951-53
Boonthai Otaganonta	Civil Engineer	-1st Grade Eng., Tech. Division	MSA 1 year	1951-53
Sara Subharngkasen	Mechanical Engineer	-1st Grade Eng., Royal Irrigation Department	MSA 1 year	1951-53
Tamon Kandhasingha	Irrigation Engineer	-2nd Grade Technician, Bhumiphol Dam Const. Proj.	MSA 1 year	1951-53
Bhochana Panyadhibya	Irrigation Engineer	-Chief, Economic & Statistical Research Unit	MSA 1 year	1951-53
Udom Rakcharya	Irrigation Engineer	-1st Grade Technician, Tech. Division	MSA 1 year	1951-53
Om Dibbabadya	Irrigation Engineer	-1st Grade Technician, Construction Division	MSA 1 year	1951-53
Amphan Poonagunta	Irrigation Engineer	-1st Grade Technician, State Irrigation Division	MSA 1 year	1951-53
Vira Suttipongse	Civil Engineer	-	MSA 1 year	1951-53

Above group sponsored by MSA for 1 year program. Additional 1 year program with USBR sponsored by Royal Irrigation Dept.

G.2 Tabulation of Trainees and Visitors during the Calendar Year 1957

TABULATION VISITORS & TRAINEES
Calendar Year 1957

Country	Trainees	Observers	Visitors	Tourists	Totals
Afghanistan	1	-	-	3	4
Argentina	-	-	1	2	3
Australia	11	2	11	8	32
Austria	-	-	-	2	2
Bolivia	-	-	-	17	17
Brazil	1	-	2	10	13
Burma	1	-	-	-	1
Cambodia	-	-	1	-	1
Canada	-	-	-	2	2
Ceylon	-	1	1	-	2
Chile	-	5	2	4	11
Colombia	-	-	-	1	1
Costa Rica	-	2	-	1	3
Ecuador	-	-	-	2	2
Egypt	-	-	1	-	1
El Salvador	-	-	-	1	1
England	-	-	3	5	8
Ethiopia	-	-	-	4	4
Formosa	3	10	12	3	28
French W. Africa	-	0	1	-	1
France	-	-	-	8	8
Germany	-	-	-	5	5
Ghana	-	-	-	1	1
Greece	-	-	-	1	1
Guatemala	-	-	-	2	2
Hungary	-	-	-	1	1
Iceland	-	-	-	1	1
India	2	6	3	12	23
Indonesia	-	-	4	1	5
Iran	-	-	3	1	4
Iraq	3	4	1	1	9
Israel	-	1	1	3	5
Italy	-	-	1	3	4
Japan	-	1	3	14	18
Kenya	-	-	1	1	2
Korea	2	4	14	-	20
Kuwait	-	-	-	6	6
Lebanon	-	2	-	1	3
Netherlands	-	-	5	-	5
New Zealand	-	-	2	3	5
Nigeria	-	-	-	1	1
Northern Rhodesia	-	-	-	1	1
Norway	-	-	2	3	5
Pakistan	4	2	2	2	10

Calendar Year 1957

ountry	Trainees	Observers	Visitors	Tourists	Totals
anama	1	-	-	1	2
eru	-	-	-	7	7
ilippines	5	2	-	5	12
erto Rico	-	-	2	-	2
otland	-	-	-	1	1
ain	-	6	19	7	32
Rhodesia	-	1	-	1	2
dan	1	-	-	2	3
eden	-	-	2	1	3
itserland	-	-	-	16	16
ria	-	-	-	1	1
ailand	2	6	1	2	11
rkey	12	3	1	7	23
ion S. Africa	-	-	3	-	3
uguay	-	-	3	-	3
nesuela	1	-	2	9	12
etnam	-	1	-	-	1
goslavia	-	4	-	1	5
tal Countries 62	50	63	109	198	420

Carry Over Trainees 60
Grand Total 480

G.3 Foreign Training and Visitor Section "Report of Participants Associated with the Bureau of Reclamation E&R Center during the Month of June 1978"

DIVISION OF FOREIGN ACTIVITIES
Training and Coordination Branch
Foreign Training and Visitor Section

Report of Participants Associated with the Bureau of Reclamation E&R Center during the:

MONTH OF JUNE 1978

COUNTRY	NAME	TITLE & ORGANIZATION	STATUS	DATE
Australia	Dr. Kathleen H. Skinner	Sr.Res Scientist/CSIRO	Visitor	June 15-
Austria	*****Dag Bergloff	--	same	June 15
Brazil	No Team (Ms.Ana Maria Araujo	Cf/Human Res/CODEVASF	Observer	June 21-
	(Sergio A.A. Gozende	Mgr/Monitoring Dept/same	same	same
	(Gerlado de Souza Araujo	Director/3d Reg Dir/same	same	same
Canada	*****Jacques Desbaillets	--	Visitor	June 15
Chile	Alberto Cortinez	Mining Engr/MNConsulting	same	June 13
Colombia	Oscar O. Estrada	Mech Engr/Integral Ltd	same	June 14
	Francisco Roman	same	same	same
	****Julio Carrizosa	Dir/Natl Inst Nat Res	same	June 26
Costa Rica	Ronald Salazar	Engineer/ICE	Observer	June 5-2
	Carlos Picado	same	same	same
Egypt	*Dr.Saad El-Demerdashe	Asst Prof Soils/Desert Inst	Visitor	June 5-6
England	John F. Cunningham	Irrig Engr/World Bank	same	June 12
Ethiopia	George Goitom	Civ Engr/CSU Student	same	June 16
France	Jean Pierre Chabal	Proj Engr/Coyne&Bellier	same	June 22
	Georges Post	Technical Manager/Same	same	same
Germany	Prof J. Raabe	Technical University	same	same
Guinea	****Abdoul G. Diallo	Prof/National Univ	same	June 26
Honduras	CORRECTION:Raul Flores	Engr/ENEE	Trainee	June 23
	Marco A.W. Ramos	Civ Engr/ENEE	Visitor	June 7
India	**Shri O.P. Najhawan	Dep Dir/Min of Energy	Trainee	June 1-3
	**Dr.Shivram S.Patil	Res Off/Cen Wtr Comm	Trainee	June 5-3
Israel	Giora Yanai	CfEngr/Israel Atomic Comm	Observer	June 7-1
Japan	Prof Michihiro Nishi	Asst Prof/Kyushu Institute	Visitor	June 8
	Takashi Kubota	Cf Hyd Engr/Fuji Elec Co	same	same
	11 Farmers accompanied by Yoko Robertson, Interpretor	Local farmers	Visitors	June 28-
Korea	Chon Myong Kim	Cf/2d Engr Sec/Agr Dev Corp-	Observer	June 1-2

(over)

.esotho	*Habofanoe E.Makhooane	Engr Tech/Thaba Sousiu Proj-	Trainee	June 1-30
1ali	***Sekou Haidara	Wtr Division Engr	Obs Team	June 8-9
	***Mamadou Kone	Magistrate, Legal Dept	same	same
1auritania	***Yedali Ould Cheikh	Secty Gen/Legal Affrs	same	same
	***Mohamedou Baba	National Coord for OMVS	same	same
1exico	****Arturo Hauser	Asst Dir/Min Agric	Visitor	June 26
1epal	****Manmohan Dhoj Joshi	Dir-Gen/Min Forestry	same	same
1ew Zealand	Murray D. Wilson	Dir/NZ Cement Holdings	same	June 13-14
1orway	*****Erik Fidji	--	same	June 15
	*****Arthur Teigland	--	same	same
'akistan	Mohammad Ashraf	Cf Engr/Wtr&Pwr Authority	Observer	June 5-23
'eru	Caesar Zapata accompanied by John Ismert, ECI, Denver	Civ Engr/Electroperu Inc	Visitor	June 28
'hilippines	Cipriano B. Billones	Asst Proj Mgr/Magat Irrig	Observer	June 1-30
'ortugal	Rui M.S.G.Henriques	Research Engr/LNEC	same	June 19-30
	J. L. Serafim	Consultant/COBA	Visitor	June 26
;enegal	***Mohamed A.Saibott	Attorney Advisor	Obs Team	June 8-9
	***Ibrahima Ba	Transportation Div Chief	same	same
	***Mamadou Konate	Div Cf/Energy&Mines	same	same
	***Moctar M'Backe	Legal Advisor	same	same
	***Mamadou Dieye	Director/Rural Equipment	same	same
;omalia	****Abdillahi A.Kerrani	Gen Mgr/Natl Range Agency	Visitor	June 26
South Africa	Jack A. Caldwell	Geotech Engr/Robertson/ & Kurston Consultg Engrs	same	June 22-23
;udan	*Siddig A. Mohamad	Cf Engr/Rahad Corp	Trainee	June 12-30
;witzerland	Dr.Henri B. Meier	Vice Pres/Motor Columbus Consulting Engineers Inc	Visitor	June 6
	*****Klaus Holler	--	same	June 15
	*****Bruno Barp	--	same	same
Thailand	Sunthorn Ruanglek	Dep Dir Gen/RID	same	June 22
	Charin Atthayodhin	RID	same	June 19
	Prahaiprook Srutanond	Dir Proj Plang/RID	same	same
	Apichai Karoonyavanich	Cf Plng/Min Agriculture	same	same
United Kingdom	Prof A.N. Schofield	Cambridge Univ/Dept Engr	same	June 22
Upper Volta	Georges Sola	Asst Hydrologist, Inter-African Committee for Hydraulics Studies	Observer	June 5-30

Country	Name	Position	Status	Date
U̇SSR	V. M. Krakovets	Counselor of Irrigation/Embassy of USSR,Washington,DC	Visitor	June 7-8
	*****Dr.Lev A. Zolotov	Research Deputy Director Scientific Research Center Hydroproject Institute	same	June 14
	*****Dr.Leren P. Mikhailov	Dir General/All-Union Des/ Surveying&Research Instit Hydroproject/Ministry of Power and Electrification	same	same
	*****Boris G. Kartelev	Sr Research Officer/Laboratory of Hydroturbine Blocks/ VNIIG	same	same
Venezuela	Elias Sanchez	Engr/Caralsolso Univ	same	same
	****Mrs.Catalina B.Arias	Hd/Soil&Wtr Div/Environ Dept	same	June 26
	****Ms.Maria A. Balza	Coord Land Use Plng/same	same	same
Yemen	Mohamed Saleh El-Bakili	--Program Administrator Tihama Dev Auth	Observer	June 29-30
Zambia	****Emmanuel N. Chidumayo	Conservator of Natural Resources, Min Wtr&Natural Resources	Visitor	June 26

* Aid sponsored

** United Nations sponsored

AID-sponsored observation program for 9-man team from Senegal River Basin Development Authority (OMVS) and Basin States.
Escort: Henry G. Doyle, USBR
Interpretor: Mr. Jean Ramseyer

**** Multi-country Natural Environmental Team, escorted by Bill Tager, IIE, Denver

***** Toured the USBR Hydraulics Laboratory. Attended the Joint Symposium on Design and Operation of Fluid Machinery at CSU, Ft. Collins, Colorado

Trainees	5
Observers	21
Visitors	53
Total	79

BIBLIOGRAPHY

Manuscript Collections

Andrew Denny Rogers Papers. American Heritage Center. University of Wyoming. Laramie.

"Carter's Hit List." Binder of Primary Source Documents. Bureau of Reclamation History Program. Denver, Colorado.

Floyd E. Dominy Papers. American Heritage Center. University of Wyoming. Laramie.

Helen Gahagan Douglas Papers. Carl Albert Research Center. University of Oklahoma. Norman.

John L. Savage Collection. American Heritage Center. University of Wyoming. Laramie.

Julius A. Krug Papers, 1936-1950. Library of Congress. Washington, D.C.

Office of International Affairs. United States Department of the Interior, Bureau of Reclamation. Washington, D.C.

Record Group 48. Records of the Secretary of the Interior. Entries 7, 768, and 779. National Archives and Records Administration. College Park, Maryland.

Record Group 59. Records of the Department of State. Decimal Files 1940-1944. National Archives and Records Administration. College Park, Maryland.

Record Group 115. Records of the Bureau of Reclamation. Office of the Chief Engineer, Accession Numbers 8NS-115-95-089, 8NS-115-95-090, "General Files", 8NN-115-92-130, "Project Histories", 8NS-115-95-122, "An Index of Cultural Resources Reports Located at the National Archives", 8NS-115-95-007, "Phoenix Office Reports", 8NN-115-85-008, "Project Histories", NRG-115-00-265, "General Correspondence 1933-1989", NRG-115-05-088, "Yangtze and Three Gorges Dam", 8NS-115-97-434, "Teton Dam Records Related to Failure, 1976-1981"; Entries 3, 7, 10. National Archives and Records Administration. Lakewood, Colorado.

Stewart Udall Papers. Western Waters Digital Library. University of Arizona. Tucson.

William E. Warne Papers. American Heritage Center. University of Wyoming. Laramie.

Secondary Sources

"$112 Million Contract for Third Powerplant: Reclamation's Largest—World's Largest," *Reclamation Era*, 56 (May 1970): 11-2.

"1963—First Payoff on the CRSP," *Reclamation Era*, 49 (August 1963): 57-60, 70-73.

"A. W. Simonds Aids New Zealand Development." *Reclamation Era*, 36 (April 1950): 84.

Adams, Darrell. "Wanted: Public Opinion." *Reclamation Era*, 63:1 and 2 (1977): 56-60.

Adas, Michael. *Dominance by Design: Technological Imperatives and America's Civilizing Mission*. Cambridge, Massachusetts: Harvard University Press, 2006.

"Additional Changes in Reclamation Personnel." *Reclamation Era,* 39 (October 1953): 174.

Anderson, Nancy K. "'Curious Historical Artistic Data:' *Art History and Western American Art*." In *Discovered Lands, Invented Pasts: Transforming Visions of the American West* by Jules David Prown, et al., chap. 1. New Haven: Yale University Art Gallery, 1992.

Anderson, Robert T. "Indian Water Rights, Practical Reasoning, and Negotiated Settlements." *California Law Review* 98:4 (August 2010): 1133-63.

"Are We Becoming a 'Have-Not' Nation?" *Current Events*, 45 (March 25-29, 1946): 201-2.

Arthur, Harold G. *Oral History Interview*. Transcript of tape-recorded Bureau of Reclamation Oral History Interviews conducted by Brit Allan Storey, senior historian, Bureau of Reclamation, during 1994 and 1995, in Denver, Colorado. Transcription by Barbara Heginbottom Jardee, Jardee Transcription, Tucson, Arizona, edited by Brit Allan Storey.

Arthur, William R. "MVA–Its Background and Issues." *Congressional Record* 29:1 (January 1950): 13-4, 32.

August Jr., Jack L. *Vision in the Desert: Carl Hayden and Hydropolitics in the American Southwest.* Fort Worth: Texas Christian University Press, 1999.

Banks, Frank A., Supervising Engineer, Columbia Basin Project. "New Empire in the Northwest." *Reclamation Era*, 32 (September 1946): 203-4.

The Barrier Between. United States Bureau of Reclamation, videocassette, 1960s.

Barsam, Richard. *The Vision of Robert Flaherty: The Artist as Myth and Filmmaker.* Bloomington: Indiana University Press, 1988.

Beard, Daniel P. Commissioner of Reclamation, "Blueprint for Reform: The Commissioner's Plan for Reinventing Reclamation," November 1, 1993.

———. "New Directions for the Bureau of Reclamation." In *Water Law: Trends, Policies, and Practices.* Kathleen Marion Carr and James D. Crammond, editors, 357-359. Chicago: ABA Section of Natural Resources, Energy, and Environmental Law, 1995, 357.

———. "Opening Statement of the Honorable Daniel P. Beard Commissioner of Reclamation before the Subcommittee on Energy and Water Development of the Senate Committee on Appropriations," May 2, 1995. Federal Document Clearing House Congressional Testimony, LexisNexis Congressional. http://0-web.lexis-nexis.com.

———. *Oral History Interview*, Transcript of tape-recorded Bureau of Reclamation Oral History Interviews conducted by Brit Allan Storey, senior historian, Bureau of Reclamation, from 1993 to 1995, in Washington, D.C. Transcription by Barbara Heginbottom Jardee, Jardee Transcriptions, Tucson, Arizona. Edited by Brit Allan Storey.

———. *Passage of the Central Valley Improvement Act, 1991-1992: The Role of George Miller.* Oral History Interview conducted by Malca Chall in 1995. Berkeley: Regional Oral History Office, Bancroft Library, University of California, 1996.

Berkman, Richard L. and Kip Viscusi. *Damming the West: The Nadar Task Force Report on the Bureau of Reclamation.* Washington, D.C.: Center for the Study of Responsive Law, November 1971.

Betsky, Aaron. "Emptiness on the Range: Western Spaces." In *Crossing the Frontier: Photographs of the Developing West, 1849 to the Present* by Sandra Phillips,

Richard Rodriguez, Aaron Betsky, and Eldridge M. Moore, 54-65. San Francisco: San Francisco Museum of Modern Art; Chronicle Books, 1996.

Biggs, David. "Reclamation Nations: The Bureau of Reclamation's Role in Water Management and Nation Building in the Mekong Valley, 1945-1975." *Comparative Technology Transfer and Society* 4 (December 2006): 225-46.

———. *Quagmire: Nation Building and Nature in the Mekong Delta.* Seattle: University of Washington Press, 2010.

Billington, David P., Donald C. Jackson, Martin V. Melosi. *The History of Large Federal Dams: Planning, Design, and Construction in the Era of Big Dams.* Denver, Colorado: United States Department of the Interior, Bureau of Reclamation, 2005.

Bingham, Jonathon B. *Shirt-Sleeve Diplomacy: Point 4 in Action.* New York: J. Day Company, 1954.

Black and White Construction Film–Hoover Dam. United States Bureau of Reclamation, videocassette, no date.

Black, Eugene R. *Alternative in Southeast Asia.* New York: Frederick A. Praeger, Publishers, 1969.

Blanchard, C. J. "The Call of the West: Homes are being made for Millions of People in the Arid West." *National Geographic Magazine*, 20 (May 1909): 403-37.

———. "Current Comments." *Reclamation Record*, 6 (June 1915): 248-9.

———. "Current Comments." *Reclamation Record*, 7 (February 1916): 51-2.

———. "Current Comments." *Reclamation Record*, 7 (July 1916): 290-1.

———. "Current Comments." *Reclamation Record*, 12 (June 1921): 253-4.

———. "Current Comments: Gathered from the Project Press and People." *Reclamation Record*, 9 (September 1918): 410-3.

———. "Current Comments: Gathered from the Project Press and People." *Reclamation Record*, 11 (August 1920): 358.

———. "Current Comments: Gathered from the Project Press and People." *Reclamation Record*, 13 (September 1922): 214.

———. "The Government Uses Motion Pictures." *Scientific American Supplemental*, 80 (August 1915): 120-1.

———. "Home-Making by the Government: An account of Eleven Immense Irrigation Projects to be Opened in 1908." *National Geographic Magazine*, 19 (April 1908): 250-87.

———. "Millions for Moisture: An Account of the Work of the U. S. Reclamation Service." *National Geographic Magazine*, 18 (April 1907): 217-43.

———. "Reclaiming the Desert." *The Mentor*, 6 (October 15, 1918): 1-11.

———. "The Spirit of the West: The Wonderful Agricultural Development Since the Dawn of Irrigation." *National Geographic Magazine*, 21 (April 1910): 334-60.

———. "Winning the West: An Account of the Marvelous Progress of Our Reclamation Service in Reclaiming the Desert." *National Geographic Magazine*, 17 (February 1906): 82-99.

Blonk, Hu. "A Braggart's Delight: The Startup of the Third Powerplant." *Reclamation Era*, 61 (Autumn 1975): 1-6.

Bloodworth, Gina and James White. "The Columbia Basin Project: Seventy-Five Years Later." *Yearbook of the Association of Pacific Coast Geographers*. Volume 70, 2008, 96-111.

Boffey, Philip M. "Teton Dam Verdict: A Foul-Up by the Engineers." *Science*, 195:4275 (January 21, 1977): 270-2.

The Boise Project: Gift of Life. United States Bureau of Reclamation, videocassette, 1983.

Bonner, Robert E. *William F. Cody's Wyoming Empire: The Buffalo Bill Nobody Knows*. Norman: University of Oklahoma Press, 2007.

Boulder Dam advertisements: *Fortune*, 11:4 (April, 1935): 5; *Fortune*, 14:4 (October 1936): 5.

"Boulder Tourist Travel." *Reclamation Era*, 25 (August 1935): 160.

"Boulder Dam Tourist Accommodations." *Reclamation Era*, 27 (July, 1937): 149.

Broder, Patricia Janis. *The American West: The Modern Vision.* Boston: A New York Graphic Society Book; Little, Brown and Company, 1984.

Brohier, R. L. *Ancient Irrigation Works in Ceylon*, 3 parts. Columbo: The Ceylon Government Press, 1934, reprinted 1949.

Brooks, Karl Boyd. *Public Power, Private Dams: The Hell's Canyon High Dam Controversy.* Seattle: University of Washington Press, 2006.

———, editor. *The Environmental Legacy of Harry S. Truman.* Kirksville: Truman State University Press, 2009.

Brower, David. *For Earth's Sake: The Life and Times of David Brower.* Salt Lake City: Peregrine Smith Books, 1990.

———. Forward in *The Place No One Knew: Glen Canyon on the Colorado* by Eliot Porter. San Francisco: Sierra Club, 1963.

"The Bureau is The Bureau … Again." *Reclamation Era*, 66:2 and 3 (Spring/Summer 1981): 1.

Burton, Lloyd. *American Indian Water Rights and the Limits of the Law.* Lawrence: University Press of Kansas, 1991.

Busch, Leo A. "Experiencing the Teton Dam Failure." *International Water Power & Dam Construction*, (August 4, 2002): 38-41.

Bustard, Bruce I. *Picturing the Century: One Hundred Years of Photography from the National Archives.* Washington, D.C.: National Archives and Records Administration in Association with the University of Washington Press, 1999.

Buttehdahl, Oscar, J. "Corralling the Colorado: Part II–Empire Builder." *Reclamation Era*, 32 (October 1949): 218-220, 229.

Cadillac Desert. Produced by KTEH/San Jose, Jon Else, writer, director, and producer. Deborah Hoffman, editor, videocassette, 1997.

California Department of Fish and Game. Citizens Advisory Committee on Salmon and Steelhead Trout. *Report on California Salmon and Steelhead Trout: The Time is Now!*, 1975.

Calkins, Hugh G. Human Land Problems in Three Regions." *Proceedings of the Inter-American Conference on Conservation of Renewable Natural Resources*,

Denver, Colorado, September 7-20, 1948. Department of State, Publication 3382, International Organization and Conference Series II, American Republics 4. Division of Publications, Office of Public Affairs.

"Camera-Eye View of Water Development." *Reclamation Era*, 47 (February 1961): 23.

Cammerer, Arno B. Director, National Parks Service. "Building a Playground at Boulder Dam." *Reclamation Era*, 28 (November 1938): 222-4.

Cannon, Brian Q. "Creating a 'New Frontier' Opportunity: World War II Veterans and the Campaign for Western Homesteads." Paper presented at the Pacific Coast Branch of the American Historical Association, Honolulu, Hawaii, August 2003.

———. *Remaking the Agrarian Dream: New Deal Rural Resettlement in the Mountain West*. Albuquerque: University of New Mexico Press, 1996.

———. *Reopening the Frontier: Homesteading in the Modern West*. Lawrence: University Press of Kansas, 2009.

———. "'We Are Now Entering a New Era:' Federal Reclamation and the Fact Finding Commission, 1923-1924." *Pacific Historical Review* 66 (May 1997): 185-207.

Carle, David. *Water and the California Dream: Choices for the New Millennium*. San Francisco, California: Sierra Club Books, 2000.

Carothers, Steven W. and Bryan T. Brown. *The Colorado River through the Grand Canyon: Natural History and Human Change*. Tucson: University of Arizona Press, 1991.

Carr, Kathleen Marion and James D. Crammond, editors. *Water Law: Trends, Policies, and Practices*. Chicago: ABA Section of Natural Resources, Energy, and Environmental Law, 1995.

Carrels, Peter. *Uphill Against Water: The Great Dakota Water War*. Lincoln: University of Nebraska Press, 1999.

Carson, Donald W. and James W. Johnston. *Mo: The Life and Times of Morris K. Udall*. Tucson: University of Arizona Press, 2001.

Carter, C. H. "Key Contract Awarded for Colorado River Storage Project." *Reclamation Era*, 43 (August 1957): 53-5, 73-4.

Carter, James E. *Public Papers of the Presidents of the United States.* Book I—January 20 to June 24, 1977. Washington, D.C.: United States Government Printing Office, 1997.

Carter, Luther J. "Water Projects Dispute: Carter and Congress Near a Showdown." *Science*, 196:4296 (June 17, 1977): 1303-1315.

———. "Dams and Wild Rivers: Looking Beyond the Pork Barrel." *Science*, 158:3798 (October 13, 1967): 233-6, 241-2.

Cash, W. J. *The Mind of the South.* New York: A. A. Knopf, 1941.

Caylor, Ann. "'A Promise Deferred': Federal Reclamation of the Colorado River Indian Reservation." *Pacific Historical Review* 69:2 (May 2000): 193-215.

A Century of Water for the West, 1902-2002. United States Department of the Interior. Bureau of Reclamation, videocassette, 2002.

"The Changing Years." *Reclamation Era*, 32 (May 1946): 113.

Chapman, Oscar L. "From Nebraska to the Sea." *Reclamation Era*, 36 (March 1950): 45-7.

Checchio, Elizabeth and Bonnie G. Colby. *Indian Water Rights: Negotiating the Future.* Tucson: Water Resource Center, University of Arizona College of Agriculture, June 1993.

"CINE—What it is. What it does." www.cine.org/history.

Clark, Jeanne Nienaber, and Daniel McCool, *Staking Out the Terrain: Power Differentials Among Natural Resource Management Agencies.* Albany: State University of New York, 1985.

Coate, Charles. "'The New School of Thought': Reclamation and the Fair Deal, 1945-1953." *Journal of the West* 22:2 (April 1983): 58-63.

———. "'The Biggest Water Fight in American History': Stewart Udall and the Central Arizona Project." *Journal of the Southwest* 37:1 (Spring 1995): 79-101.

Coffeen, William R. "The Effects of the Central Arizona Project on the Fort McDowell Indian Community." *Ethnohistory* 19:4 (Fall 1972): 345-77.

Colby, Bonnie G., John E. Thorson, Sarah Britton. *Negotiating Tribal Water Rights: Fulfilling Promises in the Arid West*. Foreword by David Getches. Tucson: University of Arizona Press, 2005.

Collins, Robert M. *More: The Politics of Economic Growth in Postwar America*. New York: Oxford University Press, 2000.

"Colorado River Storage Project Heads New Reclamation Authorization." *Reclamation Era*, (May 1956): 54-5.

Colorado River Water Users Association. *Proceedings, Second Annual Conference, Colorado River Water Users Association*. Salt Lake City, Utah, February 11, 12, and 13, 1946.

"Columbia Basin Program Speeded." *Reclamation Era*, 32 (July 1946): 148.

Committee on the Future of Irrigation in the Face of Competing Demands, Water Science and Technology Board, Commission on Geoscience, Environment, and Resources, National Research Council. *A New Era for Irrigation*. Washington, D.C.: National Academy Press, 1996.

"Commissioner Reappointed." *Reclamation Era*, 47 (February 1961): 2.

Condit, Doris M. *The Test of War, 1950-1953*. Alfred Goldberg, editor, *History of the Office of the Secretary of Defense*. Volume II. Washington, D.C.: Historical Office, Office of the Secretary of Defense, 1988.

Cone, Joseph and Sandy Ridlington, editors. *The Northwest Salmon Crisis: A Documentary History*. Corvallis, Oregon: Oregon State University Press, 1996.

Cone, Joseph. *A Common Fate: Endangered Salmon and the People of the Pacific Northwest*. Corvallis, Oregon: Oregon State University Press, 1996.

Conkin, Paul K. *Tomorrow a New World: The New Deal Community Program*. Ithaca, New York: Cornell University Press, 1959.

Contreras, Belisario R. *Tradition and Innovation in New Deal Art*. Lewisburg: Bucknell University Press, London and Toronto: Associated University Presses, 1983.

Cooper, Dana Sebren and D. Michael Harvey. "An Upstream Swim: The Crafting and Passage of the Central Valley Project Improvement Act." In *Water Law: Trends, Politics, and Practices*, Kathleen Marion Carr and James D. Crammond, editors, 253-64. Chicago: ABA Section of Natural Resources, Energy, and Environmental Law, 1995.

"Corfitzen Goes to Greece." *Reclamation Era*, 33 (October 1947): back of front cover.

Corfitzen, William E. "Ideas Go Traveling." *Reclamation Era*, 32 (May1964): 102-3.

Crampton, C. Gregory *Anthropological Papers: Outline History of the Glen Canyon Region, 1776-1922*, Charles E. Dibble, editor, Number 42 (Glen Canyon Series Number 9) University of Utah, Department of Anthropology, September 1959.

———. *Ghosts of Glen Canyon: History Beneath Lake Powell*. St. George, Utah: Publishers Place, Inc, 1986.

———. *Historic Sites in Glen Canyon Mouth of San Juan River to Lee's Ferry*, Charles E. Dibble, editor, Number 46 (Glen Canyon Series Number 12) University of Utah, Department of Anthropology, June 1960.

———. "History in Glen Canyon." *Reclamation Era*, 45 (May 1959): 41-2, 52.

Cronon, William. Foreword in *Making Salmon: An Environmental History of the Northwest Fisheries Crisis* by Joseph E. Taylor III. Seattle: University of Washington Press, 1991.

———. "Telling Tales on Canvas: *Landscapes of the Frontier Change*." In *Discovered Lands, Invented Pasts: Transforming Visions of the American West* by Jules David Prown, et al., chap. 2. New Haven: Yale University Art Gallery, 1992.

Cullather, Nick. "Damming Afghanistan: Modernization in a Buffer State." *Journal of American History* 89 (September 2002): 512-37.

———. *The Hungry World: America's Cold War Battle Against Poverty*. Cambridge, Massachusetts: Harvard University Press, 2010.

"The Dam." *Fortune*, 8:3 (September 1933): 74-88.

Dean, Robert. "'Dam Building Still Had Some Magic Then': Stewart Udall, the Central Arizona Project, and the Evolution of the Pacific Southwest Water Plan, 1963-1968." *Pacific Historical Review* 66:1 (February 1997): 81-98.

Dean, William Harper. "The Master Movie Maker." *Technical World Magazine*, 22 (January 1915): 730-2, 788.

de Chassey, Eric, editor. *American Art 1807-1947 from Winslow Homer to Jackson Pollock*. Paris: Reunion des Musees Nationaux; New York: Distributed by Harry N. Abrams, 2002.

DeGroot, Gerard. *The Bomb: A Life* (Cambridge, Massachusetts: Harvard University Press, 2005).

Dejong, David H. "'Abandoned Little by Little:' The 1914 Pima Adjudication Survey, Water Deprivation, and Farming on the Pima Project." *Agricultural History* 81:1 (Winter 2007): 36-69.

———. "'The Sword of Damocles?' The Gila Indian Community Water Settlement Act of 2004 in Historical Perspective." *Wicazo Sa Review* 22:2 (Fall 2007): 57-92.

DeLuna, P. R. "Bureaucratic Opposition as a Factor in Truman's Failure to Achieve a Columbia Valley Authority." *Historical Papers/Communications historiques* 10:1 (1975): 231-56.

De Luna, Phyllis Komarek. *Public versus Private Power during the Truman Administration: A Study of Fair Deal Liberalism*. New York: Peter Lang Publishing, Inc., 1997.

de Roos, Robert. *The Thirsty Land: The Story of the Central Valley Project*. Stanford, California: Stanford University Press, 1948.

"Design for Modern Farm Living." *Reclamation Era*, 38 (May 1952): 103, 112.

"Design Work Decentralized." *Reclamation Era*, 33 (October 1947): 227.

De Voto, Bernard. "Shall We Let Them Ruin Our National Parks." *Saturday Evening Post*, 223 (July 22, 1950): 17-19, 42-4, 46, 48.

———. *The Western Paradox: A Conservation Reader*. Douglas Brinkley and Patricia Nelson Limerick editors, with a Foreword by Arthur Schlesinger, Jr. New Haven: Yale University Press, 2000.

DeWitt, John. "Reclamation Launches Art Program." *Reclamation Era*, 56 (February 1970): 5-8.

———. "Reclamation and the Creative Artist." *Reclamation Era*, 59 (Spring 1973): 12-15.

Dippie, Brian W. "Western Art and the Dynamics of Change." In *Discovered Lands, Invented Pasts: Transforming Visions of the American West* by Jules David Prown, et al., chap. 3. New Haven: Yale University Art Gallery, 1992.

Doerksen, Harvey R. *Columbia River Interstate Compact, Politics of Negotiations.* Pullman, Washington: State of Washington Water Research Center, Washington State University, and the University of Washington, August 1972.

Dominy, Floyd E. "A New Power Giant Materializes on the West Coast." *Reclamation Era*, 51 (August 1965): 63-67.

———. *Oral History Interviews.* Interviews conducted by Brit Allan Storey, senior historian, Bureau of Reclamation, April 6, 1994 and April 8, 1996, at Bellevue Farm in Boyce, Virginia.

"Dominy Named Associate Commissioner." *Reclamation Era*, 44 (May 1958): 32.

"Dominy Named Reclamation Assistant Commissioner for Legislative Liaison." *Reclamation Era*, 43 (November 1957): 84.

Donahue, John M. and Barbara Rose Johnston, editors. *Water, Culture, and Power: Local Struggles in a Global Context.* Washington, D.C.: Island Press, 1998.

Dondero, George A, U.S. Representative from Michigan and Chairman of the Public Works Committee. "Wanted: A New Federal Policy." *Public Utilities Fortnightly*, 42 (September 9, 1948): 333-42.

Dorman, Robert L. *Revolt of the Provinces: The Regionalist Movement in America, 1920-1945.* Chapel Hill: University of North Carolina Press, 1993.

Downey, Sheridan, United States Senator. *They Would Rule the Valley.* San Francisco: Sheridan Downey, 1947.

Drennen, Jr., William M. Introduction in *FDR's Moviemaker: Memoirs & Scripts* by Pare Lorentz. Reno: University of Nevada Press, 1992.

Duck, Donald J., Deputy Director of Design and Construction. *Construction of Grand Coulee Third Powerplant.* Denver, Colorado: United States Department of the Interior, Bureau of Reclamation, 1975.

DuMars, Charles T., Marilyn O'Leary, Albert E. Utton. *Pueblo Indian Water Rights: Struggles for a Precious Resource.* Tucson: University of Arizona Press, 1984.

Dunham, Dick and Vivian. *Flaming Gorge Country: The Story of Daggett County, Utah*. Denver: Richard R. Dunham; printed by Eastwood Printing and Publishing Company, 1997.

Duvall, C. Dale. *Oral History Interview*. Transcript of tape-recorded Bureau of Reclamation Oral History Interview conducted by Brit Allan Storey, senior historian, Bureau of Reclamation, January 23, 1993, in Washington, D.C. Edited by Brit Allan Storey.

Eastman, Adam R. "The Central Utah Project, Vernal Unit." Denver: Bureau of Reclamation History Program, 2006.

———. "From Cadillac to Chevy: Environmental Concerns, Compromise, and the Central Utah Completion Act," Master's Thesis, Brigham Young University, August 2006.

"ECA Sponsors Greek Engineering Visit to Reclamation Area." *Reclamation Era*, 35 (June 1949): 137.

Echeverria, John D., Pope Barrows, Richard Roos-Collins. *Rivers at Risk: The Concerned Citizens Guide to Hydropower.* Foreword by Stewart L Udall. Washington, D.C.: Island Press, 1982.

"Echo Park-Split Mountain Dams Approved." *Reclamation Era*, 36 (August 1950): 148.

Edmonds, Carol. *Wayne Aspinall: Mr. Chairman.* Lakewood, Colorado: Great American Printing Co., Crown Point, Inc., Publishers, 1980.

Ellis, Don D. "The First Week—Saving Crops." *Reclamation Era*, 62 (Autumn 1976): 14-8.

Ekbladh, David. *The Great American Mission: Modernization and the Construction of an American World Order*. Princeton, New Jersey: Princeton University Press, 2010.

———. "'Mr. TVA': Grassroots Development, David Lilienthal, the Rise and Fall of the Tennessee Valley Authority as a Symbol of U.S. Overseas Development." *Diplomatic History* 26:3 (Summer 2002): 335-74.

Espeland, Wendy Nelson. *The Struggle for Water: Politics, Rationality, and Identity in the American Southwest*. Chicago: University of Chicago Press, 1998.

Etulain, Richard W. "Gerald D. Nash and the Twentieth-Century American West." In *The American West in 2000: Essays in Honor of Gerald D. Nash*. Richard W. Etulain and Ferenc Morton Szasz, editors, 186-198. Albuquerque: University of New Mexico Press, 2003.

——— and Ferenc Morton Szasz, editors. *The American West in 2000: Essays in Honor of Gerald D. Nash*. Albuquerque: University of New Mexico Press, 2003.

"Extraordinary Photographer Captures History in the Making." *Salt River Project Pulse*, (January 28, 1982): 4.

Farmer, Jared. *Glen Canyon Dammed: Inventing Lake Powell and the Canyon Country*. Tucson: University of Arizona Press, 1999.

Ferrell, John. "Developing the Missouri: South Dakota and the Pick-Sloan Plan." *South Dakota History* 19:3 (1989): 306-41.

Field, Alexander J. "The Most Technologically Progressive Decade of the Century." *American Economic Review* 93 (September 2003): 1399-1413.

Finsterbusch, Kurt, Jasper Ingersoll, and Lynn Llewellyn, editors. *Methods for Social Analysis in Developing Countries*. Boulder, Colorado: Westview Press, 1990.

"First Concrete Placed at Glen Canyon." *Reclamation Era*, 46 (August 1960): 72.

"A First for the First Lady." *Reclamation Era*, 50 (November 1964): 77-8.

Firth, Robert E. *Public Power in Nebraska: A Report on State Ownership* (Lincoln: University of Nebraska Press, 1962).

"Flaming Gorge Power is Signaled by President Kennedy." *Reclamation Era*, (November 1963): 106.

Ford, Gerald R. "583 Remarks at a Meeting to Discuss the Collapse of the Teton Dam in Idaho, June 8, 1976. Public Papers of Gerald R. Ford. American Reference Library—Primary Source Documents, 2001. Database: MasterFILE Premier.

———. "758 Remarks Upon Signing the Teton Dam Disaster Assistance Bill, September 7, 1976." Public Papers of Gerald R. Ford. American Reference Library—Primary Source Documents, 2001. Database: MasterFILE Premier.

"Foreign Activities." *Reclamation Era*, 37 (December 1951): 291.

Fradkin, Philip L. *A River No More: The Colorado River and the West*. New York: Alfred A. Knopf, 1981.

Francis, Rell G. *The Utah Photographs of George Edward Anderson*. Lincoln: University of Nebraska Press, 1979.

———. "Views from Mormon Country: The Life and Photographs of George Edward Anderson." *American West* 15 (November/December 1978): 14-29.

Frederick, Kenneth D. Foreword in *Markets for Federal Water: Subsidies, Property Rights, and the Bureau of Reclamation* by Richard W. Wahl. Washington, D.C.: Resources for the Future, 1989.

——— and Roger A. Sedjo, editors. *America's Renewable Resources: Historic Trends and Current Challenges*. Washington, D.C.: Resources for the Future, 1991.

Freeland, Kathleen B. "Examining the Politics of Reclamation: The 1944 Acreage Limitation Debate in Congress." *The Historian* 67 (Summer 2005): 217-33.

Friedman, Martin. *Charles Sheeler*. New York: Watson-Guptill Publications, 1975.

Friedman, Norman. *The Fifty-Year War: Conflict and Strategy in the Cold War*. Annapolis, Maryland: Naval Institute Press, 2000.

"Fundamentals of Irrigation." *Reclamation Era*, 30 (March, 1940): 83.

Garone, Philip. *The Fall and Rise of the Wetlands of California's Great Central Valley*. Berkeley: University of California Press, 2011.

Gelburd, Diane. "Reclamation's Cultural Resource Legacy." *Reclamation Era*, 63:1 and 2 (1977): 78-81.

Getches, David H. "Indian Water Rights Conflicts in Perspective." In *Indian Water in the New West*, Thomas R. McGuire, William B. Lord, and Mary G. Wallace, editors, 7-26. Tucson: University of Arizona Press, 1993.

Glaha, Ben D. "The People Who Have Gone Away." *Reclamation Era*, 32 (April 1942): 51-3.

Glaser, Leah S. *Electrifying the Rural American West: Stories of Power, People, and Place*. Lincoln: University of Nebraska Press, 2009.

———. “The Navajo Indian Irrigation Project.” Denver: Bureau of Reclamation History Program, 1998.

Glass, Todd G. “The 1992 Omnibus Water Act: Three Rubrics of Reclamation Reform.” *Ecology Law Quarterly* 22:1 (February 1995): 143-211.

“Glen Canyon Reservoir Named Lake Powell.” *Reclamation Era*, 46 (February 1960): 27.

Goldberg, Alfred, editor. *History of the Office of the Secretary of Defense*. Volume II. Washington, D.C.: Historical Office, Office of the Secretary of Defense, 1988.

Goldschmidt, Walter. *As You Sow: Three Studies in the Social Consequences of Agribusiness*. Foreword by Gaylord Nelson (Montclair, New Jersey: Allenheld Osmun, 1978).

Goldsmith, Edward and Nicholas Hildyard. *The Social and Environmental Effects of Large Dams*. Volume I—A Report to the European Ecological Action Group (ECOROPA). Camelford, Cornwall, United Kingdom: Wadebridge Ecological Centre; Wadebridge, Cornwall: Quintrell & Co. Ltd., 1984.

Golzé, Alfred R. *Reclamation in the United States*. New York: McGraw-Hill Book Co., Inc., 1952.

Goodrich, Lloyd. Foreword in *The American Artist and Water Reclamation* by Bureau of Reclamation. Washington, D. C.: Government Printing Office, 1973.

Gopalakrishnan, Chennat. “The Doctrine of Prior Appropriation and Its Impact on Water Development: A Critical Survey.” *American Journal of Economics and Sociology* 32:1 (January 1973): 61-72.

The Great Web of Water: The Central Valley Project. United States Department of the Interior. Bureau of Reclamation, videocassette, early 1980s.

The Green Echo of Snow. United States Bureau of Reclamation and the Northern Colorado Conservancy District, videocassette, 1970s.

Green Fields. Thomas J. Fields, producer, videocassette, 1963.

Gulick, Luther. *Administrative Reflections from World War II*. Birmingham: University of Alabama Press, 1948.

Hadeed, S. J. “Several Water Resources Projects Survive Carter Cutback.” *Journal (Water Pollution Control Federation)* 49:5 (May 1977): 727-30.

Hammond, Anne. *Ansel Adams: Divine Performance*. New Haven: Yale University Press, 2002.

Hansen, Oskar J. W. "From Bones of Water Pipe and Wood: The Sculptures at Boulder Dam, Part III." *Reclamation Era*, 32 (April 1942): 78-9.

———. "A Split Second Petrified on the Face of the Universal Clock: The Sculptures at Boulder Dam—Part II." *Reclamation Era*, 32 (March 1942): 57-59.

———. "With the Look of Eagles: The Sculptures at Boulder Dam–Part I." *Reclamation Era*, 32 (February 1942): 30-32.

Harden, Blaine. *A River Lost: The Life and Death of the Columbia*. New York: W. W. Norton and Company, 1996.

Harding, S. T. *Water in California*. Palo Alto, California: N-P Publications, 1960.

Harper, William Dean. "The Master Movie Maker." *Technical World Magazine*, 22 (January 1915): 730-2, 788.

Harris, Jonathan. *Federal Art and National Culture: The Policies of Identity in New Deal America*. New York: Cambridge University Press, 1995.

Harvey, Mark W. T. "Battle for Wilderness: Echo Park and the Birth of the Modern Wilderness Movement." In *A Sense of the American West: An Anthology of Environmental History*, James E. Sherow, editor, 181-200. Albuquerque: University of New Mexico Press, 1998.

———. "North Dakota, the Northern Plains, and the Missouri Valley Authority." *North Dakota History* 59:3 (1992): 28-39.

———. *A Symbol of Wilderness: Echo Park and the American Conservation Movement*. Seattle: University of Washington Press, 1994.

———. *Wilderness Forever: Howard Zahniser and the Path to the Wilderness Act*. Foreword by William Cronon. Seattle: University of Washington Press, 2005.

Hays, Samuel P. "From Conservation to Environment: Environmental Politics in the United States since World War II." *Environmental Review* VI (1982): 14-41.

Healing the Water. A KNPB public television special. Sue Pearson Atkins, producer/writer. Produced by the Water Education Foundation, videocassette, 1997.

Healy, Paul F. "Our Most Arrogant Bureaucrat." *Saturday Evening Post*, 224 (April 19, 1952): 46-7, 145-50.

"Here and There—Plan Doubling Third Powerplant at Grand Coulee," *Reclamation Era*, 56 (May 1967): 52.

Higginson, R. Keith. "New Direction for Reclamation." *Reclamation Era*, 63:1 and 2 (1977): 48-52.

———. *Oral History Interviews*. Transcript of Tape-recorded Bureau of Reclamation Oral History Interview conducted by Brit Allan Storey, senior historian, Bureau of Reclamation, March 22, 1995 and April 19, 1995.

Hogan, Harry J. *The Acreage Limitation in the Federal Reclamation Program*. Arlington, Virginia: National Water Commission, 1972.

Hogan, Michael J. *The Marshall Plan: America, Britain, and the Reconstruction of Western Europe, 1947-1952*. New York: Cambridge University Press, 1987.

Hollearin, Thomas M. "Construction Advances on the Pacific Northwest-Pacific Southwest Intertie." *Transmission & Description*, (July 1967): 66-70.

The Hoover Commission Report: On Organization of the Executive Branch of the Government. New York: McGraw-Hill Book Company, Inc., 1949.

Huffaker, Ray G. and B. Delworth Gardener, "The Distributions of Economic Rents Arising from Subsidized Water When Land is Leased." *American Journal of Agricultural Economics* 68:2 (May 1986): 306-12.

———. "The 'Hammer' Clause of the Reclamation Reform Act of 1982." *Natural Resources Journal* 26:1 (Winter 1986): 41-67.

Hughes, Thomas P. *Human-Built World: How to Think About Technology and Culture*. Chicago: The University of Chicago Press, 2004.

Hundley Jr., Norris. "The Dark and Bloody Ground of Indian Water Rights: Confusion Elevated to Principle." *Western Historical Quarterly* 9:4 (October 1978): 454-82.

———. *Dividing the Waters: A Century of Controversy Between the United States and Mexico*. Berkeley and Los Angeles: University of California Press, 1966.

———. *The Great Thirst: Californians and Water, A History.* Revised edition. Berkeley: University of California Press, 2001.

———. "The Politics of Water and Geography: California and the Mexican-American Treaty of 1944." *Pacific Historical Review* 36:2 (May 1967): 209-26.

———. *Water and the West: The Colorado River Compact and the Politics of Water in the West.* Berkeley: University of California Press, 1975.

———. "The West Against Itself: The Colorado River—An Institutional History." In *New Courses for the Colorado River: Major Issues for the Next Century*, Gary D. Weatherford and F. Lee Brown, editors, 9-49. Foreword by Governor Bruce Babbitt. Albuquerque: University of New Mexico Press, 1986.

———. "The 'Winters' Decision and Indian Water Rights: A Mystery Reexamined." *Western Historical Quarterly* 13:1 (January 1982): 17-42.

Hurt, R. Douglas. *Indian Agriculture in America: Prehistory to the Present* (Lawrence: University Press of Kansas, 1987).

"Impending Reclamation Disaster May Be Averted." *New Reclamation Era*, 15 (May 1925): 68.

"India Studies Reclamation." *Reclamation Era*, 32 (July 1946): 143.

Ingram, Helen. "Water Rights in the Western States." *Proceedings of the Academy of Political Science* 34:3 (1982): 134-43.

"Irrigation Warfare Renewed: Old Disputes over U.S. Limitations on Size of Tracts Eligible for Water." *Business Week,* (November 17, 1945): 19-20.

Isé, Sabrina and David L. Sunding. "Reallocating Water from Agriculture to the Environment under a Voluntary Purchase Plan." *Review of Agricultural Economics* 20:1 (Spring-Summer 1998): 214-26.

"Its Hoover Dam." *Reclamation Era*, (June 1947):143.

Jacobsen, Judith Eva. "A Promise Made: The Navajo Indian Irrigation Project and Water Politics in the American West." PhD diss., University of Colorado, 1989.

Jacobson, C. B. "Largest Reclamation Development Underway." *Reclamation Era*, 43 (February 1957): 1-3.

"Jean R. Walton Named Construction Engineer on Flaming Gorge." *Reclamation Era*, 42 (November 1956): 92.

Jennings, Jesse D. *Glen Canyon: An Archaeological Summary*. Foreword by Don D. Fowler. Salt Lake City: University of Utah Press, 1998.

Johnson, Rich. *The Central Arizona Project, 1918-1968*. Tucson: University of Arizona Press, 1977.

"Joint Investigation: Columbia Basin Project." *Reclamation Era*, 30 (August 1940): 219.

Kathka, David Arlin. "The Bureau of Reclamation in the Truman Administration," Ph.D. diss., University of Missouri, Columbia, 1976.

Keys III, John W. *Oral History Interviews*. Transcript of Tape-recoded Bureau of Reclamation Oral History Interview conducted by Brit Allan Storey, senior historian, Bureau of Reclamation, from 1994 to 2006, in Denver, Colorado; Boise, Idaho; Washington, D.C.; and Moab, Utah.

Kinsey, Joni Louise. *Thomas Moran and the Surveying of the American West*. Washington, D.C.: Smithsonian Institution Press, 1992.

Kirkendall, Richard S. "Social Science in the Central Valley: An Episode." *California Historical Society Quarterly* 43:3 (September 1964): 195-218.

Klein, Christina. *Cold War Orientalism: Asia in the Middlebrow Imagination, 1945-1961*. Berkeley: University of California Press, 2003.

Kociumbas, Jan. "Myths, Men and History: The Making of the Snowy Mystique." *Journal of the Royal Australian Historical Society* 88:1 (2002): 1-19.

Koppes, Clayton R. "Oscar Chapman: A Liberal at the Interior Department, 1933-1953." PhD diss., University of Kansas, 1974.

———. "Public Water, Private Land: Origins of the Acreage Limitation Controversy, 1933-1953." *Pacific Historical Review* 47:4 (November 1978): 607-36.

Kotok, E. I. Forest Service, U.S. Department of Agriculture, Washington, D.C., "The Ecological Approach to Conservation Programs," 472 in *Proceedings of the Inter-American Conference on Conservation of Renewable Natural Resources*. Denver, Colorado, September 7-20, 1948. Department of State, Publication 3382,

International Organization and Conference Series II, American Republics 4: Division of Publications, Office of Public Affairs.

Krug, Julius A. "My New Job." *Reclamation Era*, 32 (May 1946): 93-4.

Kupel, Douglas E. *Fuel for Growth: Water and Arizona's Urban Environment.* Tucson: University of Arizona Press, 2003.

Kuzmiak, D. T. "The American Environmental Movement." *The Geographic Journal* 157:3 (November 1991): 265-78.

"Labor's Memorial to Its Dead at Boulder Dam." *Reclamation Era*, 25 (July 1935): 143.

Lamar, Howard. "Looking Backward, Looking Forward: Selected Themes in Western Art Since 1900." In *Discovered Lands, Invented Pasts: Transforming Visions of the American West* by Jules David Prown, et al., chap. 6. New Haven: Yale University Art Gallery, 1992.

Lang, William L. "Failed Federalism: The Columbia Valley Authority and Regionalism," in *The Great Northwest: The Search for Regional Identity*, edited by William G. Robbins. Corvallis, Oregon: Oregon State University Press, 2001, 66-79.

——— and Robert C. Carriker, editors, *Great River of the West: Essays on the Columbia River.* Seattle: University of Washington Press, 1999.

Latham, Michael E. *Modernization as Ideology: American Social Science and "Nation Building" in the Kennedy Era.* Chapel Hill: University of North Carolina Press, 2000.

Lawson, Michael L. *Dammed Indians Revisited: The Continuing History of the Pick-Sloan Plan and the Missouri River Sioux.* Pierre: South Dakota: South Dakota State Historical Society Press, 2009.

———. *Dammed Indians: The Pick-Sloan Plan and the Missouri River Sioux.* Norman: University of Oklahoma Press, 1982.

Lawson, Sidney D. "The Meaning of Power Utilization." *Reclamation Era*, 32 (December 1946): 266-267.

Leach, William. *Country of Exiles: The Destruction of Place in American Life.* New York: Pantheon Books, 1999.

Lee, Lawrence B. "California Water Politics: Opposition to the CVP, 1944-1980." *Agricultural History* 54:3 (July 1980): 402-23.

Leebaert, Derek. *The Fifty-Year Wound: How America's Cold War Shapes Our World* (Boston: Back Bay Books, 2002).

Leighninger Jr., Robert D. *Long-Range Public Investment: The Forgotten Legacy of the New Deal.* Columbia, South Carolina: University of South Carolina Press, 2006.

Leveritt, John A. "Camp Life at Davis Dam." *Reclamation Era*, 32 (September 1946): 296-7.

Leydet, Francois. *Time and the River Flowing: Grand Canyon.* Edited by David Brower. San Francisco: Sierra Club; New York: Ballantine Books, 1968.

"L. F. Wylie Heads Glen Canyon Dam Job." *Reclamation Era*, (August 1956): 77.

Lichatowich, Jim. *Salmon without Rivers: A History of the Pacific Salmon Crisis.* Washington, D.C.: Island Press, 1999.

Lindholdt, Paul. "From Sublimity to Ecopornography: Assessing the Bureau of Reclamation Art Collection." *Journal of Ecocriticism* 1:1 (2009): 1-25.

Linenberger, Toni Rae. "Flaming Gorge Unit: Colorado River Storage Project." Denver: Bureau of Reclamation History Program, 1998.

Long, Jr., Paul V. *Archaeological Excavations in Lower Glen Canyon, Utah, 1959-1960.* Foreword by Edward B. Danson. Museum of Northern Arizona Bulletin no. 42—Glen Canyon Series no. 7. Flagstaff, Arizona: The Northern Arizona Society of Science and Art, Inc., 1966.

Look, David W. and Carole L. Perrault. *The Interior Building Its Architecture and Its Art.* Washington, D.C.: U.S. Department of the Interior, National Park Service Preservation Assistance Division, 1986.

Looking Back on a Century of Water. United States Department of the Interior, Bureau of Reclamation, videocassette, 2002.

Lorentz, Pare. *FDR's Moviemaker: Memoirs & Scripts.* Reno: University of Nevada Press, 1992.

Lucic, Karen. *Charles Sheeler and the Cult of the Machine.* Cambridge: Harvard University Press, 1991.

Lugon, Olivier. "The Machine Between Cult Object and Merchandise: Photography and the Industrial Aesthetic in the United States During the Interwar Years." In *American Art: 1807-1947 from Winslow Homer to Jackson Pollock*, Eric de Chassey, editor, chap 4. Paris: Reunion des Musees Nationaux; New York: Distributed by Harry N. Abrams, 2002.

Lyons, Eugene. *Herbert Hoover: A Biography*. Garden City, New York: Doubleday and Company, Inc., 1964.

MacAgy, Douglas. Introduction to *The American Artist and Water Reclamation* by the Bureau of Reclamation. Washington, D. C.: Government Printing Office, 1973.

MacCann, Richard Dyer. *The People's Films: A Political History of U. S. Government Motion Pictures*. New York: Hasting House, Publishers, 1973.

MacDonald, Lawrence J. *Facilitating Voluntary Transfers of Bureau of Reclamation-Supplied Water*. Volume I. Boulder, Colorado: Natural Resources Law Center, July 1991.

Man is a Giant: The Story of Boulder Dam. United States Department of the Interior. Bureau of Reclamation, sound recording, n.d., ca. 1935.

Martin, Russell. *A Story that Stands Like a Dam: Glen Canyon and the Struggle for the Soul of the West*. Salt Lake City: University of Utah Press, 1999.

Marts, M. E. "Upstream Storage Problems in the Columbia River Development." *Annals of the Association of American Geographers* 44: 1 (March 1954): 47-8.

——— and W. R. D. Sewell. "The Conflict between Fish and Power in the Pacific Northwest." *Annals of the Association of American Geographers* 50:1(March 1960): 42-50.

McCool, Daniel C. *Command of the Waters: Iron Triangles, Federal Water Development, and Indian Water*. Tucson: University of Arizona Press, 1994.

———. *Native Waters: Contemporary Indian Water Settlements and the Second Treaty Era*. Tucson: University of Arizona Press, 2002.

———. "Negotiating Water Settlements: Ten Common Themes." In *Indian Water in the New West*. Thomas R. McGuire, William B. Lord, and Mary G. Wallace, editors, 89-102. Tucson: University of Arizona Press, 1999.

———. "Politics, Water, and Utah." In *Water of Zion: The Politics of Water in Utah*, Daniel C. McCool, editor, chap. 1. Salt Lake City: University of Utah Press, 1995.

———, editor. *Water of Zion: The Politics of Water in Utah.* Salt Lake City: University of Utah Press, 1995.

McCully, Patrick. *Silenced Rivers: The Ecology and Politics of Large Dams.* London: Zed Books, 1996.

McGuire, Thomas R. "Indian Water Rights Settlements: A Case Study in the Rhetoric of Implementation." *American Indian Culture and Research Journal* 15:2 (1999): 139-69.

———, William B. Lord, and Mary G. Wallace, editors. *Indian Water in the New West.* Tucson: University of Arizona Press, 1993.

McPhee, John. *Encounters with the Archdruid.* New York: Farrar, Straus, and Giroux, 1971.

"The Meaning of Power Utilization." *Reclamation Era*, 32 (December 1946): 266-7.

Merrill, Dennis, editor. *Documentary History of the Truman Presidency.* Vol. 27, *The Point Four Program: Reaching Out to Help the Less Developed Countries.* Frederick, Maryland: University Publications of America, 1999.

Michener, James A. *Caravans.* New York: Random House, 1963.

———. "Afghanistan: Domain of the Fierce and the Free." *Reader's Digest,* 67 (November 1955): 161-72.

Mighetto, Lisa and Wesley J. Ebel. *Saving the Salmon: A History of the U.S. Army Corps of Engineers' Effort to Protect Anadromous Fish on the Columbia and Snake Rivers.* Prepared for the U.S. Army Corps of Engineers, North Pacific Division, Portland and Walla Walla Districts. Seattle, Washington: Historical Research Associates, Inc., September 6, 1994.

Milazzo, Paul Charles. *Unlikely Environmentalists: Congress and Clean Water, 1945-1972.* Lawrence: University Press of Kansas, 2006.

Miller, Char, editor. *Water in the West: A High Country News Reader.* Corvallis: Oregon: Oregon State University Press, 2000.

"Model of a Giant." *Reclamation Era*, 32 (July 1946).

Moley, Raymond. *What Price Federal Reclamation?* Washington, D.C.: American Enterprise Association, Inc., 1955.

——— and Senator Arthur Watkins. *The Upper Colorado Reclamation Project.* Washington, D.C.: American Enterprise Association, Inc., 1956.

Moncrieff, Hugh H. "Columbia Basin's Showplace." *Reclamation Era*, 38 (August 1952): 214-7, 225.

Morgan, Arthur E. *Dams and Other Disasters: A Century of the Army Corps of Engineers in Civil Works.* Boston: Porter Sargent Publishers, 1971.

"Motion Pictures Show Project Development." *New Reclamation Era*, 17 (July 1926): 125.

Moving a River: History of the Strawberry Valley Project. Roger Hansen, writer, director, and, producer. J. T. V. Productions, videocassette.

Murdock, John R. "Veterans—Here's Your Farm." *Reclamation Era*, 32 (May 1946): 95-6.

Murphy, Francis J., Project Manager, Kiewit-Judson Pacific-Murphy. "Highway in the Sky." *Reclamation Era*, 44 (November 1956): 102-4.

Naef, Weston J. and James N. Wood. *Era of Exploration: Rise of Landscape Photography in the American West, 1860-1885.* Buffalo: Albright-Knox Art Gallery; Boston: Distributed by the New York Graphic Society, 1975.

National Research Council, Committee on Water. *Water and Choice in the Colorado Basin: An Example of Alternatives in Water Management.* Washington, D.C.: National Academy of Sciences, 1969.

———. *Water Transfers in the West: Efficiency, Equity, and the Environment.* Washington, D.C.: National Academy Press, 1992.

Nash, Gerald. *The American West Transformed: The Impact of the Second World War.* Bloomington: Indiana University Press, 1985.

Nash, Roderick. *Wilderness and the American Mind.* Third Edition. New Haven: Yale University Press, 1982.

National Water Commission. *Water Policies for the Future: Final Report to the President and to the Congress of the United States by the National Water Commission*. Washington, D.C.: United States Government Printing Office, June 1973.

Netboy, Anthony. *The Columbia River Salmon and Steelhead Trout: Their Fight for Survival*. Seattle: University of Washington Press, 1980.

"New Commissioner of Reclamation." *Reclamation Era*, 45 (May 1959): 29, 38, 56.

"New Name Change for Bureau of Reclamation." *Reclamation Era*, 65:2 (1979): 1.

"News Round-Up: Bureau Engineers Loaned to Foreign Government." *Reclamation Era*, 33 (March 1947): 70-1.

Nordyke, Lewis. "River Doctor." *Saturday Evening Post*, 218 (September 8, 1945): 19, 92-4.

Ogden, Gerald R., compiler. "The Excess Lands Provisions of Federal Reclamation Law: A Bibliography and Chronology." Agricultural History Center, University of California, Davis, July 1980, transcript.

Opie, John. *The Law of the Land: Two Hundred Years of American Farmland Policy*. Lincoln: University of Nebraska Press, 1994.

———. *Nature's Nation: An Environmental History of the United States*. Fort Worth, Texas: Harcourt Brace College Publishers, 1998.

Pacific Constructors, Inc., *Shasta Dam and Its Builders*. Shingletown, California: M.V. Allen; Happy Camp, California: Naturegraph Publishers (second printing), 1987.

Palmer, Tim. *Endangered Rivers and the Conservation Movement*. Berkeley: University of California Press, 1986.

———. *The Wild and Scenic Rivers of America*. Washington, D.C.: Island Press, 1993.

Palmquist, Peter E., editor. *Once Upon a Dam Site: Howard Colby's Shasta Dam Photographs*. Redding, California: Redding Museum and Art Center, 1987.

"Panama-Pacific Prizes." *Reclamation Record*, 6 (September 1915): 394.

Pearson, Byron E. *Still the Wild River Runs: Congress, the Sierra Club and the Fight to Save the Grand Canyon.* Tucson: University of Arizona Press, 2002.

———. "'We Have Almost Forgotten How to Hope': The Hualapai, the Navajo, and the Fight for the Central Arizona Project, 1944-1968." *Western Historical Quarterly* 31:3 (Autumn 2000): 297-316.

Peeler, David P. *The Illuminating Mind in American Photography: Stieglitz, Strand, Weston, Adams.* Rochester, New York: University of Rochester Press, 2001.

Peterson, Keith C. *River of Life, Channel of Death: Fish and Dams on the Lower Snake.* Corvallis, Oregon State University Press, 1995.

Pfaff, Christine E. *The Bureau of Reclamation's Architectural Legacy: 1902-1955.* Denver, Colorado: United States Department of the Interior, Bureau of Reclamation, October 2007.

Phillips, Sandra, Richard Rodriguez, Aaron Betsky, and Eldridge M. Moore. *Crossing the Frontier: Photographs of the Developing West, 1849 to the Present.* San Francisco: San Francisco Museum of Modern Art; Chronicle Books, 1996.

"Photographic Activities Consolidated." *Reclamation Era*, 29 (August 1939): 217.

Pisani, Donald J. "The Bureau of Reclamation and the West, 1945-2000." In *The American West in 2000 Essays in Honor of Gerald D. Nash.* Richard W. Etulain and Ferenc Morton Szasz, editors, 52-68. Albuquerque: University of New Mexico Press, 2003.

———. "Federal Reclamation and the American West in the Twentieth Century." *Agricultural History* 77: 3 (Summer 2003): 391-419.

———. *Water and American Government: The Bureau of Reclamation, National Water Policy, and the West, 1902-1935.* Berkeley: University of California Press, 2002.

———. "Water Rights and the Betrayal of Indian Allotment." *Environmental Review* 10:3 (Autumn 1986): 157-76.

Pitzer, Paul C. "A 'Farm-in-a-Day': The Publicity Stunt and Celebrations That Initiated the Columbia Basin Project." *Pacific Northwest Quarterly* 82:1 (January 1991): 2-7.

———. *Grand Coulee: Harnessing A Dream*. Pullman, Washington: Washington State University Press, 1994.

Plantinga, Carl R. *Rhetoric and Representation in Nonfiction Film*. Cambridge: Cambridge University Press, 1977.

"Plaque Erected at Boulder Dam Memorializes Workers." *Reclamation Era*, 28 (January 1938): 6-7, 10.

The Plow that Broke the Plain. Pare Lorentz, writer and director. Culver City, California: Zenger Video, 1936.

Porter, Eliot. *The Place No One Knew: Glen Canyon on the Colorado*. Foreword by David Brower. San Francisco: Sierra Club, 1963.

Pratt, George O. "Foreign Activities of the Bureau of Reclamation, 1953," typescript. Office of International Affairs, Bureau of Reclamation. Washington, D.C.

Price, Wesley. "What You Can Believe About MVA." *Saturday Evening Post*, 218 (January 19, 1946): 20-4, 124-32.

"Progress of Work." *Reclamation Record*, 2 (April 1911): 154-5.

Prown, Jules David, et al. *Discovered Lands, Invented Pasts: Transforming Visions of the American West*. New Haven: Yale University Art Gallery, 1992.

Quinn, Frank. "Water Transfers: Must the West be Won Again." *Geographic Review* 58:1 (January 1968): 108-32.

Rasmussen, Stan. "Adventures in the Glen Canyon of the Colorado." *Reclamation Era*, 44 (May 1958): 41-5.

"Reclamation Abroad: New Promise in the Holy Land." *Reclamation Era*, 32 (June 1946): 151.

"Reclamation Art Exhibit." *Reclamation Era*, 58 (August 1972): 24.

"Reclamation Milestones ... President Kennedy Visits." *Reclamation Era*, 48 (November 1962): 98-99.

"Reclamation Organization Activities: Glaha Complimented." *Reclamation Era*, 25 (July 1935): 152.

"Reclamation Organization Activities and Project Visitors." *Reclamation Era*, 26 (June 1936): 155.

Reisner, Marc. *Cadillac Desert: The American West and Its Disappearing Water*. Revised and Updated. New York: Penguin Books, 1993.

——— and Sarah Bates. *Overtapped Oasis: Reform or Revolution for Western Water*. Washington, D.C.: Island Press, 1990.

Rey, Samuel S., Administrative Officer Third Powerplant Construction Office. "The Third Powerplant ... Its Impact on People." *Reclamation Era*, 58 (May 1972), 18-25.

Rhinehart, Julian. "The Grand Dam." http://www.usbr.gov/lc/hooverdam/History/articles/rhinehart1.html.

Rice, Ross R. *Carl Hayden: Builder of the American West*. Latham, Maryland: University Press of America, 1994.

The River. Farm Security Administration, USDA, Pare Lorentz, writer and director. Washington, D.C.: U.S. Department of Agriculture; Capital Heights, Maryland: distributed by National Audiovisual Center, 1937.

Robbins, William G., editor. *The Great Northwest: The Search for Regional Identity*. Corvallis: Oregon State University Press, 2001.

Roberts, Jr., Frank H. H. "River Basin Surveys: The First Five Years of the Inter-Agency Archeological and Paleontological Salvage Program," from the Smithsonian Report for 1951, 351-83. Washington, D.C.: United States Government Printing Office, 1952.

Robertson, Charlotte. "*Peterson v. Department of the Interior*: Are Contracting Rights Ever Property Rights Under the Reclamation Reform Act?" *Golden Gate University Law Review* 21:1 (Spring 1991): 175-237.

Robinson, Michael C. "Water for the West: The Bureau of Reclamation 1902-1977." *Reclamation Era*, 63:1 and 2 (1977): 1-13.

Rogers, Jedediah. "Glen Canyon Unit." Denver: Bureau of Reclamation History Program, 2006.

The Roosevelt Dam, Carolyn Gentry, editor, produced by the Roosevelt Memorial Association Film Library, located at the Library of Congress, American Memory, http://memory.loc.gov/ammem/collections/roosevelt_film.

Rosen, Michael D. and Richard J. Sexton. "Irrigation Districts and Water Markets: An Application of Cooperative Decision-Making Theory." *Land Economics* 69:1 (February 1993): 39-53.

Rosenberg, Emily. *Spreading the American Dream: American Economic and Cultural Expansion, 1880-1945.* New York: Hill and Wong, 1982.

Rowley, William D. *The Bureau of Reclamation: Origins and Growth to 1945.* Volume I. Denver, Colorado: Bureau of Reclamation, United States Department of the Interior, 2006.

Sadler, Richard W. and Richard C. Roberts. *The Weber River Basin: Grass Roots Democracy and Water Development.* Logan: Utah State University Press, 1994.

Samuels, Peggy and Harold. *Illustrated Biographical Encyclopedia of Artists of the American West.* Garden City, New Jersey: Doubleday, 1976.

Samuelson, Paul Anthony. *Economics: An Introductory Analysis* (New York: McGraw-Hill, 1951).

Sandweiss, Martha A. *Print the Legend: Photography and the American West.* New Haven: Yale University Press, 2002.

———. "The Public Life of Western Art." In *Discovered Lands, Invented Pasts: Transforming Visions of the American West* by Jules David Prown, et al. chap 4. New Haven: Yale University Art Gallery, 1992.

Sauder, Robert A. *The Yuma Project: Irrigation, Indian Allotment, and Settlement along the Lower Colorado River.* Reno: University of Nevada Press, 2009.

Sax, Joseph L. "Selling Reclamation Water Rights: A Case Study in Federal Subsidy Policy." *Michigan Law Review* 64:1 (November 1965): 13-46.

Schoots, Hans. *Living Dangerously: A Biography of Joris Ivens.* Amsterdam: Amsterdam University Press, 2000.

Schulte, Steven C. *Wayne Aspinall and the Shaping of the American West.* Boulder: University Press of Colorado, 2002.

Scobie, Ingrid W. *Center Stage: Helen Gahagan Douglas, A Life*. New York: Oxford University Press, 1992.

Scott, James C. *Seeing Like a State: How Certain Schemes to Improve the Human Condition Have Failed*. New Haven, Connecticut: Yale University Press, 1998.

Serafim, J. Laginha. "Disaster in 1976 (The Teton Dam Case)," (Desastres de Barragens em 1976 (O Caso de Teton). Translated from Portuguese: Article in the engineering review *Tecnica*, reprint from issue No. 443, pp. 7-17, published in Lisbon by the Student Association of Higher Technical Institute, 1977. This article was based on a talk given at the National Civil Engineering Laboratory on December 30, 1976. Division of Design, Engineering Services Branch, Code 280, USBR Translation Group, Denver, Colorado, May 1978.

Shafer, Alexandra M. "The Reclamation Reform Act of 1982: Reform or Replacement." *University of Pittsburgh Law Review* 45:3 (Spring 1984): 647-72.

Shepherd, Harold. "Conflict Comes to Roost! The Bureau of Reclamation and the Federal Indian Trust Responsibility." *Environmental Law* 31:4 (2001): 901-49.

Sherow, James E., editor. *A Sense of the American West: An Anthology of Environmental History*. Albuquerque: University of New Mexico Press, 1998.

Shurts, John. *Indian Reserved Water Rights: The Winters Doctrine in Its Social and Legal Context, 1880s-1930s*. Norman: University of Oklahoma Press, 2000.

Simison, Herbert E. "Action on the San Juan River." *Reclamation Era*, 48 (February 1960): 85-8.

Simonds, Wm. Joe. "The Columbia Basin Project." Denver: Bureau of Reclamation History Program, 1998.

Sluiter, Liesbeth. *The Mekong Currency: Lives and Times of a River*. Utrecht, The Netherlands: International Books, 1993.

Smith, Charles D. *Islam and the Search for Social Order in Modern Egypt: A Biography of Muhammad Husayn Haykal*. Albany: State University of New York, 1983.

———. *Palestine and the Arab-Israeli Conflict*. Boston: Bedford/St. Martin's, 2001.

Smith, Thomas G. *Green Republican: John Saylor and the Preservation of America's Wilderness*. Pittsburgh: University of Pittsburgh Press, 2006.

———. "John Kennedy, Stewart Udall, and New Frontier Conservation." *Pacific Historical Review* 64:3 (August 1995): 329-62.

The Snake: Jewel of the Gem State. United States Bureau of Reclamation, video-cassette, 1987.

Snow, Mitchell, "A Quantum Leap Forward: The Reclamation Reform Act of 1982." *Reclamation Era*, 67:4 (1982): 20-9.

Snyder, Robert L. *Pare Lorentz and the Documentary Film*. Reno: University of Nevada Press, 1994.

"Southern Reclamation Conference Makes Plans for the Future." *New Reclamation Era*, 19 (January 1928): 4.

Sowards, Adam M. "William O. Douglas's Wilderness Politics: Public Protest and Committees of Correspondence in the Pacific Northwest." *Western Historical Quarterly* 37:1 (Spring 2006): 21-42.

Spaulding, Jonathan. *Ansel Adams and the American Landscape: A Biography*. Berkeley: University of California Press, 1995.

Stamm, Gilbert. "After Teton." *Reclamation Era*, 62 (Autumn 1976): 1-3.

———. "Bureau of Reclamation International Technical Assistance in Development of Arid Lands," International Conference on Arid Lands in a Changing World, Tucson, Arizona, June 3-13. Washington, D.C.: Bureau of Reclamation, 1969.

Stegner, Wallace, editor. *This is Dinosaur: Echo Park Country and Its Magic Rivers*. New York: Alfred A. Knopf, 1955.

Steinberg, Ted. *Down to Earth: Nature's Role in American History*. New York: Oxford University Press, 2002.

Steinberg, Theodore. "'That World's Fair Feeling': Control of Water in 20th-Century America." *Technology and Culture* 34:2 (April 1993): 401-9.

Stene, Eric A. "Hungry Horse Project." Denver: Bureau of Reclamation History Program, 1994.

———. "Teton Basin Project." Denver: Bureau of Reclamation History Program, 1996.

Stessman, J. Neil. *Oral History Interviews*. Transcript of Tape-recorded Bureau of Reclamation Oral History Interviews conducted by Brit Allan Storey, senior historian, Bureau of Reclamation, from 1994 to 1996, in Billings, Montana.

Stevenson, Adlai. "Will India Turn Communist." *Look*, 17 (July 14, 1953).

"Stewart L. Udall—New Secretary." *Reclamation Era*, 47 (February 1961): 1.

Stinger, Leslie, and Bobbie Ferguson, "Portraits of Reclamation," CRM No. 4 – 1999, www.usbr.gov/museumproperty/art/.

The Story of Hoover Dam. United States Department of the Interior, videocassette, n.d.

The Story of Hoover Dam. United States Department of the Interior, videocassette, c. early 1960s.

Strand, Paul. "Photography and the New God." In *Photographers on Photography: A Critical Anthology*, Nathan Lyons, editor chap. 22. Englewood Cliffs, New Jersey in Collaboration with the George Eastman House, Rochester, New York, 1996.

Straus, Michael W. "Reclamation Street—26,000 Miles Long." *Reclamation Era*, 37 (May 1951): 95-8.

———. *Why Not Survive?* New York: Simon and Schuster, 1955.

Sturgeon, Stephen C. "Operation Blackmail: Upper Basin Reaction to the Central Arizona Project," an unabridged excerpt from the doctoral dissertation "Wayne Aspinall and the Politics of Western Water," an unpublished paper presented at the American Society of Environmental History Conference, Tucson, Arizona, April 15, 1999.

———. *The Politics of Water: The Congressional Career of Wayne Aspinall*. Tucson: University of Arizona Press, 2002.

Swain, Donald C. "The Bureau of Reclamation and the New Deal, 1933-1940." *Pacific Northwest Quarterly* 61:3 (July 1970): 137-46.

Sylten, Harold L. "Pioneering in the Missouri Basin." *Reclamation Era*, 32 (May 1946): 98-9.

Tarlock, A. Dan. "Tribal Justice and Property Rights: The Evolution of *Winters v. United States*." *Natural Resources Journal* 50:2 (Spring 2010): 471-99.

Taylor III, Joseph E. *Making Salmon: An Environmental History of the Northwest Fisheries Crisis*. Foreword by William Cronon. Seattle: University of Washington Press, 1991.

Taylor, R. G., A. M. Michelsen, and R. Huffaker. "Water Pressure: Why the Price for Federally-Developed Irrigation Water Doesn't Promote Conservation." *Choices*, (Third Quarter, 2000): 13-6.

"Technicolor Films Available." *Reclamation Era*, 32 (June 1946): 119.

Thi Dieu, Nguyen. *The Mekong River and the Struggle for Indochina, Water, War, and Peace*. Westport, Connecticut: Praeger Publishers, 1999.

Tilden, Freeman. *The National Parks: What They Mean to You and Me*. Introduction by Newton Drury, Director National Park Service, 1940-1951. New York: Alfred A. Knopf, 1951.

Townley, John M. *Turn This Water Into Gold: The Story of the Newlands Project*. Reno: Nevada Historical Society, 1998.

True, Allen Tupper. "Color and Decoration at the Boulder Dam Power Plant." *Reclamation Era*, 26 (January, 1936): 12-3, 25.

Turn This Water Into Gold: The Story of the Newlands Project. A production of the Churchill County Museum Association. Jane Pieplow, executive producer. Mark Gandolfo and Jane Pieplow, directors, videocassette, 1998.

Tyler, Dan. *The Last Water Hole in the West: The Colorado-Big Thompson Project and the Northern Colorado Water Conservancy District*. Boulder: University Press of Colorado, 1992.

Udall, Stewart L. *The Quiet Crisis*. Introduction by John F. Kennedy. New York: Holt, Rinehart and Winston, 1965.

———. *Oral History Interview: Stewart L. Udall*. William M. Moss, interviewer. Comprised of eight interviews between January and September 1970. Washington, D.C.: John F. Kennedy Library, 1981.

———. "Pausing at the Past: Reflections of a Native Son." In *Beyond the Mystic West*, Stewart L. Udall, Patricia Nelson Limerick, Charles F. Wilkinson, John M. Volkman, William Kittredge, editors, chap. 1. Salt Lake City: Peregrine Smith Books; in association with the Western Governors' Association, 1990.

Underwood, Dennis B. *Oral History Interview.* Transcript of Tape-recorded Bureau of Reclamation Oral history Interviews conducted by Brit Allan Storey, senior historian, Bureau of Reclamation, from 1995 to 1998, in Los Angeles and Ontario, California, Edited by Brit Allan Storey.

United States Department of Agriculture, Bureau of Agricultural Economics. Mary Montgomery and Marion Clawson. *History of the Legislation and Policy Formation of the Central Valley Project.* Berkeley, California: March 1946.

United States Department of Energy, Bonneville Power Administration. *Background: The Columbia River Treaty Revisited.* Portland, Oregon: Bonneville Power Administration, March 1989.

———, *Columbia River Power for the People: A History of Policies of the Bonneville Power Administration.* Portland, Oregon: United States Department of Energy, 1980.

United States Department of the Interior. Julius A. Krug, *Annual Report of the Secretary of the Interior: Fiscal Year Ended June 30, 1946.* Washington, D.C.: Government Printing Office, 1946.

———. Oscar L. Chapman, *Annual Report of the Secretary of the Interior: Fiscal Year Ending June 30, 1949.* Washington, D.C.: Government Printing Office, 1949.

United States Department of the Interior. Teton Dam Failure Review Group. *Failure at Teton Dam: Final Report*, January 1980.

United States Department of the Interior, Bureau of Indian Affairs. *Navajo Project New Mexico: Feasibility Report, Appendix D, Agricultural Economic Financial Analysis.* Window Rock, Arizona: Navajo Agency, January 1955.

United States Department of the Interior, U.S. Bureau of Reclamation, A. B. West, *Supplemental Report Central Arizona Project: Project Development Report*, January 1963.

———. *The American Artist and Water Reclamation.* Washington, D. C.: Government Printing Office, 1973.

———. *Assessment '87... A New Direction for the Bureau of Reclamation.* 1987.

———. *The Bureau of Reclamation: History Essays from the Centennial Symposium* Volumes I and II. Denver, Colorado: Bureau of Reclamation, U.S. Department of the Interior, 2008.

———. *The Colorado River, "A Natural Menace Becomes a Natural Resource": A Comprehensive Report on the Development of the Colorado River Basin for Irrigation, Power Production, and Other Beneficial Uses in Arizona, California, Colorado, Nevada, New Mexico, Utah, and Wyoming*. March 1946.

———. *The Columbia River: Final Letters of Transmittal and Comments of the Affected States and of Federal Agencies*, Volume I, H. Doc. 473, 81st Cong., 2nd sess., February 1950.

———. *The Columbia River: Supplemental Reports on the Bitterroot Valley, Cambridge Bench, Canby, Council, Crooked River, The Dalles (West Unit), Hells Canyon, Mountain Home (Payette Unit), Upper Star Valley, and Vale (Bully Creek Extension)*. Volume 2, H. Doc. No. 473, 81st Cong., 2nd sess., March 1950.

———. *Columbia Basin Joint Investigations; Character and Scope United States Department of the Interior, Harold L. Ickes, Secretary. Bureau of Reclamation, John C. Page, Commissioner.* Washington, D.C.: U.S. Government Printing Office, 1941.

———. *Eleventh Annual Report of the Reclamation Service, 1911-1912*. F. H. Newell, Director (Washington, D.C.: Government Printing Office, 1913).

———. *Federal Reclamation and Related Laws Annotated: Reclamation Reform Act Compilation 1982-1988*. Paul B. Smith, editor. Denver, Colorado: United States Government Printing Office, 1988.

———. *Federal Reclamation and Related Laws Annotated*. Volume I. Richard K. Pelz, editor. Washington, D.C.: United States Government Printing Office, 1972.

———. *Federal Reclamation and Related Laws Annotated*, Volume II. Richard K. Pelz, editor. Washington, D.C.: United States Government Printing Office, 1972.

———. *Federal Reclamation and Related Laws Annotated*, Volume III. Richard K. Pelz, editor. Washington, D.C.: Bureau of Reclamation, 1972.

———. *Federal Reclamation and Related Laws Annotated*. Volume IV 1967-1982. Louis D. Mauro and Richard K. Pelz, editors. Denver, Colorado: United States Government Printing Office, 1989.

———. *Federal Reclamation and Related Laws Annotated (Preliminary)*. Volume V 1983-1998. Donald Walker, editor. Denver, Colorado: United States Government Printing Office, 2001.

———. *Lake Powell: Jewel of the Colorado.* Washington, D.C.: U.S. Government Printing Office, 1965.

———. *Nineteenth Annual Report of the Reclamation Service, 1919-1920.* Arthur P. Davis, Director. Washington, D.C.: Government Printing Office, 1920.

———, Region 1. "Third Powerplant, Columbia Basin Project, Washington: Definite Plan Report." Boise, Idaho, September 1967.

———, Region 4. *Central Utah Project, Initial Phase: Bonneville Unit, Definite Plan Report.* Salt Lake City: Region 4, August 1964.

———. *Sixteenth Annual Report of the Reclamation Service: 1916-1917.* Washington, D.C.: Government Printing Office, 1917.

———. *They Subdued the Desert: The Story of Irrigation as Told to Barrow Lyons by the men who apply water, till the land and feed their flocks and herds*, Barron Lyons, editor, typescript, August 1947.

———. *Thirty-First Annual Report of the Bureau of Reclamation, 1931-1932.* Elwood Mead, Commissioner. Washington, D.C.: Government Printing Office, 1932.

———. "Water 2025: Preventing Crisis and Conflict in the West," May 5, 2003.

———, Upper Colorado Region. *Central Utah Project, Bonneville Unit: Municipal and Industrial System, Summary Statement*, August 1979.

———. *Jensen Unit, Central Utah Project, Definite Plan Report*, December 1975.

United States Department of the Interior, Bureau of Reclamation, and U.S. Agency for International Development, U.S. Department of State, "The Social Feasibility of Pa Mong Irrigation: Requirement and Reality," by Jasper Ingersoll, typescript mimeograph, 1969.

United States Department of the Interior, U.S. Geological Survey, Herbert M. Wilson. *Irrigation in India*, Water Supply and Irrigation Paper No. 87. Washington, D.C., 1903.

———. Water and Power Resources Service. "Navajo Indian Irrigation Project." May 1980.

———. *Project Data.* Denver: United States Government Printing Office, 1981.

———. Upper Colorado Region, *Ute Indian Unit: Central Utah Project, Concluding Report*, May 1980.

United States Federal Power Commission. Bureau of Power. *Power Market Survey: Colorado River Upper Basin.* San Francisco Regional Office, September 1947.

United States House of Representatives, *Committee Conference Report: Authorizing the Secretary of the Interior to Construct, Operate, and Maintain the Colorado River Storage Project and Participating Projects.* H. Rpt. No. 1950, 84th Cong., 2nd sess., March 27, 1956. Washington, D.C.: United States Government Printing Office, 1956.

———, Committee on Expenditures in the Executive Departments, House Subcommittee. *Investigation of Bureau of Reclamation, Department of the Interior*, H. Rpt No. 2458, 80th Cong., 2nd sess., August 7, 1948. Washington, D.C.: Government Printing Office, 1948.

———, Committee on Indian Affairs. *Right of Way through Yakima Indian Reservation: Report.* Report No. 714, 53rd Cong., 2nd sess., April 17, 1894.

———, Committee on Interior and Insular Affairs. *Amending the Act of June 12, 1962 (76 Stat. 96), with Respect to the Navajo Indian Irrigation Project*, H. Rpt. No. 91-1331, 91st Cong., 2nd sess., July 27, 1970.

———, Committee on Interior and Insular Affairs. *Central Utah Completion Act and Reclamation Projects Authorization and Adjustment Act of 1990.* H. Rpt. No. 101-764, 101st Cong., 2nd sess., September 27, 1990.

———, Committee on Interior and Insular Affairs. *Grand Canyon Protection Act of 1990.* H. Rpt. 101-641, 101st Cong., 2nd sess., July 30, 1990.

———, Committee on Interior and Insular Affairs. *National Wild and Scenic Rivers System.* S. Rpt. No. 491, 90th Cong., 1st sess., August 4, 1967.

———, Committee on Interior and Insular Affairs. *National Wilderness Preservation System: Message from the President of the United States, Transmitting National Wilderness Preservation System.* H. Doc. No. 79, 89th Cong., 1st sess., February 8, 1965.

———, Committee on Interior and Insular Affairs. *Providing for a National Scenic Rivers System, and for other Purposes*. H. Rpt. No. 1623, 90th Cong., 2nd sess., July 3, 1968.

———, Committee on Interior and Insular Affairs. *Report: Authorizing the Construction of the Navajo Indian Irrigation Project and the Initial Stage of the San Juan-Chama Project as Participating Projects of the Colorado River Storage Project*. H. Rpt. No. 685, 87th Cong., 1st sess., July 10, 1960. Washington, D.C.: United States Government Printing Office, 1960.

———, Committee on Interior and Insular Affairs, Subcommittee on Energy Research and Water Resources. *Teton Dam*, Hearing. 94th Cong., 2nd sess., June 15, 1976. Washington, D.C.: United States Government Printing Office, 1976.

———, Committee on Interior and Insular Affairs, Subcommittee on Indian Affairs, *Increasing the Appropriation for the Completion of the Construction of the Irrigation and Power Systems of the Flathead Indian Irrigation Project, Montana*. Hearing, 87th Cong., 2nd sess., August 27, 1962, unpublished hearing.

———, Committee on Interior and Insular Affairs, Subcommittee on Irrigation and Reclamation. *Central Arizona Project*, Hearings, 88th Cong., 2nd sess., November 9, 1964, Phoenix, Arizona. Washington, D.C.: United States Government Printing Office, 1965.

———, Committee on Interior and Insular Affairs. Subcommittee on Irrigation and Reclamation. *Lower Colorado River Basin Project*. Hearing, 89th Cong., 1st sess. Washington, D.C.: United States Government Printing Office, 1965.

———, Committee on Interior and Insular Affairs, Subcommittee on Irrigation and Reclamation. *Navajo Indian Irrigation Project: Hearings on S.203 and H.R. 13001 to Amend the Act of June 12, 1962, with Respect to the Navajo Indian Irrigation Project*, 91st Cong., 2nd sess., June 11,1970. Washington, D.C.: United States Government Printing Office, 1970.

———, Committee on Interior and Insular Affairs. Subcommittee on Irrigation and Reclamation. *San Juan-Chama Reclamation Project and Navajo Indian Project*. Hearings, 87th Cong., 1st sess., April 24-26, and June 1, 1961. Washington, D.C.: United States Government Printing Office, 1962.

———, Committee on Interior and Insular Affairs, Subcommittee on Irrigation and Reclamation, *Third Powerplant, Columbia Basin Project, Washington*, Hearings, 89th Cong., 1st sess., September 9 and 10, 1965. Washington, D.C.: United States Government Printing Office, 1965.

———, Committee on Interior and Insular Affairs. Subcommittee on Water and Power Resources. *Implementation of the Coordinating Agreement (COA): Implications for Water Quality in San Francisco-San Joaquin Delta and San Francisco Bay*. Hearings, 100th Cong., 1st sess., April 3, 1987, Concord, California. Washington, D.C.: United States Government Printing Office, 1987.

———, Committee on Interior and Insular Affairs, Subcommittee on Water, Power, and Offshore Energy Resources. *Management and Operation of the Central Valley Project*. Hearings, 102nd Cong., 2nd sess., May 14, 1992. Washington D.C.: United States Government Printing Office, 1992.

———, Subcommittee of the Committee on Government Operations. *Teton Dam Disaster: Thirtieth Report by the Committee on Government Operations*, H. Rpt. No. 94-1667, 94th Cong., 2nd sess., September 27, 1976. Washington, D.C.: United States Government Printing Office, 1976.

———, Subcommittee on Appropriations. *Public Works Appropriations for 1962*. Hearings, 87th Cong., 1st sess. Washington, D.C.: Government Printing Office, 1961.

———. Subcommittee on National Security Policy and Scientific Developments, *The Mekong Project: Opportunities and Problems of Regionalism*. by Franklin P. Huddle. Washington, D.C.: Government Printing Office, 1972.

———. *Supplemental Appropriations for Reimbursement for Damages Caused by Failure of Teton Dam: Communication from the President of the United States*. H. Doc. No. 94-523, 94th Cong., 2nd sess., June 11, 1976. Washington, D.C.: United States Government Printing Office, 1976.

———. *Third Powerplant Columbia Basin Project, Washington, Communication from the President of the United States, the Report of the Secretary of the Interior Concerning Economic and Engineering Feasibility of a Third Powerplant at Grand Coulee Dam on the Columbia River, and a Draft of Authorizing Legislation*, H. Doc. No. 142, 89th Cong., 1st sess. Washington, D.C.: United States Government Printing Office, 1965.

———. *Umatilla Irrigation Company: Report*. Report. No. 3379, 51st Cong., 2nd sess., January 6, 1891.

United States President. *Inaugural Addresses of the Presidents of the United States from George Washington, 1789 to Harry S. Truman, 1949*. Washington, D.C.: United States Government Printing Office, 1952.

United States Senate. *Colorado River Storage Project: Letter from the Assistant Secretary of the Interior Transmitting the Seventh Annual Report of the Department of the Interior on the Status of the Colorado River Storage Project and Participating Projects Pursuant to Public Law 485 of the Eighty-Fourth Congress (70 stat. 105) Together with a Statement and Pertinent Comments thereto by Senator Anderson of New Mexico*. S. Doc. No. 64, 88th Cong., 2nd sess. Washington, D.C.: United States Government Printing Office, 1964.

———, Committee on Energy and Natural Resources. *Columbia River, Washington*. S. Rpt. No. 100-368, 100th Cong., 2nd sess., May 27, 1988.

———, Committee on Energy and Natural Resources. *Reclamation Reform Act of 1979*. S. Rpt. No. 96-235, 96th Cong., 1st sess., June 21, 1979.

———, Committee on Energy and Natural Resources. *Reclamation Reform Act of 1982*. S. Rpt. No. 97-373, 97th Cong., 2nd sess., April 29, 1982.

———, Committee on Interior and Insular Affairs. *Central Arizona Project and Colorado River Rights: Hearings before the Committee on Interior and Insular Affairs*. 81st Cong., 2nd sess., on S. 75 and S.J. Res. 4. Washington, D.C.: United States Government Printing Office, 1949.

———, Committee on Interior and Insular Affairs. *Colorado River Storage Project: Hearings before the Subcommittee on Irrigation and Reclamation,* June 28, 29, 30, July 1, 2, 3, 1954, 83rd Cong., 2nd sess. Washington, D.C.: United States Government Printing Office, 1954.

———, Committee on Interior and Insular Affairs. *Colorado River Storage Project: Report, A Bill to Authorize the Secretary of the Interior to Construct, Operate, and Maintain the Colorado River Storage Project and Participating Projects in the States of Arizona, Colorado, New Mexico, Utah, and Wyoming*. S. Rpt. No. 128, 84th Cong., 1st sess., March 30, 1955.

———, Committee on Interior and Insular Affairs. *Designating a Segment of the Missouri River in the State of Montana as a Component of the National Wild and Scenic River System*. S. Rpt. No. 94-502, 94th Cong., 1st sess., December 1, 1975.

———, Committee on Interior and Insular Affairs. *Report: Authorizing the Secretary of the Interior to Construct, Operate, and Maintain the Colorado River Storage Project and Participating Projects*. S. Rpt. No. 1893, 83rd Cong., 2nd sess., July 26, 1954. Washington, D.C.: United States Government Printing Office, 1954.

———, Committee of Interior and Insular Affairs, Subcommittee on Water and Power Resources. *Navajo Indian Irrigation Project.* Hearing on S. 203, 91st Cong., 1st sess., July 15, 1969. Washington, D.C.: United States Government Printing Office, 1969.

———, Committee on Interior and Insular Affairs. *Third Powerplant, Grand Coulee Dam*, Hearing, 89th Cong., 1st sess., April 13, 1965. Washington, D.C.: United States Government Printing Office, 1965.

———. *Glen Canyon-Bridge Canyon Dam Project and Arizona High-Line Canal: Summary of the Arguments Made by Senator Fred T. Colter, President of the Arizona Highline Association, on S. 3414, A Bill Providing for the Building of the Glen-Bridge Dams and Arizona High-Line Canal, Introduced in the United States Senate March 1, 1926, as a Substitute for the Swing-Johnson Boulder-Black Dam and Compact Bill.* S. Doc. No. 113, 69th Cong., 1st sess. Washington, D.C.: United States Government Printing Office, 1926.

———, Select Committee on Small Business. *Will the Family Farm Survive in America?: Federal Reclamation Policy*, S. Rpt. No. 95-702, 95th Cong., 2nd sess. Washington, D.C.: United States Government Printing Office, 1978.

———, Senate Special Committee to Study the Survey Problems of American Small Business. *Small Business and the Community: Study in the Central Valley of California on Effects of Scale Operations*, 79th Cong., 2nd sess. Washington, D.C: United States Government Printing Office, 1946. Committee Print.

"Upper Colorado Compact Ok'd." *Reclamation Era*, 36 (June 1949): 140.

Upper Colorado River Commission. *First Annual Report.* Grand Junction, Colorado: March 20, 1950.

______. *Second Annual Report.* Grand Junction, Colorado: March 19, 1951.

______. *Seventh Annual Report.* Grand Junction, Colorado: April 1, 1956.

Vedder, Kurt. "Water Development in Salt Lake Valley, 1847-1985." In *Water of Zion: The Politics of Water in Utah*, Daniel C. McCool, editor, chap. 2. Salt Lake City: University of Utah Press, 1995.

Vetter, Carl P. "Corralling the Colorado." *Reclamation Era*, 32 (September 1946): 190-192.

Vilander, Barbara. *Hoover Dam: The Photographs of Ben Glaha*. Tucson: University of Arizona Press, 1999.

Voggesser, Garrit. "The Flathead Project: The Indian Projects." Denver, Colorado: Bureau of Reclamation History Program, 2001.

———. "The Indian Projects: The Crow Project." Denver, Colorado: Bureau of Reclamation History Program, 2001.

———. "The Indian Projects: Blackfeet Project." Denver, Colorado: Bureau of Reclamation History Program, 2001.

Wade, Stephan G. "The First Days." *Reclamation Era*, 62 (Autumn 1976): 4-13.

Wahl, Richard W. *Markets for Federal Water: Subsidies, Property Rights, and the Bureau of Reclamation*. Washington, D.C.: Resources for the Future, 1989.

Waltrip, John Richard. *Public Power during the Truman Administration*. New York: Arno Press, 1979.

Walton, Jean R., Construction Engineer. "Flaming Gorge." *Reclamation Era*, 45 (August 1959): 69-72, 80.

Ward, Evan R. *Border Oasis: Water and the Political Ecology of the Colorado River Delta, 1940-1975*. Tucson: University of Arizona Press, 2003.

Warne, William E. *The Bureau of Reclamation*. New York: Praeger Publishers, 1973.

———. "Corralling the Colorado, Part III." *Reclamation Era*, (November 1946): 240-243, 256.

———. *Legal and Economic Problems Associated with Excessive Withdrawal from Ground-Water Sources*. Denver: Colorado: Inter-American Conference on Conservation of Renewable Resources, 1948.

Water for a Desert Dream: The Newlands Project. Dennis Meyers and Erik Westby, producers. Curt Daniels, director, videocassette, 1996.

"'Water in the West' Goes Global." *Reclamation Era*, 38 (June 1953): 196.

"'Water in the West' Honored." *Reclamation Era*, 38 (May 1953): 107.

"Water Planning Pays Off." *Reclamation Era*, 32 (July 1946): 152-3.

"Water—The Key to Sustaining Growth." *Reclamation Era*, 53 (August 1967): 57-60.

Wathen, Albert H. "Indian Irrigation." *Reclamation Era*, 31 (December 1941): 322-323.

Watkins. T. H. *Righteous Pilgrim: The Life and Times of Harold L. Ickes, 1874-1952.* New York: Henry Holt and Company, 1990.

Weakly, Ward. "Reclamation's Roots Discovered through Archaeological Studies." *Reclamation Era*, 63:1 and 2 (1977): 82-5.

Weatherford, Gary D. and F. Lee Brown, editors. *New Courses for the Colorado River: Major Issues for the Next Century*. Foreword by Governor Bruce Babbitt. Albuquerque: University of New Mexico Press, 1986.

"Weeds." *Reclamation Era*, (May, 1940): 143.

Wehr, Kevin. *America's Fight Over Water: The Environmental and Political Effects of Large-Scale Water Systems*. New York: Routledge, 2004.

Weinberg, Marca, Catherine L. Kling, James E. Wilen. "Water Markets and Water Quality." *American Journal of Agricultural Economics* 75:2 (May 1993): 278-91.

Wescoat, James L. "Conservation and Reform of the Prior Appropriation Doctrine in Colorado." *Economic Geography* 61:1 (January 1985): 3-24.

White, Peter. "The Mekong River: Terror and Hope." *National Geographic*, 134 (December 1968).

White, Richard. *The Organic Machine: The Remaking of the Columbia River*. New York: Hill and Wang, a division of Farrar, Straus and Giroux, 1995.

Who Was Who in America Vol. IV, 1961-1968. Chicago: Marquis—Who's Who Inc., 1968.

"Why Teton Dam Failed." *Science News*, 111:4 (January 22, 1977): 60.

Wiley, Peter and Robert Gottieb. *Empires of the Sun: The Rise of the New American West*. New York: G. P. Putnam's Sons, 1982.

Willis, Roxanne. *Alaska's Place in the West: From the Last Frontier to the Last Great Wilderness*. Lawrence: University Press of Kansas, 2010.

Wilson, Herbert M. *Irrigation in India.* Water Supply and Irrigation Paper No. 87 (Washington, D.C.: Department of the Interior, United States Geological Survey, Government Printing Office, 1903), 7-10.

Winegar, Paul. "The First Months—Rebuilding Lives." *Reclamation Era*, 62 (Autumn 1976): 20-6.

Wood, Stanley. "Boulder Dam: A Portfolio of Water Colors." *Fortune*, 9:5 (May 1934): 92-100.

Worster, Donald. *Rivers of Empire: Water, Aridity, and the Growth of the American West.* New York: Oxford University Press, 1985.

Wrobel, David M. *Promised Lands: Promotion, Memory, and the Creation of the American West*. Lawrence: University Press of Kansas, 2002.

Wylie, L. F. "Roads to Scenic Treasures." *Reclamation Era*, 44 (August 1958): 57-9.

The Yakima Project–From the Cascades to the Columbia United States Bureau of Reclamation, videocassette, 1984.

Zarbin, Earl. *Let the Record Show ...: Gila River Indian Reservation Water Rights and the Central Arizona Project.* Foreword by Thomas C. Clark, General Manager Central Arizona Project. Tempe, Arizona: Earl Zarbin, 2004.

INDEX